Geological and Landscape Conservation

This volume is dedicated to the memory of Chris Stevens (1954–93),
whose energy and vision have made an indelible imprint on geological conservation
in Great Britain and internationally.

Geological and Landscape Conservation
Proceedings of the Malvern International Conference 1993

EDITED BY

DES O'HALLORAN
Joint Nature Conservation Committee, Peterborough, UK
(Present address: English Nature, Peterborough, UK)

CHRIS GREEN
Department of Geography, Royal Holloway, University of London, UK

MIKE HARLEY
English Nature, Peterborough, UK

MICK STANLEY
Hull City Museums & Art Galleries, Hull, UK

AND

JOHN KNILL
Joint Nature Conservation Committee, Peterborough, UK

1994
Published by The Geological Society, London

THE GEOLOGICAL SOCIETY

The Society was founded in 1807 as the Geological Society of London and is the oldest geological society in the world. It received its Royal Charter in 1825 for the purpose of 'investigating the mineral structure of the Earth'. The Society is Britain's national society for geology with a membership of 7500 (1993). It has countrywide coverage and approximately 1000 members reside overseas. The Society is responsible for all aspects of the geological sciences including professional matters. The Society has its own publishing house, which produces the Society's international journals, books and maps, and which acts as the European distributor for publications of the American Association of Petroleum Geologists and the Geological Society of America.

Fellowship is open to those holding a recognized honours degree in geology or cognate subject and who have at least two years' relevant postgraduate experience, or who have not less than six years' relevant experience in geology or a cognate subject. A Fellow who has not less than five years' relevant postgraduate experience in the practice of geology may apply for validation and, subject to approval, may be able to use the designatory letters C Geol (Chartered Geologist).

Further information about the Society is available from the Membership Manager, The Geological Society, Burlington House, Piccadilly, London W1V 0JU, UK.

Published by the Geological Society from:
The Geological Society Publishing House
Unit 7
Brassmill Enterprise Centre
Brassmill Lane
Bath BA1 3JN
UK
(*Orders*: Tel 0224 445046
Fax 0225 442836)

First published 1994

The publisher and the Joint Nature Conservation Committee make no representation, express or implied, with regard to the accuracy of the information contained in this book and cannot accept any legal responsibility for any errors or omissions that may be made. The opinions expressed are those of the individual authors and do not necessarily reflect those of the publisher or the Joint Nature Conservation Committee.

© The Geological Society 1994. All rights reserved. No reproduction, copy or transmission of this publication may be made without written permission. No paragraph of this publication may be reproduced, copied or transmitted save with the provisions of the Copyright Licensing Agency, 90 Tottenham Court Road, London W1P 9HE. Users registered with the Copyright Clearance Center, 27 Congress Street, Salem, MA 01970, USA: the item fee code for this publication is 1897799098/94/$07.00.

British Library Cataloguing in Publication Data

A catalogue record for this book is available from the British Library ISBN 1-897799-09-8

Typeset by EJS Chemical Composition,
Midsomer Norton, Bath, Avon

Printed by Alden Press, Oxford, UK

Distributors

USA
  AAPG Bookstore
  PO Box 979
  Tulsa
  OK 74101-0979
  USA
  (*Orders*: Tel (918) 584-2555
  Fax (918) 548-0469)

Australia
  Australian Mineral Foundation
  63 Conyngham Street
  Glenside
  South Australia 5075
  Australia
  (*Orders*: Tel (08) 379-0444
  Fax (08) 379-4634)

India
  Affiliated East-West Press pvt Ltd
  G-1/16 Ansari Road
  New Delhi 110 002
  India
  (*Orders*: Tel (11) 327-9113
  Fax (11) 326-0538)

Japan
  Kanda Book Trading Co.
  Tanikawa Building
  3-2 Kanda Surugadai
  Chiyoda-Ku
  Tokyo 101
  Japan
  (*Orders*: Tel (03) 3255-3497
  Fax (03) 3255-3495)

Contents

Preface	ix
Board of Patrons	
Organizing Committee	
Special responsibilities	
Sponsorship	
Introductory address Sir John Knill	xi
Opening address Lord Selborne	1

Theme 1: Sustainability

Sustaining the Earth's resources

Sustaining the Earth's resources R. Crofts	7
Soil conservation: an international issue E. M. Bridges	11
Conservation of peatlands: the role of hydrogeology and the sustainable development principle D. Daly	17

Planning for sustainability

International Union of Geological Sciences and the environment F. C. Wolff	23
The geotope concept: geological nature conservation by town and country planning B. Stürm	27
Sustainable development: the implications for minerals planning M. M. Scott	33
A model of consensus in aggregate mining and landscape restoration: science–industry–conservation C. Schlüchter	39
The role of local government in geological and landscape conservation E. A. Jarzembowski	43

Sustainability in practice

The building of an international airport in an area of outstanding geological diversity and quality L. Erikstad	47
The justification for a soil conservation policy for the UK T. R. E. Thompson & P. Bullock	53

Sustainability and the fossil heritage

Towards a definition of the Spanish palaeontological heritage L. Alcalá & J. Morales	57
Fossil collecting: international issues, perspectives and a prospectus D. B. Norman	63
The protection of Australia's fossil heritage through the Protection of Movable Cultural Heritage Act 1986 P. Creaser	69

The World Heritage List

The World Heritage List and its relevance to geology J. W. Cowie & W. A. P. Wimbledon	71
Australia's World Heritage fossil nomination P. Creaser	75

Sustainable use and mineral resources

Environmental aspects of mineral resource conservation in southwest England C. M. Bristow	79
The conservation of Quaternary geology in relation to the sand and gravel extraction industry D. Bridgland	87
Damage caused by opencast and underground coal mining J. Domas	93
Development and conservation at Parys Mountain, Anglesey, Wales N. J. G. Pearce	99
The Newer Volcanic Province of Victoria, Australia: the use of an inventory of scientific significance in the management of scoria and tuff quarrying N. J. Rosengren	105

Theme 2: Landscape Conservation

Evaluating the landscape
Conservation of geomorphological landscapes in Taiwan   *Shin Wang,*
   Ling-Yuh Sheu & Hsiao Yu Tang   113
Origin and use of the term 'geotope' in German-speaking countries   *F. W. Wiedenbein*   117
Natural Areas: an holistic approach to conservation based on geology   *K. Duff*   121

Conserving river systems
Earth science assessments for heritage waterways: the French River and others
   in Canada   *D. I. McKenzie*   127
The challenges of geomorphological river system conservation for the 1990s   *L. McEwen*   133
Evaluation of river conservation sites: the context for a drainage basin approach
   P. W. Downs & K. J. Gregory   139
The establishment and revegetation of vegetation zones in rural river landscapes in
   SW Finland   *J. Hietaranta*   145
Conservation management of dynamic rivers: the case of the River Feshie, Scotland
   V. Brazier & A. Werritty   147

The national park experience
Resource development, landscape conservation and national parks in Norway   *F. C. Wolff*   153
Landscape conservation and the national parks of Romania   *D. Teodorascu*   157
Conservation, access and land management conflict in upland glaciated areas of the
   Snowdonia National Park: a preliminary survey   *K. Addison & S. Campbell*   161
The national park system in Germany   *A. Grube*   175
Landscape parks and other protected areas: the Polish experience of landscape
   conservation   *K. Wojciechowski*   181

Geomorphological systems
Geomorphological systems: developing fundamental principles for sustainable
   landscape management   *J. E. Gordon, V. Brazier & R. G. Lees*   185
Strategies for conserving and sustaining dynamic geomorphological sites   *J. M. Hooke*   191
Geological conservation of Holocene peatlands in the national parks of England & Wales
   D. J. Charman   197
Conserving the Holocene record: a challenge for geomorphology, archaeology and
   biological conservation   *P. C. Buckland, B. C. Eversham & M. H. Dinnin*   201

Karst features: their conservation and management
Malta: a model for the conservation of limestone regions   *A. Spiteri*   205
Caves as unique conservation education resources   *G. N. Huppert*   209
Sixty-five years of legislative cave conservation in Austria: experience and results
   H. Trimmel   213
Protection of limestone pavement in the British Isles   *H. S. Goldie*   215
Karst and environment: a Romanian approach   *E. Silvestru*   221

National experiences
Protected volcanoes in Iceland: conservation and threats   *G. Thorvardardottir*
   & T. F. Thoroddsson   227
Development and management of geological and geomorphological conservation
   features within the urban areas of the Western Ukraine   *Y. Zinko*   231
Geological protected areas and features in Estonia   *R. Raudsep*   237
Conservation of national geological sites in China   *Pan Jiang*   243
Earth science conservation in Bulgaria   *T. A. Todorov*   247
Geological conservation in Hungary   *T. Cserny*   249
Geological and landscape conservation in India   *K. N. Prasad*   255
Geological and environmental mapping as an aid for landscape conservation in Lithuania
   J. Satkunas   259

Geological conservation in New Zealand: options in a rapidly eroding environment *J. S. Buckeridge*	263
Nature conservation and beach management: a case study of Köycegiz–Dalyan Specially Protected Area *E. Ergani & P. Kamiş*	271
Environmental geology maps for national parks and geomorphological reserves in Lithuania *J. Valiunas*	273
Country sports as an element in landscape conservation *D. B. Bragg*	279
Scale problems related to the use of the criterion 'naturalness' in the context of landscape protection and road construction in an apparently pristine Arctic environment *S. Smith-Meyer & L. Erikstad*	283
Quarry slope stability and landscape preservation in the Malvern Hills, UK *J. Knill*	287
Geological and landscape conservation in Hong Kong *D. R. Workman*	291
Some problems of geological heritage in the Russian commonwealth *S. A. Vishnevsky*	297

Theme 3: Local Conservation and Community Initiatives

Social perspectives

Legislation and attitudes to geological conservation in Queensland, Australia: past, present and future *P. G. L. Harlow*	303
The role of voluntary organizations in Earth science conservation in the UK *C. P. Green*	309
The RIGS (Regionally Important Geological/geomorphological Sites) challenge – involving local volunteers in conserving England's geological heritage *M. Harley*	313
Conservation and management of geological monuments in South Australia *R. Swart*	319
Two geological monuments in the Netherlands: De Zândkoele and Wolterholten *G. P. Gonggrijp*	323
The legal framework and scientific procedure for the protection of palaeontological sites in Spain: recovery of some special sites affected by human activity in Aragón (eastern Spain) *G. Melendez & M. Soria*	329
Naked rock and the fear of exposure *W. J. Baird*	335

Local experience in the UK

The proposed Cuilcagh Natural History Park, County Fermanagh: a locally based conservation initiative *J. Gunn, C. Hunting, S. Cornelius & R. Watson*	337
The work of the Lothian and Borders RIGS Group in Scotland *N. E. Butcher*	343
Urban site conservation – an area to build on? *C. D. Prosser & J. G. Larwood*	347
Local conservation and the role of the regional geological society *A. Cutler*	353
A person on the inside – opportunities for geological conservation in local engineering projects *G. Worton*	359

The role for computerization

Conservation, communication and the GIS: an urban case study *C. Reid*	365
Developing geological site recording software for local conservation groups *C. J. T. Copp*	371

Theme 4: Site Conservation and Public Awareness

Public awareness

The protection of geological heritage and economic development: the saga of the Digne ammonite slab in Japan *G. Martini*	383
The American Cave and Karst Museum and the work of the American Cave Conservation Association *G. N. Huppert*	387
The private sector – threat or opportunity? *R. de Bastion*	391
The La Crosse River Marsh: development vs. preservation *R. G. Wingate*	397
International support for conservation of geological, landscape and historical sites: the homeland initiative *G. McKenzie*	403

Site-based approaches

Conservation through on-site interpretation for a public audience *P. Keene*	407
National parks and geological heritage interpretation – examples from North America and applications to Australia *G. Markovics*	413
Conservation system of geological sites in the old salt mine of Wielicza (south Poland) *Z. Alexandrowicz & M. Gonera*	417
A consensus approach: industry and geoscience at the National Stone Centre, UK *I. A. Thomas & J. E. Prentice*	423
Interpreting Earth science sites for the public *T. Badman*	429
Information signs for geological and geomorphological sites: basic principles *K. N. Page*	433
Mineral collectors as conservationists *B. Young*	439
The protection of Lower Palaeolithic sites in southern Britain *C. S. Gamble & J. J. Wymer*	443
Effective management of our Earth Heritage sites: the Earth Science Site Documentation Series in Scotland *C. C. J. MacFadyen, J. A. McCurry & S. Keast*	447
Telling the story of stone – assessing the client base *T. A. Hose*	451

Involving others

Reading the landscape *A. McKirdy & R. Threadgould*	459
Involving landowners, local societies and statutory bodies in Shropshire's geological conservation *P. Toghill*	463

Education and conservation

The geo-ecological education and geological site conservation in Romania *D. Grigorescu*	467
Earth science conservation: the need for education and training *R. C. L. Wilson*	473
The changing nature of Earth science fieldwork within the UK school curriculum and the implications for conservation policy and site development *J. A. Fisher*	477
Communication of Earth science to the public – how successful has it been? *C. V. Burek & H. Davies*	483

Inspiring awareness

New Milestones: sculpture, community and the land *S. Clifford*	487
Rock poems, rock music: using poetry and the arts to interpret geology *J. Carter & T. Badman*	493

Theme 5: International Convention

Defining geological conservation *C. Stevens*	499
Geological conservation: notion, necessity and nicety *C. H. Holland*	503
Identifying geological features of international significance: the Pacific Way *E. B. Joyce*	507
Workshop 1	515
Workshop 2	519
Resolution	523
Index	525

Preface

The Malvern International Conference on Geological and Landscape Conservation (18–24 July 1993) was convened by the Joint Nature Conservation Committee of the United Kingdom on behalf of English Nature, Scottish Natural Heritage and the Countryside Council for Wales, in association with the Geological Society of London and the Geologists' Association.

The scope of the conference was determined in consultation with a Board of Patrons and its planning and management executed by an Organizing Committee, certain members of which had special responsibilities. Additionally, staff of the Malvern Hills District Council and members of the Malvern Hills Conservators provided invaluable help and support during the meeting.

Board of Patrons

Professor (now Sir) John Knill, Chairman	Joint Nature Conservation Committee and Natural Environment Research Council
The Earl of Selborne	Joint Nature Conservation Committee
The Earl of Cranbrook	English Nature
Mr Magnus Magnusson	Scottish Natural Heritage
Mr Michael Griffith	Countryside Council for Wales
Mr Guy Martini	Digne Symposium
Professor Charles Curtis	Geological Society of London
Dr Eric Robinson	Geologists' Association
Dr Peter Cook	British Geological Survey
Mr Richard Fox	European Federation of Geologists
Dr George Black	European Working Group on Earth Science Conservation

Organizing Committee

Dr Des O'Halloran, Chairman	Joint Nature Conservation Committee (now of English Nature)
Professor Sir John Knill	Board of Patrons
Mr Chris Stevens (the late)	Joint Nature Conservation Committee
Mr Mike Harley	English Nature
Mr Alan McKirdy	Scottish Natural Heritage
Dr Stewart Campbell	Countryside Council for Wales
Mr Mick Stanley	Geological Society of London
Dr Chris Green	Geologists' Association
Mrs Margaret Phillips	The Company

Special responsibilities

Technical content:	Dr Des O'Halloran
Preparation of papers:	Mr Mike Harley
Field excursions:	Dr Stewart Campbell
Registration and administration:	Mrs Margaret Phillips

Sponsorship

The conference was primarily sponsored by Great Britain's governmental conservation agencies – English Nature, Scottish Natural Heritage and the Countryside Council for Wales – acting jointly through their Joint Nature Conservation Committee.

Publication of these conference proceedings was sponsored by the Geological Society of London, and assistance with overseas travel was provided by the Geologists' Association.

In addition, selective sponsorship of individual delegates was received from the following organizations:

Clyde Petroleum
The Curry Fund of the Geologists' Association
Geological Survey of Norway
Institute of German Technical Assistance
International Science Foundation
Joint Nature Conservation Committee
The Royal Society

Introductory address

JOHN KNILL
*Highwood Farm, Shaw-cum-Donnington,
Newbury RG16 9LB, UK*

The United Kingdom has made a long and distinguished contribution both to the study of geology as a discipline, and to geological conservation. A dozen or so of our predecessors in the eighteenth and nineteenth centuries might lay some claim to being amongst the founding fathers of our subject in that they provided at least a part of the seminal input to the understanding of geology as a science. I will not attempt to argue their respective claims, but I would comment that eight were born in the eighteenth century and six came from Scotland.

And it is to Scotland that we must turn for the first hard evidence of the conservation of a geological site in Britain. However, prior to that event, to which I will return, James Hutton a farmer, scientist, philosopher and leading member of the Scottish Enlightenment who lived between 1726 and 1797 published in 1788 his *Theory of the Earth* which is widely regarded as the foundation of modern geology. Many of the localities used by Hutton to develop his ideas are in Edinburgh and these have since been recognized and conserved not only for their scientific importance in their own right but also for their contribution to the early days of our science. Anyone visiting Edinburgh today can barely fail to avoid a sight of the Castle Rock and Arthur's Seat both of which are relict volcanoes of Carboniferous age. It is not surprising, therefore, that Hutton should have been able to use these spectacular locations for arguing the fundamental tenets of his science. Hutton's Section, at the southern end of the dolerite sill of Salisbury Craigs on Arthur's Seat, demonstrates that the igneous rock was emplaced in a molten form and not deposited as a sediment in the sea. The Plutonists overcame the Neptunists. And it was Hutton who placed geology in a firm context of time by his recognition of the unconformity at Siccar Point in Berwickshire.

But it is at a later date, over forty years after Hutton's death, that the Swiss geologist, Agassiz, identified glacial striae on Blackford Hill, also in Edinburgh, and thereby showed for the first time that Scotland had been covered by ice in the past. Agassiz Rock, as it is now called, was preserved by the City Council of those days and, although it was suffered from rockfalls since, it is still protected from the public by iron railings. This was the first site to be protected for the purposes of geological conservation in Great Britain.

Later that century the City's Parks Department of Glasgow, 60 km to the west of Edinburgh, excavated a group of Carboniferous *Lepidodendron* stumps and prostrate trunks in Victoria Park in 1887 and, not to be out-done by their sister City, constructed a protective building. The need for the public awareness and understanding of science was as much in the institutional mind then as it is now.

In both that century and this, many geological sites assumed special significance whether for strictly scientific or teaching purposes, or for the historical, cultural or even musical interest such as the White Cliffs of Dover and Fingals Cave! We will be visiting later today the Gullet Quarry in the Malvern Hills, a locality that has long been one of the classic teaching localities in this country but has still yielded information of research importance. As we will hear later in the week, the Malvern Hills was one of the very first areas in this country to be 'conserved' in the sense that we use this word today.

However, it was not until the 1940s that the UK government began to take a serious interest in the systematic conservation of geological features. In 1942 the Committee on Land Utilisation in Rural Areas reported that 'geological parks' should be established as a form of nature reserve. This led directly to the appointment of the Geological Reserves Sub-Committee by the Nature Reserves Investigation Committee in 1944, who were given the responsibility of drawing up a list of Geological Parks and Geological Monuments in England and Wales. The Sub-Committee published their report in 1945 identifying 390 localities thought worthy of conservation. In retrospect it appears curious that Scotland was excluded at this stage but that is a reflection, even at that day, of the powers of national devolution. However, Professor J. G. C. Anderson took the initiative of preparing a list of 60 of the most important Scottish localities amongst the wealth that are potentially available.

No further action was taken until the Nature

Conservancy was established in 1949: the government-funded body charged with the task of identifying, and providing protective measures for, sites of conservation importance. The first geologist within the Nature Conservancy, Dr W. A. MacFadyen (self-evidently a Scot) was appointed at that time. Geological and geomorphological Sites of Special Scientific Interest (SSSIs) began to be designated in 1951 using the lists prepared several years earlier. The Scottish list was extended by a further 110 localities as a result of consultation with research workers between 1951 and 1954. SSSIs are selected on the grounds of their research importance, the primary criterion being that they be of national or international importance. Land ownership remains with the owners so that access to an SSSI can generally only be achieved with their consent.

The portfolio of Sites of Special Scientific Interest in Britain is spectacular, ranging from the dramatic landscapes of Precambrian rocks in the northwest Highlands of Scotland, through stratotypes of immense international importance, and igneous complexes where the processes of intrusion were first established, to spectacular localities illustrating glacial and recent processes.

During the next two decades the procedures for geological and geomorphological conservation were progressively refined, with for example the addition of 50 cave sites in 1956. The Nature Conservancy Council initiated a systematic programme to reassess and revise the geological site coverage in 1964, and started a similar exercise for geomorphological sites in 1972.

The Council of the Nature Conservancy Council then authorized the undertaking of a total review of all key Earth Science sites of national and international importance in 1974 and this project, the Geological Conservation Review (known as the GCR), was begun in 1977. As well as the process of site selection and formal designation, there was a continuing role for geologists and geomorphologists to be involved in the protection of the existing portfolio of sites in relation to planning proposals, as well as the wider dissemination through trail guides and other publications.

The GCR was carried out by dividing the country and stratigraphic column into 97 manageable blocks, and then inviting active research workers to identify the key sites within that block. The 'GCR contributors' were a team of some 200 scientists, but the advice of many more including those from overseas was sought where there was, and is, a site of international importance. The GCR was the most thoroughgoing and costly review of the total geological conservation resource of Great Britain ever carried out. Over 3000 GCR sites were identified and about 2200 SSSIs resulted, recognizing that some sites contained more than a single interest. A central feature of the GCR was that the sites should not simply be a series of disconnected locations where features were exceptionally well displayed, but rather that they would form a coherent role, not only being 'jewels in the crown' but forming the crown as well. Such networks of sites would, therefore, provide a rationale to a particular geological period, event or process such as a period of regional igneous activity or mountain building. This approach is important in relation to planning proposals in that the value of a site which may be threatened by development is protected not only in its own right scientifically but as part of a coherent whole.

The GCR is being published as a set of 51 volumes, together with an introductory volume, unfortunately more slowly, and at greater cost, than most of us would choose. You will see examples of the GCR volumes on display at this Conference, and I hope will be impressed by their quality and the detailed level of approach. It is important, my Lord Chairman, that as much of the GCR as possible should be made publicly available soon, otherwise it will date and lose its scientific and strategic value. I know that this is an issue that your Comittee has under active consideration at the present time.

Although the GCR has been described as the Domesday Book of geological conservation, I feel that this is a misnomer. The science evolves and so the relative importance of sites will adjust with time. Indeed some of the sites originally designated as geological SSSIs were denotified because they failed to meet the GCR criteria. Sites will also be destroyed by infill and sedimentation, or be gained by excavation during quarrying and construction. It, therefore, should be no surprise that the successor nature conservation agencies to the NCC regard the GCR as a job well done, but still to be fully placed in the public domain, and recognize the need to move on to new canvases.

Geological conservation in the UK has been dominated by site selection through the quality and coherence of the scientific importance of the site. However, the practical controls of site designation on the use of land can be offensive to land owners who may regard themselves and their forebears as good custodians of their estate. The list of Potentially Damaging Operations (PDOs) which is required to be sent to any owner of an SSSI on designation can be an unhelpful introduction to what should be a partnership. It is only subsequently that, as a management agreement is settled, many of the items on the PDO list fall away. Geological sites are inherently robust and are safe from most agricultural practices which might endanger the wildlife resource. Many areas of geological importance, such as a well-exposed coastline, a remote igneous complex or a landscape form, can be huge and it may seem to some that very different conservation designation processes and procedures should be applied to such sites as compared to those applied to localities where wildlife is on the verge of extinction and the approach to conservation must, necessarily, be different.

Some geological sites require to be protected because of the presence of a limited resource or a vulnerable feature. A debate continues in this country regarding the extent to which a limited site should be open for extraction of fossil, or mineral, material. However, geological sites are, in the main, selected to be used. We have gone further in this direction through the development of Geological Nature Reserves such as Wrens Nest which some of you will be visiting on Wednesday. We have also seen the publication of guides to areas of geological importance which are readily accessible to the public, as is illustrated by the students guide to Malvern geology given to all delegates. I hope that we will see in due course a much greater spectrum of categories of site designation in this country for Earth science sites, recognizing vulnerability of the geological resource, and the importance of sites both for science and for teaching, as well as for public awareness. In particular the importance of landscape has been undervalued, although not under-recognized. Greater use could be made of viewpoints as a means of displaying the underlying geology.

It is of course here that organizations such as the Geologists' Association, and local geological societies, have shown much of the way in supporting a wider view of the importance of geological conservation. This has been recognized in part by the establishment of so-called 'second tier' sites, those that fail the GCR criteria but nevertheless are of immense importance nationally for research, for teaching or for access by amateur groups. These Regionally Important Geological Sites (RIGS) have created new partnerships between publicly and privately funded organizations, and people on the ground and in the field.

Some of these issues which I am now addressing were coming to the fore in our thinking in the late 1980s as we saw life beyond the completion of the fieldwork and designation phase of the Geological Conservation Review. In 1990 the NCC published *A strategy for earth science conservation* which identified six themes:

- defending the existing SSSI network;
- building up a RIGS network;
- identifying gaps in applied conservation practice;
- developing and improving documentation on sites;
- increasing the profile of earth science conservation;
- develop links and interactions with international bodies.

It is Theme 6 which brings us here today, because one of its recommendations was the organization of an international conference in Britain.

We have chosen not to make this just another conference on the science or practice of geological site conservation, but to take a wider, more fundamental, view of the role of geology and landscape in conservation, and in the public eye. The result, as you see, has led us to themes such as the sustainability of geological resources, landscape conservation, the role of local and community initiatives, and public awareness. Our themes underline the duty of care that the geological profession has in conserving the geological resource, contributing to the preservation of the environment, and enhancing the quality of life.

Nevertheless, we faced the challenge of achieving more than an enjoyable conference which would leave warm memories of stimulating papers and discussions, and of the British hospitality. To justify bringing you all here we needed to add value to the occasion in a form which would have permanence.

We therefore came to the view that our fifth theme at this conference should examine the possible need for, and contents of, an international convention. As you will see this theme continues throughout our meeting. We start with keynote presentations later today and follow with workshops tomorrow which will be built on further on Wednesday. By Thursday I hope we will begin to have a consensus on some conclusions.

I would stress that the organizers of this conference and our sponsoring organizations have no views, one way or the other, on the justification for such a convention or its possible contents. I would stress that the UK Government, although aware of these discussions, has not given any consideration, formally or informally, to the matter. Nevertheless, the recently published consultation document on the UK strategy for sustainable development does identify 'the rising demand for sand, gravel and rock quarries and pits which harm wildlife, landscapes and communities' as one of the seven deadly environmental sins! There is no hidden agenda to these discussions at the conference. We genuinely believe that this is an issue which should be explored within that part of the geological profession which is concerned with all aspects of Earth science conservation. If our conclusion on Thursday is not to take the matter forward, the very process of exploration of the subject will have helped in advancing our understanding of geological conservation. If we find there is cause to take the idea of a convention forward, then there is a long hard track ahead of us in persuading others of its importance. It would not be an easy task.

I am delighted to see so many of you here today and so early. The time of 8.30 am is not a conventional one in this country to kick off such an international event, but then I hope that this will be a rather special conference.

References

HUTTON, J. 1788. *Theory of the Earth*. The Royal Society of Edinburgh.

NCC. 1990. *A strategy for earth science conservation*. Nature Conservancy Council, Peterborough.

Opening address

LORD SELBORNE
Joint Nature Conservation Committee, Peterborough, UK

Chairman, ladies and gentlemen.

I echo your words to delegates, Mr Chairman, and I am grateful for the opportunity to preface the work of Conference with a short address. On behalf of the nature conservation agencies in Great Britain, I too would like to bid you the warmest of welcomes and hope that your deliberations this week will be fruitful and stimulating. Certainly from my own contacts with the geological profession, I have little doubt that you will enjoy yourselves and you can be assured that we have made every effort to provide a programme which you will remember.

My task today – in opening the Malvern International Conference on Geological and Landscape Conservation – is a most pleasant one, for a variety of reasons.

First and foremost, the Malvern Conference is the most important international initiative that I have helped instigate during my tenure as Chairman of the Joint Nature Conservation Committee.

It really is splendid to see so many delegates – from over 40 countries – gathered here today. Your presence is a clear confirmation that the need to cherish and protect our geological features and landscapes is of global concern. Possibly geological conservation has not grasped the hearts and minds of the conservation movement, dominated as it is by wildlife issues, to the extent it should. This conference is one attempt to place the geological dimension of conservation on a sounder, international basis. And I take great heart from the support for that intent which is so clearly demonstrated by the enthusiasm with which this Conference has been greeted.

The landscape and its geological foundation provide the essential framework upon which all other conservation issues are built, controlling such critical features as topography, microclimate, water distribution and soil type. Natural features are a vital part of the world's heritage and geological conservation is a means of ensuring that we can pass them on to future generations.

The task is immense and many of us have only taken the first tentative steps down that road. Not only do we need to understand geology and landscape in their own right, but we also need to provide effective connections with other aspects of conservation. I would suggest that the geological profession needs to be even more proactive in creating pervasive associations with conservation organizations. Success will only come if you see your efforts as being complementary to those whose interests barely penetrate the soil surface!

None the less I am certain on one point. If we are to succeed we must build links – across oceans, continents and national borders, as well as across geological time – as it is only through concerted international action that we can hope to rise to the challenge. The immense public interest in the film *Jurassic Park* confirms that there is immense fascination about the geological past and hearts and minds could be won.

There is a particularly apt reason for locating this conference here in Malvern. The Malvern Hills were the first British landscape to receive statutory protection – in the 1920s – with nature conservation and landscape preservation as the intended aims. However, more than this, the Malverns have an enduring place within the British geological scene. They lie as a north–south line separating the Mesozoic rocks of the Severn valley and the Cotswold Hills from the Lower Palaeozoic rocks of Wales. The Malvern line, therefore, is an important structural entity which extends north and south well beyond the actual limits of the Malvern Hills themselves. The geology of the Welsh Borderlands has had a continuing importance and it contains, of course, the rocks upon which much of the original description of the Silurian System was based. Ludlow and Much Wenlock are as much a part of geology as they are a part of the cultural fabric of this fascinating area lying between England and Wales. The Malverns have always been a major training ground for geologists at school or university, and their rocks must be amongst the most hammered in this country. What better place to gather in Britain to examine issues of geological and landscape conservation?

The final reason for my pleasure in being here today is a more personal one. I never willingly miss an opportunity to revisit the Malvern Hills and I am delighted to find out that the geological profession shares my own delight. They offer, to my eye at least, some of the most attractive vistas of lowland Britain, yet with more than a hint of the promise of the uplands. I am delighted that you will, this afternoon, get an early opportunity to get out and see them for yourselves. There is hardly a more beautiful sight in Britain than summer sunrise over the misty Welsh Borderlands as seen from the hills above Great Malvern; I leave that as a challenge to you all.

Conservation in the UK

I would like now to briefly outline for you some of the key characteristics of the UK conservation scene – and most particularly the place of geological conservation within it.

The UK is fortunate to have a particularly strong conservation community, reflected in both statutory public bodies, and voluntary organizations. As a result, it has a respected, and well-articulated, voice which is heard in both government and more widely in society.

The weight given by Government to nature conservation is substantial as is reflected by recent planning decisions.

There are statutory agencies which are organized on a country basis, so that England, Scotland and Wales each have an independent government-funded agency charged with promoting nature conservation and enforcing conservation legislation. Responsibilities in Scotland and Wales include the wider issue of countryside conservation, but this is held by a different agency in England.

The respective bodies are English Nature, Scottish Natural Heritage and the Countryside Council for Wales.

This country-based approach ensures that nature conservation takes place close to the people, and enables a country agency to be innovative and flexible in addressing the particular issues that are special to the heritage and culture of that country. Whereas there is a continuum of conservation priorities throughout Britain, each constituent country has a mix of characteristics quite special to itself, such as the Lower Palaeozoic of Wales, the Mesozoic and Tertiary of England, and the old crystalline rocks of the Scottish Highlands.

This arrangement thus offers a diversity of conservation response – which can only be healthy. It also permits a nationwide unity when that is required.

When the country agencies – and indeed the Countryside and Wildlife Service of Northern Ireland – wish to unite in pursuing initiatives – such as the Malvern Conference – which are of Great Britain, United Kingdom or international scope, they come together through the co-ordinating medium of the Joint Nature Conservation Committee (or JNCC as it is often more conveniently referred to).

I have the honour to be the current Chairman of the JNCC and it is in that capacity that I appear before you today. Major Government support for nature conservation in Britain dates from the 1940s, and since then it has become well established both through legislation and by public interest as reflected in membership of voluntary organizations. Since then, geological conservation, as a heritage discipline in its own right, has existed alongside biological conservation.

Over recent years, as the traditional focus on protected sites selected on the basis of scientific importance alone has broadened to embrace wider countryside and landscape issues, so a greater appreciation of the role of geology has grown within the country agencies.

The influence of geology in providing the substrate for ecological habitats, and as a prime determinant of the character of the countryside, has become more widely recognized and appreciated. We see this conference not only as an opportunity for discussion and dissemination amongst the international geological community, but also as providing the wider dimension of the multiple interests in landscape conservation. It will also provide value-added to our own discussions in this country on this important issue.

Over the last five years, the number of Earth scientists employed by our nature conservation agencies has more than doubled, and now stands at almost thirty officers. They will be taking part in this meeting and ensuring that your discussions are listened to carefully.

Britain also has a rapidly-growing body of voluntary geological conservationists, who find a voice through national bodies such as the Geological Society of London and the Geologists' Association, through regional geological societies and other voluntary organizations.

They are also very active – and extremely effective – in influencing nature conservation at county level, and in safeguarding geological sites locally, notably through the Regionally Important Geological/geomorphological Sites scheme which identifies those sites which, although not in the first water, provide a vital network for research and teaching.

The two strands of geological conservation – governmental and voluntary – came together in 1990, agreed and published a joint strategy to promote geological conservation during the

1990s. This strategy underpins the national approach to geological conservation and this international conference was but one of its recommendations. The Malvern Conference is thus proof of the merits of a collaborative approach, a collaboration which we hope will now extend beyond national boundaries.

Acknowledgements and thanks

The planning for the Malvern Conference has not, of course, happened overnight, and it has involved many organizations and individuals. I cannot possibly thank all concerned by name, but it is only right to pause and acknowledge the major contributors.

Firstly, I must acknowledge the unstinting support of English Nature, Scottish Natural Heritage, and the Countryside Council for Wales over the last two years in planning for Malvern.

The resources committed – in terms of staff time and finance – have been generous, and advice and help has been readily on offer at all levels of the agencies. I extend my personal thanks to Lord Cranbrook, Magnus Magnusson and Michael Griffiths – the chairman of the agencies – in this regard.

The contribution from voluntary conservationists cannot be overemphasized. The Geological Society of London and the Geologists' Association, in particular, have been to the fore. The Geological Society will absorb the costs of publishing the Conference proceedings, and the Geologists' Association has provided generous assistance to help delegates travel to Malvern.

I am delighted to see Professor Charles Curtis, President of the Geological Society, in the audience this morning, and I know that the President-elect of the Geologists' Association, Dr Chris Green, will be joining us this afternoon.

Inspiration in planning the Conference has also extended beyond these shores.

The example of our French colleagues in organizing the Digne Symposium in 1991 has been a beacon to guide us. I am delighted to see that Guy Martini – whom I know many of you hold in great affection – is with us today.

Every effort has been made to make Malvern a worthy successor to Digne, both in terms of technical content and camaraderie – although sadly we cannot promise to equal the attractions of the Provencal wines or its climate in summer! Nevertheless, I would suggest that the legendary Malvern Water, for which these hills are world famous, will be worthy of your attention.

Focusing on the town of Great Malvern, I must thank Councillor John Young, Chairman of the Malvern Hills District Council, for all the help his staff have provided. Also Mr David Judge, Clerk to the Malvern Hills Conservators, for help on many fronts and kind permission for us to invade the hills, albeit temporarily this afternoon – I trust without long-term or irretrievable environmental damage!

Closer to home, I must thank Conference Chairman, John Knill, for leading the Malvern Board of Patrons in its task of advising on conference planning. Their collective experience and wisdom has been of great help to the Organizing Committee.

I thank Des O'Halloran and the other foot-soldiers of the Organizing Committee for their hard work and enthusiasm in planning the Conference. It is their work that has got us here together today.

Lastly I would like to thank all those – too many to mention by name – who have helped in different ways, and of course all of you for coming to represent your country.

The task ahead

Looking at the Programme, I see that you have a busy and stimulating agenda, and much ground to cover, over the coming week.

I leave it to John Knill's capable hands, and those of his colleagues, to shepherd you through the week.

I wish you every success in this task and, Mister Chairman, take great pleasure in declaring this Conference open for business!

Theme 1: Sustainability

Keynote address

Sustaining the Earth's resources

ROGER CROFTS
Scottish Natural Heritage, 12 Hope Terrace, Edinburgh EH9 2AS, UK

Abstract: Using the standard definition of sustainable development from the Brundtland Report, the paper explores the sustainable use of Earth resources, both non-renewables and renewables. Founded on recent global sustainability strategies, it establishes 10 principles of sustainable use. These embrace concepts such as the limits of toleration of the environment, the limits of resource renewal and the limits of knowledge.

The concept and impact of the exploitation of non-renewable and renewable earth resources in terms of environmental sustainability are examined within this paper. Throughout the paper, the 10 principles for sustaining the Earth's resources are highlighted (Magnusson, 1992; Fig. 1).

Initially we must define clearly what is meant by sustainable use. This can best be achieved by referring to the Brundtland Commission (World Commission on Environment and Development 1987) which states that:

> ...sustainable development is development that meets the needs of the present without compromising the ability of future generations to meet their own needs.

That definition in itself only provides a very broad guiding principle; however the report goes on to amplify this in a way which can form the basis of identification of future requirements for sustainable use. It states that:

> ...at a minimum, sustainable development must not endanger the natural systems that support life on earth: the atmosphere, the waters, the soils and the living beings.

That, then, should be the fundamental basis for sustainable use of Earth resources.

The 10 principles of sustainable use of Earth resources

In this analysis, the distinction between renewables and non-renewables provides a useful starting point.

1. Non-renewable resources should only be used wisely with a rate of depletion which does not restrict future options and provides long-term benefits outweighing the depletion of the resource.

By definition, the exploitation of non-renewable resources is essentially unsustainable. Rocks and hydrocarbons, for example, are

---

1. Non-renewable resources should only be used wisely with a rate of depletion which does not restrict future options and provides long-term benefits outweighing the depletion of the resource.
2. The rate of use must be within the renewal capacity of the resource: that is, within the limits of its regeneration and natural growth.
3. Humankind should care for other species. No species should be endangered or become extinct through human misuse of the Earth's resources, and those already threatened or in decline should be restored or enhanced.
4. We should not exploit or intervene in a way which goes beyond the tolerance of the natural system.
5. All proposals for Earth resource exploitation should be subject to independent environmental assessment.
6. Decision-makers have a responsibility to ensure that exploitation of Earth resources does not cause environmental damage elsewhere.
7. Full account of environmental effects must be taken in valuation and assessment studies.
8. Adopt the precautionary principle when there is inadequate knowledge.
9. Develop means of thoroughly appraising substitutes for pristine material.
10. Through education, seek to change personal attitudes and behaviour to respect the environment and its wise use.

Fig. 1. The ten principles of sustainable use of Earth resources

From O'Halloran, D., Green, C., Harley, M., Stanley, M. & Knill, J. (eds), 1994, *Geological and Landscape Conservation*. Geological Society, London, pp. 7–10.

non-renewable within time-scales relevant to humankind; even in locations where they are being created naturally, this creation is intensely localized and is happening infinitely more slowly than the current exploitation rate. Unique geological and landscape features are also non-renewable. All exploitation of such non-renewable resources therefore reduces their stock; but it is a counsel of despair simply to demand that such exploitation must cease – even though that could be seen as the logical implication of the concept of sustainability for this and future generations. Neither is a cessation of exploitation deliverable in practice, since much of human living depends on non-renewables.

Exploitation in our modern society seems to be governed by the maxim 'there is plenty more where that came from'. However, it should be remembered that when we talk about exploitation it is not always entirely detrimental. While some fossil features may be lost, we may also gain snapshots of the Earth's history. In exploiting large mineral deposits, man-made assets which can be utilized by later generations may be created.

2. *The rate of use must be within the renewal capacity of the resource: that is, within the limits of its regeneration and natural growth.*

Turning to renewables, most people would regard renewable Earth resources as inexhaustible and able to meet the needs of future generations without positive management and, therefore, as eminently sustainable. This proposition for unconsolidated sediments, soils, peat and water needs to be considered briefly.

Unconsolidated sediments are constantly being created by the action of water, wind and ice but, nevertheless, are being used at a rate faster than they are being produced (e.g. aggregates). Furthermore, their sources are frequently in highly dynamic and inherently unstable situations in the nearshore zone, along soft coasts and in dynamic river systems, all of which are environments that humankind frequently seeks to control. Such action often results in severe disruption to the natural process concerned. These sediments are also subject to damage by increasing recreational pressures.

Soil is a key life resource for food and timber production, as well as an essential element in the ecosystem. However, increasingly often, overuse of the soil causes damage to its structure, reduction in biological productivity, and reduction in its fertility. As a result, retaining its productive capacity requires eminently unsustainable applications of materials, often from non-renewable resources. These can cause excess concentrations of minerals and nutrients which have a detrimental effect on ecosystems.

Peat is a very significant Earth resource for a wide range of reasons: it stores considerable amounts of carbon; it provides an inventory of recent changes in Earth history; it forms the basis of special habitats; and, in its various hydro-morphological expressions, it is an important landscape type. Unfortunately, it is currently being used commercially for many essentially frivolous activities, particularly in urban areas. While it is certainly still accumulating in many locations, the present rate of extraction far exceeds the rate of accumulation.

Water, in the context of geomorphological processes, is another sustainable resource but its management often has unsustainable effects on the environment. Many parts of the world are subject to drought, some perhaps as a result of human intervention. Frequently these areas are subject to irrigation regimes which, while temporarily increasing the productivity of the land, create problems through, for example, saline deposition. Furthermore, intervention in fluvial systems is now commonplace in an attempt to prevent flooding. Unfortunately, the entrainment works so beloved of engineers frequently ignore natural fluvial processes and fail to take into account the downstream effects.

So, in these terms, are renewables really sustainable? Only if the second principle of sustainability is adhered to.

3. *Humankind should care for other species. No species should be endangered or become extinct through human misuse of the Earth's resources, and those already threatened or in decline should be restored or enhanced.*

Earth resources, whether renewable or non-renewable, provide the basis for habitats and ecosystems and are an important factor in the Earth's biodiversity. They are also the foundation for perhaps less tangible features such as landscape character, and as recreation and tourism resources which are of great benefit to human society. They cannot, therefore, be viewed just as Earth resources *per se* but should be seen as a fundamental basis for biological systems and as an essential component of human life.

By interfering in natural process systems such as coastal cells widespread effects are created which are not easily reinstated. Changes in the landscape may also be caused, for example by harnessing of water energy at the coast or inland for power generation. Further, detrimental changes may be caused in ecosystems, resulting in the loss of habitats and species through, for example, sediment deposition and heavy mineral contamination.

4. We should not exploit or intervene in a way which goes beyond the tolerance of the natural system.

The level of knowledge of the disposition and state of Earth resources is still extremely patchy despite tremendous advances, for example, as a result of remote sensing technology. Equally, some inventories are being compiled purely from the standpoint of exploitation. What is required is a proper framework for resource appraisal.

The need for the resource to be removed and/or used for people's benefit must be demonstrated, bearing in mind the Brundtland definition. More specifically, there is a requirement to assess the need for the resource which is to be obtained from each specific proposed location, bearing in mind a number of factors. The dynamics of the environment at that location must be considered, in conjunction with whether it is solid or unconsolidated material which will be exploited. It is as essential to consider the sensitivity of the resource to exploitation as it is to consider its impacts on natural systems.

5. All proposals for Earth resource exploitation should be subject to independent environmental assessment.

It is also essential to consider the impacts on adjacent ecosystems, bearing in mind that changes in water levels, water chemistry, sediment load and distribution, amongst other things, can have substantially deleterious effects on those ecosystems. In addition, the impacts which exploitation and use can have on the local economy and on local communities must not be ignored.

A resource appraisal framework should also seek to identify opportunities for managing the removal of non-renewable and renewable resources, and the *in situ* use of renewable resources, within the capacity of the environment, so that there is adherence to the second principle defined above.

6. Decision-makers have a responsibility to ensure that exploitation of Earth resources does not cause environmental damage eksewhere.

If we are to develop improved resource appraisal systems, then it is essential that effective decision-making frameworks are created within which the appraisals can be utilized.

To enable proper assessments and valuations to be made, frameworks must internalize both the environmental costs and the environmental benefits, as well as other costs (e.g. to society as a whole). These costs and benefits should be considered in terms not only of the host areas but also of the receiving areas. It is self-defeating, for example, to exploit a non-renewable resource without environmental detriment in the host area if it has impacts on the area receiving that resource beyond the limits of toleration of its environment.

The decision-making frameworks should not be so myopic as to consider only a site-based approach. The landscape approach to Earth resource use and management must be adopted more thoroughly. Zones of varying sensitivity should be identified within which the sites and areas of greatest intrinsic scientific interest for their uniqueness or representativeness, and for their educational interest, should have a higher degree of protection. The development of sensitivity zonation has to be considered in relation to the overall environmental carrying capacity, or the limits of tolerance of change. It must be ensured that the site-based approach, which has often encouraged a cumulative attrition of Earth resources outside the specifically protected areas, does not have a damaging effect on these protected areas.

7. Full account of environmental effects must be taken in valuation and assessment studies.

There needs to be a vetting and approval system which recognizes these environmental parameters. In other words it must not (and cannot) be a market-led approach. Experience reveals that a market-led approach merely leads to over-consumption and exploitation of the most convenient sources of materials, irrespective of their environmental effects. It is also essential that such a vetting and approval system is open and accountable, otherwise society will continue to be at the mercy of the market and the profit motive.

8. Adopt the precautionary principle when there is inadequate knowledge.

Our state of knowledge of internal processes and the effects of exploitation are such that, even when we have appraisal and decision-making frameworks, they are unlikely to be adequate in practice. However, if the principles of sustainable use are to be adopted, then it should be admitted if the knowledge base is limited and inadequate for proper decisions to be made on the exploitation of Earth resources. There will, therefore, be occasions when it is necessary to adopt the precautionary principle and this is the eighth principle.

It has already been indicated that use and exploitation should not go beyond the tolerance of the natural system, in other words it should take the sensitivity of the system fully into account. Where knowledge is limited, rather than adopting the maxim 'hope for the best' as was the practice far too often in the past, in the future a cautious approach should be taken. It does not necessarily mean that the activity is not undertaken at all (although it may do), but that it is undertaken in a much more limited way or at a location where it can be accommodated within environmental limits. All of these factors must be taken into account in the decision-making process.

9. Develop means of thoroughly appraising substitutes for pristine material.

On many occasions we are told that the market requirements cannot be satisfied without using pristine material. However, there is a great deal of material which is often regarded as waste which, with imagination and more investigation, could be a worthwhile substitute for pristine material. Equally, there is a great deal of over-specification of material requirements, particularly in construction, to the effect that the only possible source must be a pristine source rather than the use of a waste material. These points particularly apply to transport infrastructure and the construction of modern buildings.

The author would argue that it is necessary to assess properly the availability of substitutes and recycled materials, and to assess their adaptability to a range of uses, whilst also looking critically at specifications.

10. Through education, to seek to change personal attitudes and behaviour to respect the environment and its wise use.

It is also necessary to change the mind-sets of developers, as well as of others involved in the market, to the use of substitute and recycled material. This means doing much more to educate the users and managers of Earth resources by, for example, providing them with practical advice and guidance on environmental sensitivity, and on the tolerable limits of intervention. However, this educational process needs to be made more pervasive by instilling in those who use and manage the environment a clear sense of responsibility for passing it on to future generations in a way which does not impair their needs. It is no surprise, therefore, that *Caring for the Earth* (IUCN, WNEP, WWF 1991), the global strategy document on sustainability published prior to the Rio conference, places great stress on education.

Conclusions

Clearly, our knowledge base is still relatively limited in many areas of management of Earth resources. It is, therefore, necessary to undertake more research into sensitivity to change, in a way which provides usable output for the decision-makers and their advisors. This may involve studies of active processes as well as of palaeo-environmental records. Monitoring is also an important activity and is required not only to establish the state of the resource and its variations through time, but also to understand better its sensitivity to natural change and human impacts. Monitoring will provide a means of identifying possible tolerance limits as well as wider impacts on habitat and ecosystems.

Whether an International Earth Resources Convention would achieve more sustainable use of Earth resources is something to be considered very carefully by all countries concerned.

References

IUCN, WNEP, WWF 1991. *Caring for the Earth: a strategy for sustainable living.* Gland, Switzerland.

MAGNUSSON, M. 1992. *Sense and Sustainability: homage to Frank Fraser Darling.* 1992 NERC Annual lecture, NERC, Swindon, unpublished.

UNITED NATIONS 1992. *Earth Summit '92.* Regency Press, London.

WORLD COMMISSION ON ENVIRONMENT AND DEVELOPMENT 1987. *Our Common Future.* (the 'Bruntland Commission') OUP, Oxford.

Soil conservation: an international issue

E. M. BRIDGES

*International Soil Reference and Information Centre (ISRIC),
PO Box 353, 6700 AJ Wageningen, The Netherlands*

Abstract: Soils are described as the vital link between the inanimate world of the geosphere and the living biosphere. They are essential for all natural ecosystems, as well as for all the major crops of food, fibre and timber required by human beings. The properties possessed by soils enable them to filter organic materials from rain-water, to adsorb, transform and store many chemical elements which may be released by an exchange mechanism, and to buffer the whole environment against climatic change by storing carbon, and resisting the effects of leaching. Throughout the world, soils are being degraded by erosion, chemical contamination, physical compaction and waterlogging, and the soil's biological diversity is being reduced. This is a situation which cannot be allowed to continue. Any system of sustained development must include adequate provision for the most basic of the world's resources: the soil.

The Malvern Hills, with their core of Precambrian rocks, provided an attractive backdrop to the town in which the 1993 international conference on 'Geological and Landscape Conservation' took place. In a desert country the rocks would be at the surface, forming a stark ridge unobscured by vegetation. In contrast, a combination of environmental factors in Malvern provides a cover of soil which supports the living vegetation, giving verdant, pleasant land sufficient to inspire discussions about how to provide for sustainable development, consistent with adequate landscape conservation.

This contribution is a direct reaction to seeing a conference on earth sciences, sustainable development and landscape conservation which had no mention of soil in the whole of its brochure and other background material. This oversight is a feature of current consciousness about the soil, and it mirrors the common perception of soil as 'dirt', and therefore not worthy of recognition, let alone serious study. This paper is an attempt to remedy this unfortunate situation, and to try and show that soil is too important to be ignored by conservationists. All organisms are inextricably linked to the soil and so it needs to be carefully tended (Fig. 1).

Admittedly, for the hard rock geologists, the Earth's mantle of soil restricts what can be seen of the outcrops where their main interest lies. The importance of solid and superficial geology in the development of our landscape is not in dispute. However, without the thin covering of soil, mostly developed during the Holocene in Britain, all the other ecosystems associated with particular geological formations which people seek to conserve would not be able to survive.

EUROPEAN SOIL CHARTER

1. Soil is one of humanity's most precious assets. It allows plants, animals and man to live on the earth's surface.
2. Soil is a limited resource which is easily destroyed.
3. Industrial society uses land for agriculture as well as for industrial and other purposes. A regional planning policy must be conceived in terms of the porperties of the soil and the needs of today's and tomorrow's society.
4. Farmers and foresters must apply methods that preserve the quality of the soil.
5. Soil must be protected against erosion.
6. Soil must be protected against pollution.
7. Urban development must be planned so that it causes as little damage as possible to adjoining areas.
8. In civil engineering projects, the effects on adjacent land must be assessed during planning, so that adequate protective measures can be reckoned in the cost.
9. An inventory of soil resources is indispensable.
10. Further research and interdisciplinary collaboration are required to ensure wise use and conservation of the soil.
11. Soil conservation must be taught at all levels and be kept to an ever-increasing extent in the public eye.
12. Governments and those in authority must purposefully plan and administer soil resources.

Fig. 1. The European Soil Charter (Harcourt 1990)

*From O'Halloran, D., Green, C., Harley, M., Stanley, M. & Knill, J. (eds), 1994,
Geological and Landscape Conservation.* Geological Society, London, pp. 11–15.

It is equally surprising that biological scientists, other than the occasional ecologist, have so little interest in soils; not only does soil support life, it is itself an ecosystem teeming with living creatures belonging to many different phyla. All these different forms of life, from bacteria to mammals, participate to some degree in the recirculation of chemical elements in which the soil plays a central role.

Soil is a vital support system for life on Earth (Bridges 1978a, b). It provides all terrestrial ecosystems with physical support, nutrients and moisture, including virtually all the major crops of food, fibre and timber that the human race requires for its existence. The soil is the link between the inanimate world of the geosphere and the living biosphere; without this link, life as we know it would not exist on Earth. As the parent material for soil formation, the weathering product from the various rocks is one of the most important inputs into the soil system. Surely, protection of the soil is an aspect of conservation which deserves attention in a conservation conference.

What is soil?

Soil has been described as 'the stuff in which plants grow', but in terms of scientific appreciation of soil, it is necessary to be more precise. A comprehensive definition (Joffe 1949) combines the physical, chemical and biological constituents of soils and gives the right weight to the morphology and other components of the soil. A second definition has been provided recently by a Council of Europe document (1990):

> The soil is a natural body of animal, mineral and organic constituents, differentiated into horizons of variable depth which differ from the material below in morphology, physical make-up, chemical properties and composition and biological characteristics.

> Soil is one of the earth's ecosystems and is situated at the interface between the earth's surface and the bedrock. It is subdivided into successive horizontal layers with specific physical, chemical and biological characteristics.

These definitions are provided because it is essential to establish clearly that soil is a natural entity with a structure, composition and organization. This allows different types of soils to be identified, their distribution to be mapped and their relationships to other natural systems investigated. Only with a full knowledge of these interrelationships between geology, soils, climate, hydrology, natural ecosystems and socioeconomic setting, can the most effective, sustained use for any particular piece of land be determined.

Soil properties

Soils are the product of the interplay of several different environmental factors. Five classical factors were identified by Jenny (1941) as setting the parameters for determining the nature and properties of the soil developed at any particular place. These factors are climate, organisms, relief, parent material and time. The balance of these factors and their influence varies in strength greatly from place to place throughout the world, giving rise to a complex pattern of soils. The interaction of 'active' soil formers, climate and organisms, with the 'passive' parent material influenced by relief and acting through time gives rise to recognizable three-dimensional soil bodies, referred to as pedons, each with its characteristic sequence of horizons. Any one particular soil body is located within a pattern of other soils, partly depending upon soil hydrological conditions and position in the landscape.

There are a number of inherent soil properties which have proved vital for plant life and invaluable for people's use of soils in the production of food, fibre and timber:

- filtration: soils can filter many elements from the downward percolating rain-water, transform their chemical state and recycle them;
- adsorption: the capability to retain plant nutrients against the leaching effects of rain-water;
- storage: the capability to hold moisture and nutrients and release them when needed by plants;
- buffering: the way in which soil holds certain chemical elements means that soils can resist the effects of environmental change.

These processes of filtration, adsorption, storage, and buffering are also used to good effect in the fight against degradation of the environment.

Soil degradation

The degradation of soils, unlike many other environmental problems, is not a phenomenon of recent origin. Throughout history, soils have been influenced and changed for better or worse by human beings; indeed many soil scientists would add the activities of humankind to the classic list of five soil-forming factors described previously. In the past, soil degradation consisted mainly of the loss of soil through water or wind erosion, induced by cultivation and deforestation, but at the present time problems of chemical, physical and biological degradation are of equal concern to soil scientists.

Chemical degradation

Two main problems occur in the chemical degradation of soils: depletion and contamination. Natural depletion of all soluble elements occurs from soils through the process of leaching by rain-water, but this form of degradation also occurs through overcropping or overgrazing. Depletion of plant nutrients from soils has been an obstacle to increasing food production since people began to practise settled forms of agriculture.

Chemical pollution of soils, which may be dispersed or concentrated, is usually the direct result of human activity and has increased dramatically in recent years (Bridges 1989). Dispersed pollution results from acid deposition and atmospheric fall-out of other toxic elements and compounds. Many agricultural soils have been contaminated by the addition of sewage sludge containing toxic metals such as cadmium or zinc. In their bio-toxic forms, these metals inhibit the soil fauna, decreasing the numbers of micro- and meso-fauna by up to 50%, and the effect may be seen for more than 40 years after the soil was originally contaminated (Brookes & McGrath 1984). The toxicity of metals varies greatly with pH and the redox state of soils. Use of phosphatic fertilizers containing cadmium has also been responsible for raising the concentration of this metal in arable lands.

'Natural' soils contain calcium, magnesium, potassium, sodium and other elements in cationic form, attracted to and held by the negatively-charged clay–humus complex. Several potentially toxic metals also exist in a cationic form, and these too can be retained by the soil's exchange complex. Many studies have shown that the fall-out of copper, lead, zinc and cadmium through human activities is widespread (Alloway 1990). All these elements are retained by the soil through linkage to organic matter and clay minerals. This leads to an interesting situation whereby the soil can gradually accumulate toxic metals. Up to a critical point, the soil can absorb these without harm, which is a buffering function, and so the environment is protected temporarily against the impact of the toxic metals. Unfortunately, the soil's capacity to store these toxic metals is not infinite, nor is it of a fixed amount. For example, if the pH of the soil is decreased by acidification as is widely occurring, it is possible for some metallic elements to be catastrophically released in what has become known as a 'Chemical Time Bomb' (Batjes & Bridges 1991; Stigliani 1991; Bridges 1993).

Concentrated forms of soil pollution by chemicals are usually smaller in extent being the result of accidental spillage, ignorance or deliberate dumping of toxic substances to avoid paying for their safe disposal. Regrettably, even when supposedly disposed of safely in licensed landfills, leakages have caused problems for the adjacent ecosystems and deep percolation has reached aquifers with an unfortunate impact on drinking water supplies. Although measures can be taken to reclaim polluted soils, the expense is usually great and it is better to avoid the problem in the first place (Bridges 1992).

Physical degradation

Compaction of soils has become a problem in recent years as the size and weight of farm machinery has increased. Subsoil compaction may occur below the plough-layer through frequent use of heavy machinery; this may be difficult and expensive to rectify. Nearer the surface, a combination of compaction and tractor wheel-spin may limit the extent to which plant roots can exploit the soil for moisture and nutrients and the free movement of gases between soil and the atmosphere. Such compaction is frequently met upon sites which have been restored after opencast mining, but can also occur on normal agricultural soils which have been subject to heavy traffic, particularly in wet years.

Compaction decreases infiltration capacity and porosity, and increases bulk density and resistance to plant root penetration. In soil with a reduced infiltration capacity saturation is reached quickly and rapid surface flow is generated which can cause erosion and carry pesticide residues from the soil before they have been degraded by the soil bacteria. Crusting or surface sealing can occur, especially in silty soils, where the organic matter content has been lowered by oxidation through long-term cultivation; again this results in decreased infiltration and a greater likelihood of run-off and erosion (Van Lynden 1992).

Much effort has been expended upon erosion and its measurement, duplicating work done in the 1930s when most of the effective methods for counteracting soil erosion were worked out and demonstrated to be effective. The significance of erosion is that it is frequently a symptom that something else is wrong, and so it is necessary to look at the management of the land, particularly its physical and chemical conditions, rather than spending too much time on measuring yet another example of soil erosion. If the underlying cause or causes are addressed, rather than the effects, the problem of soil erosion falls into the correct perspective.

Biological degradation

The importance of biomass production has already been commented upon, but in addition soils are effectively the agents for decomposition of organic matter and its conversion into an innocuous, semi-stable, material referred to as humus. On farms, organic wastes from domesticated animals are used to supplement the humus content of arable fields which is gradually oxidized, and until the relatively recent advent of artificial fertilizers organic wastes were the main means of returning to the soil nutrients removed by each crop. In natural ecosystems, the leaf-fall from the vegetation and the death of the faunal component of soils also returns nutrients to the soil in a similar manner. As the soil is able to filter out, incorporate and safely decompose organic wastes, highly organic solutions do not normally reach the streams to cause eutrophication or the fouling of groundwaters in aquifers.

Soil scientists are concerned particularly about the conservation of biological diversity within the soil. The significance of the soil fauna in the biogeocirculation of elements has been stressed already. The creatures involved in this activity are at risk from application of pesticides to crops, and also from residues which fall directly onto the soil. Although soil fauna is rarely obvious, since most species are of microscopic size, it plays an important role in the breakdown of organic substances. Their unspectacular lives have not attracted much current scientific interest, but it is worth remembering that it was not beneath the dignity of the great Charles Darwin to study the humble earthworm's contribution to soil formation.

Conclusions

In Europe, 72 million hectares of cultivated land, 54 million hectares of pasture land and 26 million hectares of forested land are affected by some form of soil degradation. Erosion by water accounts for 52%, wind 19%, chemical degradation 12% and physical degradation 16% of human-induced soil degradation (Oldeman *et al.* 1991; Van Lynden 1992). The complexity of the problem arises from several different factors including an increasing urban population, which demands more land for buildings, infrastructural works and recreation. All of these activities preferably take place on level, well-drained soils which are equally in demand for crop production. Further, there is pressure on the remaining cultivated land, with intensification of agriculture using increased applications of fertilizers, manures, pesticides and herbicides, as well as heavy machinery, which has led to chemical and physical problems which usually underlie the more visible effects of soil erosion. Mechanization of many agricultural activities has led to a demand for larger fields in which the risks of erosion by water and wind are greater. Increased industrialization, changes in industrial practices, and accidents have caused indirect effects on soils through acid deposition and fallout of toxic metals. Radionuclides from nuclear testing and accidents at nuclear power stations also have resulted in unforeseen long-term soil problems of contamination for certain soils.

By the same token, different soils have greater or lesser resilience to damage, and require sympathetic management in different ways. Some soils are extremely vulnerable to degradation, others are more tolerant. This concept of soil vulnerability to degradation has been gaining recognition in the European community of environmental scientists in recent months (Batjes & Bridges, 1991; Bridges 1993). The soil, through its humus content and clay minerals, is capable of holding many chemical elements in ionic form in the process of cation exchange.

Another reason why we should take note of our soils in different ecosystems is that they are deeply involved in the relationships between the terrestrial surface and the atmosphere. It is only recently that atmospheric scientists and modellers have concluded that many of the gaps in their data revealed by their climatic change models must be related to soil gaseous emissions and absorption (Bouwman 1990). Soils contain a great deal of carbon, two to three times that stored in vegetation, which in certain circumstances may be converted into carbon dioxide and released to the atmosphere; in poorly drained conditions and in the absence of oxygen, methane may be produced; nitrous oxide, another 'greenhouse' gas, also may be formed and released as bacteria interact with nitrogenous fertilizers. Freely drained soils are also a sink for methane which is oxidized by the bacteria present. Thus soils are intimately involved in the greenhouse effect and its influence on climatic change.

Three years ago the International Society of Soil Science issued a pamphlet which posed the question *Do Soils Matter?* (Sombroek 1990). In answer to this rhetorical question it outlined a strong case for soil conservation and the recognition of the importance of soils to the existence of life on Earth.

Soils are major determinants of terrestrial ecosystems throughout the world, and are sources, transformers and stores of plant nutrients. They are major support systems for human life and welfare, strongly influencing what is produced from the land. Soils are buffers and filters for contaminants and pollutants. They are important sinks and sources for biogeochemical cycles, especially those involving carbon dioxide, methane and nitrous oxide. As they cover most of the terrestrial surface, soils influence radiant and sensible heat exchange, as well as the land surface reflection characteristics, and form a significant link in the hydrological cycle. Soils are so closely integrated with the underlying rock, climate, hydrology, vegetation and relief of the land, that they should be adequately represented in any attempts at conservation and the sustainable use of the land.

The author thanks Drs J. H. U. van Baren and N. H. Batjes for helpful comments on the content and presentation of this paper.

References

ALLOWAY, B. J. (ed.) 1990. *Heavy Metals in Soils.* Blackie, London.

BATJES, N. H. & BRIDGES, E. M. (eds) 1991. *Mapping of Soil and Terrain Vulnerability to Specified Chemical Compounds in Europe at a scale of 1:5M.* Proceedings of an International Workshop organized in the framework of the Chemical Time Bomb Project of the Netherlands Ministry of Housing, Physical Planning and Environment and the International Institute for Applied Systems Analysis (Wageningen 20–23 March 1991). International Soil Reference and Information Centre, Wageningen.

BOUWMAN, A. F. 1990. *Soils and the Greenhouse Effect.* John Wiley & Sons, Chichester.

BRIDGES, E. M. 1978a. Interaction of soil and mankind in Britain. *Journal of Soil Science*, **29**, 125–139.

—— 1978b. Soil, the vital skin of the Earth. *Geography*, **63**, 354–361.

—— 1989. *Soil Contamination and Pollution.* Annual Report 1989, International Soil Reference and Information Centre, Wageningen.

—— 1992. Dealing with Contaminated Soils. *Soil Use and Management*, **7**, 25–31.

—— 1993. Soil Vulnerability with respect to Chemical Time Bombs. *In*: TER MEULEN, G. R. B., STIGLIANI, W. M., SALOMONS, W., BRIDGES, E. M. & IMESON, A. C. (eds) *Chemical Time Bombs.* Proceedings of the State of the Art Conference on Chemical Time Bombs, Veldhoven, September 2–5, 1992. RIVM, Bilthoven.

BROOKES, P. C. & MCGRATH, S. G. 1984. Effects of metal toxicity on the size of the soil microbial biomass. *Journal of Soil Science*, **35**, 341–346.

COUNCIL OF EUROPE 1990. *Feasibility Study on possible National and/or European Actions in the Field of Soil Protection.* Report presented by the Belgian delegation to the 6th European Ministerial Conference on the Environment, Strasbourg.

HARCOURT, H. 1990. Quality as well as quantity. *Naturopa*, **65**, 4–6.

JENNY, H. 1941. *Factors of Soil Formation.* McGraw-Hill, New York.

JOFFE, J. S. 1949. *Pedology.* Pedology Publications, New Brunswick, New Jersey.

OLDEMAN, L. R., HAKKELING, R. T. A. & SOMBROEK, W. G. 1991. *World Map of the Status of Human-induced Soil Degradation.* International Soil Reference and Information Centre/United Nations Environmental Programme. Wageningen.

SOMBROEK, W. G. 1990. *Do Soils Matter?* International Society of Soil Science, Wageningen.

STIGLIANI, W. M. (ed.) 1991. *Chemical Time Bombs: Definition, Concepts and Examples.* Executive Report 16. International Institute for Applied Systems Analysis, Laxenburg.

VAN LYNDEN, G. W. J. 1992. *Handbook of Soil Conservation in Europe.* Working Paper and Preprint 92/10. International Soil Reference and Information Centre, Wageningen.

Conservation of peatlands: the role of hydrogeology and the sustainable development principle

D. DALY

Geological Survey of Ireland, Beggars Bush, Haddington Road, Dublin 4, Ireland

Abstract: Traditionally, the basis for peatland conservation has focused mainly on their ecological importance. However, their conservation is also critical from an Earth science viewpoint. They are a major landscape feature and geological unit in northwest Europe. As plant 'graveyards', they can play a vital role in understanding the climate and geography of the last 10 000 years. The high water content of peat makes it a very unusual, interesting but difficult geological material. Water in peat, as with other rocks, is present as free water in the pores, but unlike other rocks it is also an essential part of the structure of the peat.

Peat cutting, drainage, afforestation and agricultural development have destroyed most of the peatlands of northwest Europe and are posing a serious threat to the few remaining relatively intact peatlands in Ireland and Britain. Consequently, their conservation is currently a vital nature conservation objective.

In the past, the basis of conservation tended to rely on the scientific value of peatlands. There is now a need to consider peatland conservation under the general principle of sustainable development, backed up by an effective public relations campaign. This implies using economics' techniques to provide a valuation of intact peatlands. It also implies a broader basis for conservation.

One of the difficulties in conserving and restoring peatlands is that they are sensitive to the amount and quality of the water available. The key elements in the conservation of peatlands are the understanding and control of the surface water and groundwater conditions, but the ability of peat to withstand changes in water regime and chemistry is poorly understood.

Peatlands (or mires) formerly covered 8% of western Europe (Goodwillie 1980) and were a dominant landscape feature in countries such as Ireland, Scotland, the Netherlands, Sweden and Finland. A peatland is a specialized biotope, with a unique flora and fauna, which is also unique as a geological material since it consists largely of water. As a result, peatlands are extremely delicate and prone to damage. Also their utilization has a considerable economic importance. As a consequence, peat cutting, drainage, afforestation and agricultural development have destroyed most of the peatlands of northwestern Europe. With the exception of about 50 ha, all natural peatlands in the Netherlands have been lost. In Britain there has been a 90% loss of blanket bogs, while 98% of raised bogs are lost (O'Connell 1993). In Ireland only 14% of the original area of blanket bog remains relatively intact, while the figure for raised bogs is 6%. Ireland remains one of the last strongholds of peatlands, although this status is now under threat from human activities. The conservation of peatlands is now the most urgent issue confronting nature conservation – including Earth science conservation – in Ireland.

Peat formation and distribution

Peatlands are accumulations of waterlogged plant organic matter topped by a surface layer of living plants. Water plays a vital role in peat formation, since it acts as a preservative against decay by excluding the entry of oxygen. Consequently, peatlands form in wet areas where the rate of plant production exceeds the rate of plant decomposition. A high water table is an essential condition for peat formation. Achieving this depends on a number of conditions, such as climate, topography, geology and hydrogeology; conditions which vary throughout Europe. As a result, peatland distribution and types vary. They formed mainly along the northwestern fringes of Europe: in Ireland, Scotland, the Netherlands, northern Germany, northern Poland, Sweden, Finland and northern Russia. Nine main types have been described (Goodwillie 1980).

In Ireland there are three types of peatland – raised bogs, blanket bogs and fens. The term 'bog', which comes from the Gaelic word meaning soft, is used to describe peatlands where the only water source is rainfall; as a result such peatlands are acidic and poor in minerals.

Fens, in contrast, are less acidic, and often alkaline, as the main water source is mineral-rich (usually calcium-rich) groundwater and surface water. Raised bogs attain their finest development in Ireland, where they cover large areas of the Midland limestone plain. Most have developed from fens – as the peat continued to accumulate, it formed a flattened dome, slightly higher than the surrounding area – hence the name. The peat of raised bogs consists largely of *Sphagnum* mosses, and can reach thicknesses of up to 15 m. Blanket bogs form where rainfall exceeds 1200 mm per annum and falls on more than 250 days per annum. They cover vast tracts, mainly in the wetter and mountainous areas of Ireland and Scotland.

Peat: an extraordinary and special geological material

Peat is a fascinating and extraordinary earth material, seldom studied by geologists in Ireland and Britain. It is the only sediment that, in its natural state, consists of living plants growing on their inanimate and accumulated ancestors – it is both alive and dead. It can be termed a plant (and animal) graveyard. Because of the ability to preserve animals, pollen and other plant remains, peatlands play a vital role in understanding the environment of the last 10 000 years. This benefit is not merely academic or historical; it can be used to assist in the understanding and prediction of present and future pollution and climatic effects.

Peatlands were a major landscape feature and geological unit in northwest Europe. In Ireland they cover 17% of the land surface – a higher proportion than any other European country with the exception of Finland. As such, the wide open spaces provided by peatlands are part of the beauty, scenery and character of Ireland.

A striking characteristic of peat is its high water content. Peat contains more water than milk contains – it is over 90% water by volume. The high water content makes it a very unusual, interesting but difficult geological material. Water in peat, as in other rocks, is present as free water in the pores, but unlike other rocks water is an essential part of the structure of the peat, bound physically, chemically, colloidally and osmotically. Only a small proportion is mobile, although this varies with the hydraulic conditions. This means that the Darcy flow equation, which is the mathematical basis for the flow of water in rocks, does not properly apply to well-decomposed peat. As a consequence, the hydrodynamics of peat are complex and not properly understood even today.

Conservation value of peatlands

Peatlands are worthy of conservation for a variety of reasons: geological, geomorphological, ecological, social and economic. These are summarized as follows:

- Peatlands are unusual and important sediments.
- Peatlands are major landscape features in countries such as Ireland, Scotland, Poland and Finland. As such, they are not only of geomorphological importance but are part of the beauty and scenery of these countries, a factor which can lead to economic benefits from tourism.
- Peatlands are unique ecosystems (Doyle 1992). They are an essential part of the biosphere, representing extreme habitats where waterlogging and, in many cases, restricted nutrient supply are important features; and they support unique combinations of plants and animals adapted to these environmental conditions.
- Peatlands are of immense educational and scientific interest. They can be used to study (i) hydrology, hydrogeology and geotechnics, being unusual geological materials which are easy to instrument; (ii) the ecology of communities and the operation of ecosystems; and (iii) the palaeogeography of the last 10 000 years.
- Peatlands have been a source of inspiration for poets, artists, writers and musicians and, for people living close to peatlands, have a folklore and tradition associated with them.

These reasons can be incorporated into three principles for justifying the conservation of peatlands (adapted from Stevens *et al.* 1994):

(1) Peatlands have inherent values themselves and should be protected or conserved irrespective of their usefulness, now or in the future, to humans.
(2) Conservation has benefits for humankind, both now and in the future.
(3) We have a duty to future generations to preserve our heritage so that it may become their's. This principle of sustainable development can link the first two and provide a coherent basis for conservation.

Peatland conservation and sustainable development

The World Commission on Environment and Development (known popularly as the Bruntland Commission) defined sustainable development as 'development that meets the needs of the present without compromising the ability of future generations to meet their own needs' (WCED 1987). If future generations are not to be made worse off by present-day activities, they require the same potential economic opportunities for improving their welfare as the current generation enjoys. The portfolio of capital assets – 'critical' natural, 'other' natural and human-made – must, therefore, be managed in such a way as to preserve this potential (Turner 1991). As peat is effectively a non-renewable resource, it can readily be argued by scientists that intact peatlands today are 'critical' natural assets. As it is not possible to pass on peatlands as a natural asset to future generations if they are cut away or drained, then under the principle of sustainable development, it can be argued that conservation of a large proportion of the remaining relatively intact peatlands in countries such as Ireland is essential.

As an abstract concept, sustainable development is attractive and has been accepted by many. Consequently, associating peatland conservation with this concept or principle is now essential. However, operationalizing sustainability has proved difficult (Convery 1992) and it is often a cliche rather than a reality. In a pragmatic world, where 'short termism' often predominates, decision-makers and the general public are likely to need convincing of the need to conserve peatlands as part of a sustainability programme, particularly at times of high unemployment and economic difficulties in Europe. The old-fashioned scientific/academic arguments in favour of conservation, often presented poorly in scientific jargon, will not be successful. However, there is a danger that conservationists will take a narrow view of sustainable development, and feel that merely quoting the term 'sustainable development' in their arguments is enough. In fact, sustainable development represents a fragile balance between environmental conservation, economics and technology (Turner 1991). In the past the utilization of our natural resources, such as peatlands, to fuel economic growth was an accepted policy. With well-being now seen to depend also on environmental quality, how can the sustainability concept be used to assist in the conservation of peatlands now and in the future?

In the author's view, three approaches are required. First, relatively intact peatlands must be given a valuation in economic terms. This valuation can be used by decision-makers in choosing strategies for sustainability and can also be used to mobilize public support for conservation. Second, good public relations, which takes account of the sustainability principle and of a broad range of views including those of local people, is needed to market the conservation of peatlands. Third, an understanding of the water aspects as the basis for proper technical measures to engineer and manage successfully conserved peatlands and to enable peatland restoration.

What value can be given to peatlands? It is easy to value their development potential for either fuel, electricity, afforestation, horticulture or agriculture. It is not easy to obtain a value for relatively intact peatlands. Three types of value constitute the 'total economic value': direct use, indirect use and non-use (Turner 1991). In Ireland the main 'direct use' value of peatlands, where human livelihoods are supported, is likely to be limited and is derived from the scientists and tourists that visit peatlands. However, with the expansion of ecotourism in recent years this may become more important. Conserved peatlands can provide an 'indirect use' value by contributing to a good environmental quality. The perception internationally of a good environmental quality in Ireland is considered to give a competitive advantage to the agricultural and food-processing sectors and consequently is an aid in increasing exports and employment. Also tourism, which is a major industry in Ireland, depends on good environmental quality. The use of peatlands for educational and scientific purposes can be considered an 'indirect use' value. So peatland conservation can be beneficial from a strict economic viewpoint, and there is the potential for economists to estimate a use-value. However, the 'non-use' value (existence, bequest and intrinsic value) of peatlands cannot readily be measured in economic terms as it is a nebulous concept. Yet it may be significant in terms of people's willingness to pay for their conservation.

These points are easy to make by people who support peatland conservation. However, they are generalized and inadequate and so it is essential that the expertise of economists should be used to evaluate the economic importance of intact peatlands and that this evaluation should be integrated into the arguments for peatland conservation and restoration, and into public and government policy.

There are, undoubtedly, dangers with attempting an economic valuation of natural peatlands. First, it may oversimplify a complex multidimensional situation. It can give an apparently precise result, while in reality much of the information is not amenable to a numerical valuation. Second, estimating the non-use value is subjective, and different methodologies may give different results. Third, the intrinsic value of peatlands and, in particular, the value of non-human organisms, given by economists may not reflect the views of conservationists. However, economic valuation is influential among decision-makers at present. Unless conservation bodies adopt this approach, there is a danger that their influence will be less effective. Also, one means of overcoming the dangers is for the conservation bodies to carry out the valuations, rather than leaving it to those who are not sympathetic to conservation.

The role of hydrogeology in the sustainable management and conservation of peatlands

Peat equals water ... almost! As a consequence, it is a delicate geological material, prone to removal of water by drainage and consequently to damage and often destruction. An example from Clara Bog, a raised bog of international importance in the Irish midlands, illustrates this. In the last 200 years a road has traversed the bog dome. As a consequence of drainage, the thickness of peat has been reduced from 12 m to 6 m at the road, subsidence has occurred for several hundred metres on each side of the road and the bog is no longer functioning properly in this area. Drainage has removed water and in the process a substantial part of the peat. It is not just drainage on a bog that can cause subsistence; bogs and fens can also be affected by activities at a distance, for instance by arterial drainage and groundwater abstraction.

Bogs, by definition, are sensitive to the quantity and quality of water available. However, the ability of bogs to withstand changes in the water balance or water chemistry are factors which remain poorly understood. Yet the key element in the actual conservation of bogs is the understanding and control of the water regime – the water in, on, around and beneath them.

This understanding can only be achieved by detailed geological, hydrological, hydrogeological, hydrochemical and ecohydrological investigation; in other words by a multidisciplinary approach. This work usually involves techniques such as geological mapping, geophysics, surface runoff measurements, rainfall measurements, evapotranspiration estimation, drilling, piezometer installation, water level monitoring, permeability determinations, hydrochemical sampling and analyses, numerical modelling and vegetation mapping. The geologists provide the local and regional framework for bogs, the hydrologists and hydrogeologists provide the local and regional water flow regime and, together with the ecologists, integrate the varying water regime with the different plant communities. Also, hydrogeologists can, with the assistance of numerical models, provide predictions on future changes. With the understanding of the water regime and the ecology provided by the integrated multidisciplinary approach, technical measures, such as water control by dams, can be initiated to enable either conservation or restoration of peatlands. It also enables the conservation authorities to assess the conservation viability (sustainability) of a particular peatland prior to purchase or prior to ranking it as a site of scientific interest or natural heritage area.

An example of this approach is provided by the Irish–Dutch Raised Bog Geohydrology and Ecology Study, which is being undertaken on the Irish side by the National Parks and Wildlife Service of the Office of Public Works and the Geological Survey of Ireland, and on the Dutch side by the Department of Nature Conservation, Environmental Protection and Wildlife Management and the National Forest Service, with assistance provided by Irish, Dutch and British third-level colleges. This study will be completed in 1994.

Conclusions

Botanists have been to the forefront of peatland conservation and, without their work and initiative, the situation in countries such as Ireland and Britain would be much worse than that which exists at present. However, there is now a need to:

- appreciate the importance of peatlands as geological entities;
- consider peatland conservation under the general principle of sustainable development rather than the more narrow scientific approach;
- use economics' techniques to put a value on intact peatlands and on peatlands with restoration potential;
- integrate all these aspects into a public relations strategy, which should have a high priority; and
- understand the water aspects of peatlands and how they relate to plant communities. This requires the involvement of people with expertise in geology, hydrology and hydrogeology as well as ecology. It enables a proper scientific and technical approach, without which successful conservation is unlikely.

In conclusion, geologists, and in particular hydrogeologists, have a critical role to play in successful peatland conservation.

The influence of discussions with the people working at the Dutch–Irish Raised Bog Geohydrology and Ecology Study are gratefully acknowledged. This paper is published with the permission of Dr P. McArdle, Director, Geological Survey of Ireland.

References

CONVERY, F. J. 1992. Economy and environment – towards sustainability in Ireland. *In*: FEEHAN, J. (ed.) *Environment and Development in Ireland.* The Environmental Institute, University College Dublin, 1–6.

DOYLE, G. 1992. Peatlands and other wetlands. *In*: FEEHAN, J. (ed.) *Environment and Development in Ireland.* The Environmental Institute, University College Dublin, 1–6.

GOODWILLIE, R. 1980. *European peatlands.* Nature and Environment Series No. 19, Council of Europe, Strasburg.

O'CONNELL, C. 1993. Irish bogs – a crisis of survival. *Technology Ireland*, **24** (10), 32–35.

STEVENS, C., ERIKSTAD, C. & DALY, D. 1994. *Fundamentals in Earth Science Conservation.* Memoire Special de la Societe Geologique de France, **165**, 209–212.

TURNER, K. 1991. Sustainable wetlands: an economic perspective. *In*: TURNER, K. & JONES, T. (eds) *Wetlands: Market and Intervention Failures.* Earthscan Publications Limited, London, 1–38.

WCED 1987. *Our Common Future.* World Commission on Environment and Development, Oxford University Press.

International Union of Geological Sciences and the environment

FREDRIK CHR. WOLFF

Geological Survey of Norway, PO Box 3006, Lade, N-7002 Trondheim, Norway

Abstract: The International Union of Geological Sciences (IUGS) has in its 30 year existence concentrated its activities around traditional geoscience topics. In 1989 it made a turn towards activities related to the environment when the new Commission on Geological Sciences for Environmental Planning – Cogeoenvironment – was created. This paper gives a report on the activities of this commission and other IUGS involvement with environmental issues.

IUGS was founded in 1960 and today has 100 member countries. During its 30 year existence, IUGS has concentrated its activities on the traditional academic fields of our science. A number of commissions have been established on academic themes such as stratigraphy, petrology, tectonics etc. Themes of more applied geology like fossil fuels and themes within modern technology such as data-acquisition and informatics have also been covered.

At the 28th International Geological Congress in Washington in 1989 the IUGS underlined the importance of geoscientific aid to environmental planning which was considered to be the most important role of geoscientists in the next decade (Cordani 1989).

As a result of this new IUGS emphasis on the environment, the Commission on Geological Sciences for Environmental Planning (Cogeoenvironment) was established.

Report on Cogeoenvironment's activities

Since its start in 1990 the commission has:

- formulated its terms of reference, its goals and objectives;
- established its organization and liaison with a number of other international organizations;
- created a number of working groups and programmes.

Goals and objectives

From the very start the commission decided to concentrate on how to increase awareness among the general public and encourage increased understanding among planners and decision-makers of the essential role of geoscience in sound planning and managing the use and protection of the environment.

Other goals are to strengthen the interest and participation of the geoscience community in the planning process, encouraging a greater awareness of the economic, social and political constraints involved; to promote research required for a better understanding of environmental processes and the development of new approaches and methods for solving environmental problems; and to improve capabilities to predict and forecast the evolution of land and of the geological processes that affect society in the short and medium term, especially in regard to environmental impact assessments and natural hazards. Several projects and working groups have been initiated to serve some of these objectives.

Finally, mention should be made of the fact that the commission has decided to try to improve communication, cooperation and the flow of information among Earth scientists and other professionals working anywhere in this field, but especially in those countries where environmental problems are most severe. This objective is tackled by the establishment of a working group for training and education with four major programmes including a list of relevant textbooks and papers related to environmental geology.

Organization

Cogeoenvironment has established an international organization which is open for different kinds of membership:

- A Board consisting of Chairman, Vice Chairman, General Secretary and one member from each continent or larger geographical region.
- Corresponding members are colleagues who are active in this field of geoscience. The commission now has more than 60 such members from more than 30 countries.
- Supporting members are private companies and governmental organizations that regard

the work of the commission as worthy of economic support.

From the very first moment, the commission has tried to ensure cooperation with relevant organizations that may be instrumental in carrying out certain aspects of our goals. Such organizations are:

- IAEG (International Association of Engineering Geology)
- IAH (International Association of Hydrogeologists)
- AGID (Association of Scientists for International Development)
- UNESCO (The United Nations Educational; Cultural and Scientific Organization)
- SCOPE (ICSU's Scientific Committee on Problems of the Environment)
- ITC (International Institute for Aerospace Survey and Earth Sciences)
- CSPG (Canadian Society of Petroleum Geologists)
- COGEOINFO (IUGS Commission on the Management and Application of Geoscience Information)
- IAMG (International Association for Mathematical Geology)

Cogeoenvironment will continuously seek cooperation with other organizations that may be relevant to our activities.

Programmes and working groups

A number of programmes and working groups have been started to cover some of the goals and objectives that have been laid down in the commission's terms of reference.

In order not 'to spread the butter too thinly' the commission has concentrated its activities on a limited number of projects so that these programmes are able to fulfil their tasks rather than starting out in too many directions. The number of goals that might be identified within environmental geology are almost endless and the commission has therefore decided to be careful in concentrating on a small number of objectives that it believes are of fundamental importance.

Information

To collect and disseminate relevant information to the general public, Earth scientists, engineers, planners and decision-makers, a project was started to build up inventories of relevant books and journals, information on relevant international organizations that are able to provide expertise, to produce a poster to advertise Environmental Geology, and a pamphlet to explain the importance of geoscientific advice in environmental planning.

Information on the activities of Cogeoenvironment is given through a semi-annual newsletter which was launched in early 1992 (de Mulder 1992).

Education and training

A working group has been established within the commission to organize training courses of different kinds and duration concerning problems, methods and techniques related to environmental planning and management.

Different types of training courses have been suggested:

(a) Development of a one- to two-month course in geology for land-use planning and management especially designed for Asian and African countries. This type of course has been approved by UNESCO and will probably be organized in 1995.
(b) An educational programme on environmental problems for developing countries. At the request of ICSU through IUGS, a proposal for a special programme has been submitted to ICSU (Wolff 1992).
(c) Short courses, with a duration of 2 to 8 weeks, which could be given anywhere. For 1991 UNESCO agreed to give financial support to a training course of this size in Czechoslovakia (Novak 1991). The course was organized by the Czech Geological Survey with participants from a number of East European countries. The course was held in early December 1992 and proved to be a great success.

Another course of this size was organized in cooperation with CARDER (Corporation Autonomia Regional De Risaralda) in Colombia in July 1992. Approximately 300 planners from nine Latin American countries participated in a three-day training course on geology for environmental planning where the board members of Cogeoenvironment acted as lecturers. A subsequent course is planned in Hungary towards the end of 1995.
(d) Academic courses, with a duration of 4 to 8 months, resulting in a Master's degree. The commission will approach the ITC in the Netherlands to take on the responsibility for organizing this type of education as part of their international training programme.

The commission will furthermore be at the disposal of other universities and organizations which want assistance to include Environmental Geology in their curriculum.

Global inventory of geoenvironmental problems

To obtain a global overview of environmental problems related to geology, the commission has started a project compiling an international inventory of geoenvironmental problems. A questionnaire has been prepared and has so far been distributed to a number of eastern European countries. This is meant as a test of the method and eastern Europe was chosen because it is well known that this region represents one of the most heavily polluted areas of the world. The first results of this test were reported at the 29th International Geological Congress in Kyoto, Japan in September last year (Mattig & de Mulder 1992).

The questionnaire contains a number of carefully designed lists of questions concerning environmental problems related to geology, for example sterilization/protection of resources and natural hazards and the eventual mitigation of such, plus an inventory of geoinformation related to planning.

Urban geology

The size and number of big cities (450) with more than one million inhabitants and megacities (25) with more than ten million inhabitants are steadily increasing (Masure 1992). The geoenvironmental problems related to these agglomerations are quite distinct and deserve very special treatment. The commission decided, therefore, to assess the interest for the creation of an International Working Group on Urban Geology and a questionnarie was dispatched to 200 colleagues in the geosciences (de Mulder 1992).

The answers proved that interest was manifest and by January 1 1992 the International Working Group on Urban Geology was established as a joint venture of Cogeoenvironment, IAEG and IAH. Contacts have also been established with the International Society of City and Regional Planners.

The main activities of this group will be centred around the preparation of an inventory of problems in large cities which might be solved by geoscientific input. Another activity will be to exchange views during meetings and to organize special seminars and workshops in conjunction with major congresses.

Activities within the field of data collection and handling through GIS, including development of special techniques for data presentation, lie also within the sphere of interest of this group.

Brochure and poster

The production of a brochure to disseminate the necessary information on the essential role of geoscience in planning and managing the human environment was one of the earliest goals of Cogeoenvironment. Through our board member for North America, a cooperative project was established with the Environmental Geology Division of the Canadian Society of Petroleum Geologists (CSPG) and the project was successfully completed when the brochure was issued in June 1992 (Berger 1992).

Since then, several thousand copies of the brochure have been distributed through the network of our corresponding members and the IUGS National Committees worldwide. A large number of the brochure was also distributed to the participants of the 29th International Geological Congress in Kyoto in September (1992).

Representing IUGS at Scope

Scope (Scientific Committee on Problems of the Environment) which is one of ICSU's scientific committees, has been active for more than 20 years without any geoscientific participation.

Due to the new trend within IUGS, Cogeoenvironment was urged to take part in the VIII General Assembly of Scope in Seville, Spain in January 1993 and to try to introduce some geoscientific projects into the Scope programmes.

After discussions with several Scope officials and members of different projects the draft of a new project on 'Effects of Human Activities on Surficial Earth Processes and Sustainability of Land Uses' was presented to the General Assembly. There was considerable enthusiasm for the proposal and it was agreed that Cogeoenvironment should develop further plans for the project.

The project will carry out an analysis of the consequences of each major type of land use, be it 'hard' like mining, urbanization, construction of infrastructure etc. or 'soft' such as forestry or agriculture.

Geoenvironmental indicators

During Cogeoenvironment's annual meeting in Colombia in July 1992 a new Working Group on Geoenvironmental Indicators was established.

In recent years the use of environmental indicators has become common within many of the natural sciences dealing with problems of the environment. Such indicators are used as tools to quantify changes in the environment. Geoenvironmental indicators should comprise a comprehensive set of measurable geoparameters which may be integrated with non-geological indicators. They should include not only monitoring mechanisms for present and future changes, but changes in the recent past as well. The outcome should be an ability to monitor changes in the natural environment produced by both natural processes and human impact.

In the geosciences there are many such indicators: sea-level transgression/regression; soil acidification; land subsidence, just to mention a few. The search for relevant geo-environmental indicators will be the primary goal of this working group.

The Rio-92 Conference

Cogeoenvironment has been involved in the activities related to the United Nations Conference on Environment and Development (UNCED) in several ways. A short report was prepared as a contribution to one of ICSU's documents on 'Land Use and Degradation'. Contribution was also made to the 'Report of the Ibero-American universities to UNCED', which presents the views of nearly 100 institutions of higher learning in about 20 countries concerning the problems of environment and development.

The IUGS president, the Cogeoenvironment's Vice-Chairman and several of its corresponding members attented the Rio Conference.

Conclusion

During the opening of the 28th IGC in Washington three years ago the IUGS president gave a strong indication of a new trend towards environmental issues within the union.

At the 29th IGC in Kyoto in September 1992 the new trend has become evident, because most geological surveys have already adapted themselves to a shift towards environmental issues, following the interests of their own governments. In addition, the undergraduate bodies in geology all over the world include students increasingly interested in environmentally related subjects.

IUGS itself has also followed this trend, through direct participation in the Global Change programme of ICSU, its strong involvement in the Decade of Natural Disaster Reduction of the United Nations, and through the creation of a special commission dealing with Geology for Environmental Planning and Management – Cogeoenvironment.

According to § 1 of the commission's terms of reference, a central activity is to increase the awareness of the general public and political authorities of the essential role of geoscience in environmental planning and conservation. The author, therefore, believes it will be of interest to both parties to establish some form of cooperation between the International Earth Science Conservation Group and Cogeoenvironment.

References

BERGER, A. R. 1992. *Planning and Managing the Human Environment. The Essential Role of Geoscience.* Brochure, Cogeoenvironment & CSPG, Calgary.

CORDANI, U. G. 1989. *Opening Address 28th IGC, Washington.* IUGS Minutes, IUGS Secretariat, Trondheim.

MASURE, PH. 1992. Managing the physical instability of megacities. Abstract. *Proceedings of the 29th IGC.* Kyoto.

MATTIG, U. & DE MULDER, E. F. J. 1992. The world geoenvironmental problems: an inquiry. *Proceedings of the 29th IGC.* Kyoto.

DE MULDER, E. F. J. 1992. *Cogeoenvironment – Newsletter No. 1.* Haarlem.

NOVAK, M. 1991. *Short Training Course; Geoscience for Environmental Planning.* Prague, Bozi Dar, Czechoslovakia, December 8–17, 1991.

WOLFF, F. CHR. 1992. The IUGS Commission on Geosciences for Environmental Planning. *In*: CENDRERO, A., LÜTTIG, G. W. & WOLFF, F. CHR. (eds) *Planning the Use of the Earth's Surface.* Lecture Notes in Earth Sciences. Springer Verlag.

The geotope concept: geological nature conservation by town and country planning

B. STÜRM

Bruggmühlestr. 11, CH-9403 Goldach, Switzerland

Abstract: Weak points and, on the other hand, possibilities are shown within the Swiss planning system by the example of an endangered landscape. Geotopes are, from the geological or geomorphological point of view, valuable and sensitive parts of the landscape. Geotopes, fixed by planning procedures, document a public and not just a scientific need. As components of approved plans, geotope protection areas affect subsequent plan settings and determine the future intervention possibilities. In this way town and country planning could provide powerful instruments by which effective site conservation could be achieved.

Existing deficiencies in the legislative framework, as well as in the planning instruments, could be sorted out with the help of an International Convention. After this convention the signatory states would be obliged to regard protection of the geological heritage as an essential town and country planning objective and ensure that this requirement is taken into account at all stages in the drawing up of plans and in the procedures for authorizing work.

Weak points in the planning measures for an endangered Swiss moraine landscape

The moraine landscape on the edge of Switzerland's central plateau, between the rivers Sihl and Lorze (Canton of Zug), serves to illustrate this point.

Albert Heim, the celebrated geologist, described this area as the most magnificent moraine landscape of Switzerland. Particularly attractive and conspicuous are the numerous hills covered with linden trees which local people refer to simply, but affectionately, as 'Höger'. This unique, impressive scenery came into being during the last Ice Age as a result of the combined action of two immense Alpine glaciers. At the time of their maximum extension the Linth/Rhine and the Reuss Glacier merged and interacted over a distance of some 5 km. Evidence of glacial advance and retreat includes moraine ramparts, drumlins, dead ice kettles, outwash fans and meltwater channels.

Situated at the edge of the Zurich conurbation this moraine landscape suffers from the increasing pressure of residential and industrial development. The boom in the building industry has caused the demand for gravel and sand to soar to unprecedented levels. Resources with respect to current authorized quarrying schemes amount to approximately $10–15 \times 10^6 \,\mathrm{m}^3$. Further resources totalling some $40–60 \times 10^6 \,\mathrm{m}^3$ have not yet been authorized for extraction. With the market demand so high this represents a huge potential source of environmental conflict.

The instruments available to the cantonal authorities to guide and coordinate the development have been improved steadily under the pressure of public opinion. To ensure an economical use of the gravel and sand resources available a statutory system has been imposed to fix the amounts of gravel that may be quarried annually. In the area of the moraine landscape, the present figure is around $800\,000 \,\mathrm{m}^3$. Under the 1988 Act of the Canton Zug on the Protection and Maintenance of the Moraine Landscape new quarrying permits may be only granted in cases where proof of a predominantly public interest can be furnished.

The areas where gravel extraction can be permitted have been strictly laid out and are an integral part of the official town and country plans. At the same time these plans also contain seemingly contradictory specifications for protected areas, such as biotopes, so that gravel extraction is prevented in some places but not in others. Unfortunately, areas of outstanding geological or geomorphological interest have not been considered worthy of special protection and thus have not been fixed in town and country plans.

Finally, this moraine landscape has been added to the Federal Register of Landscapes of National Importance. This inventory enables the Swiss Nature Conservation Commission to intervene by assessing planning applications. But as long as planning authorities do not designate areas of outstanding geological and geomorphological interest in town and country plans the influence of the inventory will remain limited and vague.

Under these circumstances the future of the moraine landscape remains uncertain insofar as the growing scarcity of gravel and sand could justify the gravel suppliers' claim to be acting in a public interest. No one knows yet whether the present protective measures will yield or not to renewed demands for further development.

The geotope concept

The geotope concept aims at the entire integration of geoconservation into town and country planning procedures. It can be understood as a general strategy or implementation programme which may lead to effective decision mechanisms within a planning system. For this purpose an adequate **spatial unit** ('geotope') and a corresponding **operative unit** ('area or zone for geotope conservation'; i.e. 'geotope protection area') have to be defined and introduced.

The definition of the spatial unit must be easy to understand and practicable for planning purposes; furthermore, it has to comprise a strategic message. The following is an abbreviated version of the definition which has already been introduced in some German-speaking countries (Stürm 1983; Lüttig 1992; Stürm 1992; Wiedenbein 1992).

> Geotopes are distinct parts of the geosphere of outstanding geological and geomorphological interest. They have to be protected against influences which could damage their substance, form or natural development.

Geotopes can occur in many different forms, e.g. they can be active like the world famous shingle bank (beach ridge) Chesil Beach which is about 30 km long on the Dorset coast of Great Britain (NCC 1990); or they may be static like the 'Höger-complex' of the moraine landscape in the Canton of Zug, Switzerland, as described above.

The term geotope (in analogy to the term biotope) is able to focus geological and geomorphological interests. It stimulates public awareness, thus fulfilling an important catalytic function within the planning and decision-making process.

In analogy to existing planning categories the operative unit has to be defined as a field of interest or a field of activity (Stürm 1983). Geotope protection areas designated in town and country plans are a result of coordination procedures and political debate. In contrast to geotope inventories, which serve as an important basis for their delimitation, they reflect both scientific and public interests. As an integral part of town and country plans, geotope protection areas determine the arrangement of other plan components and have a strong effect on subsequent planning applications and thus influence and restrict future activities.

Implementation of the geotope concept

The Swiss planning system (a brief characterization)

The general aim of town and country planning is to ensure a sustainable use of geographical space with its heritage of geo-biological evolution, ecosystems, natural resources and landscapes.

The Swiss town and country planning system, in general, fulfills the fundamental requirements of a modern planning system (Stürm in press):

- comprehensive planning of the spatial context;
- open planning allowing the population to play an important role;
- multi-level planning (national, regional, local) which is necessary for effective vertical coordination;
- adaptable planning which reacts to changing circumstances.

The relevant legislative framework is based on: the Federal Town and Country Planning Act 1972, the various cantonal town and country planning acts, the Federal Environmental Protection Act 1983, the Federal Nature Conservation Act 1966 (+ amendment 1987), the Federal Water Protection Act 1991 and the Federal Forestry Act 1991.

Types of town and country plans available

Swiss legislation provides the following main plan types:

- *Structure plans* set the guidelines for future development and are basic tools for the co-ordination of spatial interests and impacts. Once approved they are binding for all authorities.
- *Development or land-use plans* stipulate in detail the possible exploitation of the land. The territory is divided into different zones, e.g. zones reserved for building construction, zones for agricultural use, zones for the protection of groundwater resources, zones for the conservation of biotopes. Once approved development plans are binding for everybody.
- *Nature conservation ordinances* for distinct areas can be issued to regulate in detail the specific protection needed. Nature conservation ordinances are binding for authorities,

land owners and others, depending on their specific regulations.

Under present Swiss planning legislation, geotope protection areas can be designated in town and country plans but their designation is not a 'must'.

The way a planning application functions

A central question is that of the roles that town and country plans can play in the approval procedures for a project. The interdependencies can be explained by using the example of a fictive gravel pit project within the Swiss context. Before the planning authorities decide whether the project can be carried out or not they have to clarify at least the following decisive questions:

(1) Does the project conform to existing town and country plans? Does it conflict with areas or zones already designated, e.g. for the protection of groundwater resources, the conservation of biotopes or geotopes? (Federal Order on Town and Country Planning 1989, Art. 2 lit. e)
(2) What kind of alternatives or possibilities could be considered in order to minimize conflicts between development and, for example, protected groundwater resources, protected biotopes or geotopes? (Federal Order on Town and Country Planning 1989, Art. 2 lit. b)
(3) What kind of impact does the project have on the environment (including the effects on biotopes or geotopes) and what kind of measures are necessary to reduce the ensuing strain on the environment to a reasonable level? (Federal Environmental Protection Act 1983, Art. 9)

The project leader has to define the necessary studies and to provide all the information suitable to give an accurate answer to these questions. Before the final decision is made by the planning authorities, they are obliged to weigh the interests for and against the project, taking into account the results of the preceding examinations and the intended spatial development according to approved town and country plans. (Federal Order on Town and Country Planning 1989, Art. 3).

Main steps towards implementing the concept

In order to achieve a maximum consideration and effectiveness of all efforts supporting geoconservation, the geotope concept has to be put into effect within the planning system of each country. The following main steps have to be taken:

(1) Define clearly the relevant spatial unit ('geotope').
(2) Stipulate in the planning legislation that the designation of geotope protection areas in town and country plans is obligatory.
(3) Work out the necessary planning aids such as site documentation, geotope inventories, conflict maps, guidelines and so on.
(4) Delimit and designate geotope protection areas in the official town and country plans at all administrative levels.
(5) Establish ordinances with detailed regulations for sensitive geotopes or geotope complexes with overlapping activities and interests which call for a high level of coordination.

Achieved implementation stages and remaining deficiencies

To what extent has the geotope concept already been put into effect within the planning system of each country? The requirements mentioned above serve as a standard to detect existing deficiencies and to determine appropriate strategies to improve the situation.

In the case of the Swiss planning system the analysis reveals clearly that even in a progressive and relatively well-developed town and country planning system geoconservation is poorly integrated and, therefore, plays a minor role. This contrasts with the situation in other spheres of conservation such as biological nature conservation. The term biotope, for instance, was introduced into the relevant legislation long ago and is widely accepted nowadays. Areas or zones set aside for biotope conservation belong to the standard components of Swiss town and country plans, whereas the designation of areas or zones for geotope conservation is still rather rare. For the conservation of some types of biotopes of national importance Federal Ordinances based on The Federal Nature Conservation Act have been established. Equivalents for the conservation of geotopes of national importance are missing. The compilation in Table 1 provides a synopsis of the current situation.

Requirements for an International Convention

How can the existing weak points and deficiencies in the legislative framework as

Table 1. *A comparison of the integration of geoconservation and biological nature conservation into the Swiss town and country planning system*

	Spheres of conservation	
Stages of integration into town and country planning	Geological nature conservation	Biological nature conservation
(1) Spatial unit defined: ± Definition available + Definition widely accepted	± (Geotope)	+ (Biotope)
(2) Operative unit introduced: ± Introduction voluntary + Introduction statutory	±	+
(3) Specific planning aids available: ± occasionally + widespread	±	+
(4) Areas or zones for conservation designated: ± occasionally + widespread	±	+
(5) Ordinances for distinct areas established: − not yet + widespread	−	+

well as those in town and country plans and in the planning and approval procedures be eliminated?

Although geoconservation is finding more and more acceptance, its position in the planning process is still very weak. At the same time, our geological heritage, and active geosystems as well, are increasingly threatened by human impact. This unsatisfactory situation needs to be improved by integrating step-by-step geoconservation efforts within town and country planning. In order to achieve this aim, strong pressure at a supra-national level is vital. An International Convention could promote and support the implementation of the geotope concept within the planning systems of the various countries. To do this, it would have to cover at least the following issues:

(1) Definition of the relevant spatial unit (geotope).
(2) Creation of an inventory of geotopes of international importance (including criteria and procedures for their selection).
(3) Obligations of the contracting parties:
   3.1. To pay particular attention to geoconservation in planning and development policies.
   3.2. To ensure that geoconservation is taken into account at all stages, both when drawing up plans and approving plans and projects.
   3.3. To take appropriate and necessary legislative and administrative measures to ensure effective and durable geoconservation; especially to ensure that geotope protection areas are designated in town and country plans available at all administrative levels.
   3.4. To establish an inventory of geotopes of national and regional importance.
   3.5. To support and promote private initiatives in favour of geotope conservation.
(4) Establish in each country a statutory body which has to supervise, sustain and provide the implementation of the geotope concept.

References

LÜTTIG, G. W. 1992. Der Geotopschutz in der geowissenschaftlichen Kartographie. Rückblick, Sachstandsbericht, Prognose. *Materialien I/ 1993 Ökologische Bildungsstätte Oberfranken.*

NCC 1990. *Earth science conservation in Great Britain. A strategy.* Nature Conservancy Council, Peterborough.

STÜRM, B. 1983. Anwendungsmöglichkeiten und Anforderungen an geomorphologische Karten seitens der Raumplanung. *Basler Beiträge Physiographie*, **5**.

—— 1992. Geotop. Grundzüge der Begriffsentwicklung und Definition. *Materialien* **I**/1993 Ökologische Bildungsstätte Oberfranken.

—— in press. Intégration de la protection du patrimoine géologique dans l'aménagement du territoire. *1er Symposium International sur la protection du patrimoine géologique, Juin 1991, Digne-les-bains*.

WIEDENBEIN, F. W. 1992. Zielsetzungen des Geotopschutzes in Deutschland. *Materialien* **I**/1993 Ökologische Bildungsstätte Oberfranken.

Sustainable development: the implications for minerals planning

MARY M. SCOTT

Minerals Division, Department of the Environment,
2 Marsham Street, London SW1P 3EB, UK

Abstract: The UK Government is developing a strategy for the incorporation of sustainable development principles into minerals planning, and takes the view that the aim should be to:

- conserve minerals as far as possible, while ensuring an adequate supply;
- encourage efficient use of materials and recycling of wastes;
- preserve or improve the overall quality of the environment affected by extraction.

This strategy underlies the philosophy of the current review of Minerals Planning Guidance Note 6 on aggregates provision. However, future planning guidance for other minerals will also take account of these principles.

In practical terms, a sustainable approach to minerals planning means that consideration needs to be given to the following:

- that minerals should be supplied in a way which takes full account of environmental factors;
- that minerals operations should be managed to the highest standards and sensitive working practices should aim to reduce the environmental impact of quarrying and wastage of materials. There is a need for greater emphasis on appropriate restoration and preserving or enhancing the value of restored land;
- that there should also be a positive policy of conserving raw materials, ensuring their efficient use, and encouraging recycling and the use of waste materials where this has environmental advantages.

The UK Government has made clear its support for the principles of sustainable development and is striving to make it an integral part of its domestic and international policies (The Environment White Paper 1990). The Government is following the 1992 Earth Summit in Rio and the establishment of Agenda 21 with the preparation of a report on sustainable development to be submitted to the UN Commission on Sustainable Development around the end of 1993. This will set out the actions and policies intended to meet the objectives of Agenda 21 and the conventions signed in Rio. As part of the follow up to the Rio summit, the Government organized a conference in Manchester in September 1993, which was entitled 'Partnerships for Change' on the practical implementation of sustainable development.

One aspect of these efforts is the development of a strategy for the incorporation of sustainable development principles into planning for the supply of minerals.

Role of the planning system in achieving sustainable development

The planning system is a means of balancing conflicting demands on land for mineral working, agriculture, amenity, building and other uses.

Planning Policy Guidance Note 1 (PPG 1) (DoE 1992) makes it clear that the planning system, and development plans in particular, can contribute to the objectives of ensuring that development and growth are environmentally sustainable. *PPG 1* paragraph 3 says

'the sum total of decisions in the planning field, as elsewhere, should not deny future generations the best of today's environment'.

Sustainable development of mineral resources: the key principles

The UK is fortunate to have extensive indigenous mineral resources, and exploitation of minerals makes an essential contribution to national prosperity and quality of life. However, the principles of sustainability suggest that minerals should be provided in a way which avoids unacceptable environmental disturbance for the present generation, while not undermining the potential for future generations to benefit from resources and environments.

Minerals planning involves achieving a balance among various social, environmental and economic considerations, suggesting that a proper evaluation is necessary, not only of the resources themselves, but also of environmental 'resources' such as landscapes.

The EC Environmental Council agreed the Fifth Environment Action Programme in December 1992. This recognizes that:

> 'since the reservoir of raw materials is finite, the flow of substances through the various stages of processing, consumption and use should be so managed as to facilitate or encourage optimum reuse and recycling, thereby avoiding wastage and preventing depletion of the natural resource stock ... one individual's consumption or use of these resources must not be at the expense of another's; ... neither should one generation's consumption be at the expense of those following'
> (Commission of the European Communities 1992).

This suggests there should be a positive policy of conserving raw materials, ensuring their efficient use, and encouraging recycling and use of secondary materials.

It can of course be argued that very large quantities of mineral resources exist sufficient in many cases to last far into the foreseeable future. However, minerals may be viewed as limited in terms of what can be extracted in **environmentally acceptable** locations. As the County Planning Officers' report *Planning for Sustainability* (CPOs' Society 1993) says,

> 'it is apparent in some areas that the resources from sites which do not have strong environmental constraints, will be completely used up in the not too distant future unless immediate action is taken'.

This recognizes the negative aspects of minerals extraction – not only that it requires a landtake otherwise available for alternative purposes (or undeveloped for future generations), but also that it imposes environmental impacts, such as those from dust, noise and transport, on localities and communities.

In addition to these extraction 'costs', there may also be considerable benefits arising, not only from the resource itself in the contribution made to economic growth, but also potential landscape enhancement, reclamation of derelict land, and creation of new habitats. Further, it should be recognized that aggregates, unlike energy materials, may not necessarily be used up during consumption. Since they have the potential to be recycled, they could be said to remain part of the total available resource base.

A sustainable development approach to minerals planning therefore means that the costs and benefits of minerals extraction need to be evaluated as fully as possible.

Aggregates planning and sustainable development

The current review of Minerals Planning Guidance Note 6, *Guidelines for Aggregates Provision in England and Wales (MPG 6)* (DoE 1993) is providing a timely opportunity for Government to examine the implications of sustainable development for aggregates planning.

Projections of demand commissioned from consultants indicate that up to 6.4 billion tonnes of primary aggregates may be required in England and Wales over the next 20 years (DoE 1993). These projections suggest that demand could be between 370 and 440 million tonnes per annum (mtpa) by 2011. This is an increase of between 40% and 60% on 1989 consumption levels. These projections are not production targets which must be met by the extraction of new materials, but indicate the potential demand brought about by a steady rate of economic growth. The projections provide a starting point for consideration of policy options and to inform the preparation of planning guidelines.

In the recently published draft *MPG 6*, Government set out its views on what sustainable development means for aggregates planning. These implications centre around two key themes:

(1) making the best and most efficient use of all available resources, so that extraction of new resources is limited to what is necessary to meet the current generation's needs;
(2) ensuring that the overall quality of the environment affected by aggregates extraction should be preserved or improved over time, so that future generations are not disadvantaged by the activities of the present one.

The draft guidelines sought views on how the supply requirements indicated by the demand projections could be met while pursuing the goals of sustainable development. They proposed that an increasing contribution to aggregate need should be met from alternative sources, including secondary and waste materials and remote coastal superquarries, thereby reducing the pressure on traditional land-won resources. Views were sought on the level of provision that can come from land-won primary aggregates in England and Wales without an unacceptable level of damage to the environment; and on the contribution which should come from coastal superquarries and secondary/waste materials.

Efficient use of resources and use of alternative materials

The Building Research Establishment (BRE 1993) has reported cases of 'overspecification' leading to unnecessary wastage of aggregate materials in construction. Such overspecification takes the form of both excessive margins of safety, and lack of suitable specifications for certain materials, particularly lower grade and secondary ones. The CPOs' Society (1993) encourages Government to adopt the concept of 'Minimum Necessary Specification' which would require that

> 'design specifications seek to achieve the most sparing use of natural resources, commensurate with safety and reasonable financial considerations'.

The plentiful supply of certain minerals and their relatively low price has, in some cases, meant that there can be considerable wastage of resources in processing and use. For example, BRE (1993) reports that up to 10% of building materials including aggregate minerals may be wasted on construction sites.

There are large supplies of waste materials such as slate, china clay sand, colliery spoil and demolition wastes, which could be put to use as aggregates. Technical, locational and economic factors currently constrain this use to approximately 10% of total requirements for aggregates (Arup Economics and Planning 1991). The Government is keen to increase their usage, while recognizing that working of waste materials may have its own environmental costs, including environmental problems common to primary minerals extraction, e.g. noise, dust, transport, energy consumption. Department of the Environment (DoE) research in progress on 'Recycling of Demolition and Construction Wastes in the UK' is aiming to identify how a greater level of recycling can be achieved in the most environmentally acceptable fashion.

The draft *MPG 6* includes policies to secure greater efficiency of use of all aggregates materials, including wastes. The DoE is now considering what more can be done in practical terms.

Demand minimization

Government has been urged to take steps towards the management of demand for aggregates (e.g. the CPOs' Society (1993); some respondents to the *MPG 6* consultation exercise including the Council for Protection of Rural England and English Nature). In preparing the draft *MPG 6*, the Government recognized that the marginal environmental cost of meeting increased demand is likely to rise over the longer term, and took the view that constraining future demand should be considered. It saw the long-term objective as seeking ways to meet demand while using less aggregate, for example, by using alternative materials and different patterns of design. These are issues which will remain on the agenda for further consideration.

Other Minerals Planning Guidance Notes (MPGs)

The pursuit of a sustainable policy for minerals does not stop at aggregates. Government will shortly be bringing forward draft revised guidelines on the development plan system (*MPG 1*), on provision of silica sand (the essential mineral for such industries as glass manufacture and foundry castings) and on opencast coal (*MPG 3*). Revised guidance on the reclamation of mineral workings (*MPG 7*) is being prepared. These, too, will consider the implications of sustainable development.

Key roles in pursuing sustainable development of minerals

The recognition of sustainable development principles in minerals planning has the following practical implications:

(1) *During extraction:* operations should be managed to the highest standards and sensitive working practices are required to reduce the environmental impact of quarrying and to reduce wastage of materials.
(2) *In restoration:* greater emphasis needs to be attached to appropriate restoration and preserving or enhancing the value of restored land.
(3) *In use of minerals:* steps should be taken to minimize wastage in use, including avoidance of overspecification and using recycled/waste materials where environmentally advantageous.

There are a variety of ways in which these aims might be pursued. The key players are:

- the **minerals industry** in its approach to site management and future identification of areas for exploitation;
- those **industries which use minerals** for their processes and end-products;
- **mineral planning authorities** through their development plans and development control;
- **central Government** in its advisory, regulatory and legislative roles;

- **conservation bodies**, particularly through their joint initiatives with industry for site enhancement.

A report for DoE on *The Minerals Industry Environmental Performance Study* (Groundwork Associates Ltd 1991) recommended that

'a system of Environmental Management should be established as an integral part of each company's organisation; this would include a corporate environmental statement, environmental site appraisals, regular monitoring of performance and periodic environmental audit/review'. (Fig. 1).

The minerals industry has recently made substantial progress in addressing the challenges of environmental management. Major advances include the Environmental Codes produced by the aggregates trade associations and the CBI.

A further study, *Environmental Management in the Minerals Industry: the Greensite Report* (Groundwork Associates Ltd 1993) provides practical advice on how to achieve the higher standards now expected. It establishes principles for environmental management on sites, and under the label of sustainability, recommends that operators undertake the following:

- audit energy consumption
- adopt waste minimization policy
- encourage recycling
- ensure that 'Duty of Care'[1] obligations are met

The report provides practical examples of how minerals' operators can work to these objectives, for example, by carrying out crushing and screening of imported wastes to help provide restoration material on site. Use of recycled packaging for mineral products and monitoring of energy consumption are also suggested as

---

[1] Part II of the Environmental Protection Act 1990 introduced a Duty of Care requirement from 1 April 1992 on all producers and holders of waste.

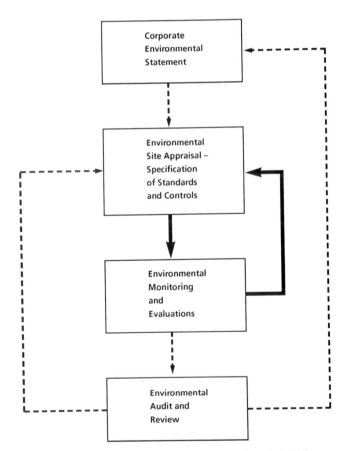

Fig. 1. Environmental Management System (from Groundwork Associates Ltd 1991).

ways of securing a more sustainable approach to operations.

Since mid-1988, the DoE has commissioned a number of research projects to provide guidance on restoration and aftercare of mineral workings (for example, *Amenity Reclamation of Mineral Workings*, Land Use Consultants 1992). Afteruses covered range from biological and geological nature conservation to sports activities. A further study on the opportunities for creating and managing nature conservation sites on derelict land and on existing mineral workings will commence during 1993.

The Government takes the view that mineral operators should aim to secure positive enhancement of sites during and following extraction. The draft Minerals Planning Guidance Note 6 suggests that in assessing planning applications, mineral planning authorities should have regard to how the proposals relate to sustainable development. In future, therefore, it can be expected that applications which demonstrate a positive commitment to sustainability should increase their chances of success.

Development plans provide an opportunity for planners to take steps to ensure that mineral supplies are obtained from areas where the environmental costs are minimized. Under the Planning and Compensation Act 1991, Mineral Planning Authorities (MPAs) are required to prepare mineral local plans and it is hoped that there will be complete coverage of these as soon as possible and by 1996 at the latest. The forthcoming revision of Minerals Planning Guidance Note 1 (*MPG 1*) will advise MPAs on sustainable development in the context of development plans.

As resources which can be worked in environmentally acceptable areas become scarcer, plan preparation will increasingly require high quality information about the nature, quantity and location of mineral resources; well-documented resource inventories will be an aid to this. The DoE's Geological and Mineral Planning Research Programme will continue to play an important role in identifying potential mineral supplies, and the planning and environmental factors relevant to exploitation.

Future steps

The Government recognizes that it needs to consider further whether a demand minimization approach is appropriate to aggregates supply, how it can practically encourage re-use and recycling of mineral resources, and how it can encourage appropriate use of high quality mineral resources and prevent their wastage in low-grade end uses.

Following on from the publication of *Policy Appraisal and the Environment* (DoE 1991), the Department of the Environment now intends to commission research on the costs and benefits of minerals extraction. This will advise on how future minerals planning policy might be developed to take account of these costs and benefits.

© Crown copyright 1994. Published with the permission of the Controller of Her Majesty's Stationery Office. The views expressed are those of the author and do not necessarily reflect the views or policy of the Department of Environment or any other government department. I am grateful to my colleagues in Minerals Division, Department of the Environment, for very helpful discussions and comments on the manuscript.

Footnote

(1) This paper was prepared in advance of the Government's consultation exercise on its strategy for sustainable development (Department of the Environment, July 1993; consultation paper, *UK Strategy for Sustainable Development*). The consultation exercise informed the preparation of two chapters on minerals for the final strategy, published in January 1994 (*Sustainable Development: the UK Strategy*, HMSO, CM 2426).

(2) This paper was prepared during the exercise to review *MPG 6*. The revised MPG was published in April 1994 (DoE 1994: *Minerals Planning Guidance Note 6: Guidelines for Aggregates Provision in England*. HMSO, London). Wales is covered in a separate guidance.

References

Arup Economics and Planning 1991. *Occurrence and Utilisation of Mineral and Construction Wastes*. HMSO, London.

Building Research Establishment 1993. *Efficient Use of Aggregates and Bulk Construction Materials*. BRE, Garston, Watford.

Commission of the European Communities 1992. *Towards Sustainability: A European Community Programme of Policy and Action in relation to the Environment and Sustainable Development*. Official Publication of the European Communities. COM (92) 23/II (Final).

CPOs' Society 1993. *Planning for Sustainability*. County Planning Officers' Society.

DoE 1991. *Policy Appraisal and the Environment: a Guide for Government Departments.* HMSO, London.
—— 1992. *Planning Policy Guidance Note 1: General Policy and Principles.* HMSO, London.
—— 1993. *Guidelines for Aggregates Provision in England and Wales: Revision of MPG 6.* Draft Consultation Document, Department of the Environment, London.
GROUNDWORK ASSOCIATES Ltd 1991. *The Minerals Industry Environmental Performance Study.* Groundwork, Macclesfield.
—— 1993. *Environmental Management in the Minerals Industry: the Greensite Report.* Groundwork, Macclesfield.
LAND USE CONSULTANTS 1992. *Amenity Reclamation of Mineral Workings.* HMSO, London.
THE ENVIRONMENT WHITE PAPER 1990. *This Common Inheritance – Britain's Environmental Strategy.* Cm 1200, HMSO, London.

A model of consensus in aggregate mining and landscape restoration: science – industry – conservation

CHRISTIAN SCHLÜCHTER

Institute of Geology, University of Bern, Baltzerstrasse 1, CH-3012 Bern, Switzerland

Abstract: The consumption of sand and gravel per capita in Switzerland is considerably above the European average. This situation is adding one more type of environmental stress to a densely populated and developed landscape. In addition to the environmental aspects of aggregate mining we need to be aware of the fact that geological information is being lost due to the high extractive capacity of this industry.

A seminal conflict over the conservation of a key Quaternary section in a pit at Jaberg in 1978 led to an awareness amongst Earth scientists, the extractive industry and planners that geoscience understanding was an important element in effective planning of aggregates extraction.

Discussions with representatives of the aggregate mining industry since 1978 on the environmental impacts of restored sites, and on the need to keep special and unique geological information preserved, have ended in the successful establishment of a 'Foundation for Aggregates and Landscape'. This joint initiative by the industry and Earth scientists seeks to find a consensus approach for resolving conflicts which avoids destructive and time-consuming confrontation. This partnership offers the prospect of a sustainable approach to managing the sand and gravel resource.

Emerging conflicts

Aggregate mining is an economically important part of Swiss industry, and through its action it is a geologically-active erosional process. The impact of this industrial activity becomes evident when two aspects are evaluated.

(a) The consumption of sand and gravel for construction purposes has averaged 14.7 tons per capita per year between 1982 and 1987 in Switzerland, compared to 6.5 tons in the former Federal Republic of Germany and 2.06 tons in the UK.

(b) The major production centres for aggregate are located in the Northern and Southern Alpine Forelands which are the densely populated areas of the country. Aggregate mining is therefore subject to the standard planning procedures which apply to restricted areas where competition with other land use activities such as forestry, agriculture, groundwater management, existing cities, traffic lines and landscape conservation exist. The major limitations to aggregate mining are the areal extent of cities, the need for groundwater production and the existence of protected areas. As a consequence, gravel mining activities are increasingly concentrated in major production centres due to such restrictions.

Domestic production of aggregate accounts for about 90% of the total volume with the remaining 10% being imported from France, Italy, Austria and the former Federal Republic of Germany. This balance needs to be carefully monitored in the future as it may develop into an important socio-economic and environmental issue. The percentage of imported material varies considerably over time and may have been as high as 15% in the past with lower values at other times in line with reduced activity in the construction industry.

The resources for the aggregate industry are the direct product of the geological processes associated with the multiple extension of the Pleistocene glaciers to the Alpine Forelands. More than 80% of the volume of deposits mined is genetically related to the Last Glaciation. The direction and focus of Quaternary geological research has been greatly influenced by this mining activity over the past 20 years. With the opening up of long vertical and longitudinal exposures. Quaternary geology evolved from a morphological study of landform surfaces to a full analysis of sedimentary facies and lithostratigraphy. This has resulted in a new and advanced understanding of geological evolution during the Quaternary ice age (Schlüchter 1976, 1988–89). However, this scientifically stimulating opportunity to explore the Quaternary landscape creates potential conflicts. One such example is briefly illustrated as follows.

The sedimentary sequences of coarse-grained outwash deposits, so typical for the Quaternary environment of the Alpine Foreland, display characteristic 'boulder beds' as an outstanding textural feature (Fig. 1). The major geological parameters of these beds are:

(a) a considerably coarser texture than beds immediately below and above with inclusions of large erratic boulders (up to several m^3);
(b) a majority of sub-rounded to rounded components;
(c) an erosional unconformity with the substratum.

However, the origin of the 'boulder beds' remained unexplained until an observation in 1978 in an active gravel pit at Jaberg. Here, the 'boulder bed' represents an erosional lag deposit of a basal till layer. The transition from the till-bed relict into the lag deposit was well exposed and its lithostratigraphic context within multiple fluvioglacial accumulative sequences was, therefore, established (Fig. 1). This simple section in a gravel pit elucidated and resolved a formerly active debate. 'Boulder beds' within fluvioglacial sequences can be interpreted as a lag deposit representing a former till bed.

With this new information on the genesis of perialpine outwash sequences available, the need to keep the important pit face intact for further geological investigations and demonstrations was obvious. Negotiations started in 1978 and the immediate result was the clearing of the pit face by the operating company prior to a visit by an international field excursion on 'Genesis and Lithology of Quaternary Deposits'. The demand for the conservation of this key site has thus been made known internationally (Schlüchter 1979). The aggregate industry, to their credit did learn from this instance about both potential future conflicts and opportunities for cooperation with the Earth science community.

'Consensus of opinion' strategy

The case history described above did not, sadly, lead to long-term conservation of parts of the artificial outcrop. The geological feature in question was mined shortly afterwards for two reasons:

(a) there were no legal possibilities to stop further development, and

Fig. 1. The lithogenetic key – the site at Jaberg with the characteristic 'boulder bed', open pit-face in 1978. (A) and (C) are coarse gravelly outwash units deposited during glacial advance (in german: *Vorstoss-Schotter*); (D) is the basal lodgement till of the Last Glaciation and (B) marks the transition between the lag deposit of the 'boulder bed' and the basal lodgement till on top of unit A.

(b) even more importantly, the industry did not yet accept geology as an integral part of its operations.

From this experience, however, it has been learned that the Earth science community does have an obligation for educating the aggregate industry on the role of geology in aggregate resource evaluations in a broad sense. This philosophy found support in the early 1980s from events within the industry itself and from a broader public awareness of the environmental impact of sand and gravel extractions. The strong support for Earth science arguments originated from two events:

(a) A major pit operation had to close down because of financial mismanagement, leaving behind a deserted area which was environmentally unacceptable. Industry found itself challenged to solve an environmental restoration problem with major geological and hydrogeological aspects and within restricted financial confines.
(b) New federal environmental protection acts (preceded by a Canton of Aargau legislation of 1981) caused substantial delays in the planning- and decision-making procedure for new pit operations and for simple extensions of active pits. The new procedures take 10 to 15 years depending on the legislation of the federal canton involved, and include (1) an initial feasibility and resource potential study, (2) an environmental impact assessment, (3) a detailed forward projection of the pit operation through to completion of site restoration.

Both events, with their strong environmental implications, have made it evident that Quaternary geosciences need to be considered at all stages of sand and gravel mining processes, and that the mining industry should accept a continuing educational input from the geological sciences. Acceptance of this simple interrelationship has resulted in a number of Diploma- and PhD-thesis works related to aggregate mining activities (e.g. Hildbrand 1990; Forster 1991; Grasmück 1992; Müller 1993).

Geology as part of the planning process and beyond

The role of Quaternary geosciences in aggregate mining can be defined at two levels.

(1) No aggregate mining is actually environmentally feasible without detailed knowledge of the geology at the site. This holds true for the full range of activities from prospecting, quality and environmental impact assessments to site restoration modelling. The recycling of a landscape without the benefits of geological understanding is irrational and unacceptable.
(2) It needs to be admitted that Quaternary sciences benefit directly from open pits and ongoing operations. The geological information is not only important but in some situations decisive for a broader understanding of geological processes. In addition, the educational value of open pits is immense. However, geological information and important sections are lost daily and this problem needs to be resolved.

A working solution

Early in the deliberations on the preservation of the Jaberg site, aggregate industry and geoscience representatives agreed to seek a solution based upon consensus of opinion rather than confrontation. The option of confrontation would have certainly backlashed on both parties. For example, scientists would loose access to new sections and industry would most probably face a doubling of the time needed to get legal permission for the proposed pit operations.

The solution based on consensus of opinion reached in the late 1980s led to the establishment of a 'Foundation for Aggregates and Landscape'. Its aims are as follows:

(a) to promote the wise planning of development, operation and restoration phases of gravel and sand mining sites; if need be, the foundation can establish contacts for project supervisions;
(b) to contribute to the development, management and coordination of environmental impact guidelines;
(c) to promote continuing education courses, mainly for teachers at all levels;
(d) to offer a consulting service in environmental impact and restoration procedures to non-members;
(e) to support research in Quaternary sciences, mainly at the Diploma level;
(f) to establish a small Documentation and Meeting Centre for Quaternary geosciences with a small exhibit in the village of Jaberg.

The financing of these activities is guaranteed through payments by member companies of contributions per unit of sand and gravel extracted. Membership of the foundation is voluntary and is attracted by the high scientific standard of its activities. At present, there are members from the federal cantons of Bern and Luzern.

In the longer term the establishment of a 'foundation of consensus of opinion' may be the only working solution to potential conflict over environmentally-sound pit operations and active site conservation. Most important of all is the need for an active programme of geoscience education to raise awareness in society at large.

References

CANTON OF AARGAU. 1981. *Dekret über den Abbau von Steinen und Erden (Abbaudekret)*. Grosser Rat des Kantons Aargau.

FORSTER, Th. 1991. *Paläomagnetische Feinstratigraphie am Thalgut-Profil (Aaretal, Kt. Bern)*. Diploma-Thesis, Institute of Geology, University of Zürich.

GRASMÜCK, M. P. 1992. Ground-penetrating radar – applications in Quaternary geology. *Eclogae Geologicae Helvetiae*, **85**, 471–490.

HILDBRAND, K. 1990. *Das Endmoränengebiet des Rhonegletschers östlich von Wangen a.A.* Diploma-Thesis, Institute of Geology, University of Zürich.

MÜLLER, B. U. 1993. *Quartärgeologie des Seeztals*. PhD-Thesis, Abt. XC, ETH-Zürich.

SCHLÜCHTER, Ch. 1976. Geologische Untersuchungen im Quartär des Aaretals südlich von Bern. *Beiträge zur geologischen Karte der Schweiz, N.F.*, **148**.

—— 1979. Contributions related to the field excursion, Introductory remarks. In (Schlüchter, Ch. ed.) *Moraines and Varves*. Balkema, Rotterdam, 377–382.

—— 1988–89. A non-classical summary of the Quaternary stratigraphy in the northern alpine Foreland of Switzerland. *Bulletin de la Société neuchâteloise de géographie*, **32–33 (1988–89)**, 143–157.

The role of local government in geological and landscape conservation

E. A. JARZEMBOWSKI

Brighton Council, Bartholomew Square, Brighton BN1 1JA, UK

Abstract: Although the Rio Summit produced many policies and commitments for national governments (most of them in Agenda 21), it is widely recognized that action based on these policies relies heavily on local partnerships involving local government, business and the voluntary sectors. Local authorities are important in determining the shape of our landscape because they oversee planning processes, establish local environmental policies and regulations, and assist in implementing national and subnational environmental policies. As the level of government closest to the people, local authorities play a vital role in educating, mobilizing and responding to the public.

In Chapter 28 of Agenda 21, local authorities are encouraged to adopt a Local Agenda 21 for their communities, in effect to define a sustainable development strategy at the local level. The role of geology in maintaining our standard of living by discovering natural resources and solving engineering problems is well established. However, this success has overshadowed the value of geology as a non-renewable resource contributing to our quality of life.

Our landscape, which is underpinned by geology, provides a panoramic medium to demonstrate the conservation value of the Earth sciences alongside living biodiversity and human heritage. The landscape also provides a ready testimony of past environmental change and that the environment cannot be taken for granted. Furthermore, landscape is valued by a wide range of interests, many of them urban-based, and including ecology, archaeology and leisure to name a few. The global process of reaching a consensus on Local Agenda 21 by 1996 is a major opportunity for world-wide action in geological and landscape conservation.

What is Agenda 21?

This was one of the main agreements at the Rio Earth Summit in June, 1992, alongside the Conventions on Climatic Change and on Biodiversity, Principles for the Sustainable Management of Forests and the Rio Declaration. Agenda 21 is a detailed environmental action plan for the twenty-first century. Its progress is monitored by the United Nation's Sustainable Development Commission. Agenda 21 recognizes (especially in Chapter 28) that local authorities are key players in fulfilling the Agenda's objectives.

Local Agenda 21

Following on from above, the Rio Earth Summit proposed that local authorities should have prepared a Local Agenda 21 by 1996 for their communities. This involves extensive public consultation in over 150 countries because environmental loss is commonly most strongly felt at a local level. Progress is monitored by the International Council for Local Environmental Initiatives (ICLEI). In essence, the Local Agenda 21 is a regional sustainable development strategy.

Sustainable development

This can be defined as:

Development that meets the needs of the present without compromising the ability of future generations to meet their needs.

Superficially, sustainability appears an abstract notion concerned with the quality of life. However, it becomes more immediate when one considers that children born in the last five years will not retire until past the year 2050, by which date the results of a number of environmental changes will be clear, e.g. global warming, habitat loss, etc.

The degree of environmental change will depend on weak or strong interpretations of sustainability by governments. Weak sustainable development takes the environment into account, but still trades it off against other objectives. Strong sustainable development identifies limits to environmental change and uses those limits as constraints or incentives in development objectives.

Geology and sustainable development

Geology is an integral and important part of the natural environment. Its function in human

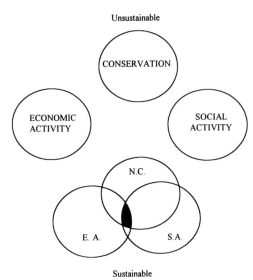

Fig. 1. Sustainable vs. unsustainable development. E.A.: economic activity; NC: nature conservation; SA: social activity.

society includes:

- provision of raw materials, energy, amenity, and life support services, e.g. place to live;
- removal of waste products, e.g. by burial.

The sustainability of geological services is constrained by non-renewability of resources, e.g. fossil fuels, and the carrying capacity of the environment, e.g. in pollution control.

Local issues asnd sustainable geology

Agenda 21 issues for UK local government can be compared with geological examples of such issues (Table 1).

We are used to conservation being interpreted as the preservation of desirable elements following geological activity. In a sustainable world, conservation begins with the more efficient use of non-renewable resources.

What can geologists do?

The Local Agenda 21 debate is just beginning, but five areas for action are already emerging:

- participation in the Local Agenda 21 consultation process, e.g. via the local community environment council, or similar forum;
- preparation of environmental statements (especially by organizations), e.g. after considering issues such as those above;
- exchange of information on environmental good practice, e.g. field work;
- supporting and resourcing geological environmental initiatives, e.g. community-based conservation schemes;
- increasing environmental awareness in geology, e.g. through training.

Local Agenda 21 will be an evolving agenda over the next three years, bringing together a range of environmental strategies, such as borough plans and environmental charters. It will become

Table 1.

Agenda 21 issues	Geological example
1 The economy	1 Mineral extraction, geotourism
2 Education and training	2 The National Curriculum, interpretation, rescue/salvage geology
3 Energy management	3 Extraction of fossil fuels
4 Environmental health	4 'Clean' technology
5 Environmental strategy and co-ordination	5 Environmental impact, cost/benefit analysis
6 Machinery of government	6 Public consultation, environmental audit
7 Housing	7 Rock extraction
8 Purchasing	8 Life cycle analysis
9 Land use planning and management (urban and rural)	9 Conservation (and enhancement) of landscape, soils, geological sites (non-statutory and statutory), palaeodiversity
10 Social welfare, equal opportunities and poverty	10 Access to information, coastal works, safety
11 Transport	11 Road building
12 Waste management	12 Landfill, reuse of materials
13 Waste and sewerage	13 Aquifer protection, groundwater pollution

increasingly prominent in monitoring reports, for example, environmental position statements. An early start in everyone's area is, therefore, desirable.

References

All the references listed here provided the background material for this paper. None are cited within the text itself.

CENTRAL AND LOCAL GOVERNMENT ENVIRONMENT FORUM. 1992. *Environment Forum Newsletter*, **1**.

COMMISSION OF THE EUROPEAN COMMUNITIES. 1992. *Towards sustainability* (EC's Fifth Action Programme on the Environment), **11**.

LEVETT, R. 1992. *Agenda 21: a guide for local authorities in the UK*, LGMB, Luton.

LOCAL AUTHORITY ASSOCIATIONS AND THE LOCAL GOVERNMENT MANAGEMENT BOARD. 1992. *Environmental practice in Local Government* (second edition), LGMB, Luton. (Loose leaf file.)

—— 1993. *The UK's report to the UN Commission on sustainable development: an initial statement/ submission by UK Local Government*, LGMB, Luton.

Additional post-MIC '93 publications:

DEPARTMENT OF THE ENVIRONMENT. 1993. UK Strategy for Sustainable Development Consultation Paper, HMSO, London.

LOCAL GOVERNMENT MANAGEMENT BOARD. 1993. *A framework for local sustainability*, LGBM, Luton.

Keynote address

The building of an international airport in an area of outstanding geological diversity and quality

LARS ERIKSTAD

*Norwegian Institute for Nature Research (NINA),
Box 1037, Blindern, N-0315 Oslo, Norway*

Abstract: Plans have been made for the new Oslo airport on the sandur surface of a raised marginal delta complex belonging to the Hauerseter substage (9800 yrs BP). The area already contains an airport, military camps and training fields. This is used as a case study to discuss the concept of sustainable development with particular reference to irreplaceable Earth science sites (geotopes). By integrating the needs of geotope conservation into the airport planning, the new airport construction will hopefully not result in a significant loss of geotope value. It may, therefore, be argued that the planning process is compatible with the concept of 'sustainable development' with respect to Earth science conservation.

Sustainable development has become a dominating term in the environmental debate. It is commonly comprehended as the principle that each generation must pass on to the next generation wealth (including environmental wealth) equivalent to the wealth that it itself inherited (Stevens *et al.* 1994). But this principle also rests on the assumption that resources are truly sustainable and allows almost indefinite exploitation in accordance with the carrying capacity of the exploitable objects. For Earth science sites (hereafter called geotopes) which are replaceable, such as some road cuttings and other man-made exposures (i.e. 'exposure sites' (Nature Conservancy Council 1990), this would imply replacement of damaged geotopes (Stevens *et al.* 1994). Most geotopes reflect values of different kinds, and it may be confusing to press them into a classification like this. Within this paper, when the term 'exposure site' is used it should be understood as an equivalent to 'exposure values'. The strategy for each site must be constructed on the basis of all the value that is present. This illustrates that although the above principle is simple and, in theory, possible to adapt, it represent a major challenge to the understanding of site values and evaluation.

This paper concentrates, however, on irreplaceable geotopes, 'integrity sites', where exploitation leads to permanent degradation of 'integrity values'. Thus, the principle of sustainable development (POSD) cannot be used in the same way. The question is, then, is it at all possible to adopt the POSD, since all modern societies imply land use practices that destroy geotopes which may be classified as 'integrity sites'?

The development of an understanding of the POSD which it is possible to apply in a normal practical planning situation, should include studies of the understanding of geological values, how they vary with time, and on what scale geotopes are defined. As a start it can be tentatively suggested that an understanding of the POSD in this context should be based on an understanding of the total value within specified management areas, and that this value should not be reduced. This would imply a thorough understanding of the geological systems, their geotopes and their value. It will not provide any easy new model for use in management theory, but calls for concentration and development of already existing systems within nature conservation, nature protection as well as integration of conservation values in the planning process at all levels.

Gardermoen: a case study

The Gardermoen area in southeast Norway consists of an ice-contact delta complex (Fig. 1) which includes the following main parts:

- a distinct ice-contact zone marking the Hauerseter substage (9800 yrs BP) in the ice recession system of the Oslofjord area (Sørensen 1983);
- two main delta formations together forming a flat sandur plain, more than 4 km wide, proximal to the ice-contact slope;

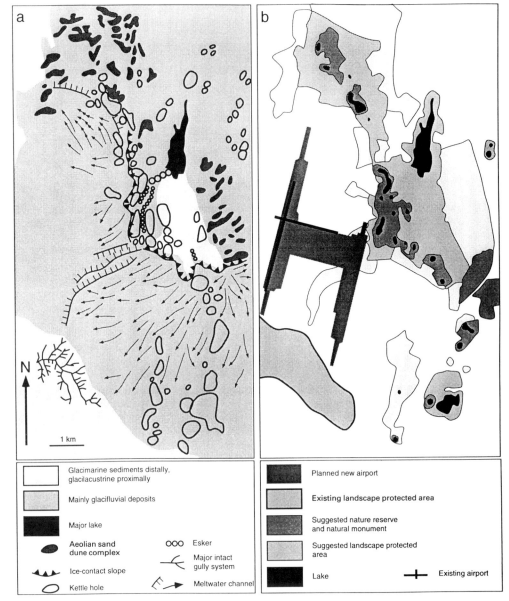

Fig. 1. (a) Glacial geology of the Gardermoen area (modified after Østmo (1976) and Longva (1987)). (b) The planned airport shown on the same map segment, together with existing and suggested nature protection areas (Erikstad & Halvorsen 1992).

- heavily gullied marine silts and clays proximal to the deltas and;
- a depression filled with glacilacustrine sediments distal to the ice-contact slope.

Additional formations are found in the area, e.g. systems of kettle holes along the ice-contact zone as well as marking the pattern of the main drainage out from the ice front when the delta was formed; a feeding esker belonging to one of the two main delta structures; meltwater channels on the sandur plain and well-developed aeolian sand dunes showing a predominant northerly wind during the time following deglaciation. The area has been mapped in detail by the Norwegian Geological Survey (Østmo 1976; Longva 1987).

The delta complex is a major aquifer. The groundwater drainage out of the delta has a major impact on the formation of gullies in the marine clays proximal to the delta. Some of the kettle holes are filled with bogs, and some with small lakes. The lakes are of two main types (Fig. 2). Some are seepage lakes which correspond with the groundwater, having a relatively rich water quality, the water level varying with the natural fluctuations of the groundwater table. Other lakes are not connected to the aquifer, even if they are situated in the same system and their depth would indicate that most of the lake is below the groundwater table. These lakes must have some sort of 'membrane' which isolates them from the surrounding aquifer. Their water level is much more stable and their water quality is poorer in nutrients since they are primarily fed by rain-water.

The nature conservation value of this area has been known for a long time (Longva 1987). Geologically it is focused on the whole suite of landforms connected to one of the youngest ice margins in south Norway, as well as on special objects such as the sandur, the gullies and the aeolian sand dunes. Limnologically it is focused on the great diversity and the great number of lakes connected to the kettle holes. The value of the area as a system is defined on an international level based both on limnological as well as geological criteria (Erikstad & Halvorsen 1992). Over the last decades several suggestions have been presented for establishing protection areas under the Nature Conservation Act. Up to now this has resulted in a $5 km^2$ landscape protected area south of the sandur which includes the best developed and undisturbed gully systems which are directly in contact with the sandur. Additionally, the local authorities have launched a time-limited restriction in

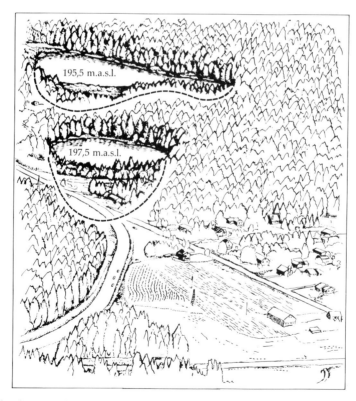

Fig. 2. Sketch showing two neighbouring lakes belonging to separate lake types in the area (Bonntjern uppermost and Svenskestutjern below). Bonntjern has contact with the groundwater (area = 4.6 ha, depth = 9 m, conductivity 5.4 mS m^{-1}, pH = 6.9 and content of calcium 10.2 mg l^{-1}). Svenskestutjern is isolated from the aquifer (area = 2.2 ha, depth = 17 m, conductivity = 1.4 mS m^{-1}, pH = 5.2 and content of calcium = 0.8 mg l^{-1}). (Erikstad & Halvorsen 1992). Sketch drawn by Bogdan Bocianowski.

establishing new gravel pits in the area, waiting for central authorities to decide what conservation actions are appropriate.

The area has already suffered from a number of severe attacks on sites with high conservation interests. Gullies have been graded as part of normal agricultural activities (Erikstad 1992); two major military training camps and training with heavy tanks (mostly on the sandur but also on the aeolian sands) and one international military airfield are situated on the sandur. In addition roads have been built and many gravel pits have been established, but major parts of the natural system are still intact.

When the new main airport for Oslo was decided to be established in the Gardermoen area, the question was raised, is some sort of 'sustainable development' at all possible, or must the conservation values of the area be considered permanently lost? During the process of environmental impact assessments an alternative location of the runway was argued for (Thommassen 1992), but without success, mostly due to aerotechnical reasons. Additionally new rail and road systems will be built and the new status of the airport will also increase the general pressure on the land in the surrounding district.

Discussion

This discussion concentrates on the physical geotopes described above and not on the groundwater aquifer as such. The relationship between the effects on the aquifer and the geotopes (and biotopes), is treated by the simple assumption that the groundwater manipulation and pollution control will work as planned and not significantly affect the described geotopes. Monitoring projects have been launched to test the reliability of this assumption and extensive research projects are in progress to gain a better understanding of the groundwater aquifer (Englund & Moseid 1992).

From a regional point of view there is no doubt that the Gardermoen area will experience major encroachments which may make a discussion of sustainable development rather hopeless. To examine this more thoroughly within the strict limits of geotope conservation, it is, however, necessary to change scale, at least to the level where the value of the area was originally defined. In this case we may split the area into sets of separate regions (or geotopes) by considering the different parts in relation to the deglaciation history (Fig. 1b).

The airport itself will only affect the sandur. Its main landform (the flat plain) will still dominate the landscape, but details such as meltwater channels and surface and subsurface materials will, of course, be altered and permanently destroyed. Communication systems will additionally affect other main geotopes on all sides of the airport. With thorough planning, however, it should be possible to limit this damage.

The main problem will be the secondary pressure on the land use that will be set up by the airport localization. Geotopes of high value are found just outside the airport on all sides and will become attractive for all sorts of development. The environmental impact assessment (Thommassen 1992), therefore, concluded that establishment of nature conservation areas around the airport was necessary to protect these valuable sites. This was later specified in a report suggesting the establishment of eight nature reserves and one natural monument within a system of three main landscape protection areas (Erikstad & Halvorsen 1992) (Fig. 1b). There already exists one such area, protecting the gullies south of the airport. In addition, special measures should be taken under area planning for protection of specified localities with aeolian sand dunes and kettle holes.

If these recommendations are followed and integrated as part of the alleviating measures in the planning process and in the construction of the airport, the total conservation value of the area will not be significantly reduced. As already pointed out, this is based on the assumption that no dramatic effects will occur in the groundwater system, and within this limit it appears possible to build the airport within the framework of sustainable development, at least with respect to Earth science conservation.

The principle of protecting defined geotopes as an integrated part of the alleviating measures in the planning procedure is a key point. If this fails, the conclusion above will be negated.

Conclusion

Sustainable development related to irreplaceable geotopes requires a well-developed understanding of geotope evaluation, environmental impact analysis and a system for conservation of geotopes both in planning and nature conservation legislation.

To promote sustainable development, the area planning system has to adopt the concept, here defined as the principle that the stock of geotope values within a management unit should not be reduced. As long as this goal applies at all planning levels, it will secure geotopes of local as well as national value. However, in order not to obstruct the planning system, it requires a well-developed system of scientific and practical input and consensus.

References

ENGLUND, J. O. & MOSEID, T. (eds) 1992. *Introductory Report, The Gardermoen Project. Literature Review and Project Catalogue.* The Norwegian National Committee for Hydrology (NHK), Oslo.

ERIKSTAD, L. 1992. Recent changes in the landscape of the marine clays, Østfold, southeast Norway. *Norsk geografiske Tidsskrift,* **46**, 19–28.

—— & HALVORSEN, G. 1992. Sites with national and international importance at the Hauerseter substage, Southeast Norway. *NINA Oppdragsmelding,* **136**, 1–28 (in Norwegian with English summary).

LONGVA, O. 1987. Ullensaker. Description of the Quaternary geological map 1915 II. *NGU-skrifter,* **76**, 1–39 (in Norwegian with English summary).

NATURE CONSERVANCY COUNCIL 1990. *Earth science conservation in Great Britain. A strategy.* Nature Conservancy Council, Peterborough.

ØSTMO, S. R. 1976. *Gardermoen, Quarternary geological map C QR 051052-20, M 1:20000.* Norges geologiske undersøkelse, Trondheim.

STEVENS, C., ERIKSTAD, L. & DALY, D. 1994. Fundamentals in earth science conservation. *Memoire de la Société Géologique de France,* **165**, 209–212.

SØRENSEN, R. 1983. Glacial deposits in the Oslofjorden area. In: (EHLERS, J. ed.). *Glacial deposits in North-west Europe.* A. A. Balkema, Rotterdam, 19–28.

THOMMASSEN, J. 1992. Environmental impact assessment, the main airport Gardermoen. *NINA Oppdragsmelding,* **140**, 1–41 (in Norwegian).

The justification for a soil conservation policy for the UK

T. R. E. THOMPSON & P. BULLOCK

Soil Survey and Land Research Centre, Cranfield University, Silsoe Campus, Bedfordshire MK45 4DT, UK

Abstract: Soil is a vital part of terrestrial ecosystems and an important ecosystem in its own right. There is evidence that land use change has led to a decline in some principal soil functions. The explicit protection afforded to geological and geomorphological features is not extended to cover soil despite its fundamental importance as an environmental resource whose processes and properties influence biota and water quality. The effects of current land use on soil are reviewed, and an argument is advanced for greater protection and conservation of soils through a three-point action plan aimed at ensuring a sustainable future for soil in England and Wales.

Soil is the weathered surface layer of the Earth's crust. It varies with space and time but generally exhibits the products of physical, chemical and biological processes. Most UK soils are the consequence of pedogenesis during the Flandrian. However, particularly in the south and east of England, soils can exhibit properties that relate to development during previous Quaternary pedogenic episodes. Soils are mostly structured, exhibit horizonation, are chemically differentiated from their parent material, contain organic matter, and exhibit biological activity. Soils possess complex biological communities ranging from viruses, bacteria, fungi and blue-green algae (microbiota) through to macro—fauna such as soil-living invertebrates.

A soil is conventionally characterized by the nature of those pedogenic processes that are most evident from a vertical profile through the soil. In addition, the texture of the soil and the nature of the material from which it has developed are used to classify soils. As well as being vertically heterogeneous, the soil kingdom varies spatially. Within England and Wales alone, 700 specific types of soil can be identified (Clayden & Hollis 1984), and these represent only a part of the world's soils. Ragg & Clayden (1973) discuss the place of British soils in the context of a world classification of soils. Avery (1980) has proposed a classification of English and Welsh soils that identifies 10 Major Soil Groups that divide consecutively into 43 Soil Groups and 118 Soil Subgroups.

Soil functions and uses

Soil is recognized as one of the primary environmental compartments (Schneider 1987) that performs a number of vital natural functions within terrestrial environmental systems. Its influence on plant growth and on animal behaviour makes its study and protection an essential part of understanding and conserving terrestrial animal and plant ecosystems. The mechanisms and patterns of retention/release of soil water, and the chemistry of soil water form a link between soil and freshwater ecosystems. Freshwater eutrophication from non-point sources and the acidification of some upland lakes and rivers are evidence of the influence of soil processes on water quality. Blum (1993) cites three ecological functions for soils and three anthropocentric functions which others (Howard *et al.* 1989) have regarded as soil uses. Soil heterogeneity results in each soil having a characteristic 'signature' in terms of its ability to fulfill the full range of functions. For example, certain soils are naturally low in fertility – acid oligotrophic peat soils of the Winter Hill Series are unable to support high rates of biomass accumulation, and shallow soils of the Powys Series over shale are unable to store large volumes of soil water. Because of this natural variability, there is an argument for viewing and protecting soil functionality at a 'kingdom' level. Attempts to derive a single 'quality' parameter for soils have their limitations (Howard *et al.* 1989) and this factor underlies the work of ecologists studying the close relationship between vegetation and soil (e.g. Tansley 1949).

Threats to soil

Few soils are untouched by the activities of man. Effects on soil can be either through complete loss of soil or via more subtle or gradual processes. Urban and industrial development covers 15% of land in England

and Wales and has continued to expand its influence at a rate of 11 000 hectares per year (Sinclair 1992). Mineral extraction can lead to long-term changes in soil even where restoration to agriculture is of a high quality. Agriculture occupies 80% of the country and 85% of that is crops and grass.

Cultivation and related practices lead to changes in the chemical, physical and biological nature of soils (Edwards 1984). The fauna of cultivated soils are characterized by lower diversity and mass. For the soils of Broadbalk field at Rothamsted, Cousins (pers. comm.) has identified a relationship between faunal diversity, as measured by body size, and topsoil organic carbon content and acidity. Long-term cultivation leads to loss of organic matter. A number of authors have reviewed potentially damaging change in soils. Some such studies have been prompted by concern for agricultural production (MAFF 1970), while others have been prompted by an interest in soil protection (Howard *et al.* 1989). This dichotomy highlights a factor that has dominated thinking on the protection of soil resources in England and Wales, the twin resources of soil and land (Thompson 1992). Concern for land protection has dominated policy and practice. The Agricultural Land Classification scheme, whereby land judged to be capable of producing consistently higher than average yields of food crops is protected from loss to development, is the only means by which the protection of land and soils explicitly enters the development control process.

Soil protection

Concerns over the protection of soil are being given considerable importance by the Council of Europe. Soil erosion, compaction, contamination, acidification and the loss of soil biological diversity are identified as significant processes affecting soils. The eutrophication, acidification and contamination of surface waters, and the contamination of groundwaters by nitrate, phosphate and biocides are associated concerns, whose solutions require changes in soil management. In England and Wales SSLRC (1993) have assessed the extent of land in the lowlands that is subject to erosion given autumn cultivation and cropping. Research into sheet erosion is currently underway, and may extend the area of land regarded as susceptible to damaging erosion. The extent of soil contamination by elements including the heavy metals is confused by lack of agreement over what constitutes a maximum acceptable concentration for each element (McGrath 1993). McGrath & Loveland (1992) report on the concentrations of 17 elements. Information on the soil concentrations of toxic organic compounds is sparse. Atmospheric sources of acidity continue to exceed buffering capacity in upland soil. Projections for traffic growth throughout Europe, and therefore NO_x exhaust emissions, raise concerns for lowland soils of low buffering capacity.

Soils and sustainability

Terrestrial ecosystems depend on soil, but there is evidence that peoples' use of the land has caused and continues to cause changes in soil.

These changes, at least in part, can be considered to represent a loss of overall quality and functional capacity (Blaikie & Brookfield 1987). Current use of soil is unsustainable in the sense that future generations will not inherit the soil capital that is currently enjoyed. Explicit conservation and protection measures should be applied to soils in order to reverse these trends.

Britain is a signatory to the European Soil Charter (Hacourt 1990) that recognizes soil as a finite and valuable resource requiring protection. Representatives of the Department of the Environment and the Ministry of Agriculture, Fisheries and Food (MAFF) are party to disccusions within the Council of Europe aimed at a revision of the Charpter (Council of Europe 1990). MAFF has drafted a *Code of Good Agricultural Practice for the Protection of Soils* (MAFF 1993). That the Code is restricted to farmed soils, and deals only with 'production-oriented' factors, highlights a major obstacle to the improvement of soil protection. Son single government department has, or recognizes, a lead responsibility for protecting soil resources. Furthermore, soil is inexplicably excluded from the responsibilities of the conservation agencies under the Countryside Act of 1968 and Wildlife and Countryside Act of 1981.

Proposed plan for the future

The authors propose the adoption of a three-point action plan aimed at encouraging the sustainable use of soils.

(1) A national review of soil quality, particularly biological quality, to establish the current status of soils and soil fauna. Contaminant levels in soils are poorly established for sites other than those that have attracted special attention. Given the national monitoring frameworks for atmos-

pheric and water quality, the establishment of such a network for the monitoring of soil quality should be considered as a second phase to the review exercise.

(2) Incorporation of soil conservation into the considerations of conservation agencies; to include a review of the need for designated and managed soil reserves to conserve particular type soils and soil biological communities. The agencies for Wales (Countryside Council for Wales) and Scotland (Scottish Natural Heritage) are in the early stages of policy determination with regard to soil conservation. With the establishment of the Environment Agency, this organization should be given equivalent duties relating to the protection of soil from degradation, particularly contamination, as are proposed for water and oil.

(3) The establishment, as part of the National Sustainability Plan, of a National Land Use Policy that takes due regard to the natural functions, heterogeneity and capacities of soil. The plan should be the result of inter-departmental discussions and would influence land use decisions across all sectors of public and private industry. The limited extent to which the environmental capacity of soil has influenced land use to date has led to widespread environmental degradation, and does not equate with the accepted principles of sustainability.

References

AVERY, B. W. 1980. *Soil classification for England and Wales*. Soil Survey Technical Monograph No. 14. Harpenden, UK.

BLAIKIE, P. & BROOKFIELD, H. 1987. *Land degradation and society*. Methuen, London.

BLUM, W. E. H. 1993. Soil protection concept of the Council of Europe and integrated soil research. *In*: EIJSACKERS, H. J. P. & HAMERS, T. (eds) *Integrated soil and sediment research: a basis for proper protection*. Kluwer, Dordrecht.

CLAYDEN, B. & HOLLIS, J. M. 1984. *Criteria for differentiating soil series*. Soil Survey Tecnical Monograph No. 17. Harpenden, UK.

COUNCIL OF EUROPE 1990. *Feasibility study on possible national and/or European actions in the field of soil protection*. Council for Europe, Strasbourg.

EDWARDS, C. A. 1984. Changes in agricultural practice and their impact on soil organisms. *In*: JENKINS, D. (ed.) *Agriculture and the Environment*. Proceedings of the ITE Symposium No. 13. NERC, Banchory.

HACOURT, H. 1990. Quality as welk as quantity. *Naturopa*, **65**, 5–6.

HOWARD, P. J. A., THOMPSON, T. R. E., HORNUNG, M. & BEARD, G. R. 1989. *An assessment of the principles of soil protection in the UK*. NERC, Merlewood.

MAFF 1970. *Modern farming and the soil*. HMSO, London.

—— 1993. *Code of Good Agricultural Practice for the Protection of Soils*. HMSO, London.

MCGRATH, S. P. 1993. Soil quality in relation to agricultural uses. *In*: EIJSACKERS, H. J. P. & HAMERS, T. (eds) *Integrated soil and sediment research: a basis for proper protection*. Kluwer, Dordrecht.

—— & LOVELAND, P. J. 1993. *The soil geochemical atlas of England and Wales*. Blackie, London.

RAGG, J. M. & CLAYDEN, B. 1973. *The classification of some British soils according to the comprehensive system of the United States*. Soil Survey Technical Monograph No. 3. Harpenden, UK.

SCHNEIDER, G. 1987. Soil protection strategy in the Community. *In*: BARTH, H. & L'HERMITE, P. (eds) *Scientific basis for soil protection in the European Community*. Elsevier, London.

SINCLAIR, G. 1992. *The lost land, land use change in England 1945–1990*. A report to the Council for the Protection of Rural England. CPRE, London.

SSLRC 1993. *Risk of soil erosion in England and Wales* (map). Soil Survey and Land Research Centre, Silsoe.

TANSLEY, A. G. 1949. *The British Isles and their vegetation*. Cambridge University Press, Cambridge.

THOMPSON, T. R. E. 1992. An approach to the setting of environmental quality objectives for soil and land. *Biologist*, **39**, 1 33–4.

Towards a definition of the Spanish palaeontological heritage

LUIS ALCALÁ & JORGE MORALES

Museo Nacional de Ciencias Naturales (CSIC), José Gutiérrez Abascal, 2, E 28006 Madrid, Spain

Abstract: The important socio-economic changes which have occurred in Spain during the past decade have impinged upon aspects of the country which have traditionally been neglected in previous development programmes. One of these aspects, concerns palaeontological heritage, which is now legally covered by two different state laws: 'Patrimonio Histórico Español' and 'Conservación de los Espacios Naturales y de la Flora y Fauna Silvestre'.

The inclusion of palaeontology within these laws is very important and positive. Nevertheless, there are many different and contradictory opinions about the definition and the limits of palaeontological heritage, reflecting the point of view of the specialist group concerned (vertebrate and invertebrate palaeontologists, archaeologists, anthropologists, etc.).

The aim of this paper is to define Spanish palaeontological heritage, in order to have a clearer and less ambiguous idea of its meaning, using three criteria: scientific, socio-cultural and socio-economic.

In contrast to the attention given to historical heritage, and the natural heritage, both of which have a long tradition of protection in Spain, Spanish palaeontological heritage has only recently begun to receive a certain amount of public interest. As a result of this recent rise in public interest, palaeontological heritage is now legally protected by two national laws. It is, however, difficult to distinguish between scientific, amateur and commercial operations within these laws.

The Spanish laws

Ley de Patrimonio Histórico Español (16/1985, 25 of June) (Spanish Historical Heritage Law)

The protection afforded by this law to palaeontological sites is ambiguous; in certain cases palaeontological finds are treated independently of any other type of heritage (Article 1.2 – components of the heritage; Article 15.4 – definition of a historical site), but generally, palaeontological resources are subordinate to those of archaeology, and palaeontological excavations are considered archaeological in nature.

Ley de Conservación de los Espacios Naturales y de la Flora y Fauna Silvestres (4/1989, 27 of March) (Law of Conservation of Natural Spaces and of the Wild Flora and Fauna)

Chapter 2 of this law is dedicated to protected natural areas or monuments, expressly stating that palaeontological sites are to be included in the protected zones:

> Natural Monuments are areas or natural features composed of elements of known uniqueness, rarity or beauty which merit being the object of special protection (Article 16.1). Geological formations, palaeontological sites, and other elements of the Earth for which there is a special interest due to uniqueness on the importance of its scientific, cultural or scenic value may also be included (Aricle 16.2).

In accordance with the Spanish Constitution of 1978, the enforcement of these laws is the responsibility of the 17 existing Autonomous Communities. These Autonomous Communities regulate the use and the protection of palaeontological heritage in very different manners, ranging from those that treat it as if it were archaeological heritage to those that have quite simply forgotten that the State has conferred upon them this responsibility (Alcalá &

Paricio 1988). Nevertheless, there is the basic that palaeontological heritage is as distinct from historical heritage as from natural heritage even though all may share certain characteristics. As a result, palaeontological heritage should require specific legal treatment.

The nature of palaeontological heritage

Palaeontological heritage shares common characteristics with historical heritage as well as natural heritage. Perhaps the most notable similarities between palaeontological and historical heritage are the existence of a strong market of collectors and the possibility of individual discoveries of great importance. In common with natural heritage, palaeontological heritage frequently has areas of interest which cover a relatively large expanse of land.

In contrast, however, to both the other types of heritage, palaeontological heritage can, at times, be considered as being comparable to renewable resource. That is, any public or private work can positively or negatively affect it and the possibilities for new discoveries may be continuous, whether by design or accident.

It is often asserted that natural processes of destruction, such as weathering and erosion, can negatively affect palaeontological heritage just as much as the activities of amateur or professional collectors. This is decidedly false. Without negating the role of geodynamic forces in the attrition of any type of heritage, the destructive effects of specimen collectors is greater, particularly at those sites where proper documentation is crucial. This documentation includes taphonomic information, which is now almost universally used at vertebrate sites and is slowly extending to other palaeontological sites.

Taking all this into consideration, it would be misleading to think that all potential or actual palaeontological sites could be protected. In a country such as Spain, with a long and diverse geological history, a well-exposed terrain and numerous very fossiliferous stratigraphic series, it would be impossible to protect the hundreds (or thousands?) of potential sites with an area of more than $300\,000\,\text{km}^2$ (or nearly three-fifths of the country).

As a result, it is essential that we determine reasonable criteria to assist the various public administrations charged with the management of Spain's palaeontological heritage and help them clearly understand the nature of their

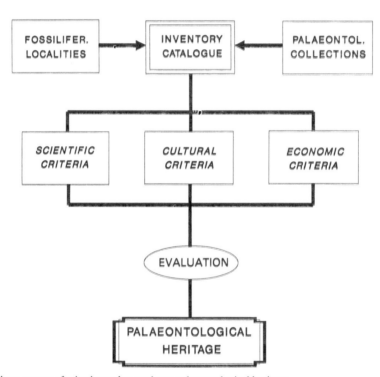

Fig. 1. The three groups of criteria used to evaluate palaeontological heritage.

responsibilities. In this paper, we present three classes of criteria which could help resolve these difficulties, and give palaeontological heritage the status it deserves (Fig. 1).

Criteria for defining the palaeontological heritage of Spain

Scientific criteria

1. Nature of fossils: Without a doubt, this is one of the fundamental criteria. Sites in which fossils of exceptional scientific importance have been found should receive special consideration.

2. Geological age of site: Age is a key factor for palaeontological sites and can be almost imseparable from the taxonomic significance of the fossils.

3. Type localities: Type localities are those from which certain species have been first recognized and defined typologically.

4. Preservation: The state of preservation of fossils is an important factor when studying the anatomical characteristics of organisms. Sites with well-preserved faunas or floras are generally therefore more important than those with poorly preserved fossils.

5. Association with archaeological remains: The history of humankind is generally valued from a scientific and socio-cultural point of view. Those palaeontological sites having archaeological remains generally possess an exceptional importance due to their direct cultural links to humankind.

6. Diversity of fossils: Sites at which a great diversity of fossilized organisms have been found (i.e. co-occurrence of vertebrates, invertebrates and plants), or in which there is simply a great diversity of taxa, are very important for palaeobiological and palaeo-environmental reconstructions.

7. Taphonomic information: Sites with unusual taphonomic processes or which represent unaltered palaeobiological communities are of very high palaeobiological importance.

8. Bio- and chronostratigraphical interest: Type localities of stages, zones etc. or which simply date important geological formations are important, often internationally.

9. Wider geological interest: Sites of wider geological interest could represent particular sedimentary environments, or show associations with volcanic activity, or permit marine-continental facies correlations, amongst many other geological phenomena.

10. Level of Knowledge: This is understood as the present level of scientific knowledge of a site and the fossils therein.

Socio-cultural criteria

1. Fragility: The fragility of a site is related to the lateral extension and potential of the fossil-bearing layer. The less extensive the layer, the more fragile the site would be.

2. Geographic situation: Sites from which are remote from urban areas and have less potential for frequent use and lower numbers of visitors, and may therefore be less vulnerable to damage.

3. Vulnerability to damage related to fossil collecting: This criterion should be applied in conjunction with the concept of fragility, as a much-visited site is most easily destroyed by collecting (Corral & Alcalá 1992). In contrast, an extensive locality which is not easily destroyed would be less vulnerable and could be opened up to private collectors (Norman 1992).

4. Historic value: Sites which form part of the history of palaeontology. In general, these will be sites which were discovered in the early days of the science (e.g. pre-1900).

5. Educational interest: The potential of a site for use in public education, university-level teaching, congresses, etc.

6. Touristic interest: Similar to the previous criterion but including sites of potential or actual value to groups interested in the fossil record.

7. Complementary value: Sites of complementary value are sites located in areas already protected for reasons of historical or natural value, and, therefore, ones that can be used in a complementary manner.

Socio-economic criteria

1. Urban value: Urban value refers to sites located in urban areas or lands potentially available for development. It is recommended that, in attempting to protect such areas, alternative solutions be sought such as rescue excavations or the incorporation of the site into parks, gardens, or *in situ* museums.

2. Mineral value: In the case of sites found in association with mineral exploitation, fossil excavation can sometimes only be realized as long as mineral extraction continues. In other cases, however, the fossil concentrations may be easily destroyed by ongoing works if the deposit is of limited extent.

3. Public works: As in the prededing case, public works can destroy the sites (Alcalá 1987) or can be potential sources of new sites (Alcalá & Morales, in press).

4. Economic value: Economic value is linked to the phenomenon of private fossil collections. The existence of a fossil market is a reality and should not be forgotten when evaluating a site (Taylor, 1988, 1992; Wood, 1988; Babin, 1992).

Conclusion

The three groups of criteria discussed here could be used to quantify the potential importance of an individual site or at least provide rigorous objective standards by which to filter and evaluate palaeontological sites, before such sites are legally declared to be part of Spanish palaeontological heritage (Fig. 2). Sites so declared should ideally be managed in such a manner as to safeguard the function or functions for which they were originally selected, leaving the rest of the country comparatively free for use by specimen collectors. Inevitably the status of known sites may change and new site discoveries can also be added to the Spanish palaeontological heritage list. In other words, a palaeontological heritage with a few well-protected and managed sites is preferable to an enormous, indefensible one which runs the risk of losing that which is truly exceptionable and irreplaceable.

However, the practical situation is complicated and still far from complying with the philosophy discussed earlier. The first point of conflict lies in the transference of the historical heritage regulations (which includes palaeontological heritage) from central government to the 17 Spanish regional administrations (or Com-

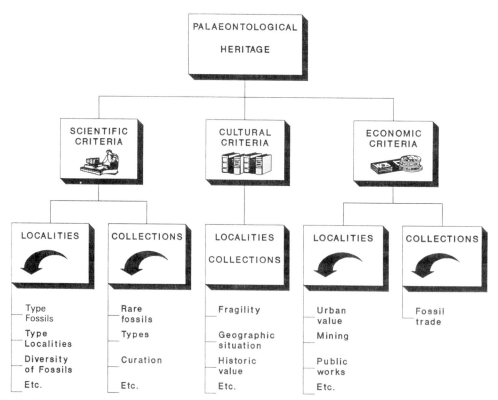

Fig. 2. Criteria proposed for definition of Spanish palaeontological heritage.

munities), in accordance with the Patrimonio Histórico Español law. Because each separate regional administration has independently interpreted the regulations concerning palaeontological heritage, without regard to any consensus of understanding, the laws are randomly enforced. As a result of this, some regions have extremely rigorous standards, while in others, there seems to be almost no regulation. Professional palaeontologists have encountered problems in those areas exercising rigorous standards because the criteria used to regulate fieldwork are highly complex and difficult to understand. The Community of Madrid is the only regional administration that has chosen to protect specific areas and localities based on the criteria discussed in this paper. The results have so far proved successful, particularly in the old 'downtown' centre of Madrid (Alcala & Morales, in press).

The current confusing situation should be dealt with in the next few years by two working groups dedicated to the protection of Spanish palaeontological heritage. One group is sponsored by the Sociedad Española de Paleontología and the other by the Sociedad Geológica de España. Both groups are trying to work together to establish the regulations laid down by the heritage laws so that all regional administrations can conserve and manage their palaeontological heritage appropriately. The main problems facing the groups are that many regional museums lack a palaeontology department, and often the department administering the regulations is composed of archaeologists and historians and not palaeontologists. The archaeologists and historians may well have interests that conflict with those of the palaeontologists, and this can lead to misinterpretation of the regulations.

With the gradual increase in appreciation of the natural heritage in Spain, it is hoped that the situation may change. The presence of a palaeontologist in regional museums and administrative departments that oversee the enforcement of palaeontological regulations may become a future requirement.

This study has been carried out under the auspices of the Comunidad Autónoma de Madrid and the Museo Nacional de Ciencias Naturales (Agreement 1992–1994 'Carta Paleontológica'). We also thank Paula Bayard for the English translation of the manuscript.

References

ALCALÁ, L. 1987. Los Aljezares de Teruel. Historia y estado actual. *Geogaceta*, **3**, 61–63.

—— & MORALES, J. (in press). The palaeontological heritage of the Community of Madrid. *Ier Symposium Interantional sur la protection du patrimoine géologique*, Digne.

—— & PARICIO, J. 1988. Protección y conservación de yacimientos palaeontológicos españoles (1984–1988). *II Congreso Geológico de España, Granada*, **1**, 253–256.

BABIN, C. 1992. Los yacimientos de fósiles, su protección y su interés cultural en algunos países de Europa. *In* (FERNÁNDEZ, S. (ed.)) *Conferencias de la Reunión de Tafonomía y Fosilización*. Editorial Complutense, Madrid, 45–62.

CORRAL, J. C. & ALCALÁ, L. (1992). El coleccionismo privado de fósiles: un paso atrás en el concepto actual de museo. *In: Paleontología y Sociedad*. Sociedad Española de Paleontología y Departamento de Estratigrafía y Paleontología de la Universidad de Granada, Granada, 69–85.

NORMAN, D. B. 1992. Fossil collecting and site conservation in Britain: Are they reconcilable? *Palaeontology*, **35**, 247–256.

TAYLOR, M. A. 1988. Palaeontological site conservation and the professional collector. *In:* CROWTHER, P. R. & WIMBLEDON, W. A. (eds) *The use and conservation of palaeontological sites*. Special Papers in Palaeontology, **40**, 123–134.

—— 1992. The local geologist 6: The fossil trade. *Geology Today*, January–February 1992, 29–33.

WOOD, S. P. 1988. The palaeontological site conservation and the price of fossils: views of a fossil hunter. *In:* CROWTHER, P. R. & WIMBLEDON, W. A. (eds) Special Papers in Palaeontology, **40**, 135–138.

Fossil collecting: international issues, perspectives and a prospectus

DAVID B. NORMAN

Sedgwick Museum of Geology, Department of Earth Sciences, University of Cambridge, Downing Street, Cambridge CB2 3EQ, UK

Abstract: The issues raised as a result of the collection of a particularly prized fossil can range from the purely personal sense of 'gain' or 'loss' by an individual, to matters of national heritage and pride. Nations and nation states differ significantly in their perception of the importance, or indeed value, of fossils and in the way that they have framed (or indeed failed to frame) legislation to deal with issues of this kind. This multiplicity of attiudes can create confusion for those scientists wishing to collect and study fossils, or for others wishing to collect fossils either as a recreational activity or as a commercial enterprise.

The legislative framework surrounding fossils is reviewed by reference to a number of nations and the policies that they have adopted. From this review it is apparent that transnational inconsistency (or inequitability) is divisive in the present day, when fossil collecting is not only a national, but in some instances an international activity. It is concluded that a generally accepted convention devoted to Earth heritage sites should also address the status of fossils. A general framework of the matters relating to fossils, that should be addressed within such a convention is suggested and put forward as the basis for wide-ranging discussion.

From the very first (Etruscan tomb relics – c. 2500 BC), fossils have been prized as items to be revered and worthy of collection (Edwards 1967). As 'collectibles' fossils have naturally attracted the attention of collectors and have commanded a price (Rolfe *et al.* 1988). Attaching a monetary value to a natural heritage item such as a fossil raises a spectrum of conflicting assessments bounded by, at one extreme, its importance to the palaeontologist as a material component of a research programme; while, at the other, it may be valued solely on the basis of its aesthetic appeal to the collector. Considering these extreme standpoints as references against which evaluations of claims for judgement by independent assessors might be made, the following issues become apparent.

Scientific assessment of the importance of fossils can be placed in an objective framework since it can be based on a set of stated criteria such as: uniqueness, being representative of a particular chronofauna, showing important anatomical features, geological and/or geographic location, or the evolutionary implications that may be associated with its discovery. While the relative importance of each of these criteria may be disputed, they nevertheless form a sound framework within which palaeontologists and neontologists can form a consensus of opinion concerning the stated object's value to science. This procedure of scientific justification, incidentally, provides an opportunity to explain the rationale behind the scientific assessment of fossil specimens and the sites from which they have been collected, to the public and legislature alike.

Aesthetic assessment is, by its very nature, a much more uncertain area of consideration. It is perhaps best summarized by the phrase 'beauty is in the eye of the beholder.' Or, as the commercial fossil collector Stanley Wood (1988; p. 137) succinctly summarized the matter: 'The "right price" for any fossil is a purely hypothetical value. When money is involved, a price must be agreed. The buyer measures in his or her mind how desirable the fossil is. This can vary from person to person, as their individual responses are quite independent'.

Conflicts arise over the issue of fossil collecting for cultural, historical, scientific and legal reasons and all four of these issues are deeply intertwined; this makes the task of mediation in the event of a dispute, should it arise, extremely difficult because quite often attitudes adopted by either party can be found to be fundamentally contradicted by their actions in the past.

As one who has been involved in attempts to mediate and set policy on the matter of fossil collecting (Norman & Wimbledon 1988; Norman *et al.* 1990a, b; Norman 1992a, b) it is of considerable concern that rather than issues becoming resolved with time, it seems that fossil collecting and the commercial sale of fossils are becoming increasingly divisive issues. The newsworthy controversies surrounding the early tetrapod 'Lizzie' (*Westlothiana lizziae* Smithson

& Rolfe 1991) which was discovered by Stanley Wood (Rolfe 1990), and eventually purchased by the Royal Scottish Museum, Edinburgh; the recent case of confiscation of a complete *Tyrannosaurus rex* skeleton by the FBI and its secure storage at the South Dakota School of Mines and Technology (USA) pending further court action (Eighth Circuit Court of Appeals 1992); and the attempts to move the Baucas Bill (Vertebrate Paleontological Resources Protection Act) through the House of Senate of the US Government, all support the view that fossils are becoming more than a local specialist issue, and indeed risk becoming a sensitive political issue which has the potential to transcend national boundaries.

Fossils and fossil collecting internationally

Attitudes toward fossils and fossil collecting vary from country to country and, at least in part, reflect the sovereign needs of those countries. This synoptic review is intended not to be exhaustive, but simply indicates the range of approaches that have been taken with regard to fossils and their status in law by various nations and nation states.

Canada

Fossil collecting is regulated by Acts of Government which are the responsibility of the Provinces (Ferguson 1988). The first such Act (The Historical Objects Protection Act) was passed in 1970 for Nova Scotia and superseded by the Special Places Protection Act (1980). The majority of Provinces – with the single exception of Prince Edward Island – have Acts which regulate the collection of palaeontological remains. The degree to which individual fossil groups are protected does seem to vary from Province to Province, and in general requires fossil collecting to be done through a permit system, which is either rigorously enforced (Alberta) or less so.

In addition to the Provincial Laws, there is a further tier of Federal Law (The Federal Cultural Property Export and Import Act [1975]), which controls the export of fossils from Canada.

Denmark

Until recently, Denmark had no specific legislation covering fossil materials and their collection. However, in 1989 the Museum Act was modified to include the concept of *Denekrae* (natural science objects – including fossils – are considered, when found, to belong to the State) (Christensen & Hald 1990). Finders are obliged by law to present such finds to natural science museums for evaluation. If considered of value to the State the object will be retained, and the finder will be offered a reward based on the rarity of the object, and the care with which it has been collected.

A 'natural sciences reference group' evaluates finds and makes a recommendation to the National Board of Museums, not only about the scientific importance of the fossil, but where it should be housed and the value of the award. Final decisions are made by the National Board of Museums.

Germany

Germany has a long history of geological conservation (Grube & Albrechts 1992) with the first geological national park being established in 1836 in the Siebengebirge. The sixteen Länderen which comprised the former German Federal Republic are governed by the Federal Monument Protection Law 1973 (Wild 1988; Grube & Albrechts 1992). The implementation of federal law varies widely across the former republic.

Bavaria, which includes in its land area such noted palaeontological sites as the Solnhofen–Eichstätt lithographic limestone quarries, has no law to protect fossils and this allows free trade in fossils to take place. By contrast, the neighbouring State of Baden–Württemberg uses its Monument Protection Law to protect fossils by considering them as cultural monuments (Wild 1986). This legislation allows for State intervention in the evaluation of fossil material once it has been discovered, and compensation to be paid to the finder if it is considered sufficiently important to be retained by the State Museum in Stuttgart. If the specimen is not considered to be of 'monument' status then it is given to the finder to dispose of in whatever way is deemed satisfactory.

Clearly such inconsistency in approaches by the German Länderen fosters conflicting attitudes to both the practice of fossil collecting and the appreciation of natural heritage items within the former Federal Republic as a whole and this cannot help in any way to promote transnational consensus with regard to the status and value of fossils.

Romania

The conservation movement in Romania has a surprisingly long history (Grigorescu & Norman 1990). In the 1930s a government commission established nature conservation areas. In 1973 the National Council for the Protection of the Environment was empowered to establish Natural Monuments (including geological sites). Recommendations for monument status are evaluated by the National Academy of Science.

At present there is no specific legislation governing the status of fossils as National Monuments, and therefore no regulation of the collection and import or export of fossils.

South Africa

The National Monuments Act (1969) protects all palaeontological, archaeological and historical sites and material older than 50 years. The legislation is sweeping and comprehensive, as the extract below demonstrates:

> 'No person shall destroy, damage, excavate, alter, remove from its original site or export from the Republic – a. any meteorite or fossil ...archaeological or palaeontological finds, materials or object, except under the authority of and in accordance with a permit issued under this section.'
>
> National Monuments Act
> (No. 28 of 1969, as amended)

Note: The appreciation of the value of the fossil heritage of South Africa, and the introduction of rigorous control of collecting and export, lies in the earlier exploitation of its extremely rich fossil heritage – notably of the remains of fossil man from the famous Plio-Pleistocene sites in the northern Transvaal (Sterkfontein, Swartkrans, Makapansgat) and of the removal of abundant fossil reptiles from the Permo-Trias of the Great Karoo Basin by individual collectors and by international expeditions.

United Kingdom

As summarized in earlier papers (Taylor & Harte 1988; Norman 1992a, b) the Wildlife and Countryside Act (1981) provides a framework within which it may be possible to regulate fossil collecting at designated Sites of Special Scientific Interest (SSSIs) or on geological National Nature Reserves (NNRs). However, there is still no law relating specifically to fossils. In the light of this, prospects for the regulation of fossil collection and trading fall within the realm of aspects of Civil and Criminal Law, combined with the general understanding (legal precedent) that fossils may be regarded as 'minerals' for the purposes of the law. While this can be explained by statutory authorities to be a **potentially** workable system, in reality it puts an enormous (and unrealistic) burden upon the site owner/occupier. A policy encouraging education of collectors and site owner/occupiers on matters relating to the law, and responsibilities for natural heritage items such as fossils, must be pursued with vigour. In addition to the absence of any specific laws concerning the status of fossils collected within the UK, it has also become evident (Rolfe 1990) that there are no UK laws relating to the sale and export of fossils, leaving them singularly vulnerable to loss.

The burden of responsibility over the matter of the legal status of fossils must be for the statutory organizations (JNCC: Joint Nature Conservation Committee; EN: English Nature; CCW: The Countryside Council for Wales; and SNH: Scottish Natural Heritage) and other geological conservation pressure groups (the Palaeontological Association, the Geological Society of London, Geologists' Association, and the British Institute for Geological Conservation) to lobby parliament in order to establish legal provisions for fossils in law.

United States of America

The issue of fossil collecting in America is one that now works on a political agenda following developments which have occurred in recent years. In 1987 there was the publication of a major document reviewing the practise of fossil collecting in the USA (Raup *et al.* 1987). In view of the wide differences in policy between States of the Union, it made a set of strong recommendations for discussion within federal government agencies, state agencies and other interested parties.

The recommendations set out the need for uniformity in national policy with regard to fossil collecting. It was recommended that fossil collecting for scientific purposes be encouraged on Federal land, but that commercial fossil collecting should be done through a permit system regulated by palaeontologists. A similar approach was to be pursued on private lands, where this was possible. And a public education programme for collectors (amateur and commercial) and land owners should be undertaken, so that the needs of palaeontologists might be better understood.

This set of strongly worded recommendations produced considerable debate and dissent (both within and outside the palaeontological community) and did not set in train legislative moves within the government.

In view of the controversy produced by the earlier recommendations a new Bill (Baucas

Bill) is being proposed to the Senate (sponsored by Senator Max Baucas) entitled the *Vertebrate Paleontology Resources Protection Act*. The key provisions of this Act are as follows:

(1) The recognition that vertebrate fossils on federal lands are a part of the heritage of the United States, and are a non-renewable resource.
(2) Recognition of the contributions of amateur collectors, and providing a means by which amateurs may lawfully collect vertebrate fossils on federal lands.
(3) Prohibition of the sale of fossils collected on Federal lands.
(4) Establishes that fossils collected on Federal lands are the property of the Nation, and should be housed in recognized public institutions, but in some circumstances these may also be retained in the collections made by amateurs.
(5) Establishes a set of financial deterrents for theft, vandalism or sale of fossils from Federal lands.

Precedent for a set of regulations such as this exists in the form of an archaeological protection bill which was passed in 1979.

The political sensitivity of this issue has been very effectively reinforced by the discovery of a complete skeleton of the dinosaur *Tyrannosaurus rex* by a commercial (fossil-selling) organization (The Black Hills Institute [BHI]) on Federal lands in South Dakota. The specimen has been seized by the FBI, following a dispute over ownership between the BHI[1] and the Cheyenne Indian Nation. The interest in the issue is not only because it concerns a dinosaur, and one that is extremely rare and well known to the public at large, but perhaps more importantly, reflects the fact that this specimen may have a commercial value somewhere in the range of 5–20 million $US.

Common themes

What emerges from this brief survey of national approaches to the collection of, and trade in, fossils, are a number of common themes, which are masked by the range and variety of stances that have been adopted. The common themes can be summarized as follows.

(1) Fossils are, without doubt, items of National Heritage. (If they are discovered, what is their status? Who owns them? Do they have the right to sell them on the open market?)
(2) Fossils (particularly vertebrate fossils) are a finite resource. (There are obviously circumstances where fossiliferous beds are laterally extensive and the 'resource' can, to all intents and purposes, be considered infinite; however this does not mean that all fossil sites are of this nature (Norman 1992*b*) and it is the vulnerable fossils that must be protected as a matter of priority.)
(3) The commercial sale of fossils has the **potential** of removing important fossils from the public arena. (This can be either by their becoming sequestered in private collections, or by being sold abroad.)
(4) Commercial sales tend to value fossils in ways that are in conflict with their value either as heritage items, or as scientific objects. (In most cases aesthetics are more important than scientific details. In some notable cases, however, scientific rarity can be used to inflate fossil prices (cf. 'Lizzie' – a singularly unappealing flattened fossil).)
(5) Commercial collecting is not necessarily carried out under scientifically rigorous conditions. (Being governed by financial parameters, the excavation techniques used by commercial fossil collectors – particularly where large fossil vertebrate skeletons are concerned – can lead to excessive haste in removal. Under such circumstances the scientific importance of fossil material in its geological context may be significantly undermined.)
(6) Regulation of fossil collecting, and its enforcement by government regulation, appears to be the ultimate sanction. (In the face of unfettered removal of important fossil material.)

Is there scope for an international convention?

Inconsistencies in approach within nation states (Germany, USA are good examples) and between nation states (see examples above), will consistently permit recognizable national heritage items such as fossils to be bought and sold, and potentially lost from the public domain entirely.

In some circumstances – such as particularly abundant fossils – this may be permissible. However, not all fossils are abundant, some are exceedingly rare and important scientifically, and in some instances even relate to national identify.

In the recent past, the majority stance from an

---

[1] I have been informed that the Black Hills Institute and various associates have been indicted on 39 counts by the South Dakota grand jury.

international perspective would seem to have been towards being in favour of the view that fossils are sufficiently abundant and not important enough to warrant legislation. What is now emerging, however, is an increasing appreciation that fossils are sometimes seen as valuable (in a variety of senses) and important items representative of the national natural heritage. Regulations relating to fossils should be in place to prevent contentious issues (relating to who owns fossils, whether there is a right to sell fossils, and whether they should be subject to export and import control) from becoming intensely political; however, it is most important that if regulations are introduced, that they should be seen to be consistent both intra- and internationally.

With this latter point in mind, the following set of proposals are presented for discussion and consideration:

(1) Each nation should establish a committee or council to consider and review the status of fossils as Scientific and National Natural Heritage items. This committee should publish its conclusions or recommendations with a view to influencing the policy of its national government.
(2) Each nation should establish a set of criteria by which fossils can be evaluated objectively as heritage items.
(3) Each nation should establish a policy for dealing with scientific, amateur and commercial fossil collecting on private and state-owned land.
(4) There should be an international forum for the consideration of transnational issues concerning the ownership of, and trade in, fossils.
(5) A set of broadly accepted guidelines should be established on the international trade in fossils which are acceptable to **all** nations.

Conclusions

The status of fossils, rights of ownership, and the marketing of some fossils are issues that are becoming increasingly divisive; they have also taken on a political profile in some local disputes. It is considered imperative that the palaeontological community (in its widest sense) be involved in, and contribute towards, a debate establishing common policy on matters concerning the status of fossils. The suggested framework outlined above allows for the potential to develop consistent national and international policies on this matter in a proactive manner, rather than having to develop policies *ad hoc* as a reaction to individual and local political initiatives.

References

CHRISTENSEN, E. F. & HALD, N. 1990. *Danekrae, et nyt begreb i dansk museumslovgivning*. Geological Museum of the University of Copenhagen. Contributions in Palaeontology No. **408**.

EDWARDS, W. N. 1967. *The early history of Palaeontology* (Handbook). British Museum (Natural History).

EIGHTH CIRCUIT COURT OF APPEALS. 1992. United States District Court, District of South Dakota, Western Division. Supplemental findings of fact and conclusions of Law. No. 92-2252SD slip op. (8th Cir. 1992). Filed 16th July 1992.

FERGUSON, L. 1988. The 'fossil cliffs' at Joggins, Nova Scotia: a canadian case-study. *In*: CROWTHER, P. R. & WIMBLEDON, W. A. (eds) *The use and conservation of palaeontological sites*. Special Papers in Palaeontology, **40**, 191–200.

GRIGORESCU, D. & NORMAN, D. B. 1990. Earth science conservation in Romania. *Earth Science Conservation*, **29**, 6–8.

GRUBE, A. & ALBRECHTS, C. 1992. Earth science conservation in Germany. *Earth Science Conservation*, **31**, 16–19.

NORMAN, D. B. 1992a. 'Collection and ownership of fossils.' Unpublished guide, issued as a ring bound booklet by Earth Sciences Branch, English Nature.

—— 1992b. Fossil collecting and site conservation in Britain: are they reconcilable? *Palaeontology*, **35**, 247–256.

——, DOYLE, P., PROSSER, C. D., DAVEY, N. D. W. & CAMPBELL, S. 1990a. Comments on C. W. Wright's 'Ideas in palaeontology: prejudice and judgement'. *Proceedings of the Geologists' Association*, **101**, 91–93.

—— & WIMBLEDON, W. A. 1988. Palaeontology in the NCC. *Geology Today*, **4**, 194–196.

——, ——, DOYLE, P., PROSSER, C. D. & PAGE, K. N. 1990b. [NCC policy on fossil collecting]. *Geologists' Association Circular*, **883**, 19–21.

RAUP, D. M. *et al.* 1987. *Paleontological collecting*. National Academy Press, Washington DC.

ROLFE, W. D. I. 1990. Export controls for valuable fossils – the trials and tribulations of "Lizzie". *Earth Science Conservation*, **27**, 20–21.

——, MILNER, A. C. & HAY, F. G. 1988. The price of fossils. *in* CROWTHER, P. R. & WIMBLEDON, W. A. (eds) *The use and conservation of palaeontological sites*. Special Papers in Palaeontology, **40**, 139–177.

SMITHSON, T. R. & ROLFE, W. D. I. 1991. *Westlothiana* gen. nov: naming the earliest known reptile. *Scottish Journal of Geology*, **26**, 137–138.

TAYLOR, M. A. & HARTE, J. D. C. 1988. Palaeonto-

The protection of Australia's fossil heritage through the Protection of Movable Cultural Heritage Act 1986

P. CREASER

Australian Cultural Development Office, Department of Communications and the Arts, Canberra, Australia

Abstract: The Protection of Movable Cultural Heritage Act 1986 protects the movable cultural heritage of Australia from indiscriminate export. The National Cultural Heritage Control List is a schedule under the Act which defines the categories of objects which are important to Australia. Palaeontological objects with a minimum value of $1000 (AUS) were included on the illegal export of objects. The status of the current Australian Federal Police investigation into alleged illegal export of fossils to South Australia and Western Australia is discussed as well as the implications of the recent Ministerial review of the Act.
One of the recommendations in the review report proposed the removal of the $1000 (AUS) minimum value on palaeontological objects. This recommendation was accepted by the government. With an increasing awareness of the Act, it is hoped there will be an increased awareness of the value of our cultural heritage, particularly the fossil heritage.

The Protection of Movable Cultural Heritage Act 1986 was enacted in 1986 and brought into operation in 1988. The principal goal of the programme is – as stated in November 1985 in the Federal Parliament by the then Minister for the Arts, Heritage and the Environment, the Hon Barry Cohen MP – 'to protect Australia's heritage of cultural objects and to extend certain forms of protection to the cultural heritage of other nations' through controls on the export and import of significant movable cultural heritage objects. A closely related goal was to enable Australia to accede to the 1970 UNESCO Convention on the Means of Prohibiting and Preventing the Illicit Import, Export and Transfer of Ownership of Cultural Property, which it did in January 1990.

Probably the most important aspect of the Act is the National Heritage Control List, which defines the categories of movable objects that are classed as 'Australian projected objects'. The Control List includes Class A objects which cannot be granted a permit for export and Class B objects that may be granted a permit for export. Class A objects include some of the most significant items of Aboriginal heritage. Class B objects include a range of other collectable items such as:

- archaeological objects
- objects of fine art; and
- natural science objects of Australian origin. This includes palaeontological specimens, minerals and meteorites.

The Control List also sets out the particular criteria defining 'Australian protected objects'

in each category controlled under the Act. Generally, the criteria include historical association, cultural significance to Australia, representation in an Australian public collection, age and current Australian market value. For example, until very recently a palaeontological object had to have an Australian market value (determined by dealers or experts in museums) of more than $1000 (AUS) before it came under the Act. However, in August 1993, the Government removed this $1000 (AUS) minimum value on palaeontological objects by amending Regulations in the Act. All fossils now require a permit if they are to be exported. Similarly all meteorites and australites are covered under the Act, but a mineral must still have a market value of $10 000 (AUS) before it comes under the Act.

The Act also provides for the establishment of the National Cultural Heritage Committee whose main function is to advise the Minister on the operation of the Act.

An application for a permit is initially assessed by Expert Examiners. These examiners are usually from collecting institutions who have a good knowledge of the item in question and similar items elsewhere in the country. The reports from the Expert Examiners are then forwarded to the Committee for consideration. The Committee, in turn, makes its recommendation to the Minister who duly decides on whether a permit should be granted. One important point to note is that the Committee does not have to abide by the reports from the examiners and the Minister does not have to accept the Committee's recommendation – i.e. it is the Minister who makes the final decision. The time taken for this process can range from a few days to a few months, particularly if the item is of major significance.

To date, over 200 applications for export permits have been received by the Department and only two items, a painting and a historic light rail engine, have been refused an export permit.

One other important aspect of the Act of interest to palaeontologists is the topic of illegal exports. The export of fossils is prohibited under the Act if a permit has not been obtained. The maximum penalty for this offence is a $100 000 (AUS) fine or imprisonment for five years, or both.

Following a request from the then Department of the Arts and Administrative Services in mid-1991, the Australian Federal Police began inquiries into the alleged theft of Ediacaran fossils from South Australia. These inquiries resulted in the identification of a number of suspects. The investigation also identified further sites in South Australia and Western Australia where fossils had been removed and in late October 1991, the Australian Customs Service in Perth intercepted a suspect allegedly attempting to smuggle fossils out of the country. A number of fossils were seized which police believe were intended for sale on overseas markets.

The police are continuing their inquiries and some fossils which have been illegally exported have been returned to Australia. It is not possible to provide all the details of this case as the investigation is still to be finalized.

This case has received considerable media coverage in Australia and has raised awareness of the importance of our fossils.

Two other recent examples which have also attracted attention involve the sale of an opalized pliosaur and the finding of an egg from the Madagascan elephant bird in Western Australia.

In the case of the opalized pliosaur, the owner was a property developer who went into liquidation. Tenders were called for so that the fossil specimen could be sold and it was feared that the successful tenderer for the specimen, which was found in South Australia in 1988, would be an overseas museum. It was estimated that the specimen was worth several hundred thousand Australian dollars. In an attempt to retain the object, the Australian Museum and the national television science program 'Quantum' launched an appeal to save 'Eric', as the opalized pliosaur had been dubbed. The appeal was successful with 25 000 donations from across Australia raising over $300 000 (AUS) for the tender bid and the bid was accepted. Similarly, the Madagascan elephant bird egg has also been retained in Australia.

A review of the Act was conducted recently to evaluate the efficiency and effectiveness of the cultural heritage export and import control programme, established by the Act and Regulations. A total of 60 recommendations were made in the report, some of which require changes to the Act. Of particular interest to palaeontologists was the recommendation to remove the $1000 (AUS) minimum market value on palaeontological objects, which has been accepted by the government.

However, it is important to note that the aim of the Act is not to restrict scientific study of fossils. There are provisions in the Act which allow recognized institutions to send fossils overseas for temporary exhibition or study without the need for a permit.

With an increasing awareness in Australia of our fossil heritage, the development of an International Convention on Geological Conservation is most timely. In developing such a convention, consideration should be given to the heritage value of fossils and the provisions of the relevant 1970 UNESCO Convention on the Means of Prohibiting and Preventing the Illicit Import, Export and Transfer of Ownership of Cultural Property.

The World Heritage List and its relevance to geology

J. W. COWIE[1] & W. A. P. WIMBLEDON[2]

[1] *Department of Geology, University of Bristol, Queens Road, Bristol BS8 1RJ, UK*
[2] *Countryside Council for Wales, Bangor, Gwynedd LL57 2LQ, UK*

Abstract: The compilation of a Global Indicative List of Geological Sites (GILGES) in recent years forms a basis for discussion within UNESCO and the ICSU family of scientific unions on geological sites to be added to the 20 or so sites alread formally agreed as on the World Heritage List. Some 250 sites are already in GILGES, with about 150 being prioritized. Sites include all categories, covering diverse aspects of geology, from fossil localities to astroblemes and fjord landscapes, from major tectonic structures to classic and historical igneous intrusions.

Present proposals are that there should be a single criterion for judging geological sites (as distinct from biological, ecological or aesthetic criteria). Sites nominated should be '...outstanding examples representing major stages of the Earth's history, significant on-going geological processes in the development of landforms, such as volcanic eruption, erosion, sedimentation etc., or significant geomorphic or physiographic features, for example volcanoes, fault scarps and inselbergs...'.

One of the considerations in putting together the GILGES has been the matter of scale. Sites suggested and considered range from whole landscapes, such as glacial or volcanic terrains, down to metre-sized outcrops with microscopic fossils. Many sites suggested as WH contenders for their intrinsic scientific merits are closer to the latter than the former in scale, but most sites presently listed as World Heritage Sites are considerably larger. The rationalization of this contrast between the national park-sized wilderness site (often with considerable geology) and the often very small, but frequently unique, site remains one of the important areas for debate. It is still not entirely clear how these gems of geological history can be accommodated in the WH system.

The World Heritage Convention came into force in 1972 and is currently signed by about 120 countries. About 100 Natural World Heritage Sites (including geological sites) have been recognized in 20 years (Cultural Sites make this total up to over 300). It seems possible that Natural Sites may increase to 200 by the year 2000. It is essential to be rigorous in the evaluation of such sites. Previously, the International Union for Conservation of Nature (IUCN) has encountered some difficulty in the evaluation of a number of the geological nominations which have been referred to it, for example Jixian, China and Lesbos, Greece.

The Global Indicative List of Geological Sites (GILGES)

The compilation of GILGES in recent years forms the basis on which the United Nations Educational, Scientific and Cultural Organization (UNESCO) and the International Committee of Scientific Unions (ICSU) can add further geological sites to the 30 already formally agreed as being on the World Heritage Site List. GILGES has been compiled from the suggestions of many organizations and individuals, put forward since the first invitation to participate was widely circulated in 1990. Under the guidance and with the involvement of representatives of UNESCO (governmental), the International Geological Correlation Programme (IGCP), the International Union of Geological Sciences (IUGS) and IUCN (the latter three being the non-governmental side of the partnership), a task force was convened in Paris to work on the refinement of a list of suggestions and the UNESCO criteria for selection that was then in use. With numerous further iterations, the present GILGES list of c. 300 sites resulted from those deliberations (Wimbledon 1993).

IUGS GEOSITES

A key development resulting from the work of the task force has been the establishment by IUGS of a new computerized global database of geological sites (IUGS GEOSITES) at Trondheim, Norway. This amplifies and supports GILGES and will gradually expand to include thousands of sites of first-class importance to global geology (including palaeobiology). It will take its place alongside the World Heritage (WH) Convention as a means of furthering

From O'Halloran, D., Green, C., Harley, M., Stanley, M. & Knill, J. (eds), 1994,
Geological and Landscape Conservation. Geological Society, London, pp. 71–73.

geological conservation and recording. Recording is, of course, important in case conservation measures are needed in the future.

Nomination of sites

Membership of UNESCO itself is not necessary to maintain the World Heritage Convention: the USA and UK are signatory countries, as are, for example, Albania, France, Germany, the former Soviet Union, Brazil, the Holy See (Vatican), Burundi and Switzerland. From the commencement of the GILGES/WH project, it has been (and remains) open to all to suggest contender localities. However, not everyone has taken the opportunity to nominate possible sites, resulting in an imbalance which is reflected in the current list, in both the topics covered and the geographical distribution of sites. It is, perhaps, too much a reflection of those countries where conservation is already well developed.

GILGES, at present, contains 300 sites. Sites include all categories, covering diverse aspects of geology, from fossil localities to astroblemes and fjord landscapes, from major tectonic structures to classic and historical igneous intrusions and global stratotypes. Currently, two sites are candidates for addition to the World Heitage List, one in Tasmania and the other in Hungary. GILGES, as it stands, shows a preponderance of European sites, but this will probably be modified.

In geology, in comparison to most other subjects, we are dealing with a backdrop of hundreds and thousands of millions of years. A recent decision is that there should be a unique criterion for judging geological sites (as distinct from biological, ecological or aesthetic criteria). Sites nominated should be

'...outstanding examples representing major stages of the Earth's history, significant ongoing geological processes in the development of landforms, such as volcanic eruption, erosion, sedimentation etc., or significant geomorphic or physiographic features, for example volcanoes, fault scarps and inselbergs...'

This provides maximum freedom for a good conspectus globally.

Problems of scale and comparison

One of the considerations in putting together the GILGES has been a matter of scale. Sites suggested and considered range from whole landscapes, such as glacial and volcanic terrains, down to metre-sized, isolated outcrops. Many sites proposed as World Heritage Site contenders for their intrinsic scientific merits are closer to the latter than the former in scale. However, most sites currently listed as World Heritage Sites are considerably larger, for example the Grand Canyon, Dinosaur Provincial Park or the Yosemite National Park in North America. The rationalization of this contrast between the national park-sized wilderness sites and the often very small, but frequently unique, sites with key evidence of, for example the evolutionary story, remains one of the important areas for further debate.

There are inherent difficulties in comparing and judging the merits of such diverse 'sites'. Wilderness areas or national parks are one form of conserved area but, from a scientific viewpoint, small sites may be of the very highest significance. It is very difficult to weigh the merits of a 'site' such as Yosemite, USA, against the best of the internationally significant fossil localities such as Durlston Bay, UK, or Holzmaden, Germany, or a locality with a single landform. The phrase 'of outstanding value' may clearly relate to all of these. It has to be admitted, however, that geological reserves and smaller sites which may be of national, rather than international, representative importance will not find a place in the World Heritage Site List or GILGES, but they may still, if of sufficient significance, be recorded in the IUGS GEOSITES computer database.

Towards the future

GILGES will continue to expand. The scientific and conservation communities must continue to co-operate and to assist in developing a better coverage of sites, correcting those imbalances that have already been identified. Thereafter, informed comparisons of sites and their relative weighting will be continuing stages in the process. It is clear that the GILGES and GEOSITES list will contain many sites – stratotypes, landscapes, localities demonstrating organic evolution – and that these lists in a refined state will be powerful tools for decision-makers in the future in their consideration of conservation priorities.

Ongoing work includes further additions to the World Heritage Site List, to GILGES and to IUGS GEOSITES. The future seems assured, with the support of over 120 governments.

Conservation, meaning physical protection, is by law a national, sovereign, responsibility; effective conservation is carried out locally. World Heritage is a reasonably well-financed, fully international convention for global geological recording and conservation, with a highly qualified, active and well-travelled professional staff working in countries within Europe, the Americas, Africa, Asia and the whole world: it is a convention which must be used.

Reference

WIMBLEDON, W. A. P. 1993. World Heritage Sites and geological conservation. Geotechnica Abstracts, Cologne.

Australia's World Heritage fossil nomination

P. CREASER

Australian Cultural Development Office, Department of Communications and the Arts, Canberra, Australia

Abstract: In the Statement on the Environment of 21 December 1992, the Prime Minister announced that Riversleigh in Queensland and Naracoorte Caves in South Australia would be nominated to the World Heritage List in 1993. The Federal Government has already provided funds for work at these sites and World Heritage nomination will ensure the protection and status they deserve. It is intended that these two sites will be part of a single 'serial' nomination, the theme of which will be the evolution of Australia's terrestrial biota featuring the evolution of marsupials. This type of nomination will allow for the addition of other related sites of international significance at a later date.

The nomination will be developed in consultation with the relevant State Governments and researchers in the field. The process for World Heritage Listing is discussed as well as the implications for sites which are inscribed on the World Heritage List.

The purpose of this paper is to give a broad overview of the World Heritage Convention, to explain Australia's participation and to give details of the first Australian Fossil Sites nomination – Murgon, Riversleigh and Naracoorte. This nomination is very relevant in the context of the development of an International Convention for Geological Conservation.

The Convention Concerning the Protection of the World Cultural and Natural Heritage, known as the World Heritage Convention, was adopted by the UNESCO General Conference in 1972. In 1974, Australia became one of the first signatories. Since then, the number of countries that are party to the Convention has risen to 139.

The World Heritage Convention aims to promote co-operation among nations to protect worldwide heritage which is of such universal value that its conservation is of concern to all people. It is intended that, unlike the seven wonders of the ancient world, properties will be conserved for all time. Member countries commit themselves to ensuring the identification, protection, conservation, presentation and transmission to future generations of their World Heritage properties. These ideas form the very backbone of the Convention.

The Convention establishes the World Heritage List, which identifies natural and cultural properties considered to be of outstanding universal value and, by virtue of this quality, especially worth safeguarding for future generations. In order to qualify for inclusion in the World Heritage List, a nominated property must meet specific criteria and integrity conditions from either a natural or cultural point of view.

The world's natural heritage offers a priceless legacy. It includes sites representing major stages in the geological and biological history of the Earth; outstanding examples of significant ongoing processes; areas of exceptional natural beauty; and significant natural habitats for the conservation of biodiversity.

Cultural heritage sites are equally priceless. Certain archaeological sites, groups of historic buildings, ancient towns, monumental sculptures and paintings whose significance transcends political or geographical boundaries constitute part of this irreplaceable heritage. Some of these treasures represent unique artisitic achievements and masterpieces of the creative genius; others have exerted great influence over a long period of time or within a cultural area of the world. Some may be the unique witness to a civilization which has disappeared or to a type of settlement that has become vulnerable under the impact of change. Others may be associated with ideas or beliefs that have left an indelible mark on the history of humanity.

A World Heritage property can be listed for both natural and cultural values. Even so, the division between natural and cultural properties is somewhat artificial, and in the last year has been addressed to some extent with the addition of cultural landscapes as a group spanning the middle ground.

To date, Australia has ten World Heritage properties, which all contribute significantly to the wealth of the tapestry. Australian World Heritage properties read like a travelogue of our most spectacular and unique places and include:

- the Great Barrier Reef, the world's largest and most complex living coral reef system was inscribed on the World Heritage List in 1981.
- the Willandra Lakes Region in western New South Wales, where oustanding evidence of the life and culture of early Aborigines has been found, was also inscribed in 1981.
- The Lord Howe Island Group was inscribed in 1982, an exceptional example of a largely intact island ecosystem.
- Uluru National Park, inscribed in 1987 contains not only Ayres Rock and the Olgas, but also a large area of desert ecosystem.
- The Wet Tropics of Queensland contains various types of tropical rainforest, including a vast array of primitive and relict plants and animals. It was inscribed in 1988.
- The East Coast Temperate and Subtropical Rainforests of Australia were inscribed originally in 1986, and re-nominated to include National Parks in Queensland in 1992. The protected areas contain examples of our temperate and subtropical rainforests.
- The Tasmanian Wilderness has also been nominated twice, taking into account an additional area that was added on to the national park. The whole area covers what has been described as 'one of the last great temperate wilderness areas of the world'.
- Shark Bay in Western Australia, inscribed in 1991 includes both marine and terrestrial areas and is the home of stromatolites, the oldest living form of life on Earth.
- Kakadu National Park has been nominated three times, to include each extension of the national park. It contains a wide range of ecosystems and also many important archaeological sites.
- The latest addition has been Fraser Island, listed in 1992 for its exceptional beauty and the geological history of the sand dunes.

All of Australia's properties have been listed for their natural values and are generally very large. Australians have the honour of having four properties listed on all four natural criteria, and three listed for both natural and cultural criteria. This puts the properties into a very elite group.

Australia has been very active in the World Heritage field, not only with the listing of its properties, but also with their protection. In 1983, the Commonwealth Government enacted the World Heritage Properties Conservation Act, which provides for the protection and conservation of World Heritage properties in Australia and its external territories.

There are now 378 sites on the World Heritage List, which includes several fossil sites:

- the Burgess Shale site in Canada, now part of the Canadian Rocky Mountains World Heritage Property, famous for its record of early soft-bodied animals from the Cambrian period;
- the Dinosaur Provincial Park, also in Canada, famous for its record of the 'Age of the Reptiles', which includes dinosaurs like *Tyrannosaurus, Triceratops* and others from the Cretaceous;
- and finally Olduvai Gorge within the Ngorongoro Conservation Area in Tanzania where the fossilized bones of many early humans from the Pleistocene were found. These included *Homo habilis*, the earliest member of our own genus.

Australia has a fossil record of great antiquity, extending from 3.5 billion years ago to the present.

On 21 December 1992, the Prime Minister, Mr Keating, announced in his Environment Statement that the fossil sites at Riversleigh and Naracoorte would be nominated to the World Heritage List in 1993. The Murgon fossil site will also be included in the nomination. Agreement has been reached with the South Australian and Queensland Governments to proceed with the nomination.

Riversleigh and Naracoorte were included on the list of Australian sites prepared for the meeting of the UNESCO Working Group Task Force on a Global Inventory of Geological and Fossil Sites held in Paris in February 1991. The two sites were allocated 'category 1' status by the meeting.

The three sites represent three key stages in the evolution of our unique Australian fauna over the last 55 million years. As sites, it would be difficult to find three places that are more different. Murgon is a lush green pasture, Riversleigh is semi-arid and tens of square kilometres in size, whilst Naracoorte is an intricate cave system.

The marsupial fossils from Murgon include a diverse suite of primitive forms, some of which appear to resemble extinct South American marsupials more closely than any previously known from Australia. Besides marsupials, many other discoveries have been made, some of the earliest known bat fossils and the tiny tooth of a very small placental mammal, which is challenging understanding about the evolutionary and biogeographic history of this region of the world. The site is about 55 million years old, and is the only link between the opalized monotreme jaw from Lightning Ridge in New South Wales and younger mammal deposits.

The Riversleigh deposits date from 25 million years ago to the present, and occur in unique freshwater limestone. With more than 20 000 specimens representing 150 faunal assemblages, Riversleigh has led to an understanding of how the environment and the animals that lived in it have changed over time from a rich rainforest community to a semi-arid grassland.

The fossil bed and Ossuaries of Victoria Fossil Cave at Naracoorte contain the remains of at least 93 vertebrate species, ranging from tiny frogs to buffalo-sized marsupials. This makes it one of the richest Pleistocene marsupial fossil deposits in the world. The sediments accumulated between 170 000 and 18 000 years ago and, therefore, over two of the Pleistocene glacial periods.

As the sites occur in two different States, the Commonwealth Government is responsible for preparing the nomination documents in close consultation with the States involved. Professor Michael Archer of the University of New South Wales has been contracted to prepare the technical chapters of the nomination, and he will be consulting with South Australian and Queensland colleagues in this work.

Nominations must be submitted to the UNESCO secretariat in Paris by 1 October each year. They are then referred to the World Heritage Bureau and thoroughly assessed. The Bureau is assisted in this task by IUCN – the World Conservation Union – which advises on natural sites, and by ICOMOS which advises on cultural sites. In addition, these organizations consult with relevant experts around the world.

The evaluations are considered by the Bureau at their meeting in the year following the submission of the nomination, and a recommendation on the listing of the property is then made to the World Heritage Committee which consists of 21 nations representing the different regions and cultures of the world. The Committee considers the recommendations from the Bureau and the evaluation from IUCN or ICOMOS and decides whether the property will be inscribed.

For the Australian Fossil Sites nomination being prepared, this timetable will mean that the nomination should have been submitted by October 1993, assessed in early 1994, considered by the Bureau in June 1994 and a decision made by the World Heritage Committee in December 1994. As can be seen, World Heritage Listing is not an instant process and usually takes about two years to complete.

This nomination seeks to tell a story of change over time, rather than just providing a snapshot of the sites. There is potential if this nomination succeeds, to nominate further sites that fill out the story of mammal evolution in Australia, or tell other unique stories of universal value. The 1993 nomination of Australian Fossil Sites will increase worldwide awareness of the unique Australian heritage. It is also hoped that this awareness will lead to an increasing understanding of the World Heritage Convention and its relevance to a Convention on Geological Conservation.

Environmental aspects of mineral resource conservation in southwest England

COLIN M. BRISTOW
Camborne School of Mines, Pool, Redruth, Cornwall TR15 3SE, UK

Abstract: In an environmentally conscious world it is sometimes overlooked that the most efficient use of the available mineral resources in the ground is an important aspect of Earth science conservation which all Earth scientists ignore at their peril.

Case histories taken from the fields of industrial minerals and constructional raw materials in southwest England will show that unthinking environmental pressures can often lead to waste and sterilization of the available mineral resources, thereby depriving future generations of the possibility of making use of these materials and often leading to increased energy usage through materials having to be brought from a greater distance to serve the markets.

The cause of Earth resource conservation is also served by ensuring that as much as possible of the material taken from the ground, besides the primary product, is utilized in order to minimize the overall impact of extractive operations; some case histories in this field will also be presented.

The value of extractive operations for use in Earth science teaching will also be discussed, as well as conflicts arising in the after use of former extractive operations between those who wish them to be entirely restored and Earth scientists who wish to see aspects of scientific interest conserved for research and teaching purposes.

'One of the guiding principles of environment management is the conservation of resources'

HRH Prince Charles, Duke of Cornwall, North Devon, 28 May 1992.

In any discussion about landscape conservation, the impact of mineral working must come up as an important factor. Indeed, the remains of ancient mineral workings may well form an integral part of some cherished landscapes in SW England, such as the north coast of the Land's End granite (tin and copper workings), Dartmoor (peat and alluvial tin working) and the Mendips (shallow lead and zinc workings). Furthermore, advanced landscape restoration techniques may enable areas of considerable industrial dereliction to be converted to landscape amenity areas, such as in the case of the re-working of coal seams by opencast methods which provide the finance for the whole clean-up operation.

Also, there is a growing awareness of the industrial archaeological significance of many areas of former mineral working, with the recent publication of a work on the china clay industry (Herring & Smith 1991) and the recent bid by the Trevithick Trust to achieve World Heritage status for the historic mining areas of Camborne–Redruth and St Just-in-Penwith being good current examples. One of the main aims of the Cornwall Regionally Important Geological/geomorphological Sites (RIGSs) Group has been to preserve old mine dumps containing rare minerals, and to preserve underground access to locations where rare minerals or types of mineralization are displayed.

However, when looked at from the point of view of the geologist working in the contemporary extractive industry; there is one aspect of conservation which most conservationists tend to ignore – the most efficient use of the available mineral resources in the ground. Conservation of mineral resources is analogous to recycling, as both have the effect of reducing the need to open up new resources of minerals to supply societies' demand for raw materials, thus conserving these resources for future generations.

Prince Charles' statement quoted above was made in North Devon last year and, taken at face value, it places a duty on those involved in the extractive industry to ensure that the most efficient use of the natural resources, which they are involved in exploiting, is made. In most cases this also makes good business sense, as the best possible use of resources is usually reflected in a good cash flow. Also, by making the best possible use of the mineral deposits currently being exploited, the area of land required for working is reduced, which postpones the need to disturb the landscape by opening new mineral workings. However, in some cases excessive environmental zeal can lead to mineral deposits being worked in a wasteful way, which in the overall interests of conservation may not be a good thing at all. As with all matters of this nature, the aim must be to strike a sensible balance.

This paper will examine some aspects of mineral conservation, with particular reference to southwest England, especially Cornwall, which has more 'hard rock' Sites of Special Scientific Interest (SSSIs) than any other English county and quite possibly has several areas which could qualify for inclusion on the GILGES/UNESCO list (cf. Cowie & Wimbledon, this volume).

The geology of southwest England

Southwest England is mainly composed of marine sediments and associated igneous rocks of Devonian and Carboniferous age, which were strongly folded and faulted in the Variscan orogeny and then intruded by a series of late Carboniferous/early Permian granites. Subsequent hydrothermal mineralization led to the formation of extensive tin and copper deposits, which resulted in this area becoming the most important producer of these metals in the world in the nineteenth century. Alteration of the granites led to the formation of extensive china clay deposits, notably in the St Austell granite. To the east, one encounters the Cretaceous chalk escarpment, which forms the western rim of the Hampshire Basin. In the Palaeogene, fault-formed basins in North and South Devon were infilled with sequences of sands, kaolinitic clays and lignites, and similar clays were also formed in the freshwater sequences in the Palaeogene succession at the western end of the Hampshire Basin. These clay deposits are today the source of much of the high grade ball clay used by Europe's ceramic industries.

Industrial minerals in SW England

China clay, or kaolin as it is known throughout most of the world, has been mined in Cornwall and Devon for over 200 years and current production is a little over 3 million tonnes, making Britain the second largest producer of kaolin in the world, after the United States. The main uses of kaolin are in paper, ceramics, plastics and paint, and SW England is the principal source of kaolin for European industries.

Ball clay is a strong, plastic white firing sedimentary clay found in fault-defined basins along the Lustleigh–Sticklepath fault zone in Devon and in the western part of the Hampshire Basin around Wareham. The Devon and Dorset ball clays are used mainly in high grade ceramics and are the principal source of high grade material for the ceramic industries of Europe. Ball clays elsewhere in Europe are not generally so suitable for high grade ceramics, and are mainly used in refractories, sewer pipes, etc. The china clay and ball clay industries also provide much needed employment in areas where unemployment is well above the UK average.

Over the last twenty years china clay has been facing increasingly tough competition from finely ground calcium carbonate made from chalk, limestone and marble, and precipitated calcium carbonate (PCC) made at the paper mill from lime and waste carbon dioxide. However, the energy audit shows that some of these calcium carbonate products (notably PCC) require far more energy to produce than china clay and a steep hike in energy prices could leave them severely disadvantaged (Bristow 1992). From a carbon dioxide emission point of view, china clay is environmentally much more acceptable than PCC.

China clay production and the environment

The china clay deposits of SW England were formed by a complex multi-stage process from the granites intruded just after the climax of the Variscan orogeny, with most of the feldspar in the granite being altered to the clay mineral kaolinite (Bristow & Exley 1994). Both hydrothermal and supergene processes appear to have contributed to the formation of the deposits seen today, with the convective circulation of fresh water of meteoric origin through the granite, driven by heat derived from the exceptionally high content of radiogenic elements, being one of the most important factors in the formation of these deposits.

As can be seen in the cross-section in Fig. 1 (top), the deposits tend to be generally funnel shaped, although there is a lot of variation in shape, with sometimes unkaolinized granite overlying kaolinized granite. The stem of the kaolinized funnel can descend to considerable depths, and in one case the kaolinization was still unbottomed in a borehole which reached a depth of 250 m below the original surface.

The St Austell granite contains the largest number of china clay workings, with around 70 km^2 of the western half of the granite occupied by a complex of pits, waste tips and processing facilities, interspersed with village settlements and small areas of farmland.

With such a massive and concentrated

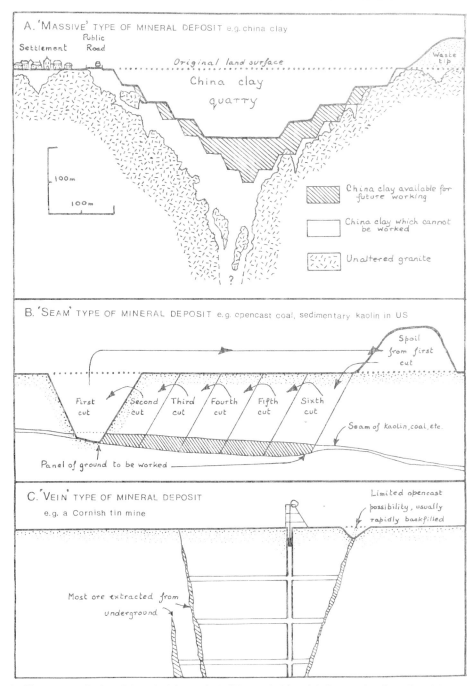

Fig. 1. Three types of mineral deposit: (a) 'massive'; (b) 'seam'; (c) 'vein'. See text for discussion. (Figure reproduced with permission of the publishers from Bristow 1993, p. 19, fig. 1.1.2.1).

environmental impact, it is not surprising that there is a lively debate over environmental issues in relation to the china clay industry. Local people tend to be on the side of the industry, not least because it provides so many people with work, but also because the dramatic and unique industrial landscape has a haunting beauty all of its own which has inspired many local poets, writers and artists. Also, many local people see it as the present-day continuation of a very Cornish tradition of mining activity which goes back to prehistoric times.

However, one of the most intractable problems facing the industry is the disposal of waste, for every tonne of china clay produced is accompanied by approximately 8 tonnes of waste in the form of residue, sand, gravel and rocks. This is the material which forms the huge tips alongside the china clay workings. Repeated studies, both by the industry and by government departments, have shown that Cornwall is too far from a major conurbation such as the London area, to be able to economically ship the material there for use in construction. A frequently heard statement by those, usually from outside the area, who are taken aback by the scale of the working is 'Why can't they backfill the pits with all the waste sand'?

The cross-section in Fig. 1 shows why this is very difficult to achieve in the case of china clay.

In the top diagram the deposit of china clay is shown, which is typically funnel shaped, and we can see that the most important resources for the future tend to lie under the bottom of the pit and that premature backfilling with waste would lead to the sterilization of this resource, thereby depriving future generations of the possibility of making use of it. This would be bad conservation practice, because it would mean that part of the resource would be wasted and a new mine would have to be opened prematurely somewhere else. That would most likely be in the forests of the Amazon Basin, where there are large resources of kaolin and there is currently much expansion of kaolin production. The global environmental disadvantage of such a move is obvious.

Of course, wherever it is possible to backfill a china clay pit, because there are no worthwhile reserves left, this is done. One of the objectives of mine planning is to create areas suitable for backfilling (see below).

By contrast the opencast coal or kaolin mine (Fig. 1, middle) can be worked as a strip mine, with progressive restoration as the seam is worked; only the section where actual extraction of kaolin or coal is taking place need be open at any one time and there are no large piles of waste material. The sedimentary kaolins of the Georgia/South Carolina kaolin producing area, currently the largest kaolin-producing area in the world, are worked in this way.

The underground mine (Fig. 1, bottom), makes even less of an impact on the landscape; although the collapse of old mine workings and shafts is a particular problem in present-day Cornwall, partly because many of the early workings are poorly documented.

In recent years the china clay industry has embarked on a series of developments, which are helping to make the maximum possible use of the material in the ground with the absolute minimum of wasted resource, whilst, at the same time, minimizing the environmental impact of the workings.

(a) The use of sophisticated software packages

Twenty of thirty years ago the development of a mine was usually preceded by a geological study based on a regular pattern of boreholes which defined what the geologist thought would be the extractable reserve. The geological report was then handed to mining engineers, who developed the mine or open pit by using their judgement and by reference to the boreholes from time to time.

In even earlier times the whole of the mining development would have been simply based on the judgement of the mining engineer-in-charge, based very largely on what was seen in the faces.

Whilst this kind of mine planning might be effective for a very small quarry or mine, it had serious drawbacks when applied to a medium-sized or large operation. In their efforts to keep to the published specification for the product, mine managers might have produced a product which, at times, was well above the quoted specification. They might then find that, as the mine developed, more and more of the faces were of substandard material, until a point was reached where a lot of the poor quality material had to be excavated onto the tip, so as to expose fresh faces of better quality material. Still worse, if they had high-graded too much in the early years of the operation, they would be left with a mine which ceased to be workable, even though much of the geologists' so-called reserves still remained in the ground. Even further complications would arise when the Marketing Department asked the mine manager to produce a new and different quality from the same orebody. A lack of proper planning by the china clay industry in the first half of the century often led to waste tips being placed on top of valuable resources of china clay.

Nowadays, in extractive industries such as china clay, the geologist, the mining engineer, the mineral processor and the marketeer all work as a single team, sharing the data, which are stored and processed on a suitable software system. All the topographic survey data, which are regularly updated, are fed directly from the field instrument into the software programme, on which is stored all the borehole information and the laboratory test results. Continuous sampling of the working faces, or 'stopes' is carried out and the laboratory tests from these are also fed directly into the computer programme.

At regular intervals, the design of the pit can be reviewed to check that the pit is being worked so as to make the most efficient use of the mineral resource in the ground. Poor quality areas can be blended away with better quality material, thereby enabling the maximum tonnage to be produced and the minimum amount of space to be taken up by waste tips.

Changes in the market which alter the ratio of the different products required, or their quality, can be easily incorporated and the production schedule from the different stopes and parts of the pit modified to take account of these changes. The positioning of major plant items such as crushers, or haul roads, can be done in such a way as to sterilize the minimum amount of kaolinized granite. The schedule of mining and tipping can be arranged to create areas suitable for residue disposal.

English China Clays use the 'Datamine' system, which has the ability to manage a huge volume of data quickly and efficiently to develop sophisticated orebody models. This makes evaluation quite straightforward and provides an overview which can be constantly updated and reviewed as circumstances dictate. The CAD facility for mine design provides an opportunity to examine many different designs and assess their advantages and disadvantages.

All in all, this leads to a much more efficient use of the mineral resource, with far less risk of finishing up with a lot of unworkable material in the pit.

(b) Better geotechnical design of pit faces

As can be seen from Fig. 1, there is a great deal to be gained, in terms of the overall efficiency of resource extraction, from steepening the angle on the sides of the pit, so that more china clay can be obtained from the deepest part of the deposit.

In many cases there is a village, or a public highway, which prevents the pit being extended to the width required to mine in depth. In other cases the configuration of the deposit may be such that much rock has to be moved in order to get at the clay in depth, so that a steepening of the sides of the pit will mean that less rock has to moved, and hence less rock is also placed on the waste tip.

Currently, rock mechanics research is in progress at Camborne School of Mines to try and discover ways of engineering steeper slopes. Dewatering of the ground behind the pit's faces is one possibility, another technique is rapid excavation, followed by tipping of waste rock against the base of the slope to ensure long-term stability. Some carefully engineered pit slopes in Rocks china clay pit have already demonstrated that stable steeper slopes are possible.

(c) Improved recoveries

After the removal of rocks and sand in the pit bottom, the clay-bearing slurry is pumped to large hydrocyclones, where the coarse residue is removed, usually referred to as micaceous residue, as it is mainly composed of mica. The clay is then dewatered and then put through a multi-stage Dorr-Oliver refining system, which produces a paper filler clay, plus another residue, which contains approximately 45% coarse kaolinite. The filler clay can then be further refined to provide a paper-coating clay, which yields yet another residue, which is composed of 70–85% kaolinite. Until recently, all these residues were pumped to large waste residue lagoons, which took up a lot of space.

In china clay matrix there are usually two types of kaolinite, a 'groundmass' type of kaolinite, occurring as pseudo-hexagonal platelets and small stacks of platelets, typically in the ½–5 μm range; and large curled stacks of kaolinite nearly 100 μm in diameter. The groundmass kaolinite has been the source of most of the china clay produced by the traditional route; the large stacks are found in the centrifuge and Dorr-Oliver residues, and to a lesser extent in the hydrocyclone residues. These large stacks are, however, much too large to use as paper filler clays and in the old days they were either put into residue lagoons or dumped straight into the local rivers, which flowed white as milk as a result, to be ultimately dispersed and lost in the sea.

However, starting about 20 years ago, all these residues have been pumped to disused pits and large dams constructed specially to retain them, and now the rivers run as clear as any other rivers in Cornwall and Devon. At the same time as the rivers were being cleaned up, research was initiated into processing these residues to see if more product could be made from them. This has been remarkably successful and by a special type of froth flotation much of the kaolinite in the centrifuge residues can now be separated from the associated mica, quartz, feldspar, etc.

This residue is then ground in special ultrafine grinders and is made into a new kind of product known as a synthetic filler. The paper industry has found that these synthetic fillers have a number of important advantages over conventional fillers and consequently they are much in demand from the paper industry. The next stage is now being tackled, with the Dorr-Oliver residues being processed by a different method.

The important point to grasp is that all these synthetic fillers are being manufactured from material which used to be thrown away, and which used to take up space in the residue lagoons. This has also had the effect of enabling the same amount of kaolin product to be produced from a much smaller quantity of raw material in the ground, thus conserving reserves for the future.

So successful have these developments been that dredges have now been put into some of the old residue lagoons to enable more material to be processed by these new methods of producing synthetic fillers.

(d) Finding uses for the waste

Whilst some progress can be reported on the re-treatment and reduction in the quantity of residues produced, the same cannot be said for the coarser waste products, and this is proving to be a far more intractable problem.

For every ton of clay product about 6–8 tons of coarse waste in the form of sand, rocks and overburden is produced. So, for a production of 3 million tons of china clay a year, over 20 million tons of coarse waste are produced; a high proportion of which is coarse sand, in part gravel by particle size definition.

The obvious use for all this sandy material is in the construction industry, but only a relatively small proportion can be readily used as a concreting sand, because the sand is generally too coarse, and the fines deficient, for this purpose. Because of its coarseness and the presence of minerals such as mica and unkaolinized felspar, concrete made from clay works sand also tends to require up to 10% more cement than conventional sands, thus creating a further disincentive for its use.

However, as a bulk fill material for the construction of embankments and road foundations, clay works sand can be used, providing the structure is properly designed for it's use. The dam wall of Colliford dam on Bodmin Moor was entirely composed of china clay sand from nearby Park china clay pit, which confirms its suitability in this sort of application. Crushed waste rock (stent) has also been used wherever possible for road foundations.

Local markets use china clay waste materials wherever possible and the local market can be said to be saturated; but this only absorbs a tiny proportion of the annual output. In order to use greater quantities, markets further afield would have to be supplied, which means taking it to large conurbations such as the London and Bristol areas.

However, as all the studies by the industry and by government have shown, the cost of transport to these markets is far more than the sand would be worth when it got there. So, currently, the main efforts are going into landscaping the tips so that they blend in with the landscape in as natural a way as possible.

Initially, the research was directed towards finding ways of rapidly establishing a vegetative cover on the tips. However, this does tend to result in tips which look rather angular and unnatural, so nowadays the tip is profiled using earthmoving equipment and then seeded. This results in surprisingly natural-looking rounded hills which can be used for grazing sheep or for amenity purposes.

However, there are some interesting possibilities which may indicate a valuable future use for china clay sand. In the past, large quantities of sand found its way into the local rivers and where these rivers reached the coast large beaches of the sand were formed, which have proved to be remarkably stable; Carlyon Bay near St Austell is a particularly good example. China clay sand therefore seems to be particularly suitable for beach make-up, because the coarseness and angularity of the particles ensure that it is retained on the beach, and not washed away, as most ordinary beach sands might be.

With rises in sea-level as a result of global warming being predicted by many climatic experts, the subject of coast protection may well become a topic of major importance in the next century. The easiest way of protecting a shore which is fronted by a beach is to make up the level of the beach so that storm energy is dissipated by the waves pounding on the beach, so perhaps china clay sand will become an important resource for coastal protection in the next century or so.

Also, the non-kaolinite component of the residues from china clay working does contain a lot of other interesting materials such as:

(1) The mica itself, when separated from the other constituents, is a valuable industrial mineral which can be used for welding and in oil-well drilling, or ground to provide special fillers for use in paint and plastics. Mica was produced in the past at Lee Moor, on the southwestern side of the Dartmoor granite and is currently produced as a by-product of kaolin working in Britanny.

(2) The St Austell granite is probably the largest resource of lithium in the European Community; much of the lithium is in the mica, the mineral which forms the bulk of the residues. Lithium is very much a space-age metal and vast quantities may be required if nuclear fusion power becomes a reality.
(3) Topaz, which could be used for fluorochemical and mullite production, occurs in abundance in the china clay waste.
(4) Small quantities of heavy minerals also occur, containing tin, wolfram, rare earths, titanium and uranium in the residues.

Whilst it is probably true to say that the recovery of these interesting minerals would not be economic for just one mineral on its own, it is nevertheless possible that an integrated plant which sought to separate and recover many, if not all, of the constituents might become economically worthwhile in the future. Designing such a plant will prove to be a considerable challenge for mineral processing engineers, but if it could be done it would mean that important resources are obtained from waste materials. In the future we will have to look increasingly at the wastes of today as the sources of many of the raw materials we will need. It is, after all, just another aspect of recycling and conservation of mineral resources.

In this context, it is important that the wastes of today are placed in such a way as to facilitate future reworking.

(e) The china clay resource as a scientific and teaching resource

Another aspect which is also often overlooked is the value of the china clay pits for scientific research and as a teaching resource.

The china clay pits in the St Austell granite offer an exceptional opportunity to see, in three dimensions, what the interior of one of the most highly altered and mineralized granites in the whole of Europe is like. As a consequence, a great deal of fundamental scientific research has been carried out which, together with the major geothermal project at Rosemanowes managed by Camborne School of Mines, has led to major advances in our understanding of how High Heat Producting (HHP) granites are emplaced, mineralized and altered.

There is undoubtably much more to be learnt from further research, which will benefit not only geologists in SW England, but also geologists working on this type of granite anywhere in the world.

It is not only research that benefits from the china clay pits; they are also an important teaching resource for undergraduate and postgraduate students. The industry hosts large numbers of parties of students from all over Britain, Europe and North America, who find that the china clay pits are an ideal location in which to instruct students in the processes of granite intrusion and mineralization.

If the china clay industry was to cease, then the pits would rapidly flood and become vegetated, and therefore be of very little value for research or teaching. The best way to conserve the value of the china clay pits for future teaching and research is to ensure that china clay working continues. This is yet another example of conservation of mineral resources.

Ball clay production and the environment

Ball clays occur is stratified sequences and are worked in open-pit quarries and sometimes in underground mines, although the latter are becoming progressively more uneconomic except for some high value clays. Ball clay mining has an even longer history than china clay, being used as far back as the seventeenth century for the manufacture of clay tobacco pipes (Rolt, 1974). However, ball clay working is on a much smaller scale than china clay and some of the smaller workings have almost the atmosphere of a cottage industry about them.

The ball clay workings are very much accepted as part and parcel of the local scene in North and South Devon, but environmental pressures are at their most intense in Dorset. Here, encroachment by housing, industry and road development has senstitized local environmental opinion to the point that any proposed development, whatever its size, is perceived as a major threat. The problem is made worse by the fact that Britain's largest onshore oilfield – Wytch Farm – lies in the same area. The value of the petroleum extracted is hundreds of times greater than the value of the ball clay, so, whereas the oil industry can afford to pay huge sums to minimize environmental disruption, the ball clay industry cannot; although it is the traditional form of extractive industry in the area and many local beauty spots, such as Blue Pool, are, in fact, former ball clay workings. The variety of species present, and hence the variety of habitats, is far greater in and around disused ball clay workings, than in the farmland that preceded them. Several flooded ball clay workinngs in South Devon have become nature reserves of national significance, and have featured in television natural history programmes.

One problem which is causing concern to the industry in Dorset is the tendency to designate every small area of heath around a ball clay working a protected area, so that ball clay working has to proceed by a patchwork of small workings to avoid the designated areas. The inevitable result is that a great deal of the resource is wasted and left in the ground and the environmental disruption affects a larger area than a single larger pit. One really must question whether the overall environmental benefit would not be served better by allowing the resource to be worked in the most efficient way possible in a single large pit. Although on a smaller scale, ball clay working utilizes the same sort of sophisticated planning methods which were described above for china clay.

Disposal of waste does not present as serious a problem to the ball clay industry as it does for china clay, because there is usually far less of it, and, because of the sedimentary geology of the deposits, most can be put back into worked-out pits. Those tips which are created can be vegetated easily and landscaped, so that they readily blend in with the landscape.

As with china clay, the ball clay pits also provide a marvellous research and teaching opportunity for undergraduate and postgraduate students. Currently, there is much exciting research into the structural evolution and clay mineralogy of these basins, which may have wide application to similar basins elsewhere in the world.

Calcium carbonate production in SW England

Yet another example of effective resource utilization is the use of chalk as a paper filling and coating material, where virtually no waste is created and 100% of what is taken out of the ground can be converted into product. English China Clays pioneered the use of this material in the late 1960s and, at that time had the world lead in this field. However, the source of special high brightness chalk at East Grimstead could not be developed for five years, because of a whole series of public enquiries caused by the clamour from local protest groups. Many irresponsible things were said by these pressure groups, who misled the public by portraying the development to be on a far larger scale than was ever envisaged, as subsequent events have so clearly shown.

Although permission to work was eventually granted, the inevitable result of the five year delay was that Britain lost the lead in this field to a Franco-Swiss company, which it has never managed to recapture.

Conclusion

So, in conclusion, these examples taken from the fields of china clay, ball clay and carbonate geology, show why it is important to think about the conservation of mineral resources, so that the maximum possible use of the resource is made, in order to conserve mineral resources for future generations to work and study, and in order to minimize the impact of mineral operations by making the maximum possible use of that which is taken out of existing holes in the ground. Conservation of mineral resources by the use of technically advanced mining and mineral processing systems is just as important as many other aspects of conservation.

The environmental lobby has a great deal of public sympathy and trust at the moment; however, it must guard against the irresponsible use of this trust by portraying misleading information about the wealth and employment-generating extractive industries and by generally overstating its case. This could lead to the public losing its confidence in the lobby in the future and also jeopardize the most efficient use of the finite mineral resources which are in the ground.

References

BRISTOW, C. M. 1992. Development of kaolin production and future perspectives. *Proceedings of the 10th Industrial Minerals International Conference, San Francisco, May, 1992*, 95–104.

—— 1993. Conservation of Mineral Resources. *In* Alfred-Wegener-Stiftung (ed.) *Die benutztg Erde, Ökosysteme, Rohstoffgewinnung, Herausforderungen*. Ernst, Berlin, 17–26.

—— & EXLEY, C. S. 1994. Historical and geological aspects of the china clay industry of southwest England. *Transactions of the Royal Geological Society of Cornwall*, **20**, 247–314.

HERRING, P. & SMITH, J. R. 1991. *The Archaeology of the St Austell China-Clay Area*. Cornwall Archaeological Unit, Cornwall County Council.

ROLT, L. T. C. 1974. *The Potters' Field*. David and Charles, Newton Abbot.

The conservation of Quaternary geology in relation to the sand and gravel extraction industry

DAVID BRIDGLAND

Earth Science Consultancy, 41 Geneva Road, Darlington, Co. Durham DL1 4NE, UK

Abstract: Many new Quaternary sites discovered during the past 20 years have been in sand and gravel workings. Few of these have been successfully conserved, in contrast to sites in gravel pits dating back to earlier years. This is primarily the result of the modern planning system for mineral extraction, which normally involves a built-in restoration programme. It is difficult to modify existing planning to incorporate geological conservation areas: often the value of gravel pits as landfill sites is important to their economic viability, and companies tend to be concerned about the adverse effect of permanent exposures on public relations.

In the 1990s gravel extraction in the UK is in decline, as a result of economic recession and the increasing use of the aggregate derived from offshore dredging or manufactured from crushed rock. This means that gravel pit exposures are at a premium, making it important that those that are available are used by the geological community to the optimum. To ensure this, a closer involvement of geological interests is needed at the planning stage. The importance of geology as part of the British natural heritage should be impressed upon planners and gravel companies alike, with the success of the archaeological community in this field being taken as a model. At present some important Quaternary sites are investigated, when threatened by development, as peripheral parts of archaeological projects. Such projects are undertaken where the occurrence of archaeological material is suspected. Only where a special scientific significance has already been noted has this type of approach been used with geological sites. It is possible to predict at the planning stage whether proposed gravel pits will provide important new geological evidence, especially where borehole records are available. Geological organizations and conservation bodies should combine to ensure that this part of our heritage is not overlooked.

In the early 1980s the author spent several years selecting sites in the Quaternary of southern and midland England as part of the Geological Conservation Review (GCR). This project was undertaken in order to update the previous coverage of Sites of Special Scientific Interest (SSSIs) in these areas, which was patchy, although comprising many important, indeed classic sites, often in rather poor condition. Former sand and gravel quarries constituted a high proportion of such sites, some of which dated back to the days of non-mechanized extraction. The scientific importance of many of these was confirmed, often after site cleaning projects carried out by the GCR Unit of the then Nature Conservancy Council in conjunction with specialist researchers from universities and museums.

Conservation of important sections in new gravel pits

At the time of the GCR, many extant gravel pits were known to have exposures of exceptional scientific interest and several new sites were discovered in such quarries while the site selection exercise was in progress. Although the sections available in these modern workings provided excellent exposures and were clearly more readily of value to the scientific community than the small sections that could be recut in old SSSIs, it proved extremely difficult to achieve long-term conservation of such exposures. This was generally possible only when interesting sections occurred at a final quarry boundary, adjacent to a road for instance, and it was often necessary to persuade operators to waive parts of landfill schemes that had already been approved, in order to leave small lengths of old quarry face uncovered.

A problem for the Earth science conservationist is that the modern planning system for mineral extraction normally involves a built-in restoration programme, so that the final form of the reclaimed land has been determined before quarrying even begins. Quarry companies are rightfully proud of their achievements in landscaping former extraction sites and returning them to agriculture or some other use. They see this side of their operations as an important one, if they are to maintain their good relationships with local communities. If the sediments exposed in the quarry prove to be of geological importance, sufficient to justify conservation, it is thus necessary to persuade the various interested parties to modify existing plants – a procedure that might prove expensive, notwithstanding the cost of lost landfill space. The revenue from landfill has been an increasingly important aspect of sand and gravel quarrying in recent years, so proposals to reduce the volume of tipping space, however insubstantial, have rarely been well received. In some cases it has proved easier to excavate purpose-dug conservation sections on adjacent land than to conserve parts of modern quarries (Wright 1993).

The recent recession in the building industry, together with tighter planning constraints, has led to a sharp decline in the number of sand and gravel quarries in Quaternary deposits. There has been a move towards the increasing exploitation of offshore aggregates and the use of crushed rock as a sustitute for river terrace and floodplain deposits, which were hitherto the most common source of sand and gravel. With the comparative rarity of gravel pit exposures in many parts of the country as a result of such trends, those that do exist may experience increasing pressure from geologists wishing to visit for research and collecting purposes, teachers wanting to bring student parties, and so on. At the same time concerns about safety, fuelled by government legislation and stipulations of industrial insurance policies, increasingly lead to permission for visits being refused. This means that the dwindling reserve of gravel pit exposures is not being used by the geological community to the full.

Comparison with archaeology

With the number of exposures in sand and gravel quarries in decline, there is a need for geologists to have a greater influence at the planning stage of mineral extraction, in order to maximize the research and teaching potential of exposures that will be created by future quarrying. A comparison with procedures followed in the related discipline of archaeology is of value to this discussion. In the 1980s and early 1990s the archaeological community has found itself in the happy position in which, thanks to a 'special relationship with planning authorities', any damage to archaeological sites as a result of development generally takes place only after a rescue investigation project funded by the developer. In addition, if there is good reason to suspect that an archaeological interest may be present at the site of a prospective development, perhaps because of proximity to known occurrences, a developer-funded evaluation is generally undertaken, to see whether there is anything worthy of a proper excavation. The mechanism behind this obviously successful system is non-statutory, but depends upon the good offices of local authority planning departments. The system has been bolstered recently by the publication of the *Planning Policy Guidance No. 16* document (PPG16: DoE 1990), in which the government requires planning authorities to pay attention to archaeological interests; this document makes archaeology a 'material consideration' to be taken into account in the planning process. The role of local authorities in archaeological resource management has been reviewed recently by Baker & Shepherd (1993). Although not all planning authorities would demand developer-funded archaeological assessment and excavation as a condition for planning consent in sensitive areas, most prospective developers now expect to provide such funding and include it in their costings for development.

Quaternary geology is sometimes a beneficiary of developer-funded work of this type, as it has a substantial overlap with Palaeolithic archaeology, the majority of Palaeolithic artefacts occurring in geological deposits. Indeed, a principal source of Lower Palaeolithic artefacts is river terrace gravel, many important assemblages of this age having been collected when gravel pits in these deposits were dug by hand. Thus a current developer-funded project at Kimbridge Farm Quarry, Hampshire, which exploits the Lower Palaeolithic Dunbridge gravel of the River Test, consists primarily of an archaeological watching brief (undertaken by 'Wessex Archaeology', Salisbury), but also includes the recording and sampling of the Quaternary deposits (Bridgland & Harding, 1993; Fig. 1). It is even possible for archaeologists to instigate research on Quaternary sediments that contain very few (if any) artefacts, on the pretext of obtaining background environmental evidence to provide a context for known archaeological assemblages. A project of this type was undertaken in 1987 at Clacton, on the site of the former Butlin's holiday camp, prior to residential development. This work involved the recording, sampling and analysis of fossiliferous sediments within the Clacton Channel Deposits but overlying the gravel that contains the type Clactonian Palaeolithic Industry (Bridgland 1994). Carried out prior to PPG16, this project was funded by English Heritage rather than the developer.

The important point about the Clacton deposits is that they are equally significant to

Fig. 1. Section at Kimbridge Farm Quarry, Dunbridge, showing the Dunbridge (Palaeolithic) gravel. This site is the subject of an archaeological watching brief, funded by the quarry company. Conservation sections have been identified in adjacent land, in older quarries at Dunbridge, the subject of a GCR excavation in 1986 (Bridgland & Harding 1987). Photograph provided by the Trust for Wessex Archaeology, Salisbury.

Quaternary geology and palaeontology as they are to archaeology; but without the archaeological interest it is probable that no investigation would have taken place at Butlin's. It is not just that no funding would have been provided as a result of the redevelopment, but also that the geological community would probably not have been informed of the opportunity were it not for the close association between archaeologists and planners. Opportunities to study important geological deposits must be missed frequently through a lack of information, as well as a lack of funding. Only when a documented geological SSSI is involved is the geological community automatically alerted to potential developments, through the various national nature conservation agencies. Thus when the Late Quaternary site at Holywell Coombe, Folkestone, was partly destroyed by the building of the Channel Tunnel, a very rare geological 'rescue excavation' was funded by the developer, Eurotunnel (Preece 1991).

One problem peculiar to mineral extraction that is faced by archaeologists and Earth science conservationists alike is that there are many planning consents already in existence, forming what is essentially a backlog of undug quarries that can be worked without recourse to modern planning procedures.

Archaeologists have, in many areas, overcome this difficulty by building relationships with the major quarry companies. Indeed, there is something of a tradition of 'rescue' work on gravel pit sites dating back to long before PPG16. In 1982 the Confederation of British Industry (CBI) published a pamphlet entitled *Archaeological investigation – Code of practice for Minerals Operators*, setting out guidelines for quarry companies on dealings with archaeology and archaeological organizations. Recently updated (CBI 1991), this code recognizes the long established good relationship between the minerals industry and archaeological groups. This has been apparent, for example, in the Upper Thames, where much archaeological rescue work has been carried out since the 1970s with the assistance of Amey Roadstone (ARC). The archaeological potential of the gravel outcrops of the Upper Thames has long been recognized; a detailed survey of aerial photographic evidence for archaeology in those parts of the valley in Wiltshire and

Gloucestershire was provided by Leech (1977), in a report that is an important source of reference in assessing the impact of future sand and gravel extraction (similar volumes were also published for other parts of the Thames valley (Benson & Miles, 1974; Gates, 1975). Unfortunately for the geologist, the archaeological interest in this area is almost entirely of Holocene age, superficial to the gravels rather than occuring within them; little new information on the sediments themselves is therefore likely to come from any future archaeological investigations.

Developer-funding of geological research and conservation

The example of Holywell Coombe shows that developer-funding can provide a valuable impetus to geological research. In this instance the routing of the Channel Tunnel through the SSSI was modified as a result of the investigation, so as to avoid destruction of the most important sediments thus benefitting earth science conservation. But should this type of rescue work occur only at sites previously scheduled as SSSIs? In the case of the sand and gravel industry, quarry operators will expect to provide funding to allow assessment and, if necessary, rescue excavation of archaeology. It seems anachronistic that our geological heritage doesn't enjoy the same treatment. Extraction frequently destroys potentially important sediments just as it destroys archaeological remains, yet there is no systematic mechanism for ensuring that the geological evidence is recorded. Rescue work by geologists would be relatively inexpensive, compared to archaeological excavation, as very much less time is involved in the field. Thus the funding of geological 'rescue' research could be seen by quarry companies and other developers as a cost-effective way of obtaining a positive public image. The geological recording, which would ideally be backed up by appropriate laboratory analyses, could be carried out by the British Geological Survey, by university or museum geologists, or by private consultants. Publications of results in a suitable journal should be a principal aim.

Assuming developers could be persuaded to fund geological conservation and rescue research, the question arises as to who would organize the necessary network of information about planned developments, such as new mineral extraction proposals. Here again a comparison with archaeology reveals great differences. Every English county now has at least one archaeological officer on its staff, responsible for the sites and monuments records in the county in question. These are backed up by a network of archaeological units, some closely allied to county councils, others independent of government funding. Geology has no such organization. Perhaps the national conservation agencies, with their close links with the planning system, are best placed to take a lead in this field and try to ensure that geological heritage, if not conserved, is fully recorded and investigated before being destroyed.

At present, the only connection that the Earth science community has with planning departments is through the government conservation agencies (English Nature, Scottish Natural Heritage and the Countryside Council for Wales), who are consulted as a statutory requirement if a geological (or any other) SSSI is affected by a planning application. There have been considerable benefits from this, as valuable information about the future availability of sections has often been passed on to researchers, who otherwise would not have known of the opportunity. However, as with developer-funded rescue work, only known sites of SSSI standard are covered by this mechanism.

Predicting the occurrence of geological interests

For geological interests to be given consideration at the planning stage of sand and gravel extraction, it would help greatly if the occurrence of scientifically important deposits could be predicted. Prediction could be based on extrapolation from previously known occurrences or the results of preliminary investigations such as borehole surveys. Such survey information is generally obtained by gravel companies prior to seeking mineral extraction consent, but it is not often available to the wider geological community for commercial reasons. Important fossil-bearing sediments are often identified directly in borehole data, particularly when mollusc remains are present. The occurrence of mammal bones and concentrations of Palaeolithic artefacts (of importance to the geologists as well as the archaeologist) are less easy to predict.

Prediction of interesting sites is another area in which the archaeological community has paved the way. An important current project, funded by English Heritage, seeks to identify those deposits with archaeological (Lower Palaeolithic) potential in southern England. This, the 'Southern Rivers Project', was designed to provide a database that catalogues the various Quaternary deposits that might contain Lower Palaeolithic assemblages, as well as compiling an exhaustive gazeteer of known artefact occurrences of this age in the areas surveyed. When completed, it will be of great value to planning departments, as it will differentiate those deposits that have a high probability of archaeological interest from those that are unlikely to contain artefacts. A similar gazeteer of potentially interesting Quaternary deposits would be of equal value, were it that planning departments could include geology as 'a material consideration in the planning process'. Unfortunately it is far from easy to predict the occurrence of important fossiliferous or other interesting sediments: in river terrace sequences these generally occur as isolated lenses within more widespread less remarkable deposits. Many interglacial sites, for example, have been discovered when such localized sediments were encountered in gravel workings.

Conclusions

Geologists need to emphasize to those in authority, and the public at large, that Britain's geological heritage is no less important than its archaeological heritage; ideally, geological interests should be treated in a similar way to archaeological ones. This is particularly pertinent to the sand and gravel industry, which currently funds considerable amounts of archaeological rescue excavation, but very little, if any, geological work. With the emphasis in Earth science conservation seemingly moving away from the rigid SSSI system, or at least to expand on this system, perhaps the time is ripe for raising the profile of geological heritage in relation to planning departments and developers, including mineral extractors. In particular, co-ordinated schemes for developer-funded rescue research on geological interests threatened by mineral extraction would be very valuable. Funding of this sort could go a long way towards bridging the gap left by cuts in funds available for work in geology from research councils.

It may be felt, by those interested in Earth science conservation, that developer-funding of work on sites that are to be damaged or destroyed is not something to place high on the agenda. After all, when such projects are necessary it means that the site is not being successfully conserved. However, there must clearly be a strong link between conservation and 'rescue' investigation: it is because the site is worthy of conservation that the rescue work is needed, and sites are being lost or damaged all the time without any research taking place beforehand. A leaf should be taken from the archaeologists' book: they regard excavation as a second-best option to intact preservation, but when the pressure for development if irresistible they are fully geared up to exploiting the developer-funding system.

References

BAKER, D. & SHEPHERD, I. 1993. Local authority opportunities *In:* HUNTER, J. & RALSTON, I. (eds) *Archaeological resource management in the UK: an introduction.* Alan Sutton, Stroud, 100–114.

BENSON, D. & MILES, D. 1974. *The Upper Thames Valley – an archaeological survey of the river gravels.* The Oxfordshire Archaeological Unit, Survey No. 2.

BRIDGLAND, D. R. 1994. *The Quaternary of the Thames.* Geological Conservation Review Series. Chapman & Hall, London.

—— & HARDING, P. 1987. Palaeolithic sites in tributary valleys of the Solent River. *In:* BARBER, K. (ed.) *Wessex and the Isle of Wight.* Field Guide. Quaternary Research Association. Cambridge, 45–57.

—— & —— 1993. Preliminary observations at Kimbridge Farm Quarry, Dunbridge, Hampshire: Early results of a watching brief. *Quaternary Newsletter,* **69**, 1–9.

CBI (CONFEDERATION OF BRITISH INDUSTRY) 1991. *Archaeological investigation – code of practice for minerals operators.* CBI. London (first published 1982).

DOE (DEPARTMENT OF ENVIRONMENT) 1990. *Planning Policy Guidance Note 16: Archaeology and Planning* (PPG16). HMSO, London.

GATES, T. 1975. *The Middle Thames Valley – an archaeological survey of the river gravels.* The Berkshire Archaeological Committee, Publication No. 1.

LEECH, R. 1977. *The Upper Thames valley in Gloucestershire and Wiltshire – an archaeological survey of the river gravels.* Committee for Rescue Archaeology in Avon, Gloucester and Somerset.

PREECE, R. C. 1991. Accelerator and radiometric radiocarbon dates from colluvial deposits at Holywell Coombe, Folkestone. *In:* LOWE, J. J. (ed.) *Radiocarbon dating: recent advances and future potential.* Quaternary Proceedings Vol. 1, Quaternary Research Association. 45–53.

WRIGHT, R. 1993. Conservation and landfill – a question of timing. *Earth Science Conservation,* **32**, 18–19.

Damage caused by opencast and underground coal mining

JAROSLAV DOMAS

Ministry of Environment, Division of Geological Environment Protection, Kodaňská 10, 101 59 Prague 10, Czech Republic

Abstract: Post-war electricity generation in the former Czechoslovakia was heavily dependent upon coal extraction. Mining and opencast extraction of brown and black coal in Northern Bohemia and Upper Silesia under a command economy have left the Czech Republic with a legacy of widespread subsidence, contaminated spoil, landscape depravation and associated industrial pollution. The work of the new government in assessing the environmental damage and initiating landscape restoration and environmental rehabilitation are discussed and reviewed.

The Czech Republic is comparatively rich in solid fossil fuels: brown coal and black coal. Since there are no other sufficient sources of energy, coal became the main source being burnt in thermal power plants. The importance of coal is evidenced by the total installed power plant output, which is around 11 000 MW. This output is produced by the following sources: coal 73%, nuclear 16%, water 11%. The amount of installed output based on coal puts high requirements on the mining industry.

North Bohemian Brown Coal Basin

In the Czech Republic brown coal extraction became important as early as during the First World War, but the decisive expansion was reached around the late 1960s–1970s. The coal is exploited in the North Bohemia Brown Coal Basin/Coal District, which covers 1450 km^2 and of which 850 km^2 is coal-bearing. The coal is of lacustrine origin (freshwater marsh), of Miocene age. There is one major coal seam with a maximum thickness of 40 m, although locally it may split into two or three smaller seams. Sulphur content is 1–4% on average. The coal crops out on margins of the basin, but can be found at a depth of 500 m in central parts.

Underground coal mining started around the year 1870 and, at its height, over 300 deep mines operated, of which just three are active today. It is estimated that approximately 50% of the coal-bearing part of the basin has been affected by this mining. The land surface has sunk about 2–7 m and the drainage pattern has been changed. Military topographic maps from the end of the eighteenth century provide information on the geomorphology of the area. Many wetlands have appeared, and are appreciated today as micro-centres of local ecological diversity. Mining subsidence on most of the ground surface area has now ceased and the land surface has stabilized, enabling the landscape to recover. Only mining experts are aware of the true extent of the historical undermining as there is little visual evidence of it. From time to time, however, a new sink hole appears at random, although construction on such surfaces is sometimes still possible.

Coal mining

In the 1960s and 1970s the needs of the economy led to an enhanced demand for energy. The government decided to obtain energy from brown coal, regardless of the impact. Several thermal power plants were constructed with power blocks of 100–200 MW. Simultaneously, sources of coal were developed and several giant opencast mines started to operate in the basin. Before long, the North Bohemian region delivered over 60% of solid fossil fuels in the former Czechoslovakia. In 1992 approximately 38% of electricity spent in the Czech Republic was produced here, and approximately 50% of all the energy produced was delivered from coal excavated in this region.

Demand for brown coal was met mostly from opencast mines. Each mine emcompasses the pit itself and an accumulation of spoil as the overburden is deposited. An average pit stretches some 1–5 km, with a width of 1–3 km. The depth can reach about 150 m depending on the thickness of beds overlying the coal seam. Each coal quarry was required to deposit the overburden on its outer reaches at the beginning of mining operations, in the process burying many square kilometres of original landscape. Pits quarrying the coal under thick overburden are obliged to spread the spoil continuously and the new profiles which result dominate the landscape.

From O'Halloran, D., Green, C., Harley, M., Stanley, M. & Knill, J. (eds), 1994, *Geological and Landscape Conservation.* Geological Society, London, pp. 93–97.

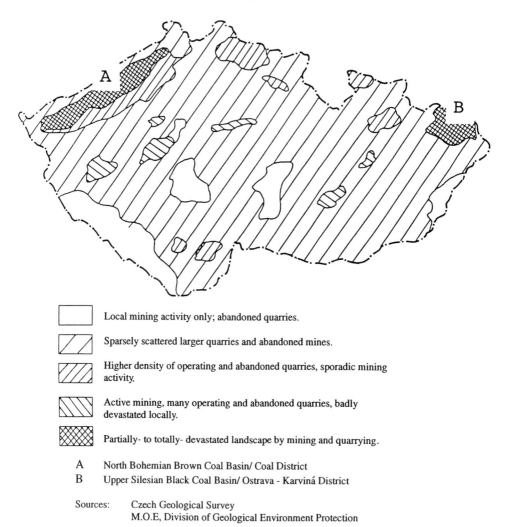

Fig. 1. Schematic map of extractive impact on the environment of the Czech Republic.

Environmental impacts

The impact of opencast brown coal mines on the landscape, the geological environment and on groundwater has been heavy. The highest coal output was in 1984 with some 74 million tons produced. In 1992, when coal demand dropped to 52 million tons, in response to political and economic change, it was necessary to strip and deposit 185×10^6 m^3 (220 million tons) of overburden. Some 260 km^2 of the land surface has been influenced by open pit mining, out of which 80 km^2 are covered by dumps. Many tens of square kilometres of landscape, including more than 70 villages and settlements have been swallowed by advancing coal quarries. Brown coal mining by the opencast method causes the greatest degree of destruction of landscape in the country. This environmental situation is, more over, exacerbated by the impact of thermal power plant pollutants which are rich in sulphur, arsenic, heavy metal, and ash content.

The mining policy of the former political regime envisaged the intended exploitation to exhaustion of the reserves of the Brown Coal Basin. Recent political change has been reflected by an increasing awareness of environmental concerns. The Ministry of the Environment, aware of the severe environmental situation in this coal basin, has introduced legislation controlling power plant pollutants. In

1991 the government also took a very important step towards landscape conservation. For each of eight operating opencast mines, obligatory spatial limits – which must not be overstepped by either quarrying or deposition of spoil – were introduced. The lifetime of pits is thereby substantially shortened, and large parts of the coal basin landscape will be preserved. Even so there will be several remaining pits with a total operating volume of $3.5 \times 19^9 \, m^3$.

The cultural landscape

Around the beginning of the 1980s the first attempts to restore the landscape disrupted by coal mining began. Work concentrated on maintenance and revitalization of both abandoned pits and spoil dumps. The smaller pits were converted into pools creating local centres of ecological diversity or were modified for swimming facilities. Spoil dumps, after morphological reshaping, were reinstated by spreading a layer of bentonite and topsoil, protected by wood chips to counter erosion, and thus were restored for farming or forestry. A large-scale study of the geomechanical properties of soil accumulated in dumps is continuing to identify dumps which can be used as sites for a range of different purposes, e.g. for a hippodrome stadium, field airport, waste site etc.

An enhanced pace of activity to promote nature conservation and to conserve the cultural landscape has become apparent in the past few years. As an important first step a study on the overall reclamation of coal basin land was completed at the end of 1992. Several new studies and projects, focused on improving environmental quality and stabilizing ecological conditions, have been launched recently by the Ministry of the Environment. An example is a project entitled 'Complex geo-ecological research into the North Bohemian Brown Coal Basin', which started in 1992. This project aims to investigate all major aspects of the environment, i.e. atmosphere, hydrosphere, pedosphere and lithosphere. Particular attention will be paid to anthropogenic accumulations (i.e. spoil dumps). This will explore ways of reshaping spoil dumps to integrate them in to the original adjacent landscape as if they were a natural feature of the area, and how they can be successfully recultivated.

Several government ministries and other organizations have begun to co-operate to find ways of reclaiming old pits. The areal extent of each of these pits can be $10-20 \, km^2$, and their depth can reach 100 m or more. Seemingly the easiest way to reclaim them is to flood them. The difficulty here is not only how to insulate the coal seam from the pit water, but also how best to separate contaminated pit water from the groundwater system. A key problem is low discharge rate of streams in the basin; even if the full discharge capacity of all streams available in the region were to be focused on a single pit only, its flooding might take at least 30 years.

When all these factors are taken into account, estimates of the cost to eliminate mining impact, restore the cultural landscape and return the area to normal ecological conditions, range from US$1–1.8 billion.

Upper Silesian Black Coal Basin

Underground black coal mining in Ostrava-Karviná Mining District, stretching over $850 \, km^2$, started in 1829. Coal seams of marine origin of Carboniferous age are deposited in a number of horizons. The thickness of individual seams is from c. 0.4 to 2.5 m, but exceptionally can reach 7 m. Coal seams are found at depths of 400–1000 m below the ground surface. The coal is of very good quality and is readily cokable.

Coal mining impacts

There are 42 active mines in the coal basin today. Mining is by the full-seam, longwall method with multiple-horizontal exploitation carried out on up to 10 floors without any back-filling. Approximately 1 billion tons of coal and 700 million tons of waste rock has already been exploited since the mining began. The mining output in 1991 was 19.5 million tons of black coal. Underground space left by mining, disregarding filling by collapsing roofs, represents a volume of $1.8 \times 10^9 \, m^3$. Some 40% of Ostrava-Karviná region, i.e. $340 \, km^2$, has been influenced by such mining. The type of multiple-horizontal exploitation has resulted in land surface subsidence. The areal extent of ground surface influenced by undermining covers 14 300 ha (35 000 acres). This figure is expected to reach 17 900 ha (44 000 acres) by the year 2050. About half of the above area will sink 2–5 m, 14% will sink more than 5 m, and 2200 ha (5400 acres) will sink down more than 10 m. In some parts of the coal basin the surface has already subsided by nearly 40 m. The original glacial and periglacial geomorphology of the landscape has been totally changed, with large areas now being flooded. In addition to the constant need to transfer and uplift roads and railways, more than 2500 houses have been destroyed.

Waste rock is accumulated on spoil dumps, although the extent of these is less than in North Bohemia. Nevertheless they cover at least 960 ha (2370 acres), and perhaps as much as 1850 ha (4570 acres). Among other coal mining impacts in Ostrava-Karviná region is the presence of settling pits infilled by waste from coal preparation plants. These pits, containing several metres of thick black mud with admixed inorganic and organic contaminants, cover approximately 816 ha (2000 acres) of land. As these pits do not have their bottoms insulated, the pullation is free to migrate into underlying geological environment and to contaminate groundwater.

Leaving aside damage caused by subsidence and to buildings, the cost of the mining impact is estimated by Associated Mining Consultants Ltd at around US$215 million. The environmental situation is worsened by air pollution arising from the relatively high concentration of heavy industry in the area, in particular from ageing coke plants and heavy metallurgy plants.

Protection and regeneration of the environment

Current government research is focused on determining the condition of the principal environmental constituents: air, soil and water. The results are alarming, for example a geochemical study of soils of the town of Ostrava and its suburbs has revealed disturbingly high concentrations of heavy metals. The same is true of the Olše River, which amounts to, in fact, little more than a waste water sewer. As a consequence, a base study on the feasibility of further mining in Karviná District has already been started and will be used for determining future land use planning for this area.

The key to regional regeneration is to decrease both mining output and the intensity of heavy industry. A coal mine closedown programme is already the subject of a government decree issued at the end of 1992. The intention is to shift coal mining from the Ostrava District to the Karviná District, where the mining economy is more profitable. This, of course, has met strong objections in Karviná, as the town itself is endangered by unrecoverable undermining. The ministries and regional representatives involved have established a special committee to try and resolve the conflicts in interest between mining on the one hand and social and conservation interests on the other. An economic consideration is the fact that extraction of 360 million tons of high quality cokable coal is at stake. It has been suggested that 'zero lines' need to be determined, beyond which no subsidence would be tolerated. Any restriction to mining will, however, need to reflect social consequences, as a number of miners and other workers would be made redundant. Moreover the tradition of being employed in the mines is deeply rooted in this region and the prospect of change is unpalatable to many people.

Land surface subsidence can be partly offset by back-filling. Waste rock is traditionally used as the filling material, but attempts to use thermal power plant ashes have been tried and $c.\ 1.5 \times 10^6\,\text{m}^3$ of ash have already been deposited underground. The application of back-filling is limited, however, by technological complications, high costs and only partial control over the subsidence which results. Objects such as buildings, endangered by subsidence, need to be either anchored or simply pulled down. Communication networks need to be relocated, or repeatedly uplifted. For instance some railways have had to be uplifted by more than 30 m. Reclamation of subsided land surface for a number of after uses can be attempted. Depressions flooded after subsidence can be infilled by waste rock, which is then covered by topsoil. If the reclaimed land is intended for forestry, the topsil is the former subsoil and plantation of 10 000 three-year-old trees per hectare follows. The original humus soil, 30 cm thick, can be used as topsoil in the case of agricultural recultivations.

Conclusions

Economic and political change in the Czech Republic are also being reflected in a new, and responsible, approach to both old and new mining impacts. The Ministry of the Environment and other responsible bodies are progressively modifying the legislation to support a process of decreasing and eliminating the negative impact of coal extraction. Mining and quarrying organizations are now obliged to fund a reclamation and recultivation plan to mitigate contemporary mining impacts. Largescale historical mining impact will need to be solved with the help of government monies. Because of the scale of this task, international support, e.g. from the World Bank, is required. Expert teams (NORWEST, NORCONSULT etc.) supported by the Canadian and Norwegian governments are currently assessing the situation. These teams will put forward proposals focused on landscape recovery and environmental remediation for consideration by the World Bank. Both of the mining districts discussed above are included in these international environmental projects, as are the Black Triangle, Silesia and Phare.

Elimination of the mining impact, so devastatingly and carelessly caused in the past, will certainly not be easy. The final resolution of the problem of necessity lies far in the future, but the first essential steps by the government and appropriate organizations have already been taken. These first steps, leading eventually to remediation and revitalization of the landscapes disrupted by mining, are encouraging and so, too, are the first successes that are becoming apparent on the ground.

Development and conservation at Parys Mountain, Anglesey, Wales

NICHOLAS J. G. PEARCE

Countryside Council for Wales, Plas Penrhos, Ffordd Penrhos, Bangor LL57 2LQ, UK
(Present address: 1 Maesyrefail, Penrhyncoch, Aberystwyth SY23 3RE, UK)

Abstract: During the late eighteenth and early nineteenth centuries, Parys Mountain, Anglesey was the world's largest copper producer, leaving a spectacular legacy of spoil heaps and mine workings. The Great Opencast Mine on Parys Mountain shows excellent exposures through the host rocks and ore body, generally accepted to be of an exhalative, volcanogenic origin (cf. Japanese Tertiary Kuroko-type deposits). The deposit is unique in the United Kingdom.

Exploration resumed at Parys Mountain in 1955, and by 1985 a reserve of approximately 5.5 million tonnes, grading 10% base metals with minor silver and gold had been proved. In January 1985 the entire Great Opencast, along with a site at Morfa Ddu to the west, was proposed as a Site of Special Scientific Interest (SSSI). The SSSI notification was opposed by the owners on the grounds that it interfered with the proposed mining operation (largely the dumping of coarse spoil). Substantial, local, unfavourable press coverage appeared regarding Earth science conservation, concerned with the loss of potential employment in a relatively depressed area should the mining application be refused. Negotiations followed and five smaller areas within the Opencast were identified showing the overall structure and the relationships between ores and host rocks. After detailed discussion with the owners (Anglesey Mining plc) and the then Nature Conservancy Council, these were notified as SSSIs. The reduced SSSI area allows mining operations to proceed unhindered and planning permission for mining was granted which specifically protects the areas of geological significance. A shaft has been sunk and when the economic climate permits the mine will generate substantial revenue for the area.

Parys Mountain (147 m above sea-level), north-eastern Anglesey, is approximately 2.5 km south of the small port of Amlwch (see Fig. 1) and shows excellent exposures of mineralized volcanics, extensive spoil heaps and abundant archaeological remains. The area is currently controlled by Anglesey Mining plc, in which Imperial Metals Corporation of Canada hold a 42% share.

The Great Opencast Mine at Parys Mountain has been selected as part of the Geological Conservation Review (GCR) 'Mineralogy of Wales Network' and designated a Site of Special Scientific Interest (SSSI) for its mineralization. The Mineralogy of Wales Network contains 22 sites covering a range of mineralization including lead–zinc, gold, manganese and copper ore deposits with sites selected to represent unique mineralogies, assemblages or styles of mineralization.

History of geological research

Greenley (1919) first reported thoroughly the geology of Parys Mountain, describing the sediments as Ordovician and Silurian, and the mineralization as younger than the 'great Post-Silurian earth movements', i.e. epigenetic. Greenley (1919) also described the mine's development and history.

Manning (1959) produced a paragenetic sequence for the mineralization in the order of

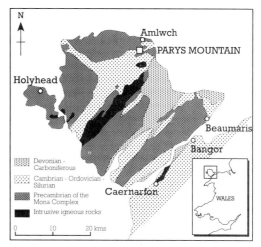

Fig. 1. Summary geology of Anglesey, showing location of Parys Mountain.

From O'Halloran, D., Green, C., Harley, M., Stanley, M. & Knill, J. (eds), 1994, Geological and Landscape Conservation. Geological Society, London, pp. 99–103.

silicification of the sediments and emplacement of quartz veins, pyritization and chalcopyritization, but did not constrain the age of the so-called 'bluestone', inter-grown galena–sphalerite, mineralization. Manning suggested that mineralization was controlled by shear zones (with little or no mineralization along faults), lithological boundaries and low grade hydrothermal alteration of the rocks to chlorite-bearing assemblages.

Hawkins (1966) first recognized the presence of Ordovician tuffs and rhyolite lavas on Anglesey, which underlie Silurian graptolitic shales and Bates (1966) placed the Parys Mountain volcanics as post-Lower Ordovician, overlying *Didymograptus bifidus* shales. Bates (1972, 1974) further described the stratigraphy and Caledonian tectonic history which generated the synclines at Parys Mountain.

The early and mid 1970s saw the major period of academic research at Parys Mountain. Thanasuthipitak (1974) described the lithological relationships and mineralization, and classified the volcanics as calc-alkaline. Sivaprakash (1977) studied the mineralogy of the ores and some of the rarer sulphosalts. Both Greenley (1919) and Manning (1959) had considered mineralization as epigenetic, whilst Wheatley (1971) considered the ores to show both epigenetic and syngenetic features. However, Thanasuthipitak (1974) concluded that mineralization was syngenetic, closely associated with the eruption of rhyolite. Ixer & Gaskarth (1975) supported this, proposing a volcanogenic model with lavas, ashes and sulphide-rich hydrothermal fluids being expelled onto the sea floor. Ixer & Gaskarth (1975) described these ores as 'Kuroko-type', by analogy with Tertiary deposits of Kuroko, western Japan. Pointon & Ixer (1980), based largely on Pointon (1979), described the host rocks and positioning of ores at Parys, referring to the deposit as 'exhalative' rather than 'Kuroko-type'. Low in the succession, ores occur as disseminations, layers and massive lenses within 'siliceous sinter' (an exhalative volcanogenic rock). Higher in the succession, ores occur as a series of lenses in shales and cherts. Pointon & Ixer (1980) described the paragenesis as dominant pyrite with chalcopyrite, sphalerite, galena, tetrahedrite-group minerals and rare lead–bismuth sulphosalts. They concluded that the deposit was the product of exhalative volcanic-sedimentary activity associated with minor rhyolitic volcanism above a subduction zone. Southwood (1984), using the minor element composition of basaltic lavas, placed the formation of these rocks in a back-arc basin associated with extensional tectonics. Pointon & Ixer (1980) placed the folding of the ore body and some remobilization as Caledonian in age.

Nutt *et al.* (1979), from K–Ar dating of core samples, proposed an alternative origin for the ores, as either entirely epigenetic or unrelated to Ordovician volcanicity. Chlorite in mineral veins records one event at 353 ± 7 Ma whilst an older event at 390 ± 10 Ma is shown by illite associated with disseminated ores. Both events are considerably younger than the Caradoc volcanics and the Silurian sediments. Nutt *et al.* invoked a late Caledonian mineralization event (390 ± 10 Ma) and a separate, unrelated Lower Carboniferous event (353 ± 7 Ma). Rundle (1981) critically challenged their interpretation, and most workers now tend to disregard these dates.

Mineralized cherts in the centre of the Great Opencast show features associated with soft sediment deformation, indicative of a syngenetic (?synvolcanic) origin (Haynes, 1993). Most recently however, a stratigraphic and structural re-evaluation of the site by Westhead (1993) has indicated that these cherts are possibly Silurian, and not Ordovician, in age.

Mineralization

The mineralization at Parys Mountain is copper–lead–zinc-based, the main sulphides being pyrite (FeS, often containing As), chalcopyrite ($CuFeS_2$), sphalerite (ZnS) and galena (PbS). Tetrahedrite group minerals ($[Cu,Fe]_{12}Sb_4S_{13}$, which may contain As in exchange for some Sb) and some rare bismuth sulphosalts (tetradymite $Bi_{14}Te_{13}S_8$; kobellite $Pb_{22}Cu_4(Bi,Sb)_{30}S_{69}$; galenobismuthinite $PbBi_2S_4$) have also been recorded. Parys Mountain is the type locality for anglesite ($PbSO_4$) first described in 1783 and named after the island, and for pisanite ($[Fe,Cu]SO_4 \cdot 7H_2O$), first recorded in 1946.

Mining history

Mining at Parys Mountain has been dated, by ^{14}C from charcoal in surface workings, to 3900 BP (Early Bronze Age, D. Jenkins, UCNW, Bangor, pers. comm.). Characteristic of these early workings are stone 'mauls' or hammers which have been found in shallow pits (Timberlake 1991). There is no direct evidence of Roman mining at the site, but folk tales, evidence of fire-setting and finds of Roman 'cakes' or ingots of copper make this likely (Manning

1959). The major period of mining commenced on March 2, 1768, after some 11 years of exploration. This unearthed the Great Golden Venture Lode, today marked by the spectacular open pits of the Great Opencast Mine (the result of the collapse of underground workings) and the Mona Mine. By the 1790s production reached 3000 tons of copper per annum (cf. Cornwall at this time producing 4434 tons annually (Greenley 1919)). During the nineteenth century mining moved further underground, reaching 174 m below sea-level. A total of 21 km of underground passages were dug during the life of the mine with access from over 100 shafts, most now sealed. By 1904, mining had ceased, although for some 60 years after, copper 'cement' was produced by precipitation on iron in large 'copper ponds'. Parys Mountain produced at least 130 000 tons of copper.

Development and conservation

Interest resumed in Parys Mountain in 1955 when the area was acquired by Anglesey Mining Exploration Ltd. The area changed hands several times, finally coming under the control of Anglesey Mining plc in 1985. Surveys since the mid 1950s have located new reserves totalling approximately 5.5 million tonnes grading 1.5% copper, 3% lead, 6% zinc, 69 grammes per tonne silver and 0.4 grammes per tonne gold.

In May 1986, Anglesey Mining plc sought planning permission to extract 350 000 tonnes of rock per annum over a 15-year period. Plans included the gradual infilling of the Great Opencast with spoil from the on-site ore processing.

In December 1986 the Great Opencast was notified as an SSSI for its geological features, in direct conflict with the dumping of waste rock material in the Opencast. As this would affect the mining operation, Anglesey Mining plc and the Imperial Metals Corporation objected to the notification in March 1987. There followed a period of discussion between Local Authority planners, the then Nature Conservancy Council and Anglesey Mining plc, with the process receiving considerable local press coverage. Press reports focused on the potential loss of employment should the SSSI designation hinder the mining operation, citing a possible workforce of 150. This would have a dramatic effect on the economy of what is a relatively depressed part of Anglesey. By July 1987 Anglesey Mining

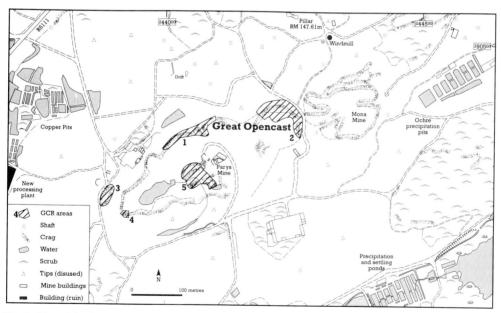

Fig. 2. Map of the Great Opencast showing the 5 areas of the SSSI. These areas show: (1) coarse, epigenetic, pyrite in siliceous volcanics; (2) cherts with sulphides along bedding planes and fractures (evidence for epigenetic remobilization) with fossiliferous Silurian shales; (3) flow-banded rhyolite; (4) core of the Parys Syncline in shales; and (5) slumped, pyrite-mineralized cherts and stockwork veins (evidence for the syn- and epigenetic origins of the ores).

plc had agreed to reduce the volume of material to be dumped into the Great Opencast to about 15% of the original estimates, and by August 1987, by a review of the GCR designation, 5 smaller conservation areas within the Great Opencast, and a further area at Morfa Ddu slightly to the east, had been identified (see Fig. 2). These are now notified as SSSIs and encompass the range of rock types and mineralization associated with this unique ore deposit. They represent a practical compromise to the original site, providing protection for exposures critical in the understanding of the deposit, whilst freeing much of the site for tipping.

In May 1988, planning permission was granted by Gwynedd County Council, who had initially voiced objections to the SSSI notification, and included some 68 clauses, one of which indicated that the operations on site 'shall include for conserving identified features of geological significance as defined by the areas making up the SSSI.'

Anglesey Mining plc immediately commenced work on a new shaft near Morfa Ddu, reaching 300 m by July 1989 (Swallow 1990). Levels have been driven through competent rocks needing minimal support, and further exploration has taken place. Coarse spoil from the shaft and levels was dumped on the flanks of the mountain, away from the Great Opencast, and by March 1990 Anglesey Mining plc considered it unlikely that any material would need to be tipped in the Opencast.

In November 1991, after showing little interest in the site during the 1980s, the Great Opencast, some of the ochre settling ponds to the east of the mine, the Windmill and Engine House were scheduled as Ancient Monuments by CADW, the statutory body for the protection of ancient monuments and historical buildings in Wales. This highly restrictive scheduling probably affords the site more legal protection than it would receive under the Wildlife and Countryside Act 1981.

Currently the mine is under care and maintenance, the main shaft and levels having been allowed to flood until it is financially viable to start production. Once operational, extracted ore will be crushed and ground on site with lead, zinc and copper concentrates being separated by froth-flotation methods. The waste slurry will be pumped into large settling ponds on the south side of the mountain.

Other conservation interest

Parys Mountain contains important biological features. At least 125 lichen species inhabit the mineralized rocks, the walls and mortar of the old mine buildings, and several of the spoil heaps, many of which are typical of old metal mine sites. Some are unique to Wales, 10 species are 'nationally scarce', and one species at least (a *Lecidea*) is unique to Britain, if not new to science. Several areas of the mine are currently proposed as SSSIs for their lichen flora. The site is host to a nationally rare liverwort (*Cladopodiella francisci*), and has the only record of stags horn club moss (*Lycopodium clavatum*) for Anglesey. Several sub-species of metal-tolerant plants have been described from the site, for example, the grass *Agrostis tenuis* Parys Mountain, widely used in the reclamation of other contaminated metal mine sites.

Bats were recorded from many of the disused shafts, including long-eared and lesser horseshoe species, the latter listed in Annex 2 of the European Community Species and Habitats Directive as an internationally endangered species. Capping of disused shafts for safety reasons may have severely affected this population.

Summary

Parys Mountain, and the Great Opencast in particular, is clearly a site where research into the processes of formation of exhalative ore bodies is still active. The understanding of such processes may assist future generations to meet their needs for base metals. Plans to infill the Opencast by the currently proposed mining would have curtailed research on the site. However, detailed discussions between the developers, local planners and conservation agencies, and the careful selection of the most important exposures, has achieved protection of geologically crucial parts of the Opencast whilst allowing the development of Parys Mountain as a mine to proceed unhindered. Careful planning has enabled land-filling and other potential threats to be accommodated (Anon, 1988) and the conservation status of the site is retained, providing, when mining starts, a valuable source of revenue for the area.

Former metal mines are commonly areas with several conservation interests, from geological and mineralogical, through floral (lichens and mosses), faunal (including bats and choughs) to a substantial archaeological heritage. The management and development of such sites must encompass the whole range of scientific and historical issues without comprising the ability of society to meet its needs for raw materials both now and in the future if conservation is to be taken seriously by the public. The conservation of the scientific features at Parys Mountain is an example where such possible conflicts have been reconciled successfully. Future mining activity, whilst providing perhaps 500 000 tons of base metals, will not detract from the spectacular nature of the Great Opencast and the associated historical buildings.

References

ANON (1988). Conservation success at Parys Mountain. *Earth Science Conservation*, **25**, 47–49.

BATES, D. E. B. 1966. The geology of Parys Mountain. *Welsh Geology Quarterly*, **2**, 27–29.

—— 1972. The stratigraphy of the Ordovician rocks of Anglesey. *Geological Journal*, **8**, 29–58.

—— 1974. The structure of the Lower Palaeozoic rocks of Anglesey, with particular reference to faulting. *Geological Journal*, **9**, 39–60.

GREENLEY, E. 1919. *The geology of Anglesey*. Memoirs of the Geological Survey of Great Britain. HMSO, London.

HAWKINS, T. R. W. 1966. Boreholes at Parys Mountain, near Amlwch, Anglesey. *Bulletin of the Geological Survey of Great Britain*, **24**, 7–18.

HAYNES, L. R. 1993. *Metallogenesis*. Geological Conservation Review. Unpublished report, Joint Nature Conservation Committee.

IXER, R. A. & GASKARTH, J. W. 1975. 'Parys Mountain – a possible Kuroko-style deposit.' Paper presented to Mineral Deposits Study Group, Geological Society of London, Session 6: Stratiform mineral deposits. Leicester, December 1975.

MANNING, W. 1959. The Parys and Mona mines in Anglesey. *In The future of non-ferrous metal mining in Great Britain and Ireland*. Institution of Mining and Metallurgy, 313–328.

NUTT, M. J. C., INESON, P. R. & MITCHELL, J. G. 1979. The age of mineralisation at Parys Mountain, Anglesey. *In*: HARRIS, A. L., HOLLAND, C. H. & LEAKE, B. E. (eds), *The Caledonides of the British Isles – reviewed*. Geological Society, London, Special Publication, **8**, 619–627.

POINTON, C. R. 1979. *Palaeozoic volcanogenic mineral deposits at Avoca, Eire; Parys Mountain, Anglesey; and in south eastern Canada – a comparative study*. PhD thesis, University of Aston in Birmingham.

—— & IXER, R. A. 1980. Parys Mountain mineral deposit, Anglesey, Wales: geology and ore mineralogy. *Transactions of the Institution of Mining and Metallurgy (Section B: Applied Earth Science)*, **89**, 143–155.

RUNDLE, C. C. 1981. Discussion on the age of mineralisation at Parys Mountain, Anglesey. *Journal of the Geological Society, London*, **138**, 755–756.

SIVAPRAKASH, C. 1977. *Geochemistry of some sulphides and sulphosalts from Parys Mountain, Anglesey*. MPhil. thesis, University of Aston in Birmingham.

SOUTHWOOD, M. J. 1984. Basaltic lavas at Parys Mountain, Anglesey: trace element geochemistry, tectonic setting and exploration implications. *Transactions of the Institute of Mining and Metallurgy (Section B: Applied Earth Science)*, **93**, 51–54.

SWALLOW, M. J. A. 1990. Renaissance of Parys Mountain. *In:* JONES, M. J. (ed.) *Metals, Materials and Industry*. Institute of Mining and Metallurgy, London, 335–342.

THANASUTHIPITAK, T. 1974. *The relationship of mineralisation to petrology at Parys Mountain, Anglesey*. Unpublished PhD thesis, University of Aston in Birmingham.

TIMBERLAKE, S. 1991. Excavations at Parys Mountain and Nantyreira. *Archaeology in Wales*, **28**, 11–17.

WESTHEAD, S. J. 1993. *The structural controls on Mineralisation at Parys Mountain, Anglesey, North Wales*. Unpublished PhD thesis, University of Wales, Cardiff.

WHEATLEY, C. J. V. 1971. *Economic geology of the Avoca mineralised belt, S. E. Ireland, and Parys Mountain, Anglesey*. Unpublished PhD thesis, Imperial College, University of London.

The Newer Volcanic Province of Victoria, Australia: the use of an inventory of scientific significance in the management of scoria and tuff quarrying

NEVILLE J. ROSENGREN
Department of Geology, La Trobe University, PO Box 199, Bendigo, Victoria, 3550, Australia

Abstract: Newer Volcanics basaltic activity in Victoria began about 7 million years (Ma) ago and has continued up to less than 6000 years ago. Many original features such as craters, cones, pyroclastic deposits, lava flows and lava caves are well preserved. However, over the past 150 years, quarrying of rock, scoria and ash from many small pits and quarries has occured, affecting the appearance of a number of the cones and craters. Quarrying has exposed features of geological interest but also degraded scientific values.

A recently completed study of the Newer Volcanic Province of Victoria presents an evaluation of the scientific significance of 354 volcanic eruption points grouped into 349 sites. The study has rated 19 sites to be of National significance; 30 of State significance; 70 of Regional significance and 112 of Local significance. Three sites are rated as of Unknown significance and 115 sites are Unassigned.

The 119 sites rated as National, State and Regional are described in detail in the report and indexed by significant rating and by local government area. A sensitivity rating is used to indicate the degree to which the scientific values could be damaged or might deteriorate by inappropriate land use or management. The study recommends that 59 sites should be included by the Australian Heritage Commission on the Register of the National Estate.

The need for continuing quarrying leads to a conflict of interest between the need for scientific conservation and the needs of local government and other groups for suitable construction material. The present study evaluates the importance of the scientific aspects of the main volcanoes. A separate study has been commissioned to assess the need for future quarrying. The two reports together will provide a rationale for planning future quarry activities across the region.

A broad belt of volcanic rock in eastern Australia comprises one of the largest intraplate volcanic provinces of the world (Fig. 1). No volcanoes are active but there are young volcanoes in north Queensland (13 000 years old) and South Australia at 6000 years old. Activity was typically intraplate effusive basaltic volcanism. The patterns of age and type are complex but models of hot spot activity applied to Australia explain the distribution, age and nature of much of this volcanism (McDougall *et al*. 1966; Wellman 1974; Wellman & McDougall 1974; Sutherland 1983, 1991; Pilger 1982).

In Victoria, volcanics younger than 7 Ma old are known as the **Newer Volcanics** or **Newer Basalts** and the area is referred to informally as the Newer Volcanic Province. There are over 350 eruption points (Fig. 2) and many craters, volcanic hills and lava flows are well preserved and the nature and sequence of volcanism and post-volcanic changes can be interpreted readily (Hills 1938). Recent radiocarbon dating has demonstrated that the youngest volcanicity is older than 20 000 years (Head *et al*. 1991).

The volcanoes are an important source of basalt rock, scoria and tuff for construction and roadmaking and in many districts are the principal source of rippable rock. As most of the eruption points are small, quarries are conspicuous and impact strongly upon both the scientific and landscape attributes of the volcanic features. Recent studies on 354 of the eruption points have evaluated their geological significance and the nature and impact of scoria and tuff quarrying (Guerin 1992; Rosengren 1992).

Type of volcanicity

The province is a distinct geomorphic/petrographic province, with extensive plains of basaltic lava, and many small, monogenetic, central vent volcanoes formed in the last stages of volcanicity (Ollier 1967).

Ollier & Joyce (1964) calculated that only 1% of the volume of the Newer Volcanics is of pyroclastic origin. The typical volcano is a low basalt hill surrounded by lava flows. The steepest and highest volcanoes are scoria cones – some with

From O'Halloran, D., Green, C., Harley, M., Stanley, M. & Knill, J. (eds), 1994, Geological and Landscape Conservation. Geological Society, London, pp. 105–110.

related to tectonic plate margins. There are at least 30 known localities where lava flows or fragmental deposits '...contain some of the richest and most diverse xenolith and magacryst occurrences known from continental volcanic provinces...' (Nicholls 1984). The inclusions are peridotites consisting of olivine and pyroxene several centimetres in diameter, with a wrapping of scoria forming the core of a volcanic bomb (Irving 1980). Megacrysts 5 mm or more in diameter are found as residuals in weathered material or soils (Dasch & Green 1975).

Geomorphology of eruption points

Table 1 classifies the geomorphology of eruption points.

Survey of geological significance

A recent study (Rosengren 1992) of the significance of eruption points of the Newer Volcanic Province examined 354 eruption points using aerial photographs, 1:25 000 topographical maps, geological maps, published and unpublished literature and field work at each site.

Selection and rating of significant sites

A **significance rating** was awarded to each site on a scale of: Unknown, Unassigned, Local, Regional, State, National. (National sites are also of international significance.)

The criteria for significance rating included:

(a) The contribution the site makes to understanding of volcanic or related geological and geomorphological features and processes or wider concepts in Earth history on a local, regional, state or national basis. National sites are assumed *a priori* to be of international significance.
(b) The extent to which it represents a type of Newer Volcanics materials or landforms.
(c) The frequency of replication, i.e. the site is a unique, rare or unusual example of a volcanic or related geological or geomorphological formation or process.
(d) The quality or clarity of the display of the volcanic material, other type of rock outcrop, volcanic structure or landform.
(e) The value of the site as a teaching or research site displaying volcanic materials or processes.
(f) The importance of the site in displaying or determining volcanic chronology and sequence.

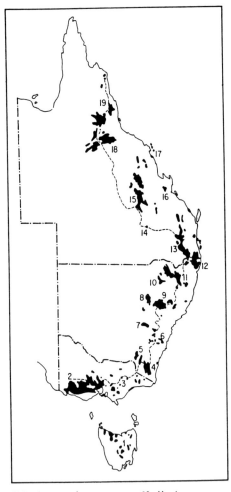

Volcanic areas of eastern Australia.
1. Tasmania.
2. Newer Volcanics of Victoria.
3. Older Volcanic of Victoria.
4. Monaro.
5. Snowy.
6. Abercrombie.
7. Orange.
8. Warrumbungle.
9. Liverpool.
10. Nandewar.
11. Inverell (Central Province).
12. Tweed.
13. Toowoomba.
14. Roma.
15. Springsure–Clermont.
16. Rockhampton.
17. Hillsborough.
18. Nulla.
19. McBride–Quincan.

Fig. 1. Cenozoic volcanic provinces of eastern Australia (from Ollier 1978).

well-preserved craters and there are a number of maars. They are comparable with Paracutin (Mexico), Surtsey (Iceland), Rangitoto (New Zealand) and the small flank eruptions on Hawaii.

The lavas are tholeiitic to alkalic lavas of continental intraplate affinity rather than

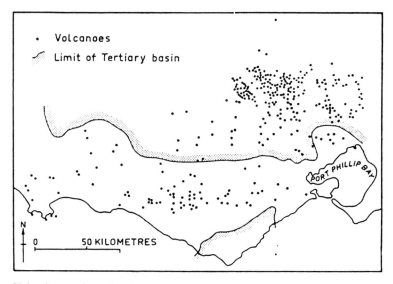

Fig. 2. Newer Volcanics eruption points in Victoria (from Ollier 1969).

(g) The environment of the site as an example of the range of terrain in which the Newer Volcanics materials were erupted. Sites, therefore, represent plains and highlands locatiions.

National (International) significance [19 sites]

These are sites that are critical to determining the character of the Newer Volcanic Province

Table 1. *Newer Volcanic Province: eruption point morphology*

	Type	Description	Number of eruption points*
Lava flow eruption points			
Lava cone with crater	1	Extended flow(s), clearly defined crater.	14
Lava shield	2	Broad, low-angle slope, overlapping multiple flows with/without crater.	29
Lava disc	3	Steep sided, without extended flow.	15
Lava hill	4	Extended flow(s), crater absent or ambiguous.	107
Pyroclastic eruption points			
Scoria cone with crater	5	Steep to moderate slope, clearly defined crater.	42
Scoria hill	6	Steep to moderate slope, rounded/flattened summit, no crater.	45
Maar-tuff ring	7	Gentle slope, crater with single eruption point.	21
Nested maar-tuff ring	8	Crater(s) with multiple eruption points in 1 km radius.	9
Miscellaneous			
Composite lava and scoria	9	Overlapping lava/scoria or scoria/lava eruptions.	50
Other	10	Other.	22

* Identified by Rosengren (1992).

and understanding volcanic or geological and geomorphological processes. The volcanoes with concentrations of xenoliths and megacrysts are of international significance for the data they provide about lithospheric and mantle composition and processes. Such features are uncommon on a comparative global basis. Other sites may be among the best national examples of a phenomenon that is either widespread or may be restricted to Australia, e.g. examples of Hawaiian and Strombolian type activity that are used as type or definitive examples in Australia. As outstanding examples of young intraplate volcanism they are accorded National/International significance.

State significance [30]

These are sites significant in defining the geological and geomorphological character and history of Victoria and supplement the National sites. They include features either limited in distribution or of significance for teaching, research or public display and interpretation.

Regional significance [70]

Sites that display the regional variation in the Newer Volcanic Province or are very clear examples of a feature with a wider distribution.

Local significance [112]

These are clear examples of relatively common features. They can be used to display local or regional character and variation.

Unknown significance [3]

Sites where there is indication of significance but insufficient data to allow a confident rating to be made.

Unassigned significance [115]

Sites that do not fall into the above categories. It does not imply that the site is not significant but indicates it does not meet the above criteria given the present state of knowledge or requirement.

As a result of the survey, it was recommended to the Australian Heritage Commission that 59 sites be nominated for listing on the Register of the National Estate.

All sites were given a **sensitivity rating** (independent of significance) on a scale of 1 (most sensitive) to 10 (robust) indicating the degree to which scientific values could be damaged by land use or other management decisions.

The geological and geomorphological significance of the region

This volcanic province is a major Australian resource for teaching and research by Earth scientists and others. Interest centres around the following attributes:

(a) It is a major example of intraplate volcanism.
(b) It contributes to understanding the causes and chronology of the volcanism of eastern Australia and in determining the later geological evolution of the continent.
(c) It relates to the theory of hot spot volcanism and provides data about the vertical and horizontal tectonics of the Australian Plate (Sutherland 1983).
(d) The megacryst and xenolith sites are of world significance as examples of minerals of deep-seated crustal and mantle origin.
(e) There are examples of diverse eruption types ranging from low lava shields to complex multiple eruption centres with maars, lava flows and scoria cones.
(f) It is an area of conspicuous volcanicity and provides some of the best examples of volcanic activity in the Australian continent.
(g) There is clear display of the varied age of volcanic events and a range of regolith types of landform development to be determined.
(h) The secondary impact of volcanic eruptions on terrain evolution can be assessed as can the erosion of volcanic products.
(i) It is a major geological and geomorphological unit covering $15\,000\,km^2$ or over 15% of the area of the state of Victoria.
(j) The region has strong cultural and economic associations with the volcanic resources, extending from Aboriginal sites and across the time of European occupation, tuff and scoria deposits continuing to be a major resource of stone both for the immediate region and beyond.

Conservation of significant sites

Aboriginal occupation had little direct physical effect on the volcanic landforms but European settlement has substantially impacted on the visibility of much of the volcanic activity.

Clearing of stones from lava flows for fencing, building or to improve land for agriculture, quarrying of basalt, tuff and scoria, urban expansion, drainage and infilling of craters, lakes and wetlands, and agricultural industries have greatly modified the volcanic landscape, natural vegetation and soils. Exotic tree plantations have been established on some volcanic hills.

Most of the region is in private ownership and is not subject to controls that are designed to protect the scientific values. Much damage to scientific value has already been achieved – probably 95% of the area of pre-European settlement vegetation has been eliminated and faunal habitat has been severely altered. Geological and geomorphological values have fared a little better than this, but only because they are generally more robust rather than as a result of considered conservative action.

Quarries and other excavations

Artificial sections (excavations, cuttings, quarries) are frequently utilized to view internal structures, sub-surface contacts and determine stratigraphy. While access for specialists may be possible, temporary exposures cannot be selected as reference sites or for ongoing studies or later revision. The most enduring exposures are road and railway cuttings, but on a flat landscape there are few of these. Many practical considerations are involved – cuttings may not be conveniently or safely accessible and geological values are often diminished by structures built to stabilize the face.

In areas of limited outcrop, active and abandoned quarries often provide the major display of a geological formation. These have the advantage of providing new and fresh exposures quicker than many natural processes. The major (almost the only) geological sections of scoria and tuff deposits and multiple lava flows are found in pits.

Exposures in operating quarries or pits are often temporary and may not be available for extended geological research. Most occur on private land and access is restricted. In pyroclastics, the exposed faces are typically over-steepened and in unstable material they are difficult to access and natural slumping obscures the detail. Combined with safety and liability issues, quarry operations and the attitude of management, the logistics of using operating quarries as geological reference sites can be formidable. Quarries also degrade some scientific values. In a region where volcanicity is no longer active, scoria and tuff are non-renewable resources and the configuration of the eruption points cannot regenerate or recover until there are more eruptions. Quarry operators may not recognize the significance of uncovered material or structures and may unknowingly destroy specimens or exposures, fill in craters or bury material with overburden or stockpile. Outcrops or topography significant in displaying volcanic history may be modified, buried or removed. Some entire deposits of scoria or tuff from small eruption points have been removed. As well as removing significant material, quarrying produces synthetic landforms. Holes and overburden mounds alter the form and slope angles of cones and mounds and may confuse future interpretation of original eruption topography and products.

Significant geological features – ranging from the unique and outstanding to mundane representative sections – are known in many abandoned and working quarries. New exposures are being created (and destroyed) in operating pits. They are a common venue of educational and specialist group excursions. While many practical issues are involved in recognizing and maintaining sites in operating pits, as a first step it should not be assumed that all disused quarries or pits need to be filled or regraded as a means of landscape rehabilitation. In quarries and other artificial sections, maintenance of the face may be necessary to remove weathered or mass movement debris, to extend or enhance the quality of the exposure or to stabilize the outcrop.

Conclusions

The need to evaluate the impact of quarries on scientific values and to take advantage of outstanding new quarry face exposures is now recognized by the Victorian Government. The department responsible for managing these operations recommends that '...the geological significance of a feature should be considered when assessing quarry applications...' and that '...investigations...to locate alternative sources of scoria and tuff where existing pits are found to be compromising significant volcanic features...' (Guerin 1992). Scientific inventories, characterized by a clear statement of the values of the geological features concerned and how these judgements were reached, are a valuable tool in reconciling Geological Conservation and quarry extraction.

References

DASCH, E. J. & GREEN, D. H. 1975. Strontium isotope geochemistry of Iherzolite inclusions and host basaltic rocks, Victoria, Australia. *American Journal of Science*, **257**, 461–469.

GUERIN, B. 1992. *Review of Scoira and Tuff Quarrying in Victoria*. Geological Survey of Victoria Report No. 96.

HEAD, L. D'COSTA, D. & EDNEY, P. 1991. Pleistocene dates for volcanic activity in western Victoria and implications for Aboriginal occupation. *In:* WILLIAMS, M. A. J., DECKKER, P. DE. & KERSHAW, A. P. (eds) *The Cainozoic of Australia: a re-appraisal of the evidence*. Geological Society of Australia Special Publication **18**, 302–308.

HILLS, E. S. 1938. The age and physiographic relationships of the Cainozoic volcanic rocks of Victoria. *Proceedings of the Royal Society of Victoria*, **51**(2), 112–139.

IRVING, A. J. 1980. Petrology and geochemistry of composite ultramafic xenoliths. *American Journal of Science*, **280**-(A), 389–426.

MCDOUGALL, I., ALLSOPP, H. D. & CHAMALAUN, F. H. 1966. Isotopic dating of the Newer Volcanics of Victoria, Australia, and geomagnetic polarity epochs. *Journal of Geophysical Research*, **71**(4), 6107–6118.

NICHOLLS, I. 1984. Inclusions of Mantle and Crustal Origin in Basaltic Deposits of the Western District, Volcanics Workshop 1984. Monash Uni. (Unpub.).

OLLIER, C. D. 1967. Landforms of the Newer Volcanic Province of Victoria. *In*; JENNINGS, J. N. & MABBUTT, J. A. (eds.) *Landform Studies from Australia and New Guinea*. ANU Press, 315–339.

—— 1969. *Volcanoes*. Australian National University Press, Canberra.

—— 1978. Tectonics and geomorphology of the Eastern Highlands. *In:* DAVIES, J. L. & WILLIAMS, M. A. J. (eds) *Landform Evolution in Australia*. Australian National University Press, Canberra, 5–47.

—— & JOYCE, E. B. 1964. Volcanic Physiolgraphy of the Western District Plains of Victoria. *Proceedings of the Royal Society of Victoria*, **77**(2), 331–376.

PILGER, R. H. 1982. The origin of hotspot traces – evidence from Eastern Australia. *Journal of Geophysical Research*, **87** (**B5**), 1825–1834.

ROSENGREN, N. J. 1992. *The newer Volcanic Province of Victoria: an inventory and evaluation of scientific significance*. Unpublished report to National Trust of Victoria (Australia).

SUTHERLAND, F. L. 1983. Timing, trace and origin of basaltic migration in eastern Australia. *Nature*, **305**, 123–126.

—— 1991. Cainozoic volcanism of eastern Australia. In: WILLIAMS, M. A. J., DECKER, P. DE., KERSHAW, P. (eds) *The Cainozoic of Australia: a re-appraisal of the evidence*. Geological Society of Australia Special Publication, **18**, 15–43.

WELLMAN, P. 1974. Potassium-argon ages on the Cainozoic volcanic rocks of eastern Victoria, Australia. *Journal of the Geological Society of Australia*, **21**(3), 359–376.

—— & MCDOUGALL, I. 1974. Cainozoic igneous activity in eastern Australia. *Tectonophysics*, **23**, 49–65.

Theme 2: Landscape Conservation

Keynote address

Conservation of geomorphological landscapes in Taiwan

SHIN WANG[1], LING-YUH SHEU[1] & HSIAO YU TANG[2]

[1]*Department of Geography, National Taiwan University, Taipei, Taiwan, Republic of China*
[2]*Resource Conservation Division, Council of Agriculture, Republic of China*

Abstract: Taiwan, known as Formosa to westerners, is an island endowed with natural beauty. With a dramatic relief, active crustal deformation and energetic erosive processes, it is obvious that nature-sculptured landscapes of very high quality are abundant.
  This paper lists the best ten features of the island's landscape and discusses the current status of their protection.

Evolution of geomorphological landscapes

Taiwan is unique on the basis of its geographical location and the Earth's tectonic background (Ho 1986). It is situated in the southeast corner of the Earth's largest continent, the Eurasian continent, and is adjacent to the largest ocean, the Pacific. In the course of the planet's evolution, the island probably did not emerge until late in the Palaeozoic. However, growth, disintegration and rebirth make the Taiwan of today vastly different from that shortly after its birth. Mountains were lifted and fault valleys continued to sink, giving rise to Taiwan's geomorphological landscapes. Every aspect from the Coastal Mountain Range, the Huatung Longitudinal Valley, the Central Mountain Range and the Pingtung Plain to the West Coast Plain is a byproduct of diastrophism.

Taiwan's top ten geomorphological wonders and their protection

1. High Mountains

The 'Five Peaks' and 'Three Pinnacles' have duly been acclaimed as such because of their towering heights and pointed tops, attesting to the orogeny in the Eurasian continent's eastern rim. Of these high mountains, the main peak of Yu Shan (Mt Morrison), rising to 3952 m, looks down upon all other peaks in East Asia.

Five Peaks: Yu Shan (3952 m), Hsueh Shan (3886 m), Hsiukuluan Shan (3824 m), Nanhuta Shan (3742 m) and Tawu Shan (3092 m).

Three Pinnacles: Central Pinnacle (3705 m), Tapa Pinnacle (3492 m) and Tafen Pinnacle (3208 m).

The preservation of the primeval appearance of these mountains is the key to their conservation. These peaks have been incorporated into national parks or nature preserves and are protected under law (Council of Agriculture and Department of National Parks 1992).

2. Taroko Gorge

Taroko Gorge (Marble Gorge) presents itself as a supreme landscape worthy of admiration with its nearly vertical marble walls and marked changes in elevation (Wang 1989*a*). It is an area with cliffs and gorges that is famous for its rare rock formations, great differences in elevation and steep slopes.
  Erosion caused by rivers is the predominant natural force carving out the face of the Earth. At Taroko Gorge, one can marvel at the multiform beauty of nature and look at ancient rocks.
  The quintessence of Taroko Gorge lies in its flora, active geomorphological processes, flowing water, awe-inspiring appearance and aboriginal peoples. As it is now part of the Taroko National Park, well-rounded measures can be drafted to preserve this invaluable landscape. The most serious threat confronting this area is the large influx of tourists and the structures required to serve them.

3. Penghu's Columnar Basalt

Column-shaped basalt makes a distinctive landscape in both geological and geomorphological terms. The columns took shape gradually as eruptions of basalt magma cooled and solidified. Scientifically speaking, the vista of columnar basalt contains important data concerning geology, geomorphology and diastrophism, by which scientists can study the evolution of the crust from the past to the present. Therefore, this distinctive geological and geomorphological landscape is valuable for tourism, education and scientific research purposes (Wang 1991).

Penghu National Scenic Area Preparatory Administration has been established, but a comprehensive conservation programme has yet to be approved. In accordance with the Cultural Heritage Preservation Law, the Council of Agriculture has designated Tingkouyu, Chishanyu and Hsiaopaisayu as nature reserves for columnar basalt, thereby banning any unsuitable land development projects (Tourism Bureau 1993).

4. Suhua Coast and Chingshui Cliff

The Suhua Coast, renowned for steep slopes and advanced wave erosion, extends from Peifangao, Ilan County to Chungteh, Hualien County. Except for a few alluvial fans, this 90-km-long shoreline is dominated by steep cliffs rising to between 300 and 1200 m.

Most of Chingshui Cliff in the Suhua Coast's southern section has been incorporated into the Taroko National Park. The greatest damage to this impressive landscape has been caused by the expansion of the Suhua Highway. Massive drilling and collapses of slopes have severely scarred the landscape. Another potential threat to the area is posed by the mining industry. Laws are needed to stipulate that all mining operations should be conducted outside the national park's domain.

5. Chihsing Shan and Shamao Shan

Chihsing Shan and Shamao Shan, independent and complete cone-shaped hills, are located in the Yanmingshan National Park, suburban Taipei, and are famous for their conical or bell-like appearance. Another feature of Chihsing Shan is its length of slope, while Shaomao Shan looks like a hat worn by officials in ancient China. In addition to these mountains' extraordinary shapes, hot springs and fumaroles are also major post-volcanic features in this region.

These mountains of volcanic origin, along with their hot springs and fumaroles, are mostly encompassed by the Yangmingshan National Park and placed under legal protection accordingly. However, activities such as mining, drawing water from hot springs and mountaineering were already widespread prior to the national park's inauguration. Damage to the landscape here has thus remained unchecked in some cases.

6. Yehliu

A famous scenic spot in northern Taiwan, Yehliu is best known for its odd-shaped hoodoo rocks. Furthermore, it is an outstanding promontory, offering views of the coast, and a cuesta landform tilting westward.

Erosion by waves has transformed the toes of the cliffs close to the sea into sea-hollowed walls. As the process continued, the rocky walls receded, giving rise to sea cliffs and shore platforms. These shore platforms along the Yehliu coast took shape on inclined layered rocks, which varied in strength. In the course of erosion, rocks of various hardness reacted differently, resulting in a slightly undulating landscape with parallel ridges and depressions. Weathering caused particles on the surface of the rocks to gradually fall off. Cracks then formed, and the elements carved the rocks into a myriad of shapes and sizes over the passage of time.

The Yehliu Scenic Zone falls under the jurisdication of the Tourism Section, Bureau of Construction, Taipei County Government. Initial planning calls for landscape preservation and interpretation. However, the administrative agency has devised various development and excavation proposals in order to boost the number of visitors. This is deplorable! Yehliu's unique landscape deserves to be designated a nature preserve similar to national monuments found in the United States. In turn, the administrative agency should be upgraded and any proposed excavation prohibited.

7. 'Small Yehliu Coast'

The Small Yehliu Coast's geomorphological and geological features make it a unique attraction in the East Coast National Scenic Area (Wang 1990), ideal for studying geology. The most distinctive characteristic of this coast is the sedimentary deformational structures of the area's rocks.

The Small Yehliu Coast is governed by the East Coast National Scenic Area Administration, which is responsible for protecting this geological landscape and executing laws and regulations when necessary. Its main responsibilities include taking precautions against visitors' insensitive behaviour, setting aside visitors' paths to prevent trampling on the rocks, and strengthening explanatory activities. Working with schools in strengthening nature eduction is another task that deserves special attention.

8. Kenting Coral Reef Coast

The coral reefs around the Hengchun peninsula can be divided into two categories: the older Hengchun limestone and the more recent coastal coral reefs. Due to the gradual uplifting of the crust, the older coral reef limestone was transformed into tableland covered with gravel and soil.

From Chialoshui southward to Haiko at the western end of the peninsula, the coast is mostly surrounded by fringing coral reefs. Sandy beaches are scattered here and there along this expanse, such as Shatao, Nanwan, Kenting and Paisha.

The single most serious threat to conservation of the Kenting Coral Reef Coast is posed by the development of fishing harbours and sightseeing resorts. In general, the area is facing fewer threats to its well-being these days as the National Park Law offers fairly sound protection.

Living coral, a rare sight for most visitors, is the Kenting National Park's most important ecological asset (Wang 1989b). Yet coral growth is largely at the mercy of the warm water discharge from a nearby nuclear power plant and other waste water released into the sea.

9. Mud Volcanoes

Mud volcanoes in Taiwan can be divided into 17 mud volcano areas. There are also several cases in which flames rise out of natural gas emissions in the absence of mud eruptions, including Kuantzeling in Tainan and Chuhuo at Hengchun. The former is a phenomenon literally known as 'water and fire sharing the same origin', while the latter has only flames.

Eruptions from mud volcanoes consist of mud along with some natural gas instead of magma. Flames can be readily triggered at mud volcanoes which have sufficiently large natural gas eruptions.

Only a few of these extraordinary natural features have been designated for protection. Some of the mud volcanoes were turned into brickyards' quarries; some were levelled to grow sugarcane and other crops; a greater number of mud volcanoes were simply confined to a small corner as nearby land was rapidly developed. A few mud volcanoes are being utilized as tourism resources, confined to limited spaces like gardens, such as the Adopted Daughter Lake in Kaohsiung County and Loshan Mud volcano in Hualien County.

10. Mt Huoyen

Mt Huoyen is a landscape with great differences of elevation and a dense pattern of ravines. It is located at Sanyi, Miaoli County, lying to the north of the Taan River and to the west of the Sun Yat-sen Freeway. Here, people can see the many pinnacle-like hills at the foot of Mt Huoyen (602 m).

One geological prerequisite for the formation of badlands such as those found at Mt Huoyen is the existence of massive gravel beds characterized by loose cementation. These gravel beds were subjected to various geological processes and also affected by day-to-day gravity. The combination of all these factors, leads to a pebble-strewn landscape of steep slopes and deep ravines.

In line with the Natural and Cultural Heritage Preservation Law, Miaoli's Mt Huoyen has been designated as a nature preserve and placed under the jurisdiction of the Forestry Administration, Taiwan Provincial Government. The agency executes management programmes on an annual basis in a bid to preserve the area's integrity. The goal of conservation is readily attainable if such programmes are carried out according to law.

References

COUNCIL OF AGRICULTURE AND DEPARTMENT OF NATIONAL PARKS 1992. *Island of Diversity: Nature COnservation in Taiwan, Republic of China*.

HO, C. S. 1986. *Geology of Taiwan*. Central Geological Survey, Ministry of Economic Affairs, Republic of China.

TOURISM BUREAU 1993. *National Scenic Areas of Taiwan, Republic of China*.

WANG, SHIN (1989a). *Geomorphology and Geology of Taroko National Park*. Taroko National Park Administration, Republic of China.

—— (1989b) *Landforms of Kenting National Park*. Kenting National Park Administration, Republic of China.

—— 1990. *Landforms of the East Coast National Scenic Area*. Tourism Bureau, Republic of China.

—— 1991. *Landforms of the Pescadores*. Tourism Bureau, Republic of China.

Origin and use of the term 'geotope' in German-speaking countries

F. W. WIEDENBEIN

*Department of Applied Geology, University Erlangen - Nuremberg,
Schlossgarten 5, D-91054 Erlangen, Germany*

Abstract: Geotope protection includes all kinds of measures of preservation, development and management of sites of Earth science heritage – geotopes. The aim of geotope protection is the preservation of geodiversity. In the framework of Earth science conservation, the term can perform the same function as the term 'biotope' does in land use planning and nature protection.

Review of development

In Germany, the term 'geotope' was introduced by Bruno Stürm for Earth science heritage sites (Stürm 1992). In its original meaning the term refers to the smallest spatial units of landscapes. Modern landscape ecology uses 'geotope' for the description of the inanimate features of biotopes, e.g. Leser (1991) in his complex physical geography.

After the Digne Symposium the following definitions were given by AG Geotopschutz, the Working Group on Geotope Protection in German-Speaking Countries, on the occasion of its constituent assembly on 6 March, 1992, in Mitwitz, Upper Franconia (Wiedenbein 1992):

> Geotopes are those parts of the geosphere which are discernible on the Earth's surface or are accessible from there; they are spatially limited and in a geoscientific sense clearly distinguishable from their surroundings. Geotopes can be endangered in their existence by anthropogenic measures and natural processes; their loss is, as a rule, irretrievable.

> Geotopes can be worthy of preservation and of protection owing to their natural characteristics (material composition, structure, form, history, content of fossils, significance for landscape ecology and cultural history), and for their beauty or rarity.

> Geotope protection includes all kinds of measures of preservation, development and management of geotopes in their natural diversity and characteristics. Geotope protection is an inherent part of the geosciences and land use planning. Geotype protection can have higher priority with respect to other objects of protection or objects open to exploitation.

After this, the meaning of the term geotope was discussed and developed in a second national meeting in April 1993 (Quasten 1993) and at the international symposium on 'Geological Heritage' connected with the Geotechnica fair at Cologne on 6 May, 1993 (Wiedenbein 1993b). The preservation of geodiversity as the main aim of geotope protection was formulated by the author in April 1993 (Wiedenbein 1993a).

Reasons for geotope protection

The preservation of diversity is distinct from the protection of monuments. The protection of natural monuments has a long tradition, which has its origin in the Age of Romanticism. When Casper David Friedrich painted the chalk cliffs of Rügen in 1819 (see Grube & Wiedenbein 1992, cover picture) he showed human individuals in a reverent dialogue with nature. At the same time, Alexander von Humboldt promoted the term 'natural monument', which had been created by the German philosopher, G. W. Leibniz, in 1691, three hundred years before the Digne Sympsoium. In his *Protogaea* Leibniz used the term for fossils (Leibniz 1691).

Modern society needs geotope protection as a complement to biotope protection, for an holistic approach to nature. Rocks, monuments, plants and animals are all threatened by acid rain and eutrophication. They need protection to their entirety. Also, the holistic approach is needed as a counter-balance against the triviality of the mass media. Geotopes are essential to show and explain to our children the development and the dynamics of the Earth. Understanding history is a prerequisite for identity and responsibility, and history, in this context, includes the history of the Earth.

Geotopes can also contain important resources, storing useful information about the history of the Earth and beginnings of life. Climatic records did in extrapolating our climatic future. Other geotopes provide information about natural disasters like earthquakes, volcanic eruptions or meteoritic impacts, particularly about their frequency in the past. Earth is not in a stable balance.

Another aspect connected with the protection of resources is the material content of the geotopes. The medical use of minerals, water and earth is as old as humankind. In Germany, mudbaths, in particular, with peat have a special tradition. Research on fossil organochemical compounds is only just beginning. While rare minerals can display special properties and can stand as models for new materials, e.g. perowskite ceramics or new zeolites. Fossils are not only records of evolution, like the *Archaeopteryx*, the famous missing link; they also allow us to have a look at the construction of living matter of past biospheres.

In the same sense we can use older fabricated geotopes as a source of information on early engineering. For example the site 'Heunesäulen', near Miltenberg on the River Main, shows us how sandstone pillars for the Romanesque cathedrals on the Rhine were made. In this case they were produced from boulders of a periglacial rock field and not in a quarry.

Another unusual example for a geotope is the so-called 'gate to hell' in the medieval town of Lüneburg. It is a nineteenth century entrance gate made of bricks with overlapping doors. The movement of the doors is a result of subrosion, here the solution of Upper Permian salt. In Lüneburg the estimation of subrosion is an important economic factor. Sites such as this gate are the prerequisite for estimating the rate of subrosion. In principle, a geo-historical assessment is necessary for all larger constructions like power plants or disposal sites.

A relatively new development in Germany is the use of geotopes for leisure time. The most spectacular example is the 'Dino-Park' in Münchehagen in Lower Saxony, arranged around a large hall with natural footprints of Upper Jurassic dinosaurians.

Naturally, special geotopes for collectors would be welcome. The collecting of minerals, fossils and rocks is a wonderful hobby which reconciles the primeval experience of searching and finding. It should be a task of geotope management to find possible locations for collectors to enjoy their hobby without damage to nature and to site which are important for teaching and research. Finally, without the reference of geotopes, the results of geoscientific research could not be checked.

The German experience

In Germany, the protection of sites is a task of the 'Länder'. In the field of nature protection and the protection of historical monuments, the Federal Government only gives general outlines for the laws of the Länder. In North Rhine-Westphalia, Hesse, Thuringia, Rhineland-Palatinate and Baden-Württemberg the protection of fossils is part of the conservation of the cultural heritage. It is often possible to protect sites by two different laws. A general overview on the differences in geotope protection was given last year in Mitwitz (Wiedenbein & Grube 1992; Öbo Mitwitz 1993).

At present, first inventory lists from nearly all German states ('Länder') are available, which all together include more than 10 000 geotopes. However, questions of use, of management and of accessibility remain mostly unsolved even after inventory. In addition, it is necessary to select sites of national and European, significance.

The main focus must be placed on a systematical approach:

- All type sections (stratotypes and parastratotypes, type profiles, petrographical and mineralogical type localities).
- Classical fossil deposits like Holzmaden, Solnhofen-Eichstätt, Messel, Öhningen, Korbach, Odernheim, Willershausen, etc.
- Extraordinary mineralogical sites (St Andreasberg, Merkers).
- Extraordinary tectonic sites (the Teufelsmauer at the Harz-North-Rim-Fault, the Bohlen near Saalfeld, the Bavarian Pfahl).
- Volcanic landscapes (the maars of the Eifel, the Laacher-See-region, the Siebengebirge, the Kaiserstuhl, the Scheibenberg).
- Meteorite craters (Nördlinger Ries, Steinheim basin).
- Karst landscapes (carbonate, sulphate and salt karst).
- Glacial landscapes (eg. Lüneburger Heide, the island of Rügen).
- Pedological sites (peatlands, palaeosoils).
- Hydrogeological sites (wells, waterfalls, etc.).
- Geomorphotopes (boulders, gorges, dunes, etc.).
- Erosion landscapes like the Saxon Sandstone Mountains.
- Ecological geotopes like the heavy metal and serpentine sites (Stolberg, Blankenrode, Bottendorf, Wurlitz, Erbendorf, etc.).
- Technical geotopes (historical mining and engineering subjects).

For the Global Indicative List of Geological Sites (GILGES), 17 proposals are given by the AG Geotopschutz (Wiedenbein 1992, 1993b):

(1) The Bavarian Pfahl;
(2) the Saxon Sandstone Mountains;
(3) the Rammelsberg;
(4) the Harz-North-Rim–Fault with the Teufelsmauer;
(5) the island of Helgoland;
(6) Holzmaden;
(7) the Kaiserstuhl;
(8) the salt dome of Lieth near Elmshorn;
(9) salt mine Merkers;
(10) Messel;
(11) Nördlinger Ries & Steinheim basin;
(12) cliff coast of Rügen;
(13) Bohlen Wall near Saalfeld;
(14) mines of St Andreasberg on the Harz;
(15) Siebengebrige with the Orachenfels;
(16) lithographic limestone of Solnhofen-Eichstätt;
(17) Mount Wilsede in the Heath of Lüneburg.

Only the Rammelsberg together with the medieval town of Goslar is protected as a World Heritage Site (since 1992). Landscapes like the Nördlinger Ries, the Steinheim basin or the Kaiserstuhl do not have any status of protection in their totality. After the saving of Messel other disposal sites are planned, e.g. in the centre of the Nördlinger Ries meteorite crater.

The protection of volcanological sites in Germany

In Germany, volcanism of Tertiary and Quaternary age documents continental intra-plate volcanism. Starting in the Eocene the first volcanoes can be found in the Upper Rhine Valley (Kaiserstuhl), which has been extensively destroyed by building so-called large terraces for viniculture. During the Neogene many volcanoes developed at the crossings of main faults like the Siebengebirge. The volcanism of the Eifel famous for its maars is mostly of Quaternary age. In particular, the caldera of Lake Laach connects a Romanesque monastery with geotopes used by humankind for five thousand years for the extraction of millstones in a unique ensemble. Also, after Schmincke et al. (1990), Lake Laach is one of the classic sites for surge deposits, best visible at the Wingertsberg. In the next few years the authorities of the Eifel will establish a volcanic park.

A unique monument of Tertiary volcanism is the halite crystal cave of Merkers in Thuringia, situated in a potassium mine several hundred metres below the surface. The halite cubes are grown up to one metre in size. The cave and the large crystals have their origin in the contact between the hot basalt intrusions with salt (Schultheiss 1992). The mine is now developed as a tourist mine, but at this moment it can only work profitably in conjunction with the extraction of potash. Therefore, after the reunion of Germany and the merger of the potash companies the future of this geotope is very uncertain. The tourist mine needs more visitors and new ideas to survive the next decades. Merkers could serve as a European centre for the documentation of salt mining and of the geology of the 'Zechstein'.

References

GRUBE, A. & WIEDENBEIN, F. W. 1992. Geotopschutz. Eine wichtige Aufgabe der Geowissenschaften. *Geowiss,* **10**(8), 215–219.

LEIBNIZ, G. W. 1691. *Protogaea*. German translation by W. V. ENGELHARDT, (1949), Kohlhammer, Stuttgart.

LESER, H. 1991. *Landschaftsökologie, Ansatz, Modelle, Methodik, Anwendung* - Uni-Taschenbücher 551. Ulmer, Stuttgart.

ÖBO MITWITZ 1993. *Ökologische Bildungsstätte Oberfranken/Naturschutzzentrum Wasserschloß Mitwitz*. Geotopschutz. Materialien 1/93. Mitwitz. Ofr.

QUASTEN, H. (HRSG) 1993. *Geotopschutz. Probleme der Methodik und der praktischen Umsetzung.* - 1. Jahrestagung der AG Geotopschutz 15–17 April 1993, Otzenhausen/Saarland. Abstracts. University de Saarlandes, Saarbrücken.

SCHMINCKE, H.-U., BOGAARD, P. V. D., FREUNDT, A. 1990. *Quaternary Eifel Volcanism*. International Volcanology Congress Mainz 1990 (IAVCEI), excursion 1AI, Workshop on explosive volcanism: Mainz.

SCHULTHEISS, T. 1992. Das Besucherbergwerk Merkers. *Geol. Bl. NO-Bayern* **42** (1/2), 163–166. Erlangen.

STÜRM, B. 1992. Geotop. Grundzüge einer Begriffsentwicklung und Definition. *In*: WIEDENBEIN, F. W. & FRUBE, A. (eds) *Geotopschutz und Geowissenschaftlicher Naturschutz*. Workshop-Abstracts. 141. (University of Erlangen-Nuremberg), Erlangen.

WIEDENBEIN, F. W. 1992. Gründung einer deutschsprachigen "Arbetsgemeinschaft Geotopschutz" in Mitwitz/Oberfranken. *Geol. Bl. NO-Bayern* **42**(1/2), 147–152. Erlangen.

—— 1993a. Ein Geotopschutzkonzept für Deutschland. *In*: QUASTEN, H. (ed.) *Geotopschutz,*

Probleme der Methodik und der praktischen Umsetzung. 1. Jahrestagung der AG Geotopschutz, Otzenhausen/Saarland, 17. University de Saarlandes, Saarbrucken.
—— (ed.) 1993b. *Geological Heritage '93. Geotope protection for Europe.* (University of Erlangen-Nuremberg), Erlangen.
—— & GRUBE, A. (eds) 1992. *Geotopschutz und Geowissenschaftlicher Naturschutz.* Workshop-Abstracts. (University of Erlangen-Nuremberg, Erlangen.

Natural Areas: an holistic approach to conservation based on geology

KEITH DUFF

English Nature, Northminster House, Peterborough PE1 1UA, UK

Abstract: In Britain, the conservation of geological and geomorphological sites is an integral part of nature conservation, and we have a long history of successful site protection. In spite of this, the links between the Earth sciences and wildlife conservation have always been rather weak, with the two activities having little in common, and occasionally coming into conflict. In England, English Nature has recently been reviewing the whole approach to wildlife and geological conservation, and has developed a system which brings these areas much closer together.

The new approach recognizes the need to set objectives and targets for wildlife and Earth science conservation, and proposes that this is best done by having a series of objectives which reflect the variety of England's landscape and wildlife. Therefore, the concept of 'Natural Areas' has been developed, which turns this idea into practice. Natural Areas are tracts of land unified by their underlying landforms, rocks and soils, displaying characteristic natural vegetation types and wildlife species, and supporting broadly similar land uses and settlement patterns. Seventy-six Natural Areas have been identified in England. The close dependency of Natural Areas on the underlying geology provides a real opportunity to bring together the causes of wildlife conservation and geological conservation. Geological and geomorphological sites, as well as having their own intrinsic interest, also allow us to see and understand the foundations from which the soils and vegetation of the Natural Areas have sprung, and the operation of the natural processes which shape the land. Links of this kind provide many opportunities to create much closer working arrangements between conservationists and land managers.

The other great strength of the Natural Areas approach is that it provides a framework within which to set joint objectives with the people who live and work in the Natural Areas. There is an opportunity to build on the strong sense of place felt by most people, especially those who have a long association with a particular area. Through interpreting the landscape evolution of an area in terms of the influence which the geology and landforms have on it, there are real possibilities for raising people's understanding of the Earth sciences.

This paper considers how stronger and more effective links can be established between the Earth sciences and wildlife conservation in England, through development and application of an approach which is based firmly on the underlying geology and landforms, and describes the actions taken by English Nature to initiate this approach. The result will be closer integration of wildlife and Earth science conservation, and one which enables us to set clear objectives for nature conservation which are based on landscape areas readily recognized by the people who live and work in them.

Inheritance

English Nature is the Government agency responsible for nature conservation in England, and has been in existence (under various different titles) since 1949. Work has tended to focus on the safeguard of National Nature Reserves (NNRs), most of which are owned or managed by English Nature, and on the safeguard of Sites of Special Scientific Interest (SSSIs), all of which are owned and managed by others. There are about 150 NNRs and 3800 SSSIs in England; about 1300 of the SSSIs are of special interest by virtue of their geological or geomorphological interest. The SSSIs cover both biological and geological sites, and over the years English Nature has been generally effective in safeguarding them against damage caused by changes in land use. There are sound statutory consultation mechanisms, which require the impact of land use change on NNRs and SSSIs to be taken fully into account before any decision is made. The survival of many classic geological and geomorphological sites around our coasts, including localities such as Hunstanton Cliffs in Norfolk, Folkestone Cliffs in Kent, and Orford Ness in Suffolk, is testimony to the success of these efforts. Biological sites have also benefited from the operation of this process.

However, the links between Earth science conservation and wildlife conservation are generally rather weak, and there have been occasions when they have come into conflict. This has tended to be when geologists have sought to re-excavate exposures which had become overgrown, and where the vegetation which had developed now supported species or plant communities of conservation interest. Lack of common ground did not help the resolution of such conflicts.

Since English Nature came into being in April 1991 it has undertaken a fundamental review of its approach to wildlife and geological conservation, to analyse the strengths and weaknesses of previous activities, and determine how best to take conservation forward in the 1990s. The analysis indicated that site protection measures had worked reasonably well in controlling changes of land use which would be deleterious to conservation interests, but that there was strong evidence that the wildlife interest of many sites was declining as a result of lack of appropriate management of the land. Heathlands, with their characteristic animal and plant communities, are gradually being overwhelmed by scrub and woodland encroachment; limestone grasslands and neutral grasslands are suffering the same fate. And upland heather moorlands are degrading as a result of over-intensive sheep grazing. So, if the overall objective of maintaining and enhancing the characteristic plant and animal communities and natural features of England is to be achieved, more effective ways of achieving the positive management of special sites, and of the wider countryside in which they occur have to be found. Ways of developing a sustainable approach to the management of the English countryside must be found, and these need to be ways which enable English Nature to carry with them the people who own and share the management of the land. Since English Nature does not own or manage the SSSIs, this is especially important.

Shared objectives

English Nature cannot secure the necessary levels of positive management on its own. The way forward is through working much more closely with landowners, land managers and occupiers towards common goals and objectives which support the wildlife and natural features of England. English Nature want to build on the sense of stewardship felt by landowners, secure shared commitments, and promote positive incentives to encourage them to manage their land in ways which support wildlife and Earth science features. This goal is achievable only if the sense of stewardship felt by landowners becomes more widespread in society, and encourages people more generally to take action to support the conservation of their heritage. There are already a number of good examples of how shared local objectives can deliver effective conservation: the Wildlife Enhancement Scheme is working well in the Culm Measures of Devon, and the Species Recovery Programme is showing how it is possible to take positive action to increase populations of some of our rarest and most vulnerable species. But how can shared objectives be developed at a nationwide scale? English Nature believes that objectives are best set locally, to build up into the overall national picture; a broad strategic view of national objectives needs to be in place to influence this. To achieve this a new way of looking at the English countryside is proposed, one which breaks away from the artificial constraints of administrative boundaries such as counties and districts, which rarely mark natural divisions in land use, land form, or natural vegetation patterns.

To make it easier to develop a shared vision of the way ahead, the concept of Natural Areas has been developed. These relate to the strong sense of place felt by people, particularly about the area in which they live. Natural Areas are tracts of land unified by their underlying geology, landforms and soils, displaying characteristic natural vegetation types and wildlife species, and supporting broadly similar land uses and settlement patterns. They are also influenced by climate, altitude and aspect. They reflect the way that land use has developed over the centuries, which is itself largely driven by the underlying geology and form of the land. The continued operation of natural geomorphological processes has also influenced their development strongly. Taking these factors together English Nature has drawn up an overall map of 76 Natural Areas in England (Fig. 1); these are supplemented by a further 11 natural coastal cells which constitute self-contained sedimentary systems (English Nature 1993).

Comparison of the Natural Areas map with a geological map of England (Fig. 2) shows the close fit between geology and Natural Areas. This is closest in southern England, where bedrock geology has not been substantially modified by glacial activity, but elsewhere there is still a reasonable fit. In the uplands, the development of extensive areas of blanket peat has tended to obscure variations in the underlying geology, and elsewhere differences of aspect and land-form have influenced the development of land use patterns. For example, the generally flat-lying Cretaceous Chalk of the Salisbury Plain area in Wiltshire, with its characteristic development of limestone grasslands managed by sheep grazing, differs markedly from the Chalk escarpment of the Chilterns, where the steep slopes have been much less heavily grazed, and a mosaic of woodlands and commons has evolved.

The definition and characterization of Natural Areas requires a considerable input from the Earth sciences, and provides us with a major opportunity to bring the practices of wildlife conservation and geological conservation much more closely together. Earth scientists must ensure that the processes which drove the development of the characteristic land management activities are well understood, and that the nature of the rocks and soils which constrain the development of the vegetation is properly appreciated. Geological exposures provide opportunities to relate Earth science to wildlife distribution, and to understand how soils have developed. They are also important in their own right as critical elements of the character of the Natural Areas, and strategies and objectives for conservation within the Natural Areas must take full account of their intrinsic interest.

Examples

Much of southern England is underlain by Cretaceous Chalk, which tends to give rounded rolling landscapes, supporting a rich agricultural economy. Natural exposures are rare, except at the coast, and quarries provide most of the opportunities to examine the rock sequence. The richness of the soil, especially when supported by the high input agriculture of the past 30 years, has encouraged the gradual change from sheep grazing to arable farming, with loss of hedgerows and small copses. The Chalk outcrop is easily recognizable from the Natural Areas map. Chalk landscapes are also characterized by high quality surface water courses, rich in wildlife, and still used locally for watercress cultivation.

In contrast, the Carboniferous Limestone of northern England is much harder and forms extensive upland areas in Derbyshire, Lancashire and Yorkshire. The calcicole vegetation has similarities to that seen on the Chalk, but altitude, aspect and land use history have led to the development of a landscape which is highly distinctive. The extensive areas of karstic limestone pavement in Yorkshire are a unique record of late glacial events in England, and support a rare association of plants and animals.

The Lower Palaeozoic rocks of the Lake District give rise to a rugged and variable topography which is very strongly influenced by the underlying geology, and by the glacial history of this centre of ice dispersal. The generally acid nature of the rocks has encouraged the development of extensive heather moorland, although excessive sheep grazing is now leading to significant erosion of this important habitat.

The late Carboniferous sediments of the Dark Peak in Derbyshire support an acidic grassland community, with the wet grassland areas providing breeding habitat for wading birds. The alternating sandstones, shales and coaly beds produce a landscape which is strikingly different from the adjacent Carboniferous Limestone, and which gives rise to extensive areas of landslipping, as at Mam Tor and Alport. Again, the close parallel between the geological extent of these sediments and the land use patterns which characterize the area are clear and distinct.

The long narrow ridge of Precambrian rocks which forms the Malvern Hills carries a range of dominantly acid grasslands and heather moorland, but with local variations which reflect the complex intermixing of rock types which has been caused by faulting. They are very clearly distinct from the low-lying Triassic and Jurassic sedimentary areas of the Severn Vale and the Vale of Evesham to the east, and from the red sandstones and mudstones of the Herefordshire plain, formed by Devonian sediments, to the west.

In contrast, the low-lying marine Flandrian sediments of the East Anglian Fens support the richest agricultural land in England, and only a few very small fragments of semi-natural vegetation remain, as nature reserves, entirely surrounded by intensively cultivated land. Fen vegetation has all but disappeared, and the extensive drainage works, initiated by Vermuyden in the seventeenth century, have lowered water levels so much that the peat itself is wasting at a rate of about 4 m per century. This is clearly demonstrated at Holme Fen, near Peterborough. Water quality in the rivers is generally low, due to the high levels of sediment, pesticide and fertiliser run off from the heavily treated agricultural land.

In the south of England, the soft clays and sands of the Tertiaries give rise to extensive heathlands, especially in Dorset, Hampshire and Surrey. Here, loss of traditional grazing has led to the heaths becoming invaded by scrub and woodland, and the few natural exposures which were once visible in the banks of streams are being overwhelmed by woodland. The nature of the sediments is well seen in coastal cliffs in Hampshire and Dorset, where the rapid erosion

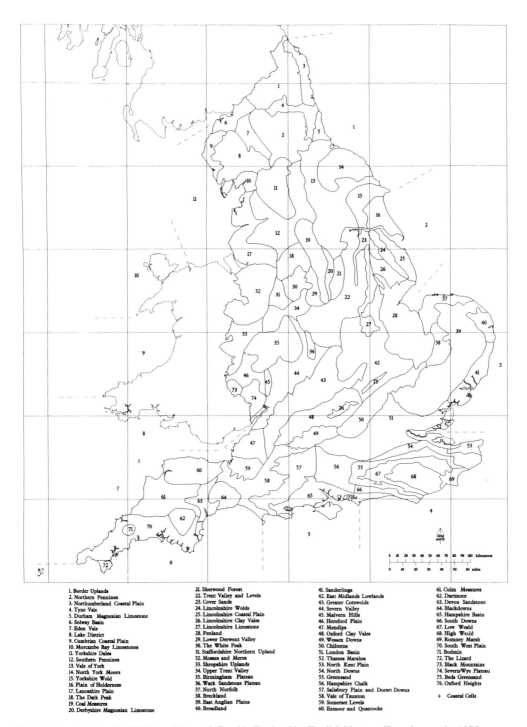

Fig. 1. Map showing the 76 Natural Areas defined in England by English Nature. (Based upon the 1975 Ordnance Survey 1 : 1 250 000 map with the permission of Her Majesty's Stationery Office. © Crown Copyright.)

1. Border Uplands
2. Northern Pennines
3. Northumberland Coastal Plain
4. Tyne Vale
5. Durham Magnesian Limestone
6. Solway Basin
7. Eden Vale
8. Lake District
9. Cumbrian Coastal Plain
10. Morcambe Bay Limestones
11. Yorkshire Dales
12. Southern Pennines
13. Vale of York
14. North York Moors
15. Yorkshire Wold
16. Plain of Holderness
17. Lancashire Plain
18. The Dark Peak
19. Coal Measures
20. Derbyshire Magnesian Limestone
21. Sherwood Forest
22. Trent Valley and Levels
23. Cover Sands
24. Lincolnshire Wolds
25. Lincolnshire Coastal Plain
26. Lincolnshire Clay Vales
27. Lincolnshire Limestone
28. Fenland
29. Lower Derwent Valley
30. The White Peak
31. Staffordshire Northern Upland
32. Mosses and Meres
33. Shropshire Uplands
34. Upper Trent Valley
35. Birmingham Plateau
36. Wark Sandstone Plateau
37. North Norfolk
38. Breckland
39. East Anglian Plains
40. Broadland
41. Sanderlings
42. East Midlands Lowlands
43. Greater Cotswolds
44. Severn Valley
45. Malvern Hills
46. Hereford Plain
47. Mendips
48. Oxford Clay Vales
49. Wessex Downs
50. Chilterns
51. London Basin
52. Thames Marshes
53. North Kent Plain
54. North Downs
55. Greensand
56. Hampshire Chalk
57. Salisbury Plain and Dorset Downs
58. Vale of Taunton
59. Somerset Levels
60. Exmoor and Quantocks
61. Culm Measures
62. Dartmoor
63. Devon Sandstone
64. Blackdowns
65. Hampshire Basin
66. South Downs
67. Low Weald
68. High Weald
69. Romney Marsh
70. South West Plain
71. Bodmin
72. The Lizard
73. Black Mountains
74. Severn/Wye Plateau
75. Beds Greensand
76. Oxford Heights

+ Coastal Cells

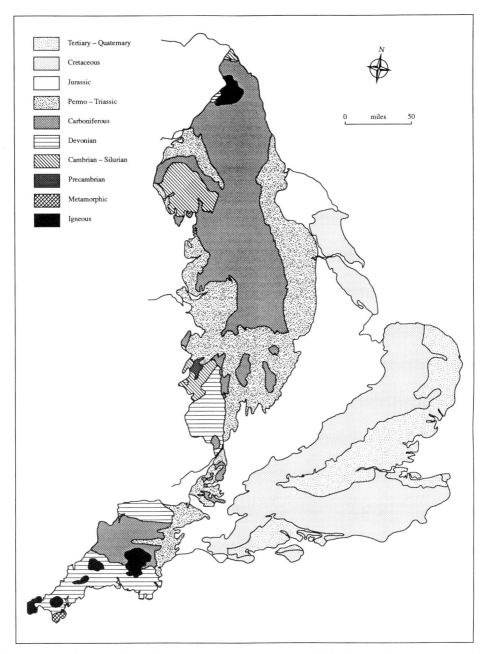

Fig. 2. Map showing the geology of England; note the close parallels with the Natural Areas in many parts of the country.

of the soft sediments by the sea is causing increased demands for sea defence works. Since several of the rock sections are of international importance as stratotypes, for example the Bartonian Stage of the Eocene, this is a matter of great concern to geologists.

In coastal conservation issues our strategy is to recognize the extent of the natural sedimentary systems which exist, and use them as the basis for coastal management and planning. In Britain there is an unfortunate history of piecemeal planning and engineering of the coast, which has resulted in a number of major engineering problems. English Nature believes, and the coastal engineering profession is increasingly taking the same view, that coastal planning should be dealt with in the context of the overall sedimentary cell; the Natural Areas approach accepts this as a fundamental principle, and recognizes 11 natural sedimentary cells around the coast of England and Wales. Eroding cliffs are a vital part of this system, since (as well as providing important exposures for geological study) they provide material which forms beaches within the systems. Once all the cliffs have been protected by engineering works, sea walls gradually become undermined as beach levels are lowered and not replenished naturally, and there is an unending cycle of reconstruction and maintenance.

Application

Natural Areas provide the framework for translating national priorities for nature conservation into local objectives and strategies, both for SSSIs and for the overall pattern of the wider environment within which they occur. The approach enables English Nature to look at the resource in an integrated way, and not just to focus on the special sites. Within each Natural Area working links will be set up with landowners and occupiers, local authorities, and the community, to work towards a shared understanding of the characteristics of the Area, and to establish jointly shared objectives for nature conservation, including geology and geomorphology. The use of a system of countryside classification which is firmly based on the broad character and form of the land should enable the development of shared ownership of plans and programmes, to which the people who live and work in the areas can easily relate. Our objectives in Natural Areas will be related to the distribution of characteristic wildlife and natural features, and will describe the 'Quality Profile' for each Natural Area. This will set objectives and targets for the high quality nature conservation resource of the area (principally comprising the SSSIs and NNRs), as well as setting broad objectives for what can be achieved at the less special level in the wider countryside as a whole. It will also set out options for what might be achieved by the enhancement of areas which have become degraded or fragmented.

Conclusions

In taking forward this approach, geological and geomorphological sites (as well as having their own intrinsic interest and importance) allow us to see and understand the foundations from which the soils and vegetation of Natural Areas have sprung, and also to understand better the operation of the natural processes which shape the land. Links like this provide many opportunities to create much closer working arrangements between geological conservationists and wildlife conservationists, and between conservationists and land managers. English Nature intends to use this process as a way of bringing geological and wildlife conservation much closer together in England, and is already encouraged by the positive reactions received from many partners, and from landowners. The future success of Earth science conservation depends upon the process becoming part of the mainstream of nature conservation, and the Natural Areas initiative is a very positive way of achieving this ambition.

Reference

ENGLISH NATURE 1993. *Natural Areas. Setting nature conservation objectives: a consultation paper.* English Nature, Peterborough.

Keynote address

Earth science assessments for heritage waterways: the French River and others in Canada

D. IAN McKENZIE

*Department of Geography and the Quaternary Sciences Institute,
University of Waterloo, Waterloo, Ontario, Canada N2L 3G1*

Abstract: Rarely, are Earth science features and processes effectively protected, conserved or managed successfully on a site-by-site basis without some supporting institutional framework. The Canadian Heritage Rivers System offers an example of a framework to provide a concatenation of Earth science features along natural lineaments and corridors. The success of protecting areas of Earth science significance lies not only with the merits of the feature itself, but the context in which the feature sits. As strange as it may sound to some geologists, an holistic, ecological approach to designation may be the most rewarding for preserving Earth science sites. Building on the strengths of the Earth science component, it is possible to superimpose other heritage values such as the life sciences, historical, cultural, and recreation and heritage appreciation, valued by residents and visitors.

Earth science features and processes can be protected in Canada in a number of ways. Components of National and Provincial legislation cover the protection of geological landscapes, for example the National Parks Act and, in Ontario, the Provincial Parks Act. Even the Canadian Wildlife Act can be used to protect Earth science features. If the feature or process cannot be designated a park, it may still find associate protection by providing habitats in coastal zones such as barrier bars and dune systems. In Ontario, on private lands and public lands not in parks, the Conservation or the Heritage Acts provide vehicles for protection (Davidson 1988). In 1978, in response to public concern that the natural resources of Canadian rivers be managed wisely, federal, provincial and territorial parks ministers asked their officials to prepare a joint proposal for a Canadian Heritage Rivers System (CHRS). A task force was set up with Parks Canada and the Province of Ontario, leading to a pilot study on the French River in 1979. Inventories were completed on the cultural, life science and Earth science components of the river (McKenzie 1979).

In 1985, a Provincial Waterway Park was created along the French River. This was necessary because some form of protection mechanism, usually a national, provincial or territorial park and suitable management plan must be in place before designation as a 'heritage river' can go ahead. In 1986, the 110 km length of the French River was designated as the first of, to date, 15 Canadian Heritage Rivers. Currently, 11 additional rivers have been nominated by the provinces and territories and expect designation over the next three years if suitable management plans are approved (Fig. 1, Table 1).

The CHRS has evolved from designating remote, prstine, 1 km-wide waterway corridors, to including the currently nominated Grand River (290 km) in populated, southern Ontario. Nomination of the Grand River is based on the human heritage and recreational component of the river. None the less, features of Earth science significance do exist along the Grand River and its tributaries, for example, the largest glacial pothole by volume according to the *Guinness Book of World Records!* Nodes of human heritage significance have been recognized and can be tied to, where possible, sites of Earth science significance.

Earth science features and processes have played a major role in developing the setting for these spectacular rivers. Formal designation as a heritage waterway serves several Earth science conservation functions. Designation provides the opportunity for Earth science features and processes to be inventoried and classified as representative of a geological region. The features and processes are evaluated as having a degree of significance at a local, regional, provincial, territorial or national level. The Earth science features and process can then

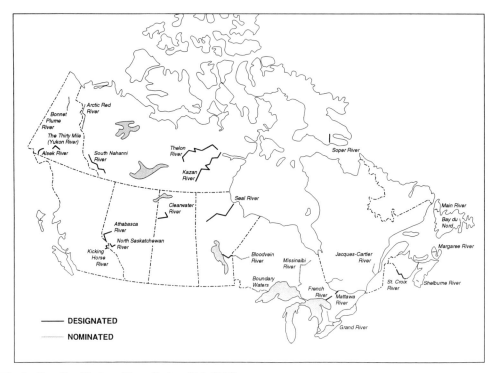

Fig. 1. Canadian Heritage Rivers System (July 1993).

provide educational and scientific opportunities, protected and managed under existing parks framework.

The Canadian Heritage Rivers System (CHRS)

The CHRS provides a national level of recognition for Earth science features and processes along waterways. Without this process many outstanding Earth science examples would not have been protected. No formal limit has been imposed on the number of rivers designated under the Canadian Heritage Rivers System. The purpose is not to provide for the conservation of all rivers of interest, importance or value, but to select only the outstanding of these from a Canadian viewpoint.

Based on guidelines proposed by Parks Canada (1984), a waterway can be selected on outstanding natural heritage values if the river environment meets one or more of the following guidelines:

(1) It is an outstanding example of river environments as they are affected by major stages and processes in the Earth's evolutionary history which are represented in Canada. This would include rivers which best represent the major periods of geological time in which the surface of the Earth underwent major changes and stream modification.

(2) It is an outstanding representation of significant ongoing fluvial, geomorphological and biological processes. As distinct from periods of the Earth's development, this focuses upon ongoing processes in the evolution and form of the river and its associated plant and animal communities.

(3) It contains along its course, habitats with rare or outstanding examples of natural phenomena, formations or features, or areas of exceptional natural beauty.

(4) It contains along its course habitats of rare or endangered species of plants and animals. This would also include areas where outstanding concentrations of plants and animals of Canadian interest and significance are found.

The guidelines for heritage rivers also include a description of 'human and recreational' values (Parks Canada, 1984).

Table 1.

River	Province/Territory	Designation date*	Length (km)
Designated Rivers			
French	Ontario (French River Provincial Park)	Feb. 1986	110
Alsek	Yukon (Kluane National Park)	Feb. 1986	90
Clearwater	Saskatchewan (Clearwater Provincial Park)	Jun. 1987	187
S. Nahanni	Northwest Territories (Nahanni National Park)	Jan. 1987	300
Bloodvein†	Manitoba (Atikaki Provincial Park)	Jun. 1987	200
Mattawa	Ontario (Mattawa River and Samuel de Champlain Park)	Jan. 1988	33
Athabasca‡	Alberta (Jasper National Park)	Jan. 1989	168
N. Saskatchewan‡	Alberta (Banff National Park)	Jan. 1989	49
Kicking Horse‡	British Columbia (Yoho National Park)	Jan. 1989	67
Kazan	Northwest Territories	Jul. 1990	615
Thelon	Northwest Territories	Jul. 1990	545
St. Croix	New Brunswick	Jan. 1991	185
Yukon (30 mile)	Yukon	Jan. 1991	48
Seal	Manitoba	Jun. 1992	260
Soper§	Northwest Territories	Jun. 1992	108
		Total	**2965**
Nominated Rivers			
Arctic Red	Northwest Territories	Sept. 1993*	450
Missinaibi	Ontario (Missinaibi Provincial Park)	Jun. 1994*	426
Jacques-Cartier‡	Québec (Jacques-Cartier Provincial Park)	Jan. 1994*	128
Grand§	Ontario	Jan. 1994*	290
Main	Newfoundland	Jan. 1994*	57
Margaree	Nova Scotia	Jun. 1994*	120
Bloodvein†	Ontario (Woodland Caribou Provincial Park)	Jun. 1995*	106
Bay du Nord	Newfoundland (Bay du Nord Wilderness Reserve)	Jun. 1995*	75
Boundary Waters	Ontario (La Verendrte/Quetico Provincial Park/Middle Falls)		250
Bonnet Plume	Yukon	Jan. 1996	350
Shelbourne	Nova Scotia	Jan. 1996	53
		Total	**2305**

* Anticipated designation date.
† Bloodvein River has been nominated in two sections, by Manitoba and Ontario.
‡ Only upper section of the river is located in the Park.
§ Length of main stem of river, excluding nominated tributaries.

The Earth science component forms the basis of the waterways' locations. The obvious is often overlooked by conservation managers. The evolution of the life science habitats is the direct product of geography. The French River, a late eighteenth century major west–east fur trading route on the southern edge of the Canadian Shield, owes its origin to faulting and pre-glacial river valley and delta. The Grand River in southern Ontario can be considered a natural corridor of time, another example of a waterway exposing and linking Palaeozoic bedrock, the first glacial meltwater channels and contemporary deposits associated with different periods in the geological development of Ontario.

The coordinating body for the CHRS is the Canadian Heritage Rivers Board, made up of members from eight provincial governments, two territory governments and the federal government. Although private citizens or groups may make recommendations to the government agency responsible for the river, nominations are made to the Canadian Heritage Rivers Board only by the participating governments. The Canadian Heritage Rivers Board reviews nominations and makes recommendation to the federal Environment Minister and to the provincial or territorial minister of the nominating government that the river meets selection criteria.

The role that public comment plays in the designation of the river is considered an important part of the procedure. This is done prior to the nomination of the river, in part to publicize the support that the river will receive. Public consultation ensures that all stakeholders concerned, from native peoples, natural and scientific interests to recreation, agriculture, forestry and mining are able to comment on the worth of the river as a Canadian Heritage River. Government officials then select a list of preferred river candidates within their jurisdiction. Input from all stakeholders is solicited because once designated, management guidelines will restrict future development or, in the case of mining, remove the potential for this activity. The number and complexity of conflicting land uses are also considered, as well as the cost of effective management. This may be contrary to the well-being of the significance of the Earth science feature. This approach may also meet a political agenda where lack of support may be related to resource use concerns. Presumably this does not imply that the complexity of the issue would prevent a river from being nominated.

The river designation process

The river designation process begins when the Board Mamber representing the nominating government agency submits a management plan to the Canadian Heritage Rivers Boards. The management plan sets out policies and practices for the government agencies reponsible to ensure development and management are consistent with objectives and guidelines of the Canadian Heritage Rivers Board. To allow time for an appropriate strategy to be developed, a management plan is presented to the board within three or four years of the nomination of the river.

After the management plans have been reviewed and accepted, the Chair of the Canadian Heritage Rivers Board advises the federal Minister of the Environment and the Ministry of the nominating agency that the requirements for designation have been met. At a key riverside location, a plaque is unveiled and a ceremonial text for the river is included into the CHRS registry. The value in formal designation ensures that at a national scale the heritage value of the river, of which the Earth science component is a part, will be managed. Responsibility of ongoing management lies with the province, territory or federal government agency responsible for jurisdiction. At this stage, it is important that the Earth science expert be re-introduced to the designation process to ensure that the Earth science values are maintained and monitored.

As part of the responsibility for management, a plan review process is required within 10 years of designation. In 1992, a plan review was called by the Ontario Parks and Natural Heritage Policy Branch for the French River (designated in 1986). Accepted in February 1993, the revised plan (Ontario Ministry of Natural Resources, 1993) reconfirmed with the public that park management and development procedures reflect current conditions for the river.

Monitoring of designated rivers

One of the key components to long-term management is a programme to monitor the feature or process designated. Within a year of the official designation of the river a list of values, including the Earth science components to be monitored, is presented to the Canadian Heritage Rivers Board by the managing government. At present, the monitoring consists largely of the dynamic life science qualities such as flora and fauna or water quality of the river. Guidelines in many protection statements do not specifically deal with the Earth sciences. Many of the Earth science features or processes within a river corridor may be seen as self-sustaining. However, this is not necessarily the case. Glacial features on a limestone surface, for example, may experience accelerated erosion due to changing atmospheric conditions.

The responsibility for monitoring Earth science features of special interest must be specifically outlined in the management plan. Guidelines for monitoring of these features needs more work. The CHRS is no exception. An assessment of the management plans for Canadian Heritage Rivers indicates that specific long-term Earth science planning could be improved. The Canadian Heritage Rivers Board requires, as part of the annual monitoring, completion of a check list of the changes in conditions of the river values. Under the natural heritage component, this includes changes in conditions both natural or human induced, to habitat alteration, wildlife, physiographic features and other natural features (Secretariat, Canadian Heritage Rivers Board 1991).

Conclusion

Earth scientists traditionally have not been resource managers. However, on the other hand, many resource managers will, on the recommendation of the Earth scientist, designate a site and then assume the feature or process to be static, not requiring management. If the feature is thousands of years old, what needs to be done to protect it other than watching to see that someone does not quarry it away? It is relatively easy to list examples of sites that only exist in books or museums or on maps and photographs but not *in situ*. The CHRS affords protection of Earth science landscapes in conjunction with other cultural and life science heritage values. It is the holistic, ecological approach that allows the sum of the parts to be greater than the whole. It makes sense from a management point of view and a political point of view to have sites protected based on a complement of strengths. The CHRS presents a framework for protecting Earth science sites. Perhaps it is the mystique of the river itself that sews all the Earth science components together to produce an ideal combination for protection.

Time (human not geological) has changed the value of the world's natural systems. The Earth sciences are a basic component of these natural systems that need to be managed for science, education, and nature appreciation. In many cases these Earth science features and processes were discovered and inventoried as part of the economic use of the resource (Davidson 1990). No doubt, many national, international and world class geological sites remain to be discovered and interpreted, often in part due to development.

Sustainable resource use and protection may require a compromise, depending on the value of the resource. This value can be geological or economic and can determine the net worth of the geological resource.

Canada, like many countries, can represent its natural and cultural history through its waterways and rivers. Beyond the national designation an opportunity exists to develop a convention at the international and global level to protect world class examples of Earth science processes and features. The strengths of developing a global convention are two fold: (1) scientists can gain the attention of world governments by demonstrating a conservation need through an holistic, ecological approach; and (2) simply, it is politically correct to support national issues that are recognized by international convention. Directions for conservation have been set at all levels in Canada. Among others, there are the Federal Government's *Green Plan*, The Canadian Environmental Advisory Council's *Protected Areas Vision for Canada*, the World Wildlife Fund Canada's campaign *Endangered Spaces/The Future of Canada's Wilderness* and the Ontario *Endangered Spaces Action Plan* (Ontario Ministry of Natural Resources 1992). The Earth sciences in general, and more specifically along corridors such as heritage waterways, will continue as important components of Canada's future.

References

DAVIDSON, R. J. 1988. *A Strategy for the Conservation of Ontario's Earth Science Heritage. Class Environmental Assessment for Timber Management on Crown Lands in Ontario*. Statement of Evidence of Panel VII. Environmental Affect. The Forest Management Unit and Resource Planning, Queen's Printer for Ontario, Toronto, Volume 2, 493–510.

—— 1990. Protecting and Managing Great Lakes Coastal Dunes in Ontario. *In: Proceedings of the Canadian Symposium on Coastal Sand Dunes, 1990*. Associate Committee on Shorelines, Natural Research Council, Ottawa, 455–71.

MCKENZIE, D. I. 1979. *French River Canadian Heritage Pilot Study, Phase 1 Assessment of Earth Science Processes and Features*. Parks Canada, Ottawa.

ONTARIO MINISTRY OF NATURAL RESOURCES 1992. *Endangered Spaces Action Plan, Protecting Ontario's Natural Heritage Areas*. Ontario Provincial Parks, Toronto.

—— 1993. *French River Provincial Park Management Plan*. Queens Printer for Ontario.

PARKS CANADA 1984. *The Canadian Heritage Rivers System: Objectives, Principles and Procedures*. Ministry of Supply and Services Canada, Ottawa.

SECRETARIAT CANADIAN HERITAGE RIVERS BOARD 1991. *Canadian Heritage Rivers System Guidelines*. Canadian Parks Service, Ottawa.

The challenges of geomorphological river system conservation for the 1990s

LINDSEY McEWEN

Department of Geography and Geology, Cheltenham and Gloucester College of Higher Education, Cheltenham, Gloucestershire GL50 3PP, UK

Abstract: The effective conservation of geomorphological sites and landform heritage for research, education and aesthetic reasons is an important area of Earth science conservation in the United Kingdom. This paper addresses the key issues and challenges facing geomorphological conservation of river systems in the 1990s at a national, regional and local level. The relative merits of different models, such as the National Parks, Sites of Special Scientific Interest (SSSIs) and Regionally Important Geological/geomorphological Sites (RIGS), for conserving fluvial landforms will be evaluated and contrasted. The scale hierarchy of conservation units needs to be matched to drainage basin functions and be sensitive to a number of key issues. For example, there are spatial and temporal disparities in river system sensitivity to artificially-induced changes at channel, catchment, regional and national levels. Other issues to be evaluated include the relationship between pure fluvial research and conservation policy and practice, the potential conflicts and compromises between site conservation and management strategies and the status of geomorphological conservation in integrated river basin management. The identified regional, national and international hierarchy of sites must also satisfy the requirements of academic researchers, educationalists and the general public.

The effective conservation of geomorphological sites and landform heritage for research, education and aesthetic reasons is an important area of Earth science conservation (Gordon 1987). Few papers, however, focus on an assessment of the problems of applying conservation principles to dynamic river environments from a geomorphological perspective (e.g. Gordon & Campbell 1991; Werritty & Brazier 1991). Currently within the UK, a hierarchy of geomorphological sites is being selected in line with the former Nature Conservancy Council's (NNC 1990) document *Earth science conservation in Britain: A strategy*. The scale hierarchy of conservation models or units encompasses National Parks, Sites of Special Scientific Interest (SSSIs), Regionally Important Geological/geomorphological Sites (RIGS) and Recorded Sites. This paper looks specifically at river systems from a geomorphological perspective to assess what characteristics make them geomorphically distinctive. It then evaluates the relative strengths and weaknesses of the different models in Earth science conservation. The problems of applying more general conservation principles to dynamic river environments are also addressed. Such sites include fluvially-derived landforms as well as integrated landscapes, which include river catchments. This paper highlights some of the challenges for river conservation from a geomorphic perspective, including problems associated with the scale of conservation units relative to drainage basin functions and the nature of potential human-induced changes in channel controls. Other issues include the relationship between research findings and the development of sound conservation policy and practice, the possible conflicts between conservation and management and the status of geomorphological conservation in integrated basin management. The extent to which sites and site documentation meet the requirements of the spectrum of potential site users and heighten geomorphological conservation awareness is also considered.

It is first necessary to assess how river basins function as distinctive geomorphic systems. For conservation purposes, the fluvial system can be subdivided into the drainage basin, which is dominated by soil through flow and the stream system, comprising the mainstream and contributing channels containing open channel flow (Gardiner 1991). Any boundaries are, however, artificially imposed. River systems provide a particular suite of challenges for geomorphological conservation. Despite the diversity of process-form relationships represented in upland and lowland catchments, the UK possesses a relatively small range of river types when viewed on a global scale (Ferguson 1981; Werritty & McEwen, in press). Sites tend to be of regional and national, rather than international, significance.

Nevertheless, there are distinct spatial and temporal variations in fluvial activity, depending on local and regional environmental controls, such as glacial or periglacial legacy. Sites range from the extensive gravel-bed braided environments of the lower River Spey to the small-scale tortuous misfit meanders in the headwaters of the River Thames basin. The geomorphic impact of floods and the stream powers required for channel change can be very different depending on the nature of sediment sources, their calibre, availability and critical thresholds for entrainment. There are, therefore, variations in landform sensitivity and robustness to externally-induced changes. Both extreme and abrupt alterations to discharge and sediment supply as well as changes to average regime conditions may potentially have a dramatic impact on channel form, depending on thresholds for change. This can lead to changes in channel form at different scales, including altered rates of channel migration and hydraulic adjustment (Gardiner 1991).

Within the catchment system, if one part is disrupted in terms of sediment availability or magnitude and frequency of the discharge regime, there are potential ramifications for channel controls in other parts of the same basin. For example, evaluation of sediment yield and the residence times of sediment in different parts of the same drainage basin necessitates assessment of sediment transfer rates at a catchment scale. The temptation is to conserve by drawing a boundary around the landform rather than maintaining the function of 'natural processes' and associated process-response systems. It is also critical to match the level and scale of conservation control with the geomorphic sensitivity of the river site. A brief critique of the different conservation models or units used in the UK is provided in the following sections.

Conservation units

(a) National Parks

National Parks were set up under the National Parks and Access to the Countryside Act, 1949 with a remit to preserve and enhance their natural beauty and to promote public enjoyment of the parks (Stedman 1993). Within the National Park concept, sites of specific geomorphological interest may be encompassed within broader areas of high conservation value. This large-scale conservation unit aims to protect, maintain and enhance the entire natural system with its landforms. The National Park model is recognized internationally. Most World Heritage Sites (International Union for the Conservation of Nature and Natural Resources; IUCN) with 'superlative natural phenomena' of geomorphological interest possess National Park status. Although within other countries such as the USA, National Parks are frequently classified as landscape complexes unaltered by human intervention, in the UK, few river systems could be categorized in this way. National Parks in the UK have the bigger problem of 'integrating the conservation of the living landscape with economic and social life in the wider countryside' (MacEwen & MacEwen 1983, p. 394). MacEwen & MacEwen (1983) point out that SSSIs in National Parks may in fact be at greater risk from human intervention recreational pressure than SSSIs elsewhere.

(b) Sites of Special Scientific Interest (SSSIs)

The NCC Earth Science Conservation Strategy (1990) aims to maintain the SSSI network. SSSIs are distinguished from RIGS by their high national or international research status as opposed to principal educational value. Issues which need addressing in the selection of river systems as SSSIs for their geomorphological interest are discussed in detail elsewhere (McEwen 1992). Considerations include the relative status of rare or unique sites against those which are typical and the differing conservation requirements for those sites which are fossil against those that are currently active. The conservation value of 'wilderness' sites needs to be compared with those which have been exposed to varying degrees of human intervention.

The logistics of defending the number of Potentially Damaging Operations (PDOs) under the Wildlife and Countryside Act 1981 and the national planning system means that site boundaries are normally kept to a minimum buffer distance from the landform of interest, and whole catchments are rarely designated unless small or containing integrated landform assemblages with conservation value.

(c) Regionally Important Geological/ geomorphological Sites (RIGS)

RIGS are sites considered worthy of protection for their educational, research, historical or aesthetic importance (NNC 1990). Research interest is, however, of lower status than for

SSSIs. All other criteria being equal, ease of site access may be a further consideration. RIGS have also been identified as useful facilitators for the wider dissemination of Earth science conservation principles. The expansion of the RIGS network was one of the key elements in the NCC (1990) Earth Science Conservation Strategy. RIGS have the potential to act as pressure sites, so relieving potential damage to SSSIs with higher-ranking research status. While the principle of expanding a network of sites of regional note targeted for educational use is sound, the temptation is to designate isolated and small landform units, frequently inappropriate for effective river conservation although this effectiveness depends on landform sensitivity to change. In addition, the methods used to classify, select and document geomorphological sites tend to be developed on a county-by-county basis rather than being presented as a nationally consistent initiative with good practice established and recognized. If terms of reference differ, it will be difficult eventually to place regionally important sites in a national context. The ultimate aim of the RIGS scheme should be a national online database of sites where information can be extracted on the basis of a number of select criteria. It should, however, be noted that the existence of a research base can add significant value to a site and there may be the possibility of upgrade from RIGS to SSSIs with future research developments.

Challenges of river conservation into the 1990s

Each of the conservation units outlined above has a distinctive function. The scale and functionality of the unit must be matched to the aims and objectives of Earth science conservation as well as the requirements of past, present and future research initiatives which investigate the nature of process-response in fluvial systems. The status of conservation in integrated river basin management and the SSSI and RIGS site documentation requirements to support research and education in Earth science conservation are also priority concerns.

Scale of unit versus conservation objectives

It is essential to match the conservation unit with the geomorphological conservation merits and requirements of a specific river site. The traditional approach associated with SSSIs and RIGS leads to the identification of highly site-specific areas of designation. A number of points support the adoption of a larger-scale approach. The residence times of sediment and water the their rates and modes of transfer are best considered at the catchment level. Pure fluvial research has undergone a shift in resolution from understanding the functioning of individual channel segments to the operation of sediment and water flows at catchment scale in time and space.

With the current emphasis on drainage basins as integrated management units for the sustainable development of land and water resources, it makes sense to attempt to implement conservation policies for river systems at this larger scale. Newson (1992a), when addressing water quality issues, has already indicated the inadequacy of delimiting areas of ecological interest and prohibiting specified PDOs, while external influences on water quality from the catchment to the site are not monitored or controlled. Similar concerns must apply to sites of geomorphological interest, which will respond to changes in sediment supply and runoff regime from areas outwith SSSI or RIGS boundaries. Similar problems should not occur if the whole catchment is situated within a National Park.

Research linked to conservation policy and practice

Research into the dynamics of river systems should determine the framework necessary for effective conservation policy and practice. In an ideal world, policy should be led by the research outcomes from pure and applied fluvial research. It is well recognized that reality differs. A number of research foci can, however, be identified which should inform river conservation policy and practice.

(i) Assessment of river channel process-form relationships Initially, there needs to be nationally consistent and appropriate categorization of both river channels and floodplain topography so that unique and typical examples can be identified with their relative geomorphological conservation status (see McEwen 1992). River systems in the UK, therefore, need to be systematically classified at the macroscale to record their planform characteristics, floodplain and valley topography, major fluvial controls and their degree of 'naturalness' (Warren 1993).

(ii) Catchment response to changing controls Research has focused on the assessment of channel response to changing external controls

at catchment level and the transfer of results from small-scale monitored catchments. The limitations of such extrapolations in different environmental settings are well recognized. There are spatial and temporal variations in the sensitivity of river systems to changes in external controls including differences in sensitivity to systems changes within different landform units in the same river system.

There may be differences in system response to a gradual adjustment in controls in contrast to an abrupt external trigger. Alterations to the land–water relationships can occur abruptly; human-induced land-use change with potential impact on river system dynamics may be due to, for example, the extensive afforestation and associated upland artificial drainage of Scotland and Wales (Newson 1990). Even sites which are categorized as 'natural' may have undergone drastic land-use changes, such as major deforestation, in recent centuries. In contrast, changing patterns of flood frequency associated with increased variability of climate provides an example of more gradual alteration to controls, which need to be considered when discussing scenarios for future change in river systems (Newson & Lewin 1991). Indirect human-induced change in fluvial controls may be at a national and international level rather than regional or local; at the larger scales, control is very much more difficult.

A sensitivity assessment should be an essential element in the evaluation of site conservation potential as different landform units can be expected to have different relaxation times after application of an external stress. The nature of complex or lagged response to system disruption or change needs to be understood, as well as how such response is reflected in past, present and future form. It is an important question as to how destabilized river systems can be effectively managed in a sensitive way without reducing the insight they might provide into the dynamics and lagged response of active river systems. Not only fluvial systems need to be considered when determining the nature of potential impacts of change, for example, in situations where there is close coupling between slope and channel systems.

(iii) Channel recovery A further area where research outcomes can be directly input into conservation assessment is in channel recovery or recreation of 'past naturalness'. With the legacy of channel training on British rivers, it is frequently necessary to assess the conservation potential of sites which have already been artificially altered, dredged or straightened, associated with land drainage and flood control. The legacy of channel training on British rivers, particularly on lowland reaches, is extensive. Research on both the extent of natural river recovery in response to human-induced change and the success of measures which artificially encourage readoption of more natural planforms is of key importance.

Such naturalization of rivers needs to address the different spatial scales of channel adjustment from the planform level (e.g. channel realignments, the reinstatement of bends to increase sinuosity; Brookes 1992) down to the recreation of bedform units (e.g. redevelopment of point bars and pool/riffle sequences). The potential of rivers for site restoration should be placed alongside the value of conservation for existing sites of regional or national note. Restoration of more natural channel dynamics has accompanying benefits for the ecological functions of both channel and floodplain (see Petts *et al.* 1992); for example, pool and riffle sequences can encourage biological diversity.

Conservation versus management?

Conservation traditionally means the prevention of all human intervention in the natural process system, although some degree of compromise might have been necessary when determining the list of PDOs for a specific SSSI. The terms 'management' and 'conservation' have frequently been considered as the opposite ends of a continuum, with a *laissez-faire* approach to conservation pitched against the more traditional structural forms of management (see Brookes 1988). 'River management works are generally incompatible with the effective conservation of active process fluvial sites' (NCC 1990; 2.9.7). Conservation has been equated with 'preservation'; already recognized as an undesirable task when dealing with active river systems. Newson (1992a, p. 52), however, notes 'Preservation of river systems is impossible because of their dynamism; however, conservation of a few wilderness rivers is equally misguided'. It is important to note that Boon's (1992) range of five management options along a spectrum of decreasing conservation value (from preservation through limitation, mitigation, restoration to dereliction) apply as much to geomorphological as to ecological conservation.

There is considerable debate as to the extent to which conservation measures can be combined with river management at channel, river corridor, valley and catchment scales. The circumstances where compatibility rather than compromise can occur needs to be determined.

Can sustainable development of catchments incorporate sound conservation practice? 'Land and water are a hydrological continuum and must be managed together' (Newson 1992b, p. 385). With the current emphasis on more integrated approaches to land and water management, it is necessary to evaluate the conservation potential of landscape assemblages at higher levels of resolution than specific SSSIs and RIGS. Basin management strategies which incorporate the trial and testing of new conservation techniques need to be developed (in a similar way to the introduction of buffer strips for dealing with nitrate-sensitive areas). It is now recognized as insufficient to only conserve selected river corridors in remoter, more 'natural' catchments; there needs to be planning control for contributing catchment areas as well. Alternative strategies might incorporate some sort of conservation land-use zoning with those activities least likely to alter sediment supply or discharge regime being located closest to the floodplain and in the most sensitive contributing catchment areas. In addition, where direct channel management is unavoidable, more environmentally sensitive biogeomorphological methods can be used or other measures which allow natural river dynamics within a controlled active area (see Werritty & Brazier 1991).

To make informed management decisions which are conservation sensitive, it is necessary to monitor and disseminate information about the success of different conservation practices, independent of whether the site is an SSSI or a RIGS. The National Rivers Authority (NRA) has responsibilities under the Water Act of 1989 to carry out all its operational and regulatory functions to 'further the conservation and enhancement of natural beauty and the conservation of flora and fauna and geological or physiographic features of special interest' (Heaton 1993, p. 301). NRA river corridor surveys now include an assessment of geomorphological features associated with both channel and floodplain.

Communicating the geomorphic significance of river sites

The improvement of site documentation and the increase in public awareness of conservation issues are key agenda items in the NCCs (1990) Earth Science Conservation Strategy. Geomorphological conservation badly needs to increase its public profile. In contrast, water quality issues, for example, possess a high national and local status. Similar, attention needs to be given to the potential impact of artificial disruption to natural river systems from a geomorphological perspective as well as ecological ramifications.

One of the benefits of the RIGS scheme is that it allows conservation education to be implemented at the local level and capitalizes on generally increased levels of environmental awareness. Conservation awareness can then be extended under control to other sites of national and international importance. To support this initiative, there is a need for up-to-date, accessible, information sources to accompany more formal site documentation for SSSIs and RIGS.

There are a number of practices which could be adopted in Britain, including site information boards, information sheets or multimedia information packs to accompany site visits. For example, Goudie & Gardiner (1988) place syntheses about regionally interesting geomorphological sites in an easily accessible format. This format could be utilized on a regional basis for RIGS support documentation. One of the problems associated with SSSIs is that such information might draw enhanced attention leading to unsympathetic use of vulnerable or monitored sites. The general public does, however, need to know why sites are valuable and what constituted Earth science conservation interest. The documentation to support the SSSI network Geological Conservation Review river volume is awaiting publication.

Conclusions

It is increasingly evident that preservation of river systems in remoter areas by site designation is not the most effective or appropriate means of river conservation for the range of river contexts found in the UK in the 1990s. For larger catchments, affected by human land-use changes and with more typical river sites of conservation interest, integrated planning for the entire catchment is necessary (Fisher 1993, p. 106). The most successful management for conservation will be where the whole catchment is included within a conservation area and where catchment conservation is integrated with ecological assessment and land and water management. SSSIs, particularly, need to be given a wider planning context if effective conservation is to be assured. At all levels, conservation can benefit from geomorphic research into the sensitivity of river systems to changes in controls at channel and catchment levels. When dealing with dynamic river systems, there is a need for compatible conservation and management measures which allow flexibility of response to spatially and temporally variable levels of robustness for system change. The establishment and dissemination of good management practice, which is also conservation sensitive, is essential. Finally, the provision of sound, accessible documentation to increase Earth science conservation awareness amongst distinct interest groups is important if the status of the different conservation units is to be more widely recognized.

References

BOON, P. J. 1992. Essential elements in the case for river conservation. *In*: BOON, P. J., CALOW, P. & PETTS, G. E. (eds) *River conservation and management*. Wiley, Chichester, 11–34.

BROOKES, A. 1988. *Channelised rivers: perspectives for environmental management*. Wiley, Chichester.

—— 1992. Recovery and restoration of some engineered British river channels. *In*: BOON, P. J., CALOW, P. & PETTS, G. E. (eds) *River conservation and management*. Wiley, Chichester, 337–353.

FERGUSON, R. I. 1981. Channel forms and channel changes. *In*: LEWIN, J. (ed.) *British rivers*. George Allen and Unwin, London, 90–125.

FISHER, R. 1993. Biological aspects of the conservation of wetlands. *In*; GOLDSMITH, F. B. & WARREN, A. (eds) *Conservation in progress*. Wiley and Sons, Chichester, 97–113.

GARDINER, J. L. 1991. *River projects and conservation*. Wiley, Chichester.

GORDON, J. E. 1987. Conservation of geomorphological sites in Britain. *In*: GARDINER, V. (ed.) *International geomorphology*. Wiley, Chichester, 583–591.

—— & CAMPBELL, S. 1991. Conserving Britain's river landforms. *Earth Science Conservation*, **29**, 10–19.

GOUDIE, A. & GARDNER, R. 1985. *Discovering Landscape in England and Wales*. Allen & Unwin, London.

HEATON, A. 1993. Conservation and the National Rivers Authority. *In:* GOLDSMITH, F. B. & WARREN, A. (eds) *Conservation in progress*. Wiley, Chichester, 301–320.

MACEWEN, A. & MACEWEN, M. 1983). National Parks: A cosmetic conservation system. *In*: WARREN, A. & GOLDSMITH, F. B. (eds) *Conservation in practice*. Wiley, Chichester, 391–409.

MCEWEN, L. J. 1992. Site assessment criteria for the conservation of fluvial systems: The Scottish experience. *In*: STEVENS, C. (ed.) *Conserving our landscape: Evolving landforms and ice-age heritage*. English Nature, Peterborough.

NCC 1990. *Earth science conservation in Britain. A strategy*. Nature Conservancy Council, Peterborough.

NEWSON, M. 1990. Forestry and water: 'Good practice' and UK catchment policy. *Land Use Policy*, **1**, 53–58.

—— 1992a. *Land, water and development; River basin systems and their sustainable management*. Routledge, London.

—— 1992b. River conservation and catchment management: a UK perspective. *In*: BOON, P. J., CALOW, P. & PETTS, G. E. (eds) *River conservation and management*. Wiley, Chichester, 385–396.

—— & LEWIN, J. 1991. Climatic change, river flow extremes and fluvial erosion – scenarios for England and Wales. *Progress in Physical Geography*, **15**, 1–17.

PETTS, G. E., LARGE, A. R. G., GREENWOOD, M. T. & BICKERTON, M. A. 1992. Floodplain assessment for restoration and conservation: Linking hydrogeomorphology and ecology. *In*: CARLING, P. A. & PETTS, G. E. (eds) *Lowland floodplain rivers: Geomorphological perspectives*. Wiley, Chichester.

STEDMAN, N. 1993. Conservation in National Parks. *In*: GOLDSMITH, F. B. & WARREN, A. (eds) *Conservation in progress*. Wiley, Chichester, 209–239.

WARREN, A. 1993. Naturalness: A Geomorphological approach. *In*: GOLDSMITH, F. B. & WARREN, A. (eds) *Conservation in progress*. Wiley, Chichester, 15–24.

WERRITTY, A. & MCEWEN, L. J. (in press) Fluvial geomorphology sites in Scotland: a review; *In*: *Fluvial Geomorphology*. Geological Conservation Review Series, Nature Conservancy Council, Peterborough and Chapman & Hall, London.

—— & BRAZIER, V. 1991. *The geomorphology conservation and management of the River Feshie S.S.S.I.* Report for the Nature Conservancy Council, March 1991.

Evaluation of river conservation sites: the context for a drainage basin approach

P. W. DOWNS[1] & K. J. GREGORY[2]

[1]*Department of Geography, University of Nottingham, University Park, Nottingham NG7 2RD, UK*
[2]*Goldsmiths' College, University of London, New Cross, London SE14 6NW, UK*

Abstract: In considering regional or national approaches to conservation sites it is necessary to consider the identification and description of the sites, their uniqueness, their sensitivity to natural and human influences, and also their value and importance. A survey of river conservation sites undertaken for English Nature in 1990 selected 85 sites according to the intrinsic value of the site, the extent to which they had been investigated in published research, and the need to embrace a range of examples. The sites were each described in terms of their key characteristics and illustrated by appropriate maps and photographs. Subsequent analysis of the national pattern of sites and their different characteristics suggested that a classification according to 30 categories could provide the basis for developing a uniqueness index similar to that originally devised by Leopold in 1969. In addition, there is a need to assess the sensitivity of the sites according to their liability to natural process adjustment and their vulnerability to direct or indirect human influences. Assessment should also test public awareness of site sensitivity to change using perception techniques. It is concluded that site evaluation needs to be seen in the context of the drainage basin and that sensitivity can be developed to provide a series of scale values that will further assist in landscape conservation techniques.

In developing regional or national approaches to river conservation sites, a four-fold approach is advocated to provide a context for analysis within a drainage basin framework. First, it is necessary to consider the identification and description of the sites. Second, a scheme of classification is needed whereby the importance and uniqueness of each site is assessed. A third need involves judging the sensitivity of the sites to natural and human influences and, due to the dynamic nature of many river conservation features, this judgement should ideally take account of the drainage basin characteristics that are responsible for, or threaten, individual features. Finally, in addition to scientific assessment, it is illuminating to investigate the perceived public value of the sites. These four sequential approaches for selection and assessment should, together, ensure that nationally rare and valued sites are chosen for conservation and that the potential for site disturbance is appreciated. The basis for this procedure is outlined for river conservation sites in the UK.

A survey of river conservation sites was undertaken in the late 1980s in response to a request from the Nature Conservancy Council (NCC, subsequently English Nature). This culminated in 1990 in the selection of 85 sites which had been chosen according to the intrinsic value of a particular site, the extent to which the sites had been investigated in published research, and the need to embrace a range of type examples. Site selection was undertaken by fluvial geomorphologists in each of five regions across England, Scotland and Wales. The areas were Scotland (A. Werrity and L. J. McEwan), Wales (J. Lewin), northwest England (A. M. Harvey), northeast England (M. G. Macklin) and central and southern England (K. J. Gregory). In each region, the first step was to determine the selection criteria from which to compile a list of possible sites. In the case of central and southern England, this initial listing was produced by reference to published information and other available research literature. This procedure gave a total of 214 possible sites, each of which was described in outline together with location details and references.

The 214 shortlisted sites for central and southern England were then classified according to a typology of 15 classes devised for the purpose of conservation, and described in Table 1. Following this stage, the distribution of the sites was considered in order to establish the spatial concentration of the class types given in Table 1 within individual counties. The classification and mapping stages thus revealed the extent to which particular sites repeated the

Table 1. *Basis of classification of river conservation features in central and southern England*

Class type	Feature
a	Terrace
b	Floodplain
c	Channel pattern
d	Process event
e	Channel change
f	Channel modification
g	Drainage network
h	River capture
i	Palaeochannel
j	Limestone fluvial landform
k	Long section/profile
l	Sediment transport
m	Stratigraphy
n	Channel study
o	Geology

characteristics of others. The next stage involved an evaluation of the quality of the sites within each of the 15 classes, and this was based upon the Nature Conservancy Council's criteria (Ratcliffe 1977) for selecting biologically important open water sites (Table 2). Together, these assessments enabled 83 sites to be chosen from the 214 sites originally shortlisted. From this list of 83 sites, a further reduction was made according to the importance of each site. Importance was determined according to the extent to which a site has been subject to geomorphological research, the degree to which the site uniquely represented a class type given in Table 1, and the intrinsic features of each site from site descriptions. The consequence of this further reduction was that 23 sites were finally selected from central and southern England as being appropriate for inclusion in the national list. This procedure, therefore, illustrates use of a progressive technique in which an initial list of conservation sites selected from the research literature can be reduced to a suitable number according to site description and simple classification, together with the representative degree of type and regional scarcity. The chosen sites can subsequently be subjected to three schemes of evaluation by considering their uniqueness, their sensitivity to natural and human influences, and their perceived value, including that by members of the public.

An important stage in nationwide evaluation is to consider the uniqueness of the chosen conservation sites. In the current example two steps were necessary. The first step, because the survey for the UK had been conducted by five groups of independent researchers, was to bring together the chosen sets of sites and to ascertain the extent to which the distinguishing characteristics of sites in one of the five areas overlapped with those in another. In the event, perhaps as a consequence of the varied rock type, topography and glacial history across the study area, comparatively little redundancy occurred and the final result led to the elimination of less than five sites from the total number chosen. The second step involved establishing a scheme to describe the characteristics of all the remaining sites, and this was achieved by using the five major categories shown in Fig. 1, together with 30 sub-categories. The major divisions reflect the variety of reasons for river site conservation in the UK context. The first category (A) reflects the significant contribution made by Quaternary development to the fluvial system; category B represents sites which demonstrate contemporary river channel activity, including those influenced by particular flood events; the third category (C) includes distinctive facets of river channel pattern and floodplain development; category D includes rivers in which historical reconstruction suggests changes in the channel dimensions; and the fifth category (E) embraces sites in which human influences are the cause of recent river channel adjustments. An important aspect is that many of the sites described include a variety of features, often relating to more than one of the five major categories and this is indicated by the key devised to represent the characteristics of each site (Fig. 1). The net result is that 52% of the sites have an association with fluvial landforms (category A), 41% with fluvial processes (B), 38% with river channel pattern and floodplain features (C), 34% with channel change (D), and 6% with human activity (E).

Although this descriptive classification allows a national pattern of conservation site characteristics to be obtained, it does not account for the uniqueness of each site. Indices of site uniqueness were developed by Leopold &

Table 2. *Classification of river conservation features based on Ratcliffe (1977)*

Class	Criteria
1	Unique feature
2	Classic feature
3	Type example
4	Good example of fluvial landform
5	Contemporary fluvial processes
6	Key facet of river evolution in the Quaternary

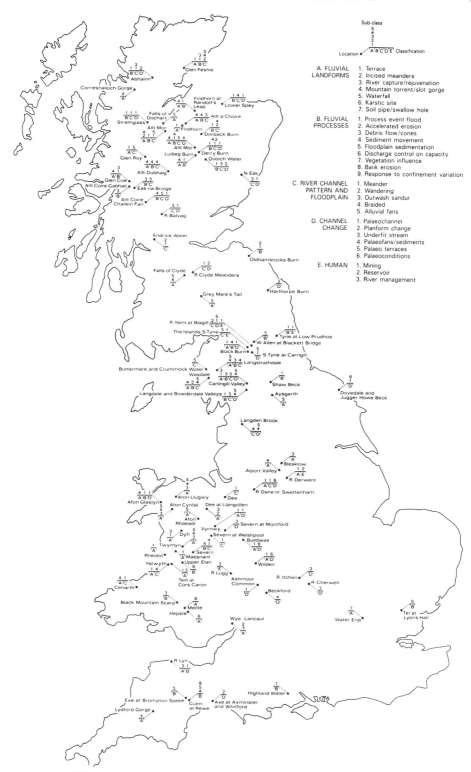

Fig. 1. Location of UK river conservation sites.

Marchand (1968) and Leopold (1969) and are summarized in Dunne & Leopold (1978). Uniqueness can be differentiated from schemes which investigate landscape value by the fact that no weighting, or preferential ranking, is assigned to the chosen landscape characteristics. Therefore, the resulting indices simply compare the rarity of the characteristics possessed by each site. The rarity value for individual features is determined by the proportion of sites which possess the feature and the overall uniqueness is given either as the arithmetic mean or the summation of individual features (Dunne & Leopold 1978). A method of uniqueness assessment was therefore devised based upon the prevalence of each of the 30 characteristics given in Fig. 1. Both the 'summation' uniqueness (total of scores obtained for each feature at a site) and the 'mean' uniqueness (score total divided by the number of separate features noted at each site) were calculated for the 85 sites. Table 3 indicates the 12 most unique river conservation sites based upon the average rank of the summation and mean uniqueness indices: the sites can be located on Fig. 1.

A further assessment that requires consideration is the sensitivity of the sites to geomorphological changes which might threaten the existence of the conservation feature. Determining sensitivity is difficult, because a generally accepted deterministic understanding of river channel systems does not exist, but is an important requirement because some river channels are very sensitive to adjustment from influences either within the reach, or transmitted from upstream or downstream of the reach. River channel sensitivity to change has been referred to in contrasted ways in the literature, prompting Downs & Gregory (1993) to suggest four hierarchical definitions which would enable sensitivity assessments to be achieved within the limitations of current knowledge. In relation to river conservation, the classification used in Fig. 1 is being developed according to a series of scale values which indicate the sensitivity of individual sites to natural changes and to human influences. When combined with the possibility of change being promoted by factors external to the site, it becomes apparent that site evaluation needs to be seen in the context of the drainage basin.

In addition to establishing the scientific value and sensitivity of river conservation sites, it is desirable to understand the value placed upon the chosen sites. There are two primary ways in which this might be achieved. The first is a development from the technique used to determine the final listing of sites and uses separate environmental criteria as a basis for categorization. Data of this type have been used frequently for instream river channel habitat analyses and river quality analyses, and a number of examples are discussed by Mosley (1987). An alternative approach is to use perception techniques to analyse the overall impression of sites as visualized by the public. In a densely populated country such as the UK where there is limited scope for designated conservation areas, increased public awareness and appreciation of the need for, and the importance of, conservation sites is necessary both to safeguard sites and to provide the foundations for wider acceptance and long-term support (NCC 1990). Understanding public perception towards the attractive-

Table 3. *Most unique national river conservation sites*

Rank	Site		Uniqueness indices			
		No. of features	'Summation' score	Rank	'Mean' score (max. = 1)	Rank
1	North Esk	2	1.077	2	0.539	5
2	R. Derwent	2	1.071	3	0.536	6
4.5	Upper Elan	1	1.000	6.5	1.000	2.5
4.5	R. Lugg	1	1.000	6.5	1.000	2.5
4.5	Ter at Lyons Hall	1	1.000	6.5	1.000	2.5
4.5	Highland Water	1	1.000	6.5	1.000	2.5
7	Oldhamstocks Burn	2	1.056	4	0.528	7
8	Culm at Rewe	3	1.476	1	0.492	12
10.5	Maesnant	1	0.500	18	0.500	9.5
10.5	Meilte	1	0.500	18	0.500	9.5
10.5	Hepste	1	0.500	18	0.500	9.5
10.5	Water End	1	0.500	18	0.500	9.5

ness of particular river channel features may also allow a complete realization of the general amenity potential of conservation areas. Previous perception studies include those of Mosley (1989), House & Sangster (1991) and Gregory & Davis (1993). In this latter example, the perception of riverscape aesthetics from a large group of respondents was examined for sites on two Hampshire rivers, one of which, the Lymington River, includes as a tributary the Highland Water, a selected river conservation site. The conclusions demonstrated that channels which are intrinsically the most natural are not necessarily the most highly valued, and a similar conclusion has been reached in New Zealand (Mosley 1989). It may, therefore, be necessary to educate public awareness of natural river features as part of conservation programmes.

It is concluded that the four procedural stages for selecting river conservation sites detailed above, namely; site selection, analysis of uniqueness, expression of sensitivity, and assessment of perceptual judgement, provide a simple approach for maximizing the scientific and public benefit of river conservation areas. However, due to the importance of current river channel processes in creating and maintaining many desirable geomorphological features, river conservation must also entail scientific investigation into the origins and retention of individual site characteristics. This will require an understanding of the hydrological, sedimentological and network characteristics of the drainage basin in order to gauge the sensitivity of the site to adjustment. Therefore, retaining river conservation features is likely to require active management within a drainage basin context, rather than simply preservation of the site itself. Currently, neither the 'exposure' or the 'integrity' approaches to Earth science conservation (NCC 1990) appear to incorporate this dimension.

References

Downs, P. W. & Gregory, K. J. 1993. The sensitivity of river channels in the landscape system. *In*: Thomas, D. S. G. & Allison, R. J. (eds) *Landscape Sensitivity*. J. Wiley and Sons, Chichester, 15–30.

Dunne, T. & Leopold, L. B. 1978. *Water in Environmental Planning*. W. H. Freeman, San Francisco.

Gregory, K. J. & Davis, R. J. 1993. The perception of riverscape aesthetics: an example from two Hampshire rivers. *Journal of Environmental Management*, **39**, 171–185.

House, M. R. & Sangster, E. K. 1991. Public perception of river corridor management. *Journal of the Institution of Water and Environmental Management*, **5**, 312–317.

Leopold, L. B. 1969. *Quantitative comparison of some esthetic factors among rivers*. United States Geological Survey, Circular **620**.

—— & Marchand, M. O. 1968. On the quantitative inventory of the riverscape. *Water Resources Research*, **4**, 709–717.

Mosley, M. P. 1987. The classification and characterisation of rivers. *In*: Richards, K. S. (ed.) *River Channels: Environment and Process*. Blackwell, Oxford, 295–320.

—— 1989. Perceptions of New Zealand river scenery. *New Zealand Geographer*, **45**, 2–13.

NCC. 1990. *Earth Science Conservation in Britain: a strategy*. Nature Conservancy Council, Peterborough.

Ratcliffe, D. 1977. *A Nature Conservation Review, Volume 1*. Cambridge University Press, Cambridge.

The establishment and revegetation of vegetation zones in rural river landscapes in SW Finland

JARI HIETARANTA
Department of Geography, University of Turku, FIN-20700 Turku, Finland

Growing interest in the cultural landscapes of Finland has recently been accompanied by demands to conserve and restore the traditional rural landscape. The Ministry of Environment has started a programme to make an inventory of valuable Finnish rural landscapes. It is paradoxical that there are both demands to maintain the cornerstones of rural landscape, e.g. supporting open farmland, livestock, traditional farmhouses and old gardens, and attempts to maintain and protect natural river channels from erosion or sedimentation. These diverse aims may lead to conflict if pursued separately.

In the study reported here, an attempt was made to link stream restoration to values associated with landscape. The idea was that the proposed green zones or erosion shields should be suitable and in harmony with the different landscape patterns found in the study zones.

The study

The study area, in the Aurajoki river valley, is situated in southwestern Finland and has a long cultural landscape heritage. The study area consists of two separate zones (5 km and 9 km in length, respectively) along the channel. To meet the needs of landscape conservation and to improve the recreational value of the river, vegetation zones adjacent to the channel were planned.

First, on the basis of the differences in landscape pattern, the study area was divided into three landscape base units. The work was done mainly from maps and by aerial-photo interpretation. During the field work, the present vegetation types of each unit were mapped to get an idea of the vegetation already present along the stream. Secondly, the morphometry of slopes adjacent to the river channel was determined with NIKON DTM-5 takymetric equipment. The slope profiles were surveyed at 50–100 m intervals. To simplify greatly: three different morphometrical slope types with different vegetation cover were identified. Finally, the revegetation and restoration of the selected zones (10–100 m wide) was introduced for nine different combinations of slopes and landscapes (three landscape base units and three morpometrical slope types).

The main aim was to preserve the individual characteristics of each landscape unit. The treatment of zones varied from tree felling to afforestation. The choice of species was important, with traditional plant species being favoured. In addition, the demands of stream protection were taken into account.

Conclusions

This study attempted to combine the requirements of landscape protection and the urgent need for river restoration. The study also tried to demonstrate that these separate goals could be pursued together. In order to achieve these aims morphometrical data were used together with landscape evaluation. The approach of this pilot study has been to treat landscape as an object for planning and conservation.

In the future, research methods should be developed further. Various disciplines are interested in landscape planning, but a consensus on concepts and methods of understanding the landscape is still missing. In particular, investigations and applications are needed at the local level.

Conservation management of dynamic rivers: the case of the River Feshie, Scotland

VANESSA BRAZIER[1] & ALAN WERRITTY[2]

[1]*Scottish Natural Heritage, Earth Science Branch, 2 Anderson Place, Edinburgh EH6 5NP, Scotland*
[2]*School of Geography and Geology, The University, St Andrews, Fife KY16 9ST, Scotland*

Abstract: Scotland has a rich variety of fluvial environments which are of prime interest to landscape, Earth science and biological conservation. However, effective conservation management of dynamic river systems, which are common in most upland areas, is complex because of the hazards associated with dynamic rivers, specifically floods and channel instability. In the case of the River Feshie in the Cairngorms, conservation of a large actively aggrading alluvial fan and two braided reaches has involved striking a balance between enabling local people to carry out some hazard mitigation works and minimizing impacts on the river system, but often at the expense of damage to ephemeral landform assemblages. In developing a sustainable management approach we have had to rely on conceptual modelling of geomorphic sensitivity of these river environments and on the documentary record of known past impacts on the river environment. This approach has enabled identification of extreme activities which would result in fundamental change in river process regime. In practice this means it is possible to identify seriously damaging activities, but it is still difficult to assess the cumulative effects of relatively small-scale interferences. The latter are the most commonly requested hazard mitigation works for dynamic river Sites of Special Scientific Interest (SSSIs), and usually require long-term and ongoing maintenance. Many of these small-scale works have little effect in reducing the hazard concerned, and in some instances actually exacerbate the original problem. The main difficulty in sustainable management of dynamic rivers is counteracting the perceived need for people to 'do something', regardless of whether or not it will successfully reduce or contain the impact of the hazard.

There is a rich variety of river environments in Scotland which contribute to the diversity of Scotland's landscape and provide important ecological habitats. Scotland's upland rivers and associated modern and ancient fluvial landforms are characteristic of formerly glaciated landscapes and include: bedrock reaches, gorges, spectacular waterfalls, boulder-bed torrents, small alluvial fans and steeper debris cones, and a variety of gravel bed river landforms from sinous to wandering and braided river reaches. The extent of lowland fluvial environments is more restricted in Scotland compared with the rest of the UK, but within these lowlands there are good examples of actively meandering sand bed and gravel bed rivers, the majority of which are uninhibited by large-scale river control works.

Internationally and nationally important examples of the whole range of fluvial landforms and environments is reflected in the selection of 28 fluvial sites in the Geological Conservation Review (GCR) (Werritty & McEwan, unpublished). The two principal criteria used in selecting these sites were: first, variations in channel type; and second, the principal characteristics which make Scottish river channels geomorphologically significant. The resulting selection of sites covers 6 distinct geomorphic characteristics (Table 1), including the geomorphic role of extreme floods. As a result of the GCR many of the 28 fluvial sites in Scotland are now designated as Sites of Special Scientific Interest (SSSIs). This paper examines the practical problems facing effective conservation of dynamic river systems, where flooding and channel instability are important attributes of the Earth science conservation interest. The paper specifically reviews some of the problems associated with the conservation management of dynamic rivers using as a case study the River Feshie, in the Western Cairngorm Mountains of Scotland.

The River Feshie

Four areas of Glen Feshie were designated as a SSSI in 1989 on the following basis:

Table 1. *Summary of significant geomorphological characteristics of Scottish rivers*

Characteristics	Examples
1. Classic sedimentary structures	River Endrick
	Allt Choire Chailein
	Allt Coire Gabhail
2. Downstream changes in fluvial controls	Allt a' Choire
	Allt Choire Chailein
	Allt Mor (Nairn)
	Allt Mor (Glenmore)
3. Interfaces between different types of geomorphic activity	Allt a' Choire
	Glen Feshie
	River Quoich
	Allt Mor (Nairn)
	Allt Mor (Glenmore)
4. Evidence of the geomorphic impact of extreme floods	River Feshie
	Allt Mor (Glenmore)
	River Quoich
5. Types and rates of fluvial adjustment over historic time	River Clyde
	River Endrick
	Abhainn an t-Srath Chuilleanaich
6. Lateglacial and Holocene fluvial adjustment	Glen Roy
	Glen Feshie
	Findhorn Terraces
	River North Esk

(1) areas of multiple river channels, characterized by rapid and frequent shifts in channel positions:
(2) associated assemblages of small-scale sedimentary structures and landforms; and
(3) a large assemblage of slope and valley floor landforms and sedimentary features, which together provide an impressive and long-term record of landscape evolution since deglaciation, *c.* 13 000 years ago.

On the basis of these criteria, four sections (A–D) of the river were identified as component parts of the River Feshie SSSI (Werritty & Brazier 1991*a*). Only Section B comprises an assemblage of relict landforms, which complement the active geomorphological interests of Sections A and C (two braided reaches) and section D (a large actively aggrading alluvial fan). These geomorphologically active sections collectively identify the dynamic reaches of the river in which parts of the valley floor are being reworked by floods. This flashy runoff regime also generates a great variety of small-scale features and sedimentary structures which typically only survive until the next major flood. Section D of the SSSI is focused upon in this paper since it affords a particularly good example of a dynamic river and one in which there is a major conflict between the interests of local landwoners and earth science conservation.

The alluvial fan which comprises Section D is located immediately downstream of Loch Insh at a very constricted site where the River Feshie flows into the River Spey (Fig. 1). This particular alluvial fan is composed of two parts: a modern actively aggrading alluvial fan inset within a much larger relict fan more than 10 000 years old. The present-day morphology and micro-topography of the fan is determined by floods capable of eroding, transporting and depositing the coarse river gravels which make up the bulk of the river's bed. Floods capable of mobilizing such material are intermittent (typically occurring less than 3–4 times per year) and thus the downstream movement of coarse sediment is highly episodic, the river being akin to a very jerky conveyor belt. As each flood wave passes through a reach bed material is locally eroded, potentially creating new channels, and moves off downstream, often in discrete sedimentary waves. Thereafter, as the velocity lessens, the coarser material is selectively deposited in shallows (riffles), bars and gravel sheets in and adjacent to the channel. With a further reduction in the flow velocity, finer sediments are deposited forming ephemeral features such as sand sheets, ripple complexes and crevasse splays. These

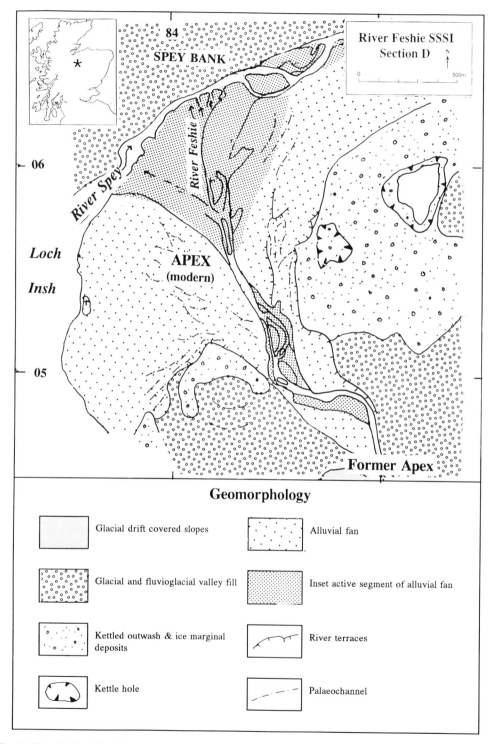

Fig. 1. River Feshie alluvial fan.

are especially significant in deciphering the record of flow immediately prior to the cessation of sedimentary movement.

The distribution of sediment accumulating on the alluvial fan surface is determined by the River Feshie changing its position on the fan over time. Such changes have been reconstructed from old maps and aerial photographs documenting the area from 1750 to the present day (Werritty & Brazier 1991a, b). The locus for major channel change (the apex), is the point at which deposition of sediment dominates over erosion and throughput of sediment. Characteristic fan-scale channel switching occurs on the active Feshie alluvial fan, and is triggered by the build-up of channel sediment near the apex, effectively raising the river bed higher than the surrounding land.

The Feshie alluvial fan is a classic example of its type and the best to be found in upland Britain (Werritty & Brazier 1991b). Its scientific significance specifically rests upon the channel pattern dynamics (repeated switching of the position of major channels), the manner in which the bed material moves downstream and the ephemeral fluvial forms it continually recreates. The integrity of the site is especially important for the opportunity it provides to investigate the processes which sustain and develop an active fan. Furthermore it acts as a modern analogue for the reconstruction of past environments based on ancient sedimentary sequences, as well as being of considerable interest as one of the last actively forming river confluence alluvial fans in Britain.

There is, however, a downstream impact of the Feshie alluvial fan in terms of constricting the valley floor immediately downstream of Loch Insh and thereby exacerbating local flooding on the River Spey. This arises because the channel capacity immediately downstream of river confluences in upland Britain is rarely capable of efficiently transporting all sediment inputs from a tributary river.

This is specifically the case with the confluence of the Rivers Feshie and Spey (Werritty & Brazier 1991b). The map and aerial photographic record shows that by 1899 a series of alternate channel bars had formed downstream of the confluence; these continued to grow and had developed into the Spey Islands by 1946 (see figs 3.6 and 3.7 in Werritty & Brazier 1991b). These islands have been progressively built up from coarse sediment, brought down by the River Feshie, that initially accumulated as tongues of sediment at mouths of the distributary channels. These tongues of sediment, or deltas, are sometimes extensive and intrude into the flow of the River Spey. They are periodically trimmed by the Spey in spate, but the sediment is only transported a short distance downstream to be deposited in shallower water (such as the area immediately upstream of the islands). This pattern of sediment transport from the Feshie into the Spey is already limiting the long-term viability of channel straightening works downstream of the confluence undertaken in recent years, because the Spey is not sufficiently competent to evacuate the Feshie-derived sediment. An estimate of 20 000 tonnes per year for bedload transport rates has been suggested for the Feshie at the confluence (Werritty & Brazier 1991b), which gives an indication of the magnitude of sediment that potentially reaches the Spey, forming the so-called 'Feshie Plug'.

The Earth science conservation interest outlined above is, however, in conflict with the local landowning interests which are damaged by the hazardous behaviour of the river. These hazards comprise flooding, bank erosion and channel changes, all of which have dramatically impinged on human activities throughout the glen during recent winters, and most notably at the confluence of the Rivers Feshie and Spey. Thus the practical conservation problems at this site centre on the question of how to strike a balance between enabling local people to carry out some hazard mitigation works at the same time as minimizing their adverse impacts on the dynamic character of the river. The situation is further complicated in that SNH has to set this site within the wider context of nature conservation interests in the whole area. In practice SNH has had to assume that Section D of the River Feshie SSSI is more 'robust' to proposed engineering schemes than the Spey and Insh Marshes SSSI.

Hazard mitigation and Earth science conservation

In the case of the River Feshie, public concern has focused on the role of the sediment load 'plugging' the Spey, and on the flood peak constricting the outflow of the Spey from Loch Insh. Both attributes of the behaviour of the River Feshie have been identified by local people as major exacerbating factors in the degree of recent winter flooding experienced in Badenoch, in Strathspey. However, there has been little appreciation in the debate on the Feshie and Spey of the increased incidence of winter flooding on many Highland rivers during the last decade (Anderson & Black 1993). Despite recommendation of the need for an integrated catchment scale approach to flood alleviation in Strathspey (Cuthbertson & Partners 1990), Highland Regional Council concluded that flood defence work could be carried out as and when required by landowners, who could seek financial assistance from the Scottish Office Agriculture and Fisheries Department (SOAFD). Thus, in keeping with current practice in Scotland, the responsibility for flood alleviation les mainly with the individual landowner.

When areas of river and land are designated as SSSIs the landowner or occupier must notify SNH of activities likely to damage the conservation interest of the site. In practice, all river engineering works on river or wetland SSSIs are notifiable activities. In the case of the confluence of the Rivers Spey and Feshie, SNH has sought to develop a sustainable approach to balance local fears about flood risk against the requirements of conservation. This has necessitated an assessment of the relative merits of different types of engineering works. Moreover, as an inevitable result of the assessment of engineering threats, SNH has had to balance the benefits of certain management practices on the Feshie confluence with the needs of the Spey and Insh Marshes SSSI. This neighbouring SSSI and RAMSAR site is recognized as an internationally important site for wetland flora, wildfowl and wildlife. Although the whole of the River Feshie SSSI has been considered to be of international importance for the Earth sciences (Werritty & Brazier 1991a), it is not possible to evaluate objectively the relative merits of one SSSI over the other.

Thus generalizing from the specific details of this case study, the main attributes of active alluvial fans which are of interest to Earth science conservation are also hazardous to the local people who live and work in the area. Flooding and channel instability cause direct damage to property and land-use, and indirectly have implications for land ownership defined by the centre of the main channel. The question faced by conservationsits is how then to best manage the conservation interest of dynamic rivers when they are threatened by engineering schemes?

Conservation management of dynamic rivers

The main objective of Earth science conservation of dynamic river systems is to maintain the integrity of the river process environment. In managing hazardous sites such as rivers prone to flooding and bank erosion, it is not always feasible to conserve individual landforms. However, conservation of the environment in which such landforms can reform should be possible without compromising essential hazard mitigation works, provided that such works are appropriately designed and can fulfil the objectives of Earth science conservation (Werritty & Brazier 1991a, b).

Geomorphological sensitivity provides a conceptual framework for assessing what types or degree of intervention in a river system are sustainable in terms of maintaining the essential character of the river system (Werritty & Brazier in press). In essence we need to know whether or not the tolerance of the river to externally driven change will be exceeded. River systems may therefore be divided into essentially **robust** systems and **sensitive** systems, for a given level of interference. However, identifying when stable robust conditions (or more probably metastable conditions) in a naturally dynamic environment are exceed by unstable sensitive conditions is perhaps one of the main unresolved problems in fluvial geomorphology (Leopold 1976; Newson 1992). At the present time all we can rely on to provide an assessment of sensitivity is to evaluate the evidence for the degree and persistence of past impacts caused by human interference on a given type of river system. Even this information is limited, in that we do not yet understand the cumulative impact of minor interferences (such as repeated bulldozing of bed material up against an eroding river bank) on an apparently robust river system. Bearing these points in mind, the following discussion outlines our current thinking on the relative robustness of the dynamic distributary system at the confluence of the Feshie and the Spey.

The impact of past flood control works on the Feshie alluvial fan

The pattern of relict features on the surface of the whole alluvial fan gives some insight into long-term patterns of sediment build-up at the confluence of the River Feshie and the River Spey. In particular, the formation of inset alluvial fans progressively downstream indicates both natural changes in the quantity of sediment supplied to the fan, and the effectiveness of past river control measures. The map and aerial photographic record for this area demonstrate that the modern active inset fan has occupied a central location on the palaeo-alluvial fan since the 1750s (figs 3.2 to 3.15, Werritty & Brazier 1991b). Flood control works on the alluvial fan date back to the nineteenth century and include flood banks, a straightened channelized reach of the river, and recently emplaced bank reinforcements and bulldozed reaches of the river. All of the flood banks crossing the active inset alluvial fan have been partly or wholly buried by the build-up of river sediments. None of these works, to date, have permanently damaged or destroyed the integrity of the site, in terms of allowing the river to behave like classic alluvial fan rivers, which have divided channel patterns and are subject to periodic switches in their main channel paths across their alluvial fans. Furthermore, none of these works on the Feshie have, to date, destroyed the range of sedimentary environments or resultant features typical of actively accumulating alluvial fans. At one level, therefore, the active alluvial fan represents a relatively robust example of an active alluvial fan environment occupying the central axis of a much larger palaeo-alluvial fan, that in total is of considerable interest to Earth science conservation and research.

However, not all hazard mitigation schemes that have been proposed for the River Feshie alluvial fan, would enable the river to maintain a divided channel pattern. Since the 1860s there have been several detailed specifications drawn up to channelize and stabilize the braided distributary reach (Werritty & Brazier 1991*b*). However, the costs of implementing and maintaining a trapezoidal straight channel have resulted in this option not being implemented. This style of intervention would destory the active alluvial fan, because the river would not be allowed to reform its dominant divided channel pattern. What is astonishing is that none of the specifications for channelizing the Feshie recognized that, if effective in engineering terms, all of the bedload of the Feshie would be discharged directly into the Spey, exacerbating problems of constriction immediately downstream of the confluence with the Spey.

The Spey Islands downstream of the present Feshie/Spey confluence have been permanently damaged since the designation of the River Feshie SSSI. The most recent incidents of this damage are a result of bulldozing river gravels to create a flood relief channel in 1991 and the realignment of the Spey channel in the autumn of 1992, which has resulted in the removal of the bulk of the island adjacent to the north bank of the Spey. Although in the very long term the Spey Islands might reform, it cannot be assumed that this will be the case.

In extreme cases then, we can identify those developments which would be detrimental to the continued survival of the river process environment. The River Feshie is robust enough to withstand modification caused by minor peripheral works, that often have a short life expectancy. However, we do not yet understand what the cumulative impact of such short-term activities will be on the overall stability of the river system. Larger-scale, and usually capital intensive, schemes that modify the channel pattern threaten the integrity of the site, and have been vigorously resisted. However, there is some scope for innovative solutions to the flood and channel change hazard that affect the confluence as part of a sustainable approach, sensitive to Earth science conservation interests and working in sympathy with river processes.

References

ANDERSON, J. & BLACK, A. 1993. Tay flooding: Act of God or Climatic Change? *Circulation*. British Hydrological Society Newsletter.

CUTHBERTSON, R. H. & PARTNERS 1990. *Flooding in Badenoch and Strathspey*. Report to Highland Regional Council (2 vols).

LEOPOLD, L. B. 1976. Reversal of erosion cycle and climatic change. *Quaternary Research,* **6**, 557–562.

NEWSON, M. 1992. Geomorphic thresholds in gravel-bed rivers. *In*: BILLI, P., HEY, R. D., THORNE, C. R. & TACCONI, P. (eds) *Dynamics of gravel-bed rivers*. John Wiley and Sons, Chichester, 3–20.

WERRITTY, A. & BRAZIER, V. 1991*a. The geomorphology, conservation and management of the River Feshie SSSI*. Report to the Nature Conservancy Council (2 vols).

—— & —— 1991*b. Geomorphological aspects of the proposed Strathspey flood alleviation scheme*. Report to the Institute of Hydrology.

—— & —— (in press). Geomorphic sensitivity and the conservation of fluvial geomorphology SSSIs. *In*: STEVENS, C., GREEN, C. P., GORDON, J. E. & MACKLIN, M. G. (eds) *Conserving our Landscape: Evolving Landforms and Ice-Age Heritage*. English Nature, Peterborough.

Keynote address

Resource development, landscape conservation and national parks in Norway

F. C. WOLFF

Geological Survey of Norway, Box 3006, N-7002 Trondheim, Norway

Abstract: Establishment of national parks and other types of landscape protection in Norway, has led to conflicts between the government's environmental authorities and developers of mineral potentials who traditionally have been active in these areas. These conflicts have created a new need for geoscientific analysis of the environmental consequences of mineral exploitation. A few examples of such analyses will be described.

Increased awareness amongst the general public has forced politicians to acknowledge the importance of protecting the natural environment. This has led to the establishment of a number of national parks and other types of protected area in Norway. Resource development such as mineral exploration has traditionally been frequent in these areas. The political opinion on utilization and protection of the natural environment is influenced by different pressure groups (Fig. 1). In one of these – frontier economics – the consideration of nature both as a mere store of natural resources and energy and as an object of exploitation prevails completely. The result of this policy may be seen in the black triangle between Poland, former East Germany and the Czech Republic. At the other extreme occur the deep environmentalists who want to keep the globe totally green. Their policy would, however, lead to starvation of the world's steadily increasing population. The role of the natural scientist will be to find a point of equilibrium between these two extremes. Some examples of conflict between resource development and natural environmental protection and the lack of balanced solutions will help to illustrate this problem.

Three case histories (Fig. 2)

(A) Conflict between the establishment of a new national park in a Lapplandian area in northernmost Norway and prospecting for resource development of a potential nickel ore.

(B) Conflict between the establishment of a less rigorous type of protected area called a 'protected landscape' and an existing talc quarry.

(C) Conflict accompanying the attempt to develop an apatite-iron ore mine near an urban area in one of the most densely populated parts of Norway.

(A) Anarjokka National Park

When an area has been declared a 'national park', prospecting is no longer legal. Prior to the declaration of this national park in December 1975, a mining company had, since 1962, invested considerable sums in geological mapping and prospecting for nickel ore in the area. The company sued the Norwegian state to try to recover their loss. The case was brought to court as a case of appraisal. The referees were all geoscience or mining experts. The court found that the company had not suffered any undue loss and could not agree to the claim of compensation. They agreed, however, to refunding the company the cost of bringing the case to the court. The reason for the conclusion was that the company owned neither the ground nor the possible ore occurrence. The rights were only the right of priority for eventual mining

* FRONTIER ECONOMICS

* RESOURCE MANAGEMENT

* SUSTAINABLE DEVELOPMENT

* SELECTIVE ENVIRONMENTALISM

* DEEP ENVIRONMENTALISM

Fig. 1. Pressure groups that may influence political opinion on the utilization and protection of the natural environment.

From O'Halloran, D., Green, C., Harley, M., Stanley, M. & Knill, J. (eds), 1994, *Geological and Landscape Conservation*. Geological Society, London, pp. 153–156.

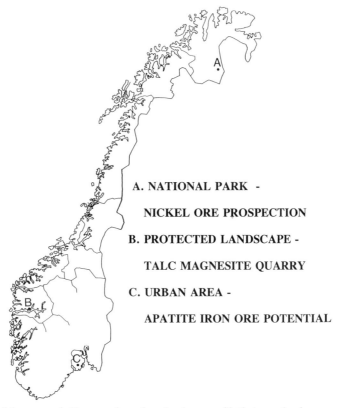

Fig. 2. Location of three areas in Norway where there has been conflict between landscape protection and resource development.

over other potential companies. The area, as well as the possible ore, remained the property of the Norwegian state and the company would in any case, according to Norwegian mining law, have had to seek the permission of the ground owner to carry out mining in the area.

(B) Vik Talc Quarry

Sogn and Fjordane county authorities submitted in 1987 a proposal for a protected landscape area of 379 km^2 within the Sognefjkord district of west Norway. Since 1981, in cooperation with a private company, the Geological Survey of Norway had developed a talc-magnesite occurrence within a marginal part of the area. The occurrence is definitely one of the largest of this kind that has been discovered in Western Europe in recent years. With respect to local and national economy, including job opportunities in the area, the Geological Survey of Norway submitted a request for a reduction of the protected area in order to have the quarry excluded.

A permission for a test mining of up to 5 years has been obtained so far.

(C) Kodalen apatite-iron ore and the analysis of the environmental consequences of an eventual mining operation

In 1977, prior to a development of this large apatite-iron ore deposit, a mining company wanted to analyse the environmental consequences of establishing a large mining operation in an urban area (Fig. 3). The most extensive impact analysis to date in Norway, was therefore carried out (Norsk Hydro *et al.* 1977). The report states that this is the only large potential resource of phosphorus in Norway. There then follows a thorough analysis of the planned open pit mining and its effect on the natural environment and the adjacent urban areas. Finally, it discusses alternative modes of operation in order to reduce the effects.

A focal point of the report is where and how

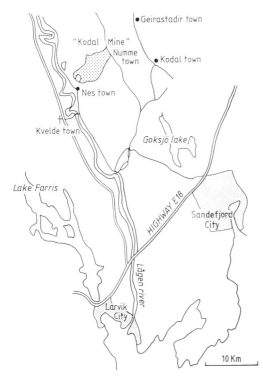

Fig. 3. Location of Kodalen apatite-iron ore in relation to the surrounding urban areas.

Another important aspect was the possible chemical and material pollution from the ore dressing plant. It was proposed that the waterborne pulp would be deposited in a nearby lake. As a result, after 15 years of mining the lake surface would have been raised by 38 m. This would mean the construction of a large dam with a total content of 20×10^6 tons of which 13×10^6 tons would consist of fine-grained sterile waste. The water from the ore dressing plant would amount to $c.$ 600 m^3 per hour and would be turbid due to suspended fine-grained rock fragments and ore remnants. It would have a high electrical conductivity, be slightly alkaline with inorganic components and dressing chemicals in relatively low concentrations. In addition to the dam, which would act as sedimentary depository for the pulp particles, a cleansing plant would have to be installed before the water could be let out into the nearby river, which is rich in trout and salmon.

The study concludes with an overview of the calculated consequences of the planned mining operation. This large iron–phosphorus resource has not yet been developed and one can assume that the cost of the minimization of the environmental consequences has discouraged the company from the exploitation of this valuable resource.

General comment

to deposit the sterile waste. The size of the dump would have extensive visual effect on the landscape (Fig. 4). Underground redepositing was, therefore, suggested.

The examples presented demonstrate clearly how legislation for environmental protection affects potential resource exploitation in areas where there are conflicts between protection

Fig. 4. View of the potential dump of sterile waste in Kodalen.

of the natural environment and industrial developments. In order to avoid or mitigate such problems in the future, the importance of geoscientific research within this field must be stressed.

To obtain concessions for exploitation of mineral resources the mining companies in the future will have to accompany their applications with a thorough analysis of the environmental impact. The developing field has, therefore, produced a new challenge to geoscientific research.

Reference

Norsk Hydro, a/s, i/s Miljøplan & Platou, F. S. a/s. 1977. *Beskrivelse og Miljøanalyse av eventuell bergverksdrift i Kodalfeltet*. Unpublished.

Landscape conservation and the national parks of Romania

D. TEODORASCU

Ministry of Waters, Forestry and Environmental Protection and Conservation, Bucharest, Romania

Abstract: The National Protected Areas Network of Romania encompasses 4.6% of the country's surface and includes three Biosphere Reserves (2.96%), 10 national and 2 natural parks (1.68%) and 571 strictly protected areas according to IUCN (the International Union for the Conservation of Nature) categories.

At the moment, a national research programme, financed by the Ministry of Waters, Forests and Environmental Protection, on landscape and biodiversity conservation is being developed. One of its goals is the reorganization of the existing National Protected Areas Network by reassessing the biodiversity which requires protection and considering proposals for its expansion. The results will be used for better classification according to IUCN categories and also for adequate management of these areas.

The first step is to reorganize the national parks of Romania, including the borders, buffer zones and international zones and also to implement management plans and park rangers to ensure that protected areas are properly used and maintained.

Romania is rich in geological remains, geomorphological elements, palaeontological deposits, erosional features, caves and other geologically interesting objects. The majority of these have been designated nature monuments or scientific/nature reserves, representing 172 protected sites. Some of them are situated within the national park boundaries. The main threats facing protected geological sites are from mining, modification of the natural speleological environment, increased fossil collecting and uncontrolled tourism.

The Ministry of Waters, Forestry and Environmental Protection (MoWFEP) was created in early 1990 as the central authority for public administration to develop, set the guidelines, and improve the activity of environmental protection on a national scale.

MoWFEP is responsible for water management, forest management, pollution control management, nature resource management, and nuclear safety. Within these areas MoWFEP drafts new legislation, elaborates administrative regulations, issues permits and has responsibility for the enforcement of the policy. MoWFEP is responsible for managing the National Protected Areas Network, and for identification and designation of new natural areas to protect. These are the main tasks of the Conservation and Protected Areas Management Unit of the Department of Environmental Protection. In fulfilling its obligations in this field, the unit co-operates with the Commission for the Protection of Natural Monuments of the Romanian Academy that, since 1932, has been responsible for identifying the areas to protect. Proposals for protected areas are submitted to Parliament for designation.

At present in Romania the operative legislation is the Environmental Law of 1973 (Law no. 9). It provides a conceptual framework for environmental regulation, including principles and duties for the protection and improvement of the environment. These principles were developed through special laws and regulations for air, water, soil and subsoil protection, forest and vegetation, land and water fauna, nature reserves and nature monuments.

However, despite the intentions set out in the law, structural weakness, including the present political situation involving transition to the market economy, land privatization, changes in all spheres of activity and lack of finances, have rendered the laws largely ineffective.

In 1991, the Ministry began the process of drafting a new general environmental law. After improvements and discussions, a draft Environmental Protection Law was submitted to the Romanian Parliament in 1992 and is expected to be approved as soon as possible.

Current situation

The first nature reserve in Romania was declared in 1932 and the first national park, namely Retezat National Park, was established in 1935.

Protected areas cover 4.8% of Romania's surface (1 140 590.3 ha of the total surface area of 23 750 200 ha) and include three Biosphere Reserves (2.96%), 12 national and 2 natural parks (1.68%) and 571 strictly protected areas conforming to protected area categories of the International Union for the Conservation of Nature (IUCN).

The list of Protected Areas in Romania as it stands at the moment will be modified. Thus, firstly, changes will be made after confirmation of the actual state of each protected area listed. It is clear that some of them will be deleted from the list because of their degradation due to non-observance of the existing environmental laws or because they were declared on a subjective basis, without having supporting scientific justification. On the other hand, it appears that the existing protected areas network fails to cover completely the full diversity of the different ecosystems that exist in Romanian territory.

Started in 1991, a major research programme financed by MoWFEP on 'Delineation of Romania's Ecoregions' has as one of its main goals the organization of the protected areas network to ensure conservation of biodiversity and to create permanent control areas, with minimal anthropogenic impact, to assist the monitoring of the direction and rate of evolution in other ecosystems. Of great importance in the achievement of this programme is a resources assessment by aerial photographic interpretation using a geographic information system (GIS). As a result of studies already done, 22 ecoregions are recognized in the territory of Romania (21 continental ecoregions and one marine ecoregion – the Black Sea). The most significant ecosystems in each of these ecoregions will be represented in the protected areas network. Thus we estimate that the total extent of protected areas will increase to about 6% of Romania's territory. The programme will end in 1994. Its main results will be gathered into the 'Atlas of the Environment of Romania'.

In early 1990, on the initiative of the Department of Forestry, 12 national parks with a total extent of 396 761 ha were hurriedly established. These areas contain only the forest ecosystems of 12 mountain massifs and their surroundings without including the alpine pastures, meadows and rocky areas. Thus a basic idea was ignored namely the structural and especially the functional unity of a mountain massif.

Assessment of national parks

In this context, MoWFEP proposed and financed, in early 1992, a project called 'Studies concerning the Protected Areas Network of Romania'. The first study was 'The assessment of the National Parks of Romania'. The aims of this project are as follows:

- scientific analysis and assessment of areas that will form parts of the national parks and their delimitation clearly in the field;
- the internal delineation in terms of management categories of the component areas;
- evaluation of the present state of the biodiversity and geological heritage.

The basic concepts regarding the establishment of the new national parks of Romania are:

- the expansion of national specific biodiversity and natural habitat conservation in large ecological units;
- the conservation of complex ecosystems in their full integrative and functional state;
- the assurance of a maximum representation of the national natural landscape, with at the same time the creation of a valuable scientific, ecological and genetic resource;
- the proper integration of the component areas within the international network of management, according to the IUCN categories;
- the inclusion of the national parks within a national system of protected areas with a unitary, complete and responsible management of its territory.

The project is developed in cooperation with the Commission for Natural Monuments of the Romanian Academy and with many research institutes as an interdisciplinary study. To obtain the best results this study involves distinguished specialists in the field of biology, geology, speleology, geography, forestry and tourism. It is believed that after the end of the project essential modifications to the national parks of Romania will be made through their reorganization and even by cancelling those which do not correspond to the established criteria.

The establishment of national parks involves large land areas and consideration of the type of property (state or private), land use, economic activities and not least the opinion of local populations.

Land redistribution is a serious threat for Eastern Europe, especially in national parks where the change in land tenure threatens their management or where development pressure on the ecological components puts in question the viability of the park's existence. Fortunately in Romania, as a consequence of Law 18, which stipulates that all protected areas are public land, land privatization does not represent a threat.

Generally, as far as possible, national parks do not include localities with a large number of inhabitants, industrial objectives, intensely used roads and railways networks on private properties.

In each case, depending on the elements mentioned above, the proposed areas forming the national parks will be classified within the most appropriate IUCN management category: II – National Park; V – Protected Landscape; IX – Biosphere Reserve; as parts of the National Protected Areas Network, or will be protected at local level, with appropriate management.

Once finalized, the proposals for the establishment of national parks will be submitted to Parliament for approval, together with proper management plans, involving important financial resources to be allocated from the state budget.

Of great importance is the foundation for each national park of an administration, a scientific council and an ecological guard, responsible for their management. Obviously, for the moment, Romania is confronted with the lack of a rangers' network, well trained to ensure that protected areas are properly used and maintained, and so with insufficient control over the threatening factors.

The Danube Delta was declared a Biosphere Reserve (591 200 ha) in September 1990 (by Decree no. 983), as a 'Ramsar' site in May 1991 and over 50% of its area was placed on the World Heritage List in December 1991. A park administration has been established in the Danube Delta Reserve, including an ecological corps of rangers. There is already a training programme developed and used for the training of the Danube Delta guards. The concept of that programme involving physical training, foreign language training, courses in biodiversity, education and servicing of the public can be used also for other parks and protected areas.

A biodiversity conservation project proposed by MoWFEP is under consideration by IUCN as part of the Action Plan for Central and Eastern Europe. It involves a short-term investment for Retezat National Park management and conservation. The proper management of Retezat and its immediate surroundings will provide a model for the other national parks of Romania.

Existing national parks

The National Parks of Romania cover the most spectacular natural landscapes of the Romanian Carpathians. In the Northern Carpathians there are Rodna National Park and Caliman National Park. A part of Rodna National Park, namely Pietrosul Rodnei, was declared in 1980 as a Biosphere Reserve. The metamorphic crystalline rocks of the area were transformed by Quaternary glaciation resulting in a strongly fragmented relief, glacial cirques and 13 glacial lakes. The Caliman National Park preserves the largest and most complex volcanic structure of Romania's territory, 'Caliman Caldera'. As a result of intense volcanic activity there are many little volcanoes surrounded by pyroclastic deposits in the south and the huge stratified volcano and Caliman Caldera in the north. In this area exists the entire group of eruptive rocks in the shape of lavas, pyroclastites and intrusive bodies that appears all along the Neogene Eruptive chain of the Eastern Carpathians. Nine features are strictly protected here.

Ceahlau National Park is located in the centre of the Eastern Carpathians in the Cretaceous flysch. This geological structure is developed in thrust sheets. The special features of Ceahlau Mountain are represented by the Albian (K_1) conglomerates that form specific erosional features, and by the existence of over 60 endemic species of which 8 species are declared as Nature Monuments. Special attention is given to 15 nature sites here.

The establishment of Bicaz–Hasmas National Park, located in the limestone area of the Eastern Carpathians, refers to conservation and protection of the Bicaz gorges and other spectacular endo- and exokarstic features, such as karst-pits, caves, dolines, lapies and also forest and meadows.

The Bucegi National Park is the last in the Eastern Carpathians. In the area over 53 sites are protected as follows: 3 scientific reserves, 9 nature reserves, 2 protected landscapes and 39 Nature Monuments, representing about 32% of the total area. These sites are fossiliferous deposits in limestones and other Jurassic and Cretaceous rocks, caves, erosional features, river gorges and forest. Unfortunately there are many threats that have already affected the area of Bucegi National Park and some of them are still present: uncontrolled tourism, overgrazing, roads and tourism facilities, limestone quarries, water abstraction, barrage building and severe soil erosion.

The Southern Carpathians are represented by 7 national parks. The first of these is Cozia National Park composed of two areas separated by the River Olt. The Cozia Mountain is a great crystalline horst. The lithology of the area, especially gneiss, conglomerates, sandstones, marls and sands creates a large range of lithologically controlled relief forms: imposing cliffs, gorges, toothed ridges, sharp peaks and waterfalls.

Anina Mountains National Park includes Jurrasic and Cretaceous limestones disposed in strongly tectonized anticlines and synclines. This area is karstified presenting spectacular gorges, caves and karstic springs. About 16 sites are strictly protected in the area.

Cioclovina National Park was created to protect and conserve the unique geomorphological features of the Sureanu Mountains. The rivers here have cut spectacular gorges and defiles and the small brooks have generated karstic complexes composed of storeyed caves of different ages and supporting a variety of features. The exokarstic morphology is very complex as a result of long subaerial exposure and the evolution of limestone tablelands. About 31 sites, especially speleological, are strictly protected.

Cerna Valley National Park includes the southern Cerna Mountain and Mehedinti Mountains separated by the River Cerna. The specific features of the area are: existence of endemic Mediterranean species, thermomineral waters and impressive karstic features.

The last large protected are of the Southern Carpathians is Portile de Fier National Park ('Iron Gates') that includes the spectacular defile created by the Danube river through the mountains. The beautiful landscape contains a large biological diversity (over 4000 species of flora, of which 28 are endemic species, and 5300 species of fauna), fossiliferous deposits, caves and forest reserves.

The most representative karstic area of Romania is located in the Apuseni Mountains. Within the Apuseni National Park about 60 speleological Nature Monuments are protected, 2 scientific reserves and 23 mixed nature reserves. The karstic system is very complex including:

- 'Cetatile Ponorului' – the greatest complex of exo- and endokarstic forms in Romania;
- 'Piatra Altarului' cave – one of the most beautiful caves in the world, especially for crystalline forms;
- 'Humpleu' cave – the largest cave of Romania, with 12 halls over 100 m long;
- 'Ghetarul de la Scàrisoara' cave – containing the greatest underground fossil glacier in the world.

Cave fauna is well represented by 27 endemic taxa.

At the moment there are about 150 new localities (c. 85 000 ha) registered to be declared as protected areas after evaluation of their biodiversity and natural habitat value by the Romanian Academy.

For the present, special attention is being given to the establishment of a new national park in Dobrogea Mountains to conserve a particular landscape. The geological significance of the area is the unusual sculptural forms of the 'Pricopan' granite.

Conservation, access and land management conflict in upland glaciated areas of the Snowdonia National Park: a preliminary survey

K. ADDISON[1] & S. CAMPBELL[2]

[1] *University of Wolverhampton & St Peter's College, Oxford University, Oxford OX1 2DL, UK*
[2] *Countryside Council for Wales, Plas Penrhos, Bangor, Gwynedd LL57 2LQ, UK*

Abstract: The mountains of Snowdonia in northwest Wales contain some of the most spectacular upland glaciated landscapes in Britain. A long record of scientific investigation of these landscapes commenced during the formulation of the 'Glacial Theory' in the 1830s, and has continued unabated with a current major reappraisal focused on the Geological Conservation Review (GCR). The Ordovician rocks and Caledonian structures of Snowdonia have also received considerable recent attention. The region, therefore, constitutes a major scientific and educational resource for the study of Quaternary and Lower Palaeozoic environments – reflected by the relatively large number of GCR sites, National Nature Reserves (NNR) and Sites of Special Scientific Interest (SSSIs) found in the region.

The mountains are also the focal region of the Snowdonia National Park by virtue of the outstanding scenery created by this combination of geological forces. A substantial area is owned by the National Trust, although most remains in private ownership; there are also several public utility interests. 'National Park' designation has stimulated public awareness of the scenic and geological landscape resources, creating considerable visitor and recreational pressures.

Three separate mountain groups – Snowdon, Y Glyderau and Y Carneddau – illustrate the major problems of managing these pressures and resolving access conflicts, whilst also conserving the scenic and geological heritage. Each has distinctive Quaternary landscapes, visitor pressures and management issues. Snowdon is primarily an area of classic radial alpine glaciation with widespread and intense visitor pressure. Y Glyderau exemplifies ice-sheet glaciation with intense site-specific pressure. Y Carneddau endured alpine glaciation with permafrost ornamenting slopes and surviving pre-glacial plateaux, and are the least accessible mountains with only moderate visitor pressures. A range of policy and management solutions is discussed with respect to these nationally-important areas.

Outline regional geomorphology: the scientific resource

The mountains of Snowdonia cover an area of 1383 km^2 over 800 ft (244 m) OD in northwest Wales and comprise the highest relief of England and Wales (Fig. 1), including all 14 Welsh peaks over 3000 ft OD (915 m) and a further 38 over 2000 ft (610 m). They rise abruptly up to 1000 m above the mostly subdued ancient basement rocks of Arfon and Anglesey which lie to the northwest across the Menai Strait fracture system. The late Proterozoic and Lower Palaeozoic lithology and structure of North Wales gained international renown during the nineteenth century (Addison 1990) and feature prominently among national and international stratotypes and type sites. The contemporary revision of the regional geology, and Ordovician terrains in particular, by the British Geological Survey (Howells *et al.* 1991) and the Geological Conservation Review (GCR) (Campbell & Bowen 1989) highlight the scientific importance of the region.

The mountain core is synonymous with the outcrop of Ordovician marine sediments and resistant Caradoc (Upper Ordovician) volcanic rocks and associated igneous intrusions. Their tectonic deformation, between convergent crustal plates closing the Lower Palaeozoic Iapetus Ocean (the Caledonide Orogeny), is reflected in the principal structures of Snowdonia (Kokelaar 1988; Treagus 1992). The summit crest follows the sinuous axis of the Ordovician syncline for 45 km southwest from Conwy and major glacial breaching has exposed magnificent examples of the synclinal structure, especially in Nant Ffrancon, Cwm Idwal and the Llanberis Pass. The NE–SW Caledonian strike was exploited differentially

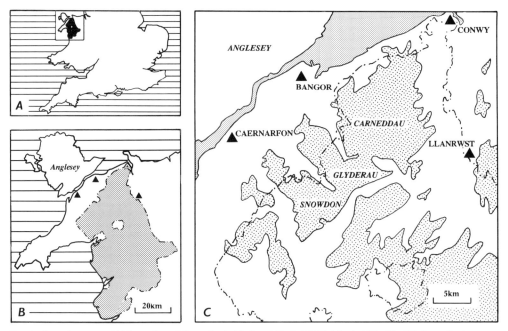

Fig. 1. Location map. (A) Location of Snowdonia National Park (SNP) (shaded black). (B) Snowdonia National Park and North Wales region. (C) Northern area of SNP and principal mountain groups of Snowdonia; land over 305 m (1000 ft) shaded.

by glacial cirque erosion (Addison 1981). These glaciated mountains contrast markedly with the subdued uplands of the Cambrian Mountains southeast of Snowdonia, formed mostly of Lower Palaeozoic (especially Silurian) marine sediments. They rarely exceed 600 m OD and have been eroded into a system of erosion plateaux and incised valleys (Brown 1960).

Snowdonia and northwest Wales are also internationally important for their Quaternary land systems, stratigraphy and palaeoenvironmental history. The development of the 'Glacial Theory' early in the nineteenth century and its application to the British Isles was strongly influenced by Trimmer's discovery of marine shells at 426 m OD on Moel Tryfan in 1831, Buckland's description of 'glacial-diluvial' landforms in Nant Ffrancon in 1842 and, in the same year, the account of the Cwm Idwal moraines and glacio-marine sediments in Arfon by Charles Darwin. Since then, well over a hundred scientific papers have been published on the Quaternary geology and geomorphology of the region, with enhanced status through the contemporary revolution in glacial sciences and concern for rapid environmental change. This was recognized by the first published GCR volume *Quaternary of Wales* (Campbell & Bowen 1989) and Quaternary Research Assocation's *The Quaternary of North Wales: Field Guide* (Addison *et al.* 1990).

The geological and geomorphological scientific resource is confirmed by 29 GCR and 2 NNR sites located within the 6 mountain groups of Snowdonia (Fig. 2). It might appear that the grand scale and abundance of individual sites in upland glaciated areas obviates the need for a conservation policy; it is, after all, the case that the geological foundation of Snowdonia has survived for 400 million years and the Pleistocene legacy for over 10 000 years. However, there are competing land-use, site maintenance and access pressures which could threaten the integrity of and access to this scientific resource in Snowdonia. This is recognized by the 'Integrity' site designation of the NCC's Earth Science Conservation Strategy (1990) and we examine here its application to the upland glaciated landscapes of Snowdonia.

Site descriptions of glaciated uplands

The mountains of Snowdonia represent the surviving elements of a broad dome, created

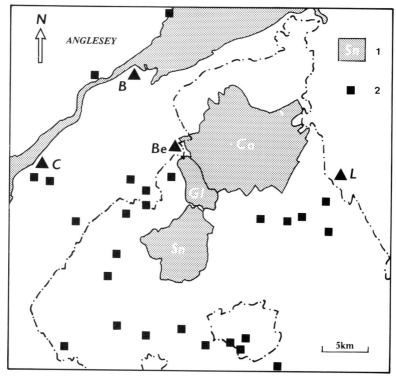

Fig. 2. Quaternary GCR sites in and around the northern Snowdonia National Park. (1) Principal GCR areas (*Ca*: Carneddau; *Gl*: Glyderau; *Sn*: Snowdon; (2) other GCR sites.

by Tertiary crustal uplift and subsequently dissected by intense Pleistocene glaciation. Major glaciated troughs, breached by over-riding ice under ice sheet conditions, fragmented the mountains into 6 discrete groups (Fig. 1). Llanberis Pass and Nant Ffrancon, the two most impressive troughs over 900 m deep, isolate the Snowdon, Glyderau and Carneddau groups from each other and provide access routes through the heart of the range. Alpine glaciation exploited the great diversity of site and aspect opened up by the excavation of the outlet glacier trough network.

Northwest Wales was glaciated several times during the last 0.5 Ma of the Mid and Late Pleistocene and probably at least twice during the most recent (Devensian) cold stage. Early Devensian Welsh ice advanced as far as Caernarfon c. 70 000 yrs BP (Addison & Edge 1992) and the region was inundated by Welsh and Irish Sea ice in the western sector of the British Late Devensian ice sheet c. 18 000 yrs BP. The emphasis then switched from ice sheet (polar) to alpine (temperate) glaciation as deglaciation proceeded, with a final minor readvance of cirque glaciers between c. 11–10 ka (Younger Dryas). Severe cold during the intervening Mid-Devensian interstadial and again during the Younger Dryas stimulated permafrost activity to ornament the glacial landscape. Although evidence of all three styles of cold stage processes occurs throughout Snowdonia, each mountain group described here is also geomorphologically distinct.

Snowdon

Physical setting and context

Snowdon is internationally important for its glacial geomorphology and stratigraphy. More a single complex massif covering 85 km^2 than a mountain group, it is an outstanding monument to alpine glaciation. Two multiple-basin cirque staircases, 3 large cirques and 4 smaller ones were excavated asymmetrically into the core pyramidal peak or horn of Yr Wyddfa (1085 m OD), virtually destroying the pre-glacial plateau. Many are ringed by impressive arêtes which isolate the satellite summits of Carnedd

Ugain, Crib Goch and Y Lliwedd; there are 25 cirque forms altogether (Unwin 1973). Cirque orientation and altitude reflect fluctuating snowline conditions during glacial stadia and recent attempts to reconstruct the glacier mass balance and dynamics of the last Snowdon glaciers focus on the impressive staircase of 4 cirques forming the 'Snowdon Horseshoe' (Addison 1987; Gray 1982a) at the core of the National Trust Snowdon Estate.

For all its classic alpine landforms and radial pattern of glaciers, transfluent ice-sheet discharge swept past Snowdon at the glacial maximum, breaching the pre-glacial watershed in Nant y Betws and the Llanberis Pass. This was recognized from striations high above the Pass as early as 1860 by Ramsay and, currently, from small-scale, abrasional landforms in the artificially-controlled waters of Llyn Llydaw (Gray 1982a). East Snowdon ice was clearly drawn northwest into the transfluent outlet glacier before resuming an easterly flow on return to alpine conditions.

Smaller-scale, more sensitive landforms – contrasting with the grand scale of the cirques – are also fundamental in establishing the late Quaternary history of Snowdon. Suites of valley and cirque moraines complete the alpine geomorphic inventory and are interpreted widely as marking either stages in the retreat of Late Devensian glaciers or the furthest advance and subsequent retreat of smaller Younger Dryas cirque glaciers (Gray 1982a). Relative dating is constrained precisely at a number of sites by the ^{14}C age of organic sediments, which accumulated in lake basins ponded by moraines and in cirque rock basins. Such sites in Snowdonia helped to establish the extent and duration of the Younger Dryas in Britain. A number of 'moraines' have been re-interpreted as protalus ramparts, indicative of nivation processes around snow banks rather than glaciers, and thus form important evidence of a late permafrost climate.

Conservation status

The Snowdon massif contains two separate GCR sites, Snowdon and Cwm Dwythwch, selected as part of the *Quaternary of Wales* network. The Snowdon (Yr Wyddfa) site is by far the larger, occupying c. 37.5 km^2. It was selected specifically for its large- and small-scale features of glacial erosion, its complexity of Younger Dryas moraines and protalus ramparts, other periglacial features and Holocene vegetation history. The site includes the principal cirques radiating from the summit of Snowdon and representative sections of the spectacular glacial troughs in the Llanberis and Gwynant valleys.

The site falls mostly within a large biological SSSI notified for a wide range of interests including arctic-alpine flora, sheep-grazed acid grassland, clubmosses, choughs and pine martens. A substantial part of the biological SSSI and GCR site fall within the Snowdon NNR, established in 1966 in recognition of this exceptional combination of biological and Earth science features. The GCR site is in multiple ownership, although substantial tracts are freely open to the public.

Y Glyderau

Physical setting and context

Separated from Snowdon by the Llanberis Pass, Y Glyderau cover 72 km^2 flanked to the east by the impressive glacial breach of Nant Ffrancon–Nant y Benglog, above which the mountain crest is punctuated by 11 parallel cirques showing remarkable structural and palaeoclimatic control. This relationship is interpreted as the composite product of transfluent, ice-sheet outlet glaciation and subsidiary (mostly younger) alpine glaciation – the latter concentrated on the sheltered, northeast face of the Glyderau where cirque glaciers exploited and exposed Caledonian structures (Addison 1988).

Despite such large-scale glacial excavation, involving the entire east face of the Glyderau and truncation of the high-level plateau overlooking Nant Peris to the west, the summit plateau appears to have survived as a nunatak. Intense frost action in this zone created tors, blockfields and stone stripes. A large debris ridge above the east flank of Nant Ffrancon, interpreted as a protalus rampart postdating ice retreat, and more recent debris flow activity in the same zone demonstrate a continuum of active slope processes from cold stage to interglacial. Many areas of glacially-oversteepened slopes are still in disequilibrium and active geomorphologically.

The northern cirques are closely associated with the Nant Ffrancon trough and its infilled lake basin, which has provided the fullest, dated sequence of progressive deglaciation and Late Devensian/Holocene environmental and vegetational changes in Snowdonia. Cwm Idwal, the largest cirque, is also probably the best known individual site in Snowdonia. Its international importance is associated with Charles Darwin's description of the Idwal moraines in 1842 and its designation as the first NNR in Wales in 1955 (Hughes *et al.* 1986; Williams *et al.* 1987).

Conservation status

Y Glyderau contain 4 GCR sites selected independently for their Quaternary (Y Glyderau), Cambrian (Cwm Graianog), Ordovician (Cwm Idwal) and Caledonian structure (Cwm Idwal) features. The Quaternary site of Y Glyderau is by far the largest, occupying 17.5 km^2 and containing individual sites in Cwm Idwal selected for their Caledonian and Ordovician interests. Further GCR sites at Llyn Peris (Quaternary), Llyn Padarn (Cambrian) and Dyffryn Mymbyr (Caledonian Structures) lie on the flanks of the massif. Like Snowdon, Y Glyderau Quaternary site was selected primarily for its glacial erosional features – particularly the series of structurally-controlled cirques along the west side of Nant Ffrancon, which provide one of the most spectacular glaciated landscapes in Britain and contrast markedly with the radial pattern of Snowdon cirques.

Cwm Idwal lies at the heart of this site and its three, independent GCR interests are listed in separate GCR site selection networks: *Caledonian Structures in Britain; Ordovician Igneous Rocks of Wales* and *Quaternary of Wales* (NCC 1990). It is also a Grade 1 Nature Conservation Review (NCR) site and Llyn Idwal has been proposed as an internationally-important wetland under the Ramsar Convention. A substantial biological interest throughout Y Glyderau is recognized by SSSI status and, at Cwm Idwal, the combined geological, geomorphological, botanical and zoological value of the site is emphasized by its designation as Wales' first NNR. Leased to CCW by the National Trust, the grazing rights on its 298 ha are tenanted and all parts of the reserve are open to the public.

However, the cwm can be reached within 10 minutes' walk of the A5 Euro-route through the mountains, subjecting it inevitably to intense educational and leisure pressures. Human trampling greatly exacerbates existing damage to sward, mires, screes and paths due to sheep-stocking levels. Site-specific conservation requirements, competing land uses and pressures in Cwm Idwal epitomize regional problems.

Y Carneddau

Physical setting and context

Y Carneddau are the largest mountain group in Snowdonia, covering 238 km^2 and are physically the most isolated. Radial ice discharge at glacial maximum via Nant Ffrancon to the west and Nant Conwy to the east relieved Welsh ice-sheet pressure on the Carneddau and only the lower eastern areas were overridden. Alpine glaciation prevailed, centred in 7 cirques on Carnedd Llewelyn (1064 m OD); consequently, the mountains were less dissected and more of the summit plateau survives with 8 peaks over 3000 ft (915 m), providing the Carneddau with their distinctive character (Campbell & Bowen 1989).

Glacial breaching was confined to parallel troughs holding the eastern lakes Cowlyd, Crafnant and Geirionydd. Elsewhere, glacial geomorphology focuses on the range of cirque, moraine and other ice-limit landforms (Gray 1982*a*) whose diversity was facilitated by the absence of a controlling ice-sheet. In this sense, Y Carneddau are the most alpine of the 3 groups but their remoteness discourages substantial scientific investigation and a large research resource remains.

The survival of extensive high-altitude plateau surfaces in Y Carneddau – the largest area over 900 m OD in Wales – was instrumental in the formation of a range of periglacial landforms (Tallis & Kershaw 1959; Ball & Goodier 1970). Moreover, frost-action continues locally with excellent patterned ground on low-angled summits on volcanic rocks. It is probable that the widespread and conspicuous summit blockfields were formed during the Devensian cold stage (Scoates 1973). With extensive younger, mostly Holocene, scree formation degrading glacially-oversteepened slopes and contemporary summit cryoturbation, the array of cryogenic forms represents an instructive continuum of processes without equal elsewhere in Wales.

Conservation status

Y Carneddau contain multiple GCR interests with the core of the massif having been selected as a *Quaternary of Wales* site for its outstanding periglacial landforms. The Quaternary site occupies an area of *c.* 22.3 km^2 and includes the smaller *Ordovician Igneous* site in Llyn Dulyn. Whereas Snowdon and Y Glyderau provide prime examples of glacial erosional and depositional landforms, more of the pre-glacial topography survives in Y Carneddau and this has become ornamented with an exceptional range of peri-glacial features including stone stripes, polygons, blockfields and solifluction lobes (Scoates 1973; Campbell & Bowen 1989). The site also provides exceptional Younger Dryas glacial landform assemblages (Cwm Dulyn and Cwm Ffynnon Llugwy) and a detailed Holocene environmental record (Cwm Melynllyn; Walker 1978). Like Snowdon and

Y Glyderau, much of Y Carneddau forms a biological SSSI and is in multiple ownership. Most of the area is freely open to the public.

Conservation, access and management conflict in the national park

Management constraints

Statutory authority for the management of the national parks in Britain lies with the respective National Park Authority (NPA) but the balance of power and influence over activities in the parks has always been complex and at times confusing. The two prime objectives established under the National Parks and Access to the Countryside Act 1949 (Section 5) of (i) preserving and enhancing the natural beauty of the areas and (ii) promoting their enjoyment by the public, were augmented and qualified by the requirement to have 'due regard to their social and economic interests, as well as to the needs of agriculture and forestry;, set out in the Countryside Act 1968 (Section 37). The full nature and extent of confusing and often conflicting influences on the purpose and management of the parks – leading to potential conflict of interest – is beyond the scope of this paper but they might be summarized as follows:

- inadequate NPA resources and powers, often divided between elected authorities;
- a recent fundamental shift from NPA's traditional role in managing development to influencing land management;
- frequent subjugation of principal park aims by other considerations, most notably the primary objectives and responsibilities of public and statutory bodies;
- land ownership concentrated in the hands of public authorities and, particularly, private landownership;
- the impact of Central Government and EC policies, especially on agriculture and forestry;
- uncoordinated and often conflicting interests between Government ministries, especially the Department of Environment and Ministry of Agriculture, over agriculture, forestry and conservation;
- the absence of any initiative on national parks or protected landscapes in EC policy;
- changing long-term economic, resource and environmental management needs and strategies;
- major shift in the nature and magnitude of social, leisure and recreational pressures on national parks;
- area-specific direct conflicts of potential land-uses.

These issues are immediately evident in the Snowdonia National Park, where only 1.2% of its 838 square miles (2171 km^2) are owned by the NPA, 1.7% by the Countryside Council for Wales (CCW), 8.9% by the National Trust compared with 15.8% by the Forestry Commission and 69.9% in private ownership. Twenty-five public bodies or agencies operate in the park where there are also 17 National Nature Reserves (NNRs), 12 Forest Nature Reserves (FNRs), 1 Royal Society for the Protection of Birds Reserve (RSPB) and 45 Sites of Special Scientific Interest (SSSIs) covering an area of 388 km^2 – the largest concentration of statutory conservation sites in any British national park, with 15.4% of the total (Snowdonia National Park Authority 1987).

Land management of this complex operating environment by the NPA is effected through the formulation of Development Plans under guidance from the Welsh Office, implementing Central Government policy, and a National Park Plan leading to 5-year programmes of implementation. These plans contain important policy and management statements for the glaciated uplands of Snowdonia. The NPA brief also extends explicitly to a regard for the social, economic and cultural well-being of the local communities, with its additional but incompletely-defined ethnic distinction of *Cymreictod* (Welshness) – unique in British National Parks – and the promotion of the public enjoyment of Snowdonia. These concerns are shared by the Report of the National Parks Review Panel (1991) noting that:

'... the environmental concerns of the National Parks must embrace more than the protection of fine scenery and should take an active interest in all environmental attributes and cultural traditions which contribute to their high quality ...'

Park Authority geological and geomorphological conservation strategy

Section 3 of the Wildlife and Countryside (Amendment) Act 1985 conferred responsibility on the national parks to:

'... prepare a map showing those areas of mountain, moor, heath, woodland, down, cliff or foreshore (including any bank, barrier, beach, flat or other land adjacent to the foreshore), the natural beauty of which, the Authority considers it is particularly important to conserve.'

The SNP Section 3 Map (Fig. 3) and accompanying Policy Statement defines the specific nature of, and its policy towards, these environmental categories in Snowdonia. The Section 3 Map covers 73.5% of the park area – the highest value for any British national park – with 52% defined as 'Mountain and Moor', 4% 'Woodland', 1.5% 'Coast' and a further 16% in a sub-category defined as 'Areas of wooded character'. Although the document acknowledges that the 'extensive wind swept uplands and rocky crags' – essentially, the glacial inheritance and its subsequent superficial modification by natural and anthropogenic processes – were the *raison d'être* for the park's designation as 'National Park' in 1950, the sole parameter alluding to the role of geological and geomorphological substrates is relief and topography. Other deciding parameters of size and scale and context are not explained, although clearly these are heavily dependent on the geological foundation of Snowdonia. Nevertheless, the primary emphasis is on vegetation and land use systems and therefore so too are most of the policy statements.

It is our case that the geomorphological integrity of Snowdonia depends on the recognition of (i) glacial and permafrost land systems at the large scale and (ii) characteristic landform and stratigraphic exposures at discrete sites. This is given only informal recognition in the Section 3 Policy Statement (Snowdonia National Park Authority, 1991) by the aim to 'Conserve the natural beauty of the mountains and moorlands' and in the *Snowdonia National Park Plan: First Review* (Snowdonia National Park Authority, 1987) 'To take measures to protect the landscape and ecological value of the Park.', in line with the 'Sandford principle' that the preservation and enhancement of the natural beauty of the area should prevail over public enjoyment and recreation.

The Conservation Document of the Plan Review is more explicit:

'The glaciated topography of Snowdonia has a marked effect on upland ecology. Steep slopes, cwms and ridges combined with

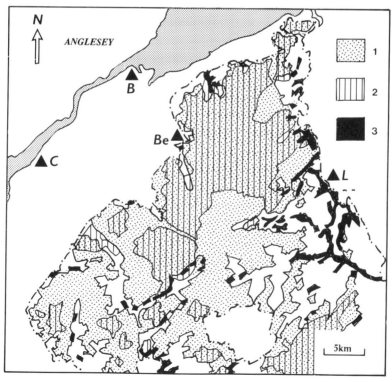

Fig. 3. Section 3 Map of the northern Snowdonia National Park. (1) Mountain and moorland; (2) statutory nature conservation sites and common land; (3) woods. B: Bangor; Be: Bethesda; C: Caernarfon; L: Llanrwst.

varying geology and soils form a variety of distinctive habitats. Higher mountains frequently have a deep layer of frost-shattered rock debris whilst lower slopes have varying layers of glacial drift. Soils on the whole however tend to be shallow.'
(6.3.2)

and makes 5 Policy Statements which assist the conservation case:

- A landscape classification map will form the basis of the NPA's decisions on policies relating to conserving and enhancing the natural beauty of the Park. [42B]
- The NPA will continue to support the Nature Conservancy Council in its efforts to conserve the habitats and landscapes of National Nature Reserves, all sites included in the Nature Conservation Review and other designated sites within Snowdonia. [42C]
- The NPA will press the NCC to utilise its powers under Section 15 of the Countryside Act 1968 to promote the enhanced conservation management of land not only on SSSIs but on all important habitats in Snowdonia. [42CH]
- The NPA will resist developments which threaten to damage the character of upland glaciated landscapes. [11A]
- The NPA whilst reiterating its demand for forestry operations to be subject to planning control will work closely with the Forest Consultative Panel to resolve conflicts between conservation, recreational and forestry interests. [45A]

Geomorphological conservation interests are also recognized specifically for the first time in Recommendation 3.1 (i) of the National Parks Review Panel (1991):

> to protect, maintain and enhance the scenic beauty, natural systems and <u>land forms,</u> and the wildlife and cultural heritage of the area; (our underlining)

Despite this support, we share NPA concern (Plan Review) that 'no fully effective method of conserving both the landscape and wildlife of the countryside exists'. The NPA identifies the most important threats as coming from continuing changes in use of the uplands from 'badly conceived but also incremental agricultural and forestry operations, and other forms of development' and – referring to SSSIs – concedes that even 'the means of safeguarding the nation's areas of nature conservation importance has its weaknesses.'

Conservation and access conflicts

With these caveats in mind, we now highlight the principal sources of conflict in the conservation of, and access to, the geomorphology of upland glaciated landscapes in the park (see also Table 1).

Forestry Large-scale afforestation, particularly with plantation conifers, is widely regarded as inimical to the conservation of those scenic and other national park resources which depend on extensive open land. In Snowdonia it is restricted to less-elevated areas of the eastern Carneddau forming part of Gwydyr Forest and in the northern Hebog group, west of Snowdon,

Table 1. *Qualitative summary of actual and potential land use conflicts hazarding the scientific resource in the 3 principal mountain groups of northern Snowdonia. Circles indicate* major *(solid),* intermediate *(half) and* minor *(open) degrees of conflict. To the right, the available extent of site interpretation and educational information – from most to least – is also shown for each group.*

	Forestry	Water resources	Hydro-power generation	Quarrying and mining	Construction	Visitor pressure	Footpath erosion	Site interpretation	Educational information
SNOWDON	◐	○	◐	○	◐	●	●	●	◐
Y GLYDERAU	○	◐	●	●	●	◐	◐	◐	◐
Y CARNEDDAU	●	●	●	◐	○	○	○	○	○

where it illustrates forest impacts on upland glaciated land systems.

The eastern Carneddau were overrun and heavily abraded by ice from the south, forming a *knoch* and *lochan* land system of parallel lake basins, bogs and large-scale *roches moutonées* rarely found in Wales. The variety of relief undoubtedly enhances forest vistas and the larger lakes (Llyn Gerionnydd and Llyn Crafnant) prevent blanket cover, but an important Quaternary resource is obscured and access to valuable mires severely restricted. Beddgelert Forest in the eastern Hebog group exemplifies the impact of plantation on cirque landforms; only higher, degraded rock slopes survive and the entire geomorphological configuration of rock basin, moraines and surviving or infilled cirque lakes is lost.

The treeless character of British uplands is the product of agricultural and other human interventions over past millennia. The upland environment can support either extensive pastoral farming systems or a return to woodland. Both are dependent on economic subsidies and agricultural pricing policies are determined increasingly at intergovernmental level – well beyond the reach of NPAs.

The introduction in 1988 of an assumption against further upland afforestation in England does not apply to Wales, although a consultative panel has existed in SNP since 1955. Updated in the form of a Forestry Agreement in 1975, it identifies areas where there are **assumptions** for and against afforestation in Snowdonia (Fig. 4). There are many specific instances of policy and planning cooperation between the NPA and public and private forestry interests and it is recognized that in many districts, sympathetic planting has enhanced scenic and recreational resources. Most land in the three mountain groups is covered by existing constraint or an **assumption** against, but the ability to resist future afforestation is far from secure.

However, in recognizing pressures in the forest industry for rapid expansion, the NPA acknowledges the land capability for commercial forestry throughout the park, that no special economic concessions apply within the

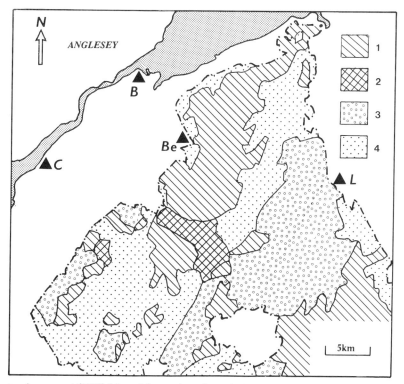

Fig. 4. Forestry Agreement (1975) Map of the northern Snowdonia National Park. (1) Areas under existing constraints; (2) areas with a strong presumption against forestry; (3) areas with a strong presumption that afforestation will be acceptable; (4) areas where afforestation proposals **might** be acceptable, although there is a strong presumption against.

national parks and that the Forestry Commission 'does not appear to recognise the special status of National Parks' (Plan Review). Further, it concedes that the principal constraint to afforestation is other land use and **expects** further rapid expansion. Consequently, the NPA Section 3 Policy Statement on forestry presumes against afforestation in mountain or moorland areas and asserts that 'The planting of mountain or moorland with either conifers or substantial areas of broad-leaved woodland will be resisted.'

It would seem unwise to rely on the essentially voluntary basis of agreement and, indeed, the NPRP Recommendation 7.1 states that 'The Government should declare that no further major coniferous afforestation should occur in national parks.' In particular, the 1975 Forestry Agreement is ambiguous with large areas of the park classed as 'Areas where, although there is a presumption against afforestation, proposals might be acceptable'; this extends to large tracts of southwest Snowdon and eastern Carneddau.

Water Resources It is inevitable that the large number of natural, glacially-excavated lakes feature prominently in the acquisition and regulation of water resources in Snowdonia, which experiences some of the highest rainfall in Britain with Snowdon summit receiving $\geqslant 5000$ mm annually (Embleton, 1962). Many lake levels are regulated by dams or weirs and their resource range embraces domestic and industrial supply (e.g. Llyn Ffynnon Llugwy, Llyn Ogwen), hydroelectric generation (Llyn Eigiau, Llyn Cowlyd, Llyn Peris, Llyn Llydaw), river compensation (Llyn Ogwen), recreational uses (Llyn Crafnant, Llyn Padarn) and – formerly – extractive industry (Llyn Glaslyn, Llyn Geirionydd). Some lakes are interconnected by leats and tunnels.

The existence of numerous natural lakes reduces the pressures for artificial reservoirs in an important source area within national water strategy. The impact is limited generally to the artificial raising of water levels by small dams (although these often obliterate glacial moraine dams), pipes and ancillary structures. However, glaciated rock basins offer considerable storage capacity and recent reservoir construction elsewhere in North Wales in less protected but less suitable terrains, and even within Snowdonia, demonstrate the vulnerability of glaciated landscapes to changing water and conservation strategies. Most Devensian and Holocene sediments and depositional landforms are located in valley floors and thus endangered by reservoir construction. Glacial lakes also form an important scenic asset.

Power generation Recent construction for electricity generation in Snowdonia illustrates the vulnerability of lake environments and the unpredictable value placed on mountain scenery. The largest pumped storage scheme in Europe opened at Dinorwic in 1984, with an average station output of 1740 MW – enough to power Wales – specifically utilizing glacial lakes in tandem. Llyn Marchlyn Mawr, occupying a cirque in the western Glyderau, and Llyn Peris in the Llanberis Pass act as the upper and lower reservoirs respectively. The scheme obliterated Cwm Marchlyn moraines and subjected Llyn Peris to diurnal fluctuations in level, but did permit access to important glaciated bedrock surfaces (Gray, 1982b) and lacustrine sediments (Tinsley & Derbyshire, 1976) during construction.

Most importantly, the potential impact of a major generating plant and ancillary works on the oustanding scenery of the western Glyderau and Llanberis Pass was recognized by the Central Electricity Generating Board and planning authorities by siting the power station underground – even though it is just outside the SNP boundary. In stark contrast, the same generator reconstructed a 2 km penstock **overland** in the Snowdon Horseshoe in 1988, serving a minor hydroelectric plant from Llyn Llydaw, despite considerable protest from conservation bodies.

A smaller pumped storage scheme, using glacial lakes in tandem but without locating the power station underground, was developed at Ffestiniog southeast of Snowdon in the 1960s, and the Dolgarrog aluminium smelter was powered from a hydroelectric scheme built in 1924, harnessing 6 glacial lakes in the Carneddau. Other sites in Snowdonia have been investigated for their hydroelectric potential and continuing changes in generating strategy may create pressure for future development.

Quarrying and mining The Lower Palaeozoic rocks of North Wales have long been a source of metal ores and construction materials although this has diminished considerably over the past few decades as reserves ran out or became uneconomic. Slate was the most extensively quarried rock, primarily from the Cambrian belts flanking the Snowdonian syncline. The SNP boundary was judiciously steered around the world's two largest quarries, in the western Glyderau, and around Nantlle and Blaenau Ffestiniog southwest and southeast of Snowdon respectively. Similarly, large and still-active igneous road-stone quarries around Penmaenmawr were excluded from the park in the northwest Carneddau.

Metal ores were mined much less extensively and mining finally ceased in 1958. The largest single area was the lead–zinc belt in the eastern Carneddau, with copper mined on Snowdon and in Nantgwynant. Abandoned mines and quarries furnish an important resource for tourism and industrial archaeology; conservation problems are associated primarily with vast spoil tips and polluted ground, and whatever geomorphological merit abandoned sites may have possessed is beyond reclamation. Extractive industry therefore appears to pose few new threats to the scientific resource and, indeed, in some instances facilitates access. However, two recent developments – reassessment of the ore body on Parys Mountain (Anglesey) and of the status of planning consent for slate quarrying on Moel Tryfan (west of Snowdon) – demonstrate the episodic nature of the industry. Although outside Snowdonia, they are indicative of the remaining potential within the park boundary.

Road construction Northwest Wales is currently in a phase of major road improvements under regional (Gwynedd County Council) and national (Welsh Office) direction, particularly with respect to the A55 Euro-route which skirts northwest Snowdonia. Ancillary improvements to assist local, tourist and regional traffic through the park area, and limited provision of visitor parking, usually involves widening schemes. These create temporary sections of potential geological interest which are usually then lost through grading. Stabilization of slopes undercut in this way has been extended recently to older road sections, particularly in Nant Ffrancon, Llanberis Pass and Nant Gwynant. Methods vary from massive concrete and masonry walls to masonry aprons and re-vegetation over a geotextile base.

Although road construction usually operates at smaller scales than other developments, and stabilization methods provoke arguments over visual amenity or intrusion, conservation opportunities may be lost. Of greater concern was the announcement that the A5 – which follows Nant Ffrancon – Nant y Benglog through Snowdonia – was marked out by the EC for upgrading to motorway standard. The scale of this would destroy the geomorphological integrity of the route and, although the proposal appears to have been rejected, it is indicative of threats posed by changing perceptions and economic strategies.

Visitor pressures The *Snowdonia National Park Plan* (1977) estimated that there were over 4.7 million visitor-nights per annum in 1975; there were no estimates for the number of day visitors. Specific, small localities in national parks known as 'honeypots' attract very large numbers of visitors and the same plan revealed that 3 such sites in the park had more than 1 million visitors in the same year, and a further 19 had more than 100 000 visitors. Snowdon summit is reached by $c.$ 200 000 people per annum with $c.$ 2000 per day during the peak season. Direct visitor pressure on the landscape occurs in the form of path and slope erosion through trampling, vegetation loss and compaction; indirect pressure is exerted through provision of structures and facilities geared to tourism and recreation.

The impact of highway and related roadside structures was referred to above; in general, the NPA adopts a restrictive policy towards building provision and is sensitive to environmental and conservation requirements. The greater threat to the geomorphological resource occurs directly, through public access to the landscape on foot and off-road vehicles (including bikes) in the pursuit of recreational enjoyment; this is part of an internationally-recognized problem of promoting greater access to the countryside.

Serious erosion associated with footpaths occurs almost everywhere on Snowdon, which has 6 major paths extending over 36 km radiating from the summit to roadside access points; on the cluster of paths focused on Cwm Idwal, Tryfan, Y Garn and the main Glyder summits in the Glyderau; and the main Carneddau ridge including access via Pen yr ole Wen. Paths generally traverse moraine and wetland sites in valley floors and screes on steeper slopes; damage often extends well beyond the path to adjacent ground, threatening general 'unravelling' of sensitive soil and vegetation cover.

Heavily-used paths on steep slopes may expose vegetated scree and remobilize unvegetated scree slopes, a problem which becomes more extensive in areas of intensive climbing activity. The upper section of the Miner's Track on Snowdon, large areas of the Glyderau slope of the Llanberis Pass beneath popular rock climbs and high-level screes around the Glyderau and Carneddau summits, in particular, have been affected in this way. The impact of remedial measures as well as primary damage threatens not only the integrity of many landforms but also future access and visual amenity. There is an emerging skill in the sensitive reinforcement of paths but pre-emptive measures are hampered by inadequate funding.

Conclusion

The impressive scale and range of cold-climate landforms in Snowdonia has been altered relatively little by the impact of both climatic amelioration during the current interglacial and by concomitant human activity during the past 10 ka. W.M. Davis (1909), a founding figure of modern geomorphology, recognized this in describing Snowdonia as '... admirably developed for field study. There is nothing in the United States east of the Rocky Mountains that can be compared with North Wales in the way of vivid exhibition of ... glacial features'. Indeed, Snowdonia provides an outstanding, compact and accessible microcosm of ice-sheet, alpine and permafrost landscapes. This, together with varying patterns of land use, ownership and public access creates a complex mosaic of management and conservation interests.

It is widely accepted that insufficient powers are vested in British National Park Authorities and that management and conservation are often subordinated to perceived national or international interests and changing economic strategy. Although the National Parks Review Panel is pressing for improvements in powers and funding and there is a more sensitive climate to environmental protection, the position is unlikely to change substantially in the short term.

A successful conservation strategy, able to respond to future conflicts of interest in particular, therefore needs to work more effectively within existing frameworks. The National Parks Review Panel has recommended that responsibility for the management of NNR sites and SSSI in national parks should be delegated to the NPA from CCW and other national conservation bodies (recommendation 4.2.2, 1991) but this would risk sacrificing in-depth scientific expertise for parochial views of a national resource. Purely in terms of an understanding of the scientific resource, it is our view that NPAs are not best placed to respond and there needs to be close liaison with the statutory conservation bodies – CCW in Wales, English Nature in England and the Countryside Commission.

The need for an integrated strategy is underlined by the qualitative assessment of the extent of conflict and hazard to the scientific resource in the 3 mountain groups, summarized in Table 1. This suggests that a greater threat is posed by alternative, large-scale land-use potential in the Carneddau than elsewhere, with visitor pressure and footpath erosion more extensive in the Glyderau and, especially, Snowdon. Greater access to the Carneddau could reduce intense visitor pressure elsewhere but would probably be resisted by the NPA who have designated the Carneddau a 'quiet zone'.

An extension of 'Zoning' in the mountains with a wider set of criteria contributed by all interested parties might emerge from an integrated strategy, and consideration should be given to allowing subject-specific integrity of some areas to be sacrificed for cast-iron protection elsewhere. This clearly requires a far better understanding of the scientific resource and an agreed differentiation between the biological and Earth science components. A 'whole landscape' approach is required and it is evident from the great diversity of landscape character, pressures and conflicts within 3 mountain groups alone, illustrated here, that blanket conservation policies are meaningless.

The voluntary sector should also be drawn fully into the pursuit of an integrated strategy. The RIGS scheme permits a wide range of user-interest to be assessed outside the statutory authorities in designating sites, initially one tier below SSSI and GCR level, which are notified to planning authorities. RIGS groups provide a marriage of professional and lay interest in conservation and provide an important bridge with educationalists and the general public for whom, after all, the national parks were designated. Education of the wider public into an appreciation of the origins of our landscapes and the purpose and significance of geomorphological conservation, completes this strategy.

The authors wish to acknowledge the assistance of the following in the preparation of this paper: Trefor Jones (formerly Principal Planning Officer, Future Planning, Snowdonia National Park Authority and now Head of Policy & Future Planning, Ynys Môn District Council); Peter Ogden (Head of Policy & Local Planning, Snowdonia National Park Authority).

References

ADDISON, K. 1981. The contribution of continuous rock-mass failure to glacier erosion. *Annals of Glaciology*, **2**, 3–10.

—— 1987. *Snowdon in the Ice Age*. Addison Landscape Publishing, Broseley.

—— 1988. *The Ice Age in Y Glyderau and Nant Frrancon*. Addison Landscape Publishing, Broseley.

—— 1990. Introduction to the Quaternary in North Wales. *In*: ADDISON, K., EDGE, M. J. & WATKINS, R. (eds) *The Quaternary of North Wales: Field Guide*. Quaternary Research Association, Coventry, 1–19.

—— & EDGE, M. J. 1992. Early Devensian interstadial and glacigenic sediments in Gwynedd, North Wales. *Geological Journal*, **27**, 181–190.

——, —— & WATKINS, R. (eds). 1990. *The Quaternary of North Wales*. Quaternary Research Association, Coventry.

BALL, D. F. & GOODIER, R. 1970. Morphology and distribution of features resulting from frost-action in Snowdonia. *Field Studies*, **3**, 193–217.

BROWN, E. H. 1960. *The relief and drainage of Wales*. University of Wales Press, Cardiff.

BUCKLAND, W. 1842. On the glacial-diluvial phenomena in Snowdonia and the adjacent parts of North Wales. *Proceedings of the Geological Society of London*, **3**, 579–584.

CAMPBELL, S. & BOWEN, D. Q. 1989. *Quaternary of Wales*. Geological Conservation Review, NCC, Peterborough.

DARWIN, C. 1842. Notes on the effects produced by the ancient glaciers of Caernarvonshire, and on the boulders transported by floating ice. *Philosophical Magazine*, **21**, 180–188.

DAVIS, W. M. 1909. Glacial Erosion in North Wales. *Quarterley Journal of the Geological Society of London*, **65**, 281–350.

EMBLETON, C. 1962. *Snowdonia*. Geographical Association, Sheffield.

GRAY, J. M. 1982a. The last glaciers (Loch Lomond Advance) in Snowdonia, N. Wales. *Geological Journal*, **17**, 111–133.

—— 1982b). Unweathered, glaciated bedrock on an exposed lake bed in Wales. *Journal of Glaciology*, **28**, 483–497.

HOWELLS, M. F., REEDMAN, A. J. & CAMPBELL, S. D. G. 1991. *Ordovician (Caradoc) marginal basin volcanism in Snowdonia (north-west Wales)*. British Geological Survey, HMSO, London.

HUGHES, R. E., GODWIN, H. & WOODHEAD, N. 1986. National Nature Reserves in Wales: a systematic survey, 7. Cwm Idwal NNR, Gwynedd. *Nature in Wales*, **3** (1 & 2), 79–90.

KOKELAAR, P. 1988. Tectonics of Ordovician arc and marginal basin volcanism in Wales. *Journal of the Geological Society*, **145**, 759–775.

NATIONAL PARKS REVIEW PANEL. 1991. Fit for the Future, (Report of the NPRP; R. Edwards, Chairman), Countryside Commission, Cheltenham.

NATURE CONSERVANCY COUNCIL. 1990. *Earth science conservation in Great Britain: A strategy*. Peterborough.

RAMSEY, A. C. 1860. *The Old Glaciers of Switzerland and North Wales*. Longman, Green, Longman and Roberts, London.

SCOATES, E. A. 1973. 'Cold climate processes in North Wales.' M. Phil thesis, University of Reading.

SNOWDONIA NATIONAL PARK AUTHORITY. 1977. *Snowdonia National Park Plan*. Gwynedd.

—— 1987. *Snowdonia National Park Plan: First Review*. Gwynedd.

—— 1991. *Section 3 (Wildlife and Countryside (Amendment Act) 1985: Conservation Map (Policy Statement)*. Gwynedd.

TALLIS, J. H. & KERSHAW, K. A. 1959. Stability of stone polygons in North Wales. *Nature*, **183**, 485–486.

TINSLEY, H. M. & DERBYSHIRE, E. 1976. Late-glacial and post-glacial sedimentation in the Peris–Padarn rock basin, North Wales. *Nature*, **260**, 234–238.

TREAGUS, J. E. 1992. *Caledonian Structures in Britain*. GCR Series 3, Chapman & Hall, London.

TRIMMER, J. 1831. On the diluvial deposits of Caernarvonshire between the Snowdon chain of hills and the Menai Strait and on the discovery of marine shells in diluvial sand and gravel on the summit of Moel Tryfane near Caernarvon, 1000' above the level of the sea. *Proceedings of the Geological Society of London*, **1**, 331–332.

UNWIN, D. J. 1973. The distribution and orientation of corries in northern Snowdonia, Wales. *Transactions of the Institute of British Geographers*, **58**, 85–97.

WALKER, R. 1978. Diatom and pollen studies of a sediment profile from Melynllyn, a mountain tarn in Snowdonia. *New Phytologist*, **81**, 791–804.

WILLIAMS, I. E., HOWELLS, M. F., LEVERIDGE, B. E., REEDMAN, A. J. & ADDISON, K. 1986. National Nature Reserves in Wales: a systematic survey, 7. Cwm Idwal NNR (cont'd). *Nature in Wales*, **5**(1 & 2), 19–32.

The national park system in Germany

ALF GRUBE

*Geological Institute, Christian-Albrechts-University Kiel,
Olshausenstr. 40, 24098 Kiel, Germany*

Abstract: The oldest national park in Germany is about 20 years old, while the oldest nature reserve dates back to 1836, and early attempts to protect caves even reach back to 1668. Today, 11 national parks exist in Germany, covering an area which is twice as large as that protected by the existing 5000 nature reserves and which is about 2.5 times the size of Luxemburg.

The largest national parks are situated along the coasts of the North Sea and the Baltic Sea, where large tidal flat areas, dune landscapes, barrier islands, graded shore lines and bodden-type landscapes can be found. The Cretaceous cliffs of Rügen belong to the most famous heritage sites of Europe. The Müritz National Park comprises a number of glacigenic forms and sediments like end-moraines, outwash plains and dead ice features. Two other national parks enclose hill country areas, one in the Bavarian Forest, the other (a large granitic pluton) in the Harz Mountains. The Sächsische Schweiz National Park, with its erosional features carved into a sandstone plateau, is a unique landscape. Last but not least, the Berchtesgarden National Park encloses part of the northern calcareous alps, with alpine areas, large alluvial cones and glacial troughs. The list of German national parks demonstrates that coastal features are well represented in comparison with petrographic sites, cuesta landscapes and other geological and geomorphological forms.

A number of other potential national parks have been discussed, the majority of which are situated in the hill country of Germany. As in other countries, the influence of geoscientists on the establishment of protection areas in Germany should be increased. National parks provide good opportunities for the protection of large landscapes, but existing laws make the establishment of parks, based mainly on geological reasons, almost impossible. 'Geological Parks' are therefore favoured and supported by many German Earth science conservationists. Concepts are currently being developed in the Eifel volcanic province.

Despite the small number and coverage of the national parks, these comprise some of the most valuable central European geosites and thus contribute to the protection of the geological heritage of Europe.

Nature protection in Germany has its roots in the Romantic movement and the main aim has, until recently, been the protection of beautiful and interesting landscapes and natural objects. The first ideas to establish national parks (NPs) in Germany were based on concepts developed in the USA. Today 11 NPs exist, of which the majority had precursors in the form of 'Nature Reserves'.

Historical aspects, legal framework and distribution

The first attempts to establish NPs reach back almost a hundred years, when for example, the famous German poet Hermann Löns (1866–1914) worked actively for the establishment of a national park in the Harz Mountains, but the oldest national park is only about 25 years old. In contrast, the oldest nature reserve dates back to 1836, and early attempts to protect caves reach even further back to 1668, when Rudolf August, Duke of Braunschweig and Lüneburg, enacted a decree for the Baumanns cave.

NPs are mentioned in §14 of the German Act on the Conservation of Nature, where it is stated that NPs should protect **unique** and **large scale** areas which have **not** been **altered** by humankind (or at least only slightly). The main goal is the 'conservation of a diverse native plant and animal population'. NPs are accessible to the public as long as no severe interference with protection exists.

Until now the responsibility for NPs as well as the individual national park legislation was handled differently in each German state. In general there is still some economic activity in parts of the German NPs (e.g. forestry, military activities, intense tourism in inner zones). The situation will hopefully improve, since the foundation of a German Federation on national parks in 1991.

Five national parks were established shortly before the reunification of Germany in the eastern 'Länder', although plans to establish NPs in the former GDR had existed since the 1950s in the regions of Elbsandsteingebirge, Oberharz, Darß and Spreewald, but these had always been obstructed by policitcal interests. Fast work and great personal effort made the protection of comparatively large parts of the country possible in 1989–90. Three of the 11 NPs have been established in Mecklenburg-Vorpommern, one of the less wealthy and relatively sparsely populated Länder, where alternative uses were of minor importance.

Though the 11 NPs cover an area twice as large as the existing 5000 nature reserves, the overall percentage makes up only about 2% of the country; still, this is 2.5 times the size of Luxemburg.

Fig. 1. National parks in Germany.

Description of the national parks

Bayrischer Wald (130 km², established in 1970, decree from 1969)

The Bayrischer Wald, the first national park in Germany, comprises a highland area in a magmatic province with crystalline schists and gneisses (Moldanubikum, Bohemian Masse). Glacigenic features also occur.

Berchtesgaden (210 km², established in 1978)

This national park encloses subalpine to alpine regions in the northern calcareous alps which comprise a number of parallel valleys and mountain chains with karst features, caves and large lakes. The highest altitude of 2713 m is reached by the Watzman. Geological sequences dominated by dolomites and limestones from the Permian to the Quaternary are exposed. During the Pleistocene, alpine glaciers changed the landscape to a large extent, producing moraines, U-valleys, cirques and so on. Large boulder fields and Holocene fans exist within the valleys. Part of the area has been under protection since 1910. The threats to geosites are only of minor importance but do still, for example exist through climbing activities.

Hochharz (59 km², established in 1990)

This area comprises a Variscan pluton in a Variscan mountain area with metamorphic zones (Hornfels). The Harz was lifted and eroded during the Alpine orogeny. Sack-like structures of granite, block fields and undisturbed bogs are typical for the area. The morphology is threatened by a railway line and military/civil structures.

Jasmund (30 km², established in 1990)

The Jasmund National Park is famous for its chalk cliffs (Fig. 2), that reach a height of 120 m (Königsstuhl) and which have been made popular by the paintings of Caspar David Friedrich and the works of various writers. The Cretaceous rocks, in which karst features occur, can be differentiated into c. 20 blocks, which are interpreted as having been ice pushed by glacier-tongues from the Baltic Sea. Other scientists believe in a dislocation by isostatic adjustment. In between fillings of Quaternary sediments occur – the so-called stripes, which comprise mainly sediments from Saalian to Holocene age. Steep valleys cut into the Chalk. All these features are mentioned in the official decree of the national park, including the possibility of active geomorphological processes.

Fig. 2. Jasmund National Park with Chalk cliffs and Quaternary sequences on the island of Rügen.

Quarrying of chalk was carried out in the area and, in the past, large quantities of boulders for construction sites have been extracted from the coast. Early protests against this exploitation led to protection as a nature reserve in 1935. In 1992 intensive local protests prevented the construction of a large dockyard on an adjacent site – a great success for conservationists in Germany.

Müritz (313 km^2, established in 1990)

The geological–geomorphological forms of Müritz are characterized by sandur plains, glacier basins and lakes, the latter being formed by glaciers, meltwater or dead ice. Lake terraces reach a width of more than 3 km and mires are also widespread. End-moraines are included in the fringe area and inland dunes are part of the sandur plains. People have altered the landscape by lowering the water level, which has led to the formation of many small lakes (>100) instead of one large one, and through military structures.

Odertal (224 km^2, temporarily protected in 1992)

This area protects large floodplains, with adjoining till plains and spillway illustrating dissection of their flanks as well as undisturbed bogs. The lower Oder Valley is the last extensive estuary in central Europe which remains in a near-natural to natural state. Of high biological value, the Odertal National Park is only of medium value for Earth science conservation (active processes, sedimentation).

Sächsische Schweiz (93 km^2, established in 1990)

Similar to the chalk cliffs of Rügen, the area of the Elbsandsteingebirge (Elbe sandstone mountains) has long since been a favourite subject of painters and poets. The beautiful landscape was the reason for early attempts at protection during the last century. Within the last four decades a number of nature reserves have been established, and these were incorporated into the national park in 1990.

The NP protects a unique erosional landscape which has been carved into Cretaceous sandstone by the River Elbe and its precursors. Various levels can be differentiated, which comprise table mountains, planation terraces, different cave types and steep valleys. The Cretaceous sandstone was overthrust by older granites at the end of the Cretaceous (Lausitzer overthurst). Single basalt dykes of Tertiary age are another geological feature. The landscape continues into the Czech Republic. With 3 million visitors per year, this national park is one of the most popular natural areas for tourism in Germany. Such popularity causes a number of problems and, in addition, in some places the landscape has been altered by quarrying popular building stone (Barock of Dresden).

Vorpommersche Boddenlandschaft (805 km^2, established in 1990)

The national park protects part of a near-natural coastline of the Baltic Sea with bodden-type landscape (drowned hilly glacigenic landscape), forming an area with more than 180 associated beach ridges, as well as lagoons, sand reefs and dunes. The 'Fischland' is seen as part of an end-moraine. Both flat shores and cliffs occur. The largest parts of the NP lie offshore. Some of the islands, consisting of glacigenic sediments, have been linked by present beach material in the Holocene. A fossil cliff from the *Littorina* transgression is also integrated. The dunes reach a height of up to 13 m. Human impact is visible in some places, e.g. the connection between the two largest islands has been closed in order to protect the coast. Part of the area was used for military training until recently.

Wattenmeer (5370 km^2, established between 1985 and 1990)

The Wattenmeer National Park comprises a large tidal flat area. A number of barrier islands, cliffs, beach ridges, lagoons, sand reefs, dune landscapes and river mouths are included in the park. The daily-changing wash-out gullies are a good example of active geomorphological processes. The NP decree clearly states the natural processes should be allowed to proceed undisturbed, but because of the size of the park and the necessity of coastal protection in the north German lowlands – a large subsidence area – some compromise has been necessary.

The biological interest is clearly dominant. The Wattenmeer National Park is one of the largest areas of its kind in the world and has major functions as a wildlife habitat, breeding site for many birds, and last but not least, as a major self-cleaning area for the North Sea. Production of oil is still being carried out, but the license for this will expire in a few years time.

National parks and Earth science conservation

NPs are a high protection category, that is also represented by nature reserves and 'natural monuments' (the latter two comprising less than 1% of the country, and both protecting smaller areas). 'Landscape Protection Areas' (more than 20%) give only superficial protection, while 'Nature Parks' (more than 20%) are mainly designed for tourism and recreation. Besides these there are biological protection categories like 'Biosphere Reserves' and possibilities for the protection of palaeontological sites (only in some 'Länder'); the laws vary between the different Länder.

Only in a minority of NPs, like Berchtesgaden, Jasmund and Sächsische Schweiz, were geoscientific issues important in the establishment of the NP. The largest NPs are situated along the coasts of the North Sea and the Baltic Sea. So, in contrast to geological and geomorphological forms of glacigenic origin, of cuesta landscapes, and of the Variscan hill countries etc., the representation of coastal features is comparatively good.

Detailed investigation of single geoscientific sites has been carried out only in a minority of NPs. The geoscientific aspect has been insufficiently integrated into the management plans of NPs and even strong supporters of the NP idea tend to neglect it.

Future perspective

The overall number and area of NPs will be increased during the coming years. The Hochharz National Park will be enlarged by the Oberharz National Park in the adjoining Lower Saxony, and an extension of the Bayrischer Wald National Park is under consideration. In the future an Unteres Elbetal (lower Elbe Valley) National Park will be established, comprising part of the large Elbe ice marginal spillway with adjoining Saalian landscapes and some of the largest inland dunes in Europe (Dömitz, Klein Schmöhlen).

A number of other areas have been considered as possible NPs (e.g. in Hessen and the Black Forest), most of which are interesting for Earth science conservation. A couple of areas are 'cultural landscapes', which can also be very interesting for Earth Science Conservation. These did not become NPs, because in Germany NPs only protect natural, and not artificial, landscapes (only smaller components to be included according to NP guidelines of IUCN). A discussion has been going on about this for many years. An example is the Lüneburger Heide Nature Protection Park in North Germany, which comprises a large area of heath, which is the result of exploitation of many hundreds of years. From a geomorphological point of view the park is valuable because it protects a typical Saalian landscape, that has only been slightly altered. However, as a nature protection park the area does not have a high protection level.

Geoscientifically orientated site interpretation is still developing in all parks. In the future it is thought there will be a strong demand for this topic. Because of the possible natural geomorphological processes in connection with natural biotic communities, some NPs offer especially good conditions for Earth science orientated educational issues.

In all, the area of NPs will never cover a very large proportion of the country and so Earth science conservationists should not expect too much of national parks. A separate protection category for geosites, such as a 'Geological Park', would be an advantage. A change in the legislation for NPs to include human-made areas, like pits, is no alternative, since this could do harm to the practical biological conservation work in NPs that has already been established. There are several possible geological parks, e.g. the Eifel-region and the Nördlinger Ries (impact crater). Some promising ideas are in preparation for two geoparks in the Eifel region.

Despite the fact that NPs in Germany are mainly orientated towards biological interests, the geoscientific component will hopefully play a more important role in the future. In spite of the small number and coverage of the NPs, these comprise some of the most valuable central European geosites and so contribute to the protection of the geological heritage of Europe.

References

All the references listed here provided the background material for this paper. None are cited within the text itself.

BIEBELRIETHER, H. 1987. *National Park Bavarian Forest*. Grafenau.

—— 1991. Nationalparke – Ihre internationale und regionale Bedeutung. *In*: Nationalpark Hochharz-Erste wiss. Tagung. NATIONALPARKFORSTAMT HOCHHARZ (ed.). Wernigerode, 7–9.

BURGER, H. (ed.) 1991. *Nationalpark (Umwelt Natur). Sonderausgabe Deutsche Nationalparke,* Grafenau.
GESELLSCHAFT ZUR FÖRDERUNG DES NATIONALPARKS HARZ 1992. *Konzept für einen Nationalpark Harz* (3rd edn). Goslar/Wernigerode.
GRUBE, A. 1992. Earth Science Conservation in Germany. *Earth Science Conservation,* **31**, 16–19.
LEMKE, K. & MÜLLER, H. 1988. *Naturdenkmale – Bäume, Felsen, Wasserfälle.* Stapp Verlag, Berlin.
LÜTTIG, G. et al. 1980. *General geology of the Federal Republic of Germany.* Schweizerbart, Stuttgart.
SCHULZ, W. (in preparation) Zur Geologie der zehn Großschutzgebiete in Mecklenburg-Vorpommern, Proc. 1. Jahrestagung AG Geotopschutz, Saarland.

SEMMEL, A. 1984. *Geomorphologie der Bundesrepublik.* Steiner, Stuttgart (= Geogr. Z., Beihefte).
SEUFFERT, O. (ed.) 1989. *Manual of field trips in and around Germany.* Excursion Guide 2nd International Conference on Geomorphology, Frankfurt. Geoöko-Forum.
STEINICH, G. 1992. *Quartärgeologie der Ostseeküste Mecklenburg-Vorpommerns.* Exkursionsführer DEUQUA '92, Exkursion A1, Kiel.
WAGENBRETH, O. & STEINER, W. 1982. *Geologische Streifzüge - Landschaft und Erdgeschichte zwischen Kap Arkona und Fichtelberg.* VEB Deutscher Verlag, Leipzig.
WEGENER, U. & SCHADACH, V. 1991. *Nationalpark Hochharz mit Brocken, Sachsen-Anhalt.* Goslar.
WILD, R. 1988. The protection of fossils and palaeontological sites in the Federal Republic of Germany. *Special Papers in Palaeontology,* **40**, 181–189.

Landscape parks and other protected areas: the Polish experience of landscape conservation

KRZYSZTOF WOJCIECHOWSKI

Earth Sciences Institute, Maria Curie Sklodovska University, Akademicka 19, 20-033 Lublin, Poland

Abstract: Following new regulations in 1991, it is now possible to organize an integrated system of protected natural areas in Poland. The elements of this system (different categories of protected area) were created independently over several decades. The current legal situation and people's ecological awareness has enabled completion of a territorial protection system consisting of national parks, landscape parks and protected landscape areas as the main components, supported by various smaller elements: nature reserves, nature monuments, protected study and research sites.

The task of completing a nationwide environmental protection system is an ongoing process, but creation of the basic framework is now well advanced in some regions, for example Lublin Region and Southeastern Poland. The environmental protection system established in Lublin Region was created on the basis of detailed multidisciplinary field studies, repeated evaluation processes and consultation with local authorities. Conservation areas were categorized according to their potential research and recreational value, their unique natural qualities and important landscape elements. A hierarchy of conservation areas was set up. The existing system now protects all main types of landscape (landforms together with traditional land-use patterns) which occur in the region.

The regional protection systems, set up by regional administrative bodies, are being slowly adapted to form larger systems – multiregional or even international. In Southeastern Poland, efforts are now directed towards finding the most effective way of integrating the conservation systems with the three neighbouring states of Slovakia, Ukraine and Poland.

Creating a legal structure in Poland to enable an integrated system of protected areas to develop has taken from 1918, when the Poles first regained their independence, until 1991. The oldest of the various systems and the areas with the highest rank of protection are the nature reserves and national parks. Landscape itself was subordinated to the nature reserves and only in the 1970s did it become a primary concern with the creation of the first landscape parks. These were planned and located independently of each other in various areas. Different regulations concerning the conservation of landscape and the separate components of the environment were developed for the landscape parks in different voivodeships (administrative districts).

It was also during those years that in the Warsaw Institute for Environmental Planning the *Ecological System for Protected Areas* was worked out (Rózycka 1977). Individual elements of this system and different classifications of various areas were to be planned for larger regions. One of the features of the proposed systems of conservation was a territorial interconnection of all elements. The new element in this concept was the appearance of conservation regions of a lower order – 'protected landscape areas' and the so-called 'ecological corridors'.

After the administrative reform of Poland in 1975, the 'new' authorities acquired the option of establishing further protected regions of a lower order: landscape parks with buffer zones as well as the 'protected landscape areas'. It was at this time that, under the auspices of the administrative authorities, several scientific centres (including the Institute for Environmental Planning and its local branches) started creating plans for landscape parks, their park complexes and the first system of conservation areas (Chmielewski 1990). By the end of the 1970s several dozen landscape parks were established in Poland; work was also initiated on further parks, complexes and systems of conservation areas. Despite tumultuous political, economic and social change in Poland during the 1980s, the process of enacting laws for environmental protection and establishing new conservation areas went on without interruption. The culmination of these efforts can be considered the passing of the Nature Conservation Act by the parliament of the Republic of Poland on October 16, 1991. The act stated the exact competencies of the particular administrative levels involved in the process of creating separate elements of the system of protected regions. The act defined already existing and newly created forms of conservation, described the responsibilities of the services involved in protecting nature and assigned the protected areas to the appropriate type of land-use unit, as listed in the land-use development plans in accordance with legislation concerning land-use planning.

The Nature Conservation Act of 1991 defines the following forms of protection (Fig. 1):

(1) the creation of national parks;
(2) designating prescribed areas as nature reserves;
(3) creating landscape parks;
(4) designating protected landscape areas;
(5) recognizing appropriate sites and objects as:
    (a) monuments of nature
    (b) research stations
    (c) ecological sites
    (d) nature/landscape complexes.

National parks and reserves are established at the level of central government, while the remaining types of conservation area can be created at lower administrative levels. A significant development is that the lowest conservation levels, points 4 and 5 in the above list, can be enacted at the level of the municipality (the only form of self-government aside from the national level in Poland at the present time).

Development of the current systems of conservation areas

The task of establishing and constantly modifying the system of regional environmental and landscape conservation must be treated as an ongoing process. At present it is possible to point out several regions in Poland where this process is fairly advanced and has led to the development of basic elements of a functional, integrated system, containing all the fundamental conservation types to facilitate protection of the most valuable natural and landscape areas. A good example of such a region is the area encompassed by the Lublin, Chelm and Zamosc voivodeships in the eastern part of Poland.

The present system of protected areas developed by degrees, initially through the creation of individual, independent elements. In this way, during the inter-war period, the first nature reserves were established. Following this, came the Roztocze National Park in the voivodeship of Zamosc, the first national park to be created in this region, and then further nature reserves in all three of the voivodeships.

Fundamental studies to develop separate systems of conservation areas were initiated in the 1980s in each of these voivodeships independently. Thus, in each of the voivodeships the process of developing conservation areas evolved differently.

Voivodeship of Chelm

The earliest system of protected areas was developed in the voivodeship of Chelm. The special natural qualities of this voivodeship, with its lakes and marshlands, had been known for some time, but were only partially protected by a few reserves. In the 70s, however, new coal mines started operating in this voivodeship, which entailed a fundamental change in land-use planning. A project for a system of conservation areas, mainly on the basis of already existing studies, was done by the Warsaw Institute for Environmental Planning. With only minor alterations, the project was accepted in 1983. It encompassed four landscape parks and four protected landscape areas. The Polesie National Park was created from one of the most valuable sections of one of the landscape parks in 1991.

Voivodeship of Lublin

Developing a 'system of ecological conservation areas' for the voivodeship of Lublin was initiated in 1983. The basis of the project was formed from numerous detailed field studies conducted by many interdisciplinary scientific teams. The results of the studies served as the foundation for evaluating the environment. This evaluation was conducted by taking into account the hierarchy of potential functions that the prospective areas would serve. Thus, the highest place in the hierarchy was reserved for those areas conserving unique natural qualities and areas of particular scientific interest. A lower place was designated to areas of a primarily recreational value but with an important natural environment and landscape. Lower still were areas of industrial or agricultural use. On the basis of this evaluation a project was developed for a system encompassing six landscape parks with their associated buffer zones and seven protected landscape areas. In 1990, the voivodeship authorities accepted the project which comprised 37.8% of the voivodeship's area.

A significant proportion of the protected area is made up of traditional farmlands, while along river valleys, which are not included in traditional forms of conservation, ecological corridors were established, which were taken into account in local planning (Wilgat 1992).

National Parks	Landscape Parks	Protected Landscape Areas
I Roztoczański	1 Kazimierski	a Annówka
II Poleski	2 Wrzelowiecki	b Pradolina Wieprza
	3 Koziowiecki	c Kozi Bór
	4 Nadwieprzański	d Dolina Ciemięgi
	5 Krzczonowski	e Chodelski
	6 Pojezierze ćęczyńskie	f Kraśnicki
	7 Poleski	g Czerniejowski
	8 Sobiborski	h Poleski
	9 Chelmski	i Chelmski
	10 Strzelecki	j Grabowiecko-Strzelecki
	11 Szczebrzeszyński	k Pawlowski
	12 Krasnobrodzki	
	13 Puszczy Solskiej	Buffer Zones -----
	14 Poludnioworoztoczański	
	15 Lasy Janowskie	Ecological Corridors

Fig. 1. Protected areas in the Lublin Region.

Voivodeship of Zamosc

The project for a system of conservation areas for the Zamosc voivodeship was already developed by 1983 and was also based on the results of field studies. In this voivodeship, where a national park already protected the regions of highest natural value, the voivodeship authorities have only established further landscape parks in recent years (at present there are three). Up until now, all the conservation areas established cluster along a forested chain of hills in Roztocze and the territories bordering upon it, which have some particularly valuable natural habitat.

Tasks for the future

Currently, work in the field of landscape and environmental protection is being focused on two tasks. Firstly, there is the problem of perfecting the administration of protected areas. They need to be provided with effectively functioning control systems and the environments need to be monitored to enable evaluation of potential threats.

Administrative bodies have been created to manage the landscape parks and their complexes by authority of the existing laws. Yet their exact competencies and establishing the appropriate forms of cooperation between the park administrative body and other agencies, e.g. the regional organs of the central government, planning institutions and local governments still requires additional legal regulation. Moreover, practical experience in this field in lacking.

For most people living within the lower-ranking protected areas, as well as most within the micro-economic units, awareness of the existence of additional restrictions is not very high nor fully accepted. So the incidence of misunderstanding or attempts at circumventing stiffer regulations are relatively frequent. It is also often very difficult for the parks' administration and staff to respond in time to actions which pose an environmental threat, as they are underfinanced, limited in numbers and inadequately equipped with the technical means of control and monitoring. They are likewise insufficiently supported by pro-ecological social organizations.

The second task is to modify the existing systems with the necessary new elements to facilitate integration of the systems created independently for each voivodeship into extra-regional ones.

In the last several years, increasing cooperation with new bordering countries has permitted international projects to be undertaken for systems of protected areas. Initial projects are already at an advanced stage in creating integrated complexes of protected areas which include the borderlands of Poland, Ukraine, Slovakia and Belorus (Wojciechowski 1993).

References

CHMIELEWSKI, T. J. 1990. *Parki Krajobrazowe w Polsce – Metody delimitacji i zasady zagospodarowania przestrzennego.* SGGW-AR Warszawa.

RÓŻYCKA, W. 1977. Propozycja formowania ekologicznego systemu obszarów chronionych w planach zagospodarowania przestrzennego. *Czlowiek i Środowisko*, **1**, 4.

WILGAT, T. (ed.) 1992. *System obszarów chronionych województwa lubelskiego.* TWWP Lublin.

WOJCIECHOWSKI, K. (ed.) 1993. *Edukacja ekologiczna i ochrona środowiska na pograniczach.* TWWP Lublin.

Geomorphological systems: developing fundamental principles for sustainable landscape management

JOHN E. GORDON, VANESSA BRAZIER & R. GEORGE LEES

Scottish Natural Heritage, 2 Anderson Place, Edinburgh EH6 5NP, UK

Abstract: This paper examines the application of geomorphology to sustainable land management and outlines an approach that addresses the requirements of both Earth science conservation in its narrow sense and also aspects of wider landscape conservation. The understanding of fundamental attributes of geomorphological systems and their sensitivity to change across a range of spatial and temporal scales allows the development of a set of principles that form an important primary contribution for geomorphology in a multidisciplinary approach to sustainable landscape management.

Scottish Natural Heritage (SNH) is the statutory body responsible for the conservation of Scotland's diverse range of natural assets. Its general aims are to secure the conservation and enhancement of the natural heritage of Scotland, and to foster understanding and facilitate the enjoyment of that heritage. The fundamental principle underlying the work of SNH is that anything done in relation to the natural heritage of Scotland is done in a manner which is sustainable. Furthermore, the focus of Scottish Natural Heritage is the conservation of the whole landscape, not only target sites or protected areas. Geomorphology has an important contribution to make in two respects to the conservation of Scotland's natural heritage. First, Scotland possesses a valuable geomorphological resource that is important in its own right as a record of past changes in the landscape and of its continued evolution in response to changing environmental conditions and human interference. Second, geomorphological processes are instrumental in the formation and perpetuation of many terrestrial and coastal habitats and hence underpin the faunal and floral communities which these habitats support. An understanding of these processes is thus fundamental to the conservation of such communities.

The geomorphological resource in Scotland

The natural heritage of Scotland contains a diverse range of landforms and deposits that record the effects of past and present environments. These landforms include many relict features formed during environmental conditions that differ from those of the present: for example, glacial and glaciofluvial landforms and deposits; flights of postglacial river terraces; a variety of landslide types that range from large rock slope failures to small hillslope debris flows; periglacial features, such as patterned ground and solifluction sheets; and raised beaches, abandoned cliff lines and dune systems. These relict landform assemblages, together with the record of environmental change preserved in the sediments and in the floral and faunal remains of peat, lacustrine and marine deposits, provide an impressive and largely intact record of Quaternary environmental change.

Modern active Earth surface process environments include highly dynamic coastal and fluvial systems. However, the rest of the landscape is by no means geomorphologically inert. For example the role of hillslope processes such as debris flows and shallow landslides can dramatically alter localized parts of the landscape. Other processes operate on such a small scale, or sufficiently slowly, that they are virtually imperceptible. Examples include nivation processes associated with semi-permanent snowbeds, or the development of small-scale patterned ground on mountain tops.

These dynamic and relict elements of the landscape are increasingly at risk from a range of human activities, such as land-use change, mineral extraction and coast protection, as well as being subject to naturally induced changes. One particular aspect of dynamic landforming processes environments that is of great interest to science is their variability of behaviour, as in the magnitude and frequency of coastal storm surges, floods and river channel changes. Such behaviour often constitutes a hazard to human activities, and consequently dynamic environments are frequently a target for hazard mitigation and management measures that significantly modify the natural processes.

From O'Halloran, D., Green, C., Harley, M., Stanley, M. & Knill, J. (eds), 1994, *Geological and Landscape Conservation.* Geological Society, London, pp. 185–189.

Geomorphology within a multidisciplinary approach to landscape and habitat conservation

Geomorphology not only forms an integral part of Scotland's natural heritage, but also has an essential contribution to play in a wider multidisciplinary approach to habitat and landscape conservation (Gordon, in press). Modern and relict landforms and landforming environments underpin the viability of many important habitats. For example, salmonoid spawning grounds are usually associated with well-oxygenated gravels in shallow waters, found in riffles in mobile gravel-bed rivers. A further example is the viability of certain dune grass communities, which is at least partly dependent upon the continuing supply and deposition of blown sand on to the dunes. More commonly the relationship between habitat and geomorphology is more subtle. For example, the sedimentology and internal structure of glaciofluvial deposits, in combination with the history of pedogenic processes, can lead to groundwater and surface discharge patterns that favour the development of specialized fen communities. However, the dependency of certain types of habitat on either landforms or dynamic process environments has been largely unexplored during the short history of Earth science conservation in Britain.

Geomorphology is therefore inextricably linked with the success of the conservation of Scotland's natural heritage, whether this is Earth science conservation or the effective protection of habitats and landscapes. In this paper we address the conceptual basis, considering two aspects in particular:

(1) conservation objectives for geomorphology;
(2) geomorphological sensitivity and its implications for sustainable land management.

Conservation, geomorphology and sustainability

At the outset, it is useful to consider the definition and objectives of conservation in so far as they relate to geomorphology. In a broad sense, conservation may be defined as 'the process of sustainable management of renewable natural resources, as an essential foundation for the human future on this planet' (Holdgate 1991, p. 2). This definition reflects recent international developments in conservation thinking (IUCN 1980; World Commission on Environment and Development 1987) and encompasses three principal elements: (1) sustainable management of natural resources; (2) protection of wildlife and physical features for scientific reasons; and (3) protection of the aesthetic value of the landscape. Nature conservation in Great Britain in the past has principally been concerned with the second of these elements and to some extent with the third (cf. Ratcliffe 1977). Earth science conservation has essentially fitted into this mould, as reflected in the emphasis on the Geological Conservation Review (GCR). The challenge for the Earth sciences now is to reflect the developments taking place in conservation, with the growing focus on landscape systems and sustainability.

We believe that the most promising way forward is to develop an approach based on the concept of sustainability and the links between geomorphological systems and the landscape across a variety of geographical and temporal scales. Such an approach would encompass the conservation of geomorphological features as well as the contribution of geomorphology to a wider multidisciplinary approach to landscape conservation. Site conservation would therefore form only one element in a broader strategy.

In order to redefine the conservation approach and objectives for geomorphology, we need to identify the fundamental elements of interest. Historically, the focus in Earth science conservation has been upon sites, with a strong emphasis on evolutionary aspects of scientific interest, as in networks of sites representing the geomorphological and environmental changes during the Quaternary. In contrast, biological approaches to conservation have involved stronger geographical and ecosystem elements.

A more focused geographical perspective, based on systems concepts, provides a potentially powerful framework for geomorphology and is worth exploring further. Such an approach has several advantages. It permits the identification of a hierarchy of different scales of land units, comprising whole landscapes (morphogenetic regions), landform assemblages encompassing groups of related landforms, and individual landforms and their constituent sediment sequences. Further, the development of a structured management framework across a range of spatial scales allows individual features to be managed in the context of their interrelationships with other components of the wider landscape matrix (cf. Franklin 1993).

In Scotland much of the landscape has been modified by human activities over many centuries. Therefore, conservation cannot only be concerned with protection of pristine habitats, features or landscapes. Sustainable conservation management implies a flexible or pragmatic approach to landscape management, where human-induced changes can be tolerated. Conservation management of the geomorphological heritage must ensure that the integrity of the record of past environmental change stored in the relict landform resource of Scotland is not wantonly or irretrievably damaged. Similarly, conservation management of active geomorphic systems must aim to ensure that the range in behaviour of dynamic process systems is not fundamentally or irrevocably changed by artificial means. These approaches include, where appropriate, the protection of unique or exceptional features, at a site-specific scale. Thus the objectives of sustainable conservation management of our geomorphological heritage can be expressed as follows:

(1) to maintain the national and regional diversity of landscapes, landform assemblages, landforms and sediment exposures, including the key localities identified in the GCR;
(2) to maintain the essential geomorphological features which determine the character of the landscape, including the form of individual landforms;
(3) to ensure that integrated landscape management is based on a sound understanding of geomorphological principles and works in sympathy with natural processes;
(4) to maintain the integrity and continuity of landform assemblages and natural landforming process systems;
(5) to maintain the integrity of landform-process links;
(6) to maintain the visibility of the geomorphological interest, and to maintain exposures and access to features of interest;
(7) to provide interpretation at appropriate levels of understanding.

These objectives relate not only to specially protected sites or areas, but equally to the wider landscape.

Geomorphological sensitivity

Geomorphological components of the landscape are either dynamic or relict. Relict features are the products of processes or conditions that no longer operate in Scotland. In effect they are fossilized and if disturbed or damaged, cannot be easily repaired or recreated. Dynamic geomorphological systems have the ability to respond to external interference in a robust or sensitive manner (Werritty & Brazier, in press). Robust behaviour is characterized by little or short-lived visible readjustment to external interference. For example, a bulldozed straightened reach of a gravel bed river may revert to a more natural planform after a moderate flood flow. However, sensitive behaviour in a dynamic system results in fundamental change, where the system is unable to readjust to either the type or level of external interference. For example, a hypothetical river system may be robust to moderate amounts of gravel extraction from the river bed, because the river has a high sediment flux. However, the same river may be highly sensitive to changes in its runoff regime, such as large-scale water abstraction. The effect of dropping discharge levels could result in a change from a sinuous channel form to a sediment-choked and smaller, divided channel pattern. These two examples are extremes, but little is known of what the cumulative impact of minor interferences could be on the sensitivity of dynamic geomorphological systems. Understanding the sensitivity of geomorphological systems, and how they will respond, is fundamental for sustainable management, and requires awareness and application of basic principles of geomorphology such as thresholds, equilibrium, frequency and magnitude relationships, and complex responses.

The concept of geomorphological sensitivity has been used to determine whether extreme examples of certain human activities are sustainable, in both an active river system and relict assemblages of landforms (cf. Schumm, 1977; Werritty & Brazier 1991; Gordon & Campbell 1992; Werritty & Brazier, in press). Relict and dynamic features differ in their sensitivities to change, both natural and anthropogenic in origin, and this has important implications for conservation objectives and sustainable management (Werritty et al. in press). Change to sites arises from a variety of activities (e.g. Gordon & Campbell 1992; Werritty et al. in press). In some cases, as in response of dynamic features to natural change, this change will form part of the intrinsic scientific and landscape interest. Such change, however, may constitute a hazard to human activity and therefore be perceived to require management (cf. Brazier & Werritty, this volume). In most cases, particularly those arising from anthropogenic origins, change will result in destructive or damaging effects on the features of interest through:

- physical loss
- fragmentation involving loss of integrity, context and interrelationships
- loss of naturalness
- change of state (e.g. activation, fossilization)

An essential first step in both site and landscape management is the identification of potentially sensitive areas of the geomorphological resource. Such zonation requires an inventory of all landform assemblages and active process environments, and the assessment of the sensitivity of each element of the landscape.

Geomorphology and conservation principles

In essence, the central requirement for effective conservation of landscape, fragile habitats and our geomorphological heritage is the need to identify the geomorphological sensitivity of the system or features of interest. Several general working principles can be suggested for developing a sustainable approach that integrates geomorphology. These are put forward as follows as a basis for discussion.

(1) It is necessary to maintain a network of protected sites to conserve the diversity of the capital resource for scientific, educational and heritage reasons. Management prescriptions will be most stringent for such sites.
(2) Conservation does not stop at protected site boundaries. We need to include the wider landscape matrix and its component landforms, landform assemblages and process systems. Management prescriptions will vary in their degree of stringency but should be based on sensitivity zonation at the appropriate spatial scale.
(3) Most of the landscape of Scotland is modified by human activity. Since the approach to conservation is generally one of management rather than protection, the concept of sustainability is fundamental. Sustainability implies according due recognition to the sensitivity of both relict and dynamic geomorphological features.
(4) In the wider context, it is important to recognize that geomorphology underpins not only habitats but also the very fabric of the landscape. Sustainable conservation of the wider natural heritage requires a multidisciplinary approach to landscape management, with geomorphology providing a primary contribution. Equally, sustainable development requires geomorphologically informed management of both the environment and human uses of it.
(5) A landscape-orientated, systems-based approach provides a potentially useful means of developing the role of geomorphology in sustainable land management.

Conclusion

Conservation solutions that are sympathetic not only to the intrinsic scientific interest of the geomorphological resource, but also to the dynamics and inherent sensitivity of geomorphological systems, are essential elements in any holistic strategy for sustainable environmental management. A sound understanding of physical processes, their sensitivity to change, their spatial and temporal variability and their interactions with habitat ecology is therefore fundamental in a multidisciplinary approach to the management of Scotland's landscapes. From a geomorphological viewpoint, the integration of site protection with sustainable landscape management across a range of spatial scales represents an important applied role for the discipline.

References

BRAZIER, V. & WERRITTY, A. (this volume). Conservation management of dynamic rivers: the case of the River Feshie, Scotland.

FRANKLIN, J. F. 1993. Preserving biodiversity: species, ecosystems, or landscapes? *Ecological Applications*, **3**, 202–205.

GORDON, J. E. (in press). Geomorphology and conservation in the uplands: framework for a developing role in sustaining Scotland's natural heritage. *In*: STEVENS, C., GREEN, C. P., GORDON, J. E. & MACKLIN, M. G. (eds) *Conserving our Landscape: Evolving Landforms and Ice-Age Heritage*. English Nature, Peterborough.

—— & CAMPBELL, S. 1992. Conservation of glacial deposits in Great Britain: a framework for assessment and protection of Sites of Special Scientific Interest. *Geomorphology*, **6**, 89–97.

HOLDGATE, M. W. 1991. Conservation in a world context. *In*: SPELLERBERG, I. F., GOLDSMITH, F. B. & MORRIS, M. G. (eds) *The Scientific Management of Temperate Communities for Conservation*. Blackwell Scientific Publications, Oxford, 1–26.

IUCN. 1980. *The World Conservation Strategy*. IUCN, Gland.

RATCLIFFE, D. A. 1977. Nature conservation: aims, methods and achievements. *Proceedings of the Royal Society of London*, **197B**, 11–29.

SCHUMM, S. A. 1977. *The Fluvial System*. Wiley, New York.

WERRITTY, A. & BRAZIER, V. 1991. *The Geomorphology, Conservation and Management of the River Feshie SSSI*. Report to the Nature Conservancy Council.

—— & —— (in press). Geomorphic sensitivity and

the conservation of fluvial geomorphology SSSIs. *In*: STEVENS, C., GREEN, C. P., GORDON, J. E. & MACKLIN, M. G. (eds) *Conserving our Landscape: Evolving Landforms and Ice-Age Heritage*. English Nature, Peterborough.

——, ——, GORDON, J. E. & MCMANUS, J. (in press). The freshwater resources of Scotland: a geomorphological perspective. *In*: MAITLAND, P. S., BOON, P. J. & MCLUSKY, D. S. (eds) *The Freshwater Resources of Scotland*. Wiley, Chichester.

WORLD COMMISSION ON ENVIRONMENT AND DEVELOPMENT. 1987. *Our Common Future*. Oxford University Press, Oxford.

Strategies for conserving and sustaining dynamic geomorphological sites

J. M. HOOKE

Department of Geography, University of Portsmouth, Portsmouth PO1 3HE, UK

Abstract: Geomorphological sites are of enormous importance not only from their scientific value and the understanding they provide of processes and development of landforms but also because of their scenic and amenity value, enjoyed widely by the public. Sustenance of ecological diversity is also dependent on conservation of sites with a range of physical attributes. Dynamic sites present particular problems through their very activeness.

Strategies for conserving and sustaining such areas should include two major strands: (1) the understanding of sites and possible management techniques through scientific analysis; (2) education of various groups of people as to the nature of such sites and possible strategies of management.

On the scientific side there are three primary information requirements. Firstly, it is essential that the characteristics of the site are understood, particularly the rates of change and their long-term development. This involves historical analysis and monitoring. Demonstration that erosion is natural and long-continued is very important. Secondly, it is essential to view the system as a whole and understand links, as exemplified in coastal work. Thirdly, all alternative management strategies should be examined and their impacts assessed adequately. Information then needs to be presented to the public, landowners and decision-makers with emphasis on: the processes involved, their rates and importance in landscape development; the interrelated nature of different sites and the transfer of impacts; the association between ecological and physical characteristics of sites; and the available techniques of management. Some problems do arise even from conflicting conservation aims. All of these points will be exemplified by our own work on rivers and coasts.

Geomorphological sites are usually classified as integrity sites as opposed to the exposure type of Earth science site (NCC 1990). A valuable subdivision of geomorphological sites is into those which are essentially fossil landforms e.g. glacial features of northern Britain, and those which are active in which dynamic processes are integral to their maintenance. These latter pose particular problems of conservation because it is often the very processes themselves which are perceived as a threat, e.g. river erosion, coastal erosion or land instability. Pressures are increasing on such sites not only because of development but because of people's expectations of protection from the effects of nature. Ironically, this is also at the same time as an apparent increased public desire to have access to 'natural-looking' places. These active geomorphological sites are important for a number of reasons:

(1) for their scientific value, where processes can be directly measured, change and reaction in the environment can be monitored, and the mechanisms of longer-term development of landforms can be understood;
(2) for their scenic and amenity value to the public; the dynamism of such areas conveys a particular characteristic of 'naturalness' and beauty;
(3) as the basis for ecological habitats; ecological diversity is maintained by the range of types of habitat created by the active processes, e.g. on active rivers, bare eroding banks, sandy bars and old cutoffs at different stages of seral succession are frequently found in close proximity (Fig. 1). Both morphological and sedimentological attributes contribute to the diversity.

Geomorphological sites range widely in size, complexity and activeness and it is the combination and juxtaposition of certain landforms that can give rise to distinctive landscapes. In Britain the Site of Special Scientific Interest (SSSI) network provides some protection for scientifically important sites but cannot be used to justify sites for their scenic beauty. National parks and Areas of Outstanding Natural Beauty (AONBs) are a recognition of landscape value for large areas. However, other areas of geomorphological value fall between these in scale and justification. The most obvious are river valleys and coasts, although strategies for management and

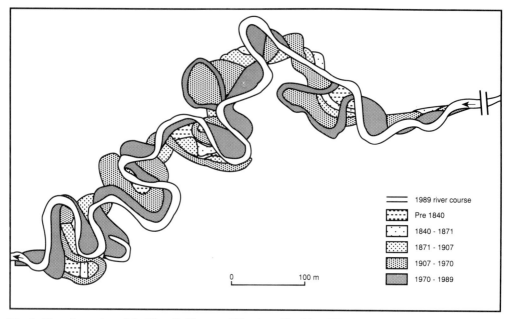

Fig. 1. Map of a section of the River Bollin, near Wilmslow, Cheshire, UK, showing zones of the floodplain dated from historical maps.

conservation are increasingly being developed by the public authorities responsible for those zones, i.e. the National Rivers Authority (NRA) and local authorities. Nevertheless numerous cases still arise of conflict of interest and where the public pressure for protection, i.e. cessation of the active processes, becomes irresistible for decision-makers. Therefore conservation of such sites and areas can only succeed with the two-pronged approach of both understanding the site and educating the decision-makers, managers and public.

Scientific requirements

Three types of scientific and technical information need to be provided for effective and sustainable management of active sites. The first is on the characteristics of the site and the nature of the processes. This should involve not only a present, on-site assessment but also a historical perspective. This is essential for measurement of long-term rates of processes, assessment of their variability over time, assessment of whether a net trend is apparent and thus for any extrapolation of changes and assessment of impacts of processes. It is also a vital clue in investigation of causes of change (Hooke & Redmond 1992). Very often the hazard posed by a 'natural' process is regarded as a threat because it is perceived that the processes are something new and have been caused by some other activity or interference in the vicinity. This is particularly so where residents are new to an area and unused to the environment. Some activity and changes may well be caused by human actions in the neighbourhood and rates could be accelerating but, in many cases, historical analysis will show that the processes have been long continued and cannot be attributed to an immediate and recent cause. Likewise, much geomorphological activity takes place in large but infrequent events. Historical evidence can help to demonstrate that particular conditions are not abnormal and, of course, can be the basis for calculation of recurrence interval. It can be easier to persuade people that sites should remain active if it can be shown that the processes have been active for centuries and that any solution would be working against the powerful forces of nature.

A second category of information that is needed is that which elucidates and demonstrates the interconnectedness of processes and the links between areas. All those trained in understanding the physical environment through systems analysis will appreciate these concepts but they need to be fully applied. The requirement is most apparent in sediment-transporting systems, i.e. in rivers and on the coast. Severe problems have arisen on much of the soft rock coast of Britain in recent years through beach depletion. To understand the causes and to find solutions that are sustainable and do not in turn cause furthe problems entails finding where sediment is coming from and where it is moving to. It necessitates understanding the nature of circulations so that impacts of interference in one area on others can be assessed. Recent work we have undertaken on the south coast of England aimed to increase that understanding and will now be exemplified.

Coastal protection is the statutory responsibility of local authorities in Britain and sea defences are the responsibility of the National Rivers Authority (NRA). In 1986 the engineers from these groups together with some other statutory authorities such as English Nature (then NCC) came together to form a coastal forum on the central south coast called SCOPAC (Standing Conference on Problems Associated with the Coastline). They did so because they realized that the works they were undertaking on one part of the coast were affecting another part. We were commissioned to identify the pattern of sediment cells and the nature of sediment processes (Bray et al. 1991). Much information was found in consultants reports, theses and other unpublished material on specific sites. The evidence was examined and synthesized to produce a map of the major cells and subcells (Fig. 2), as well as more detailed maps of sediment transfers. It emerged that the major current source of sediment naturally sustaining beaches is from cliff erosion in areas such as Dorset and on the Isle of Wight (Bray 1992). Therefore, if further problems of beach erosion are to be avoided then these sources must be maintained and the cliffs not protected. Local residents near eroding cliffs tend to lobby for protection but coastal management strategies have to be seen in a wider, cell-based context rather than the immediate site. If protection is allowed to go on increasing not only will we have no natural looking areas with varied ecological habitats but we will have difficulty even maintaining that protection.

The third type of information required concerns possible solutions to problems and the need to consider all possible strategies, including doing nothing, and to assess the full environmental impacts of each. Research into alternative methods of erosion protection has increased in recent years (Hemphill & Bramley 1989; Hydraulics Research 1991) and there is much greater awareness now of the techniques available and the advantages of soft engineering. Further

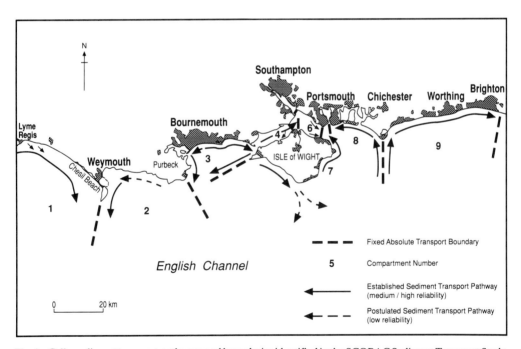

Fig. 2. Cells, sediment transport pathways and boundaries identified in the SCOPAC Sediment Transport Study (Bray et al. 1991).

research is still needed into the actual performance of such measures in different types of situation. Solutions that are wider than sites must also be considered, that are adequate for the landscape scale. An increasingly successful concept with regard to rivers is that of river corridors and catchment planning (Gardiner 1991; Newson 1992).

Education requirements

It is obvious that, having acquired the appropriate scientific information it must be communicated to the relevant groups.

Managers of geomorphological sites and areas must have a thorough understanding of the processes and landscape change at various spatial and temporal scales. The public must be given an appreciation of the dynamics of the landscape. If they understand it they are much more likely to conserve it. Managers must have access to adequate information on alternative methods and strategies including the fine detail of implementation. The public also need to understand the interconnectedness of natural systems and the consequences of not conserving sites. It is only in this way that attitudes of 'rights to protection' and belief that hard engineering is the only type of solution that will work can be altered. For example, on the River Thames the NRA intended to issue a leaflet to riparian landowners on alternative methods of bank protection to try to reduce the amount of sheet piling being emplaced and to inform them of 'softer' techniques (Hooke *et al.* 1991).

Conservation conflicts

Progress is being made in both the scientific understanding and the modification of attitudes. However, conflicts can arise even within the conservation field and two examples will illustrate this. The first is from Hurst Castle Spit, a large gravel spit at the western end of the Solent on the south coast of England. Historical evidence indicates that the neck of this spit has been migrating northeastwards for centuries. Recent problems have arisen because of a thinning of the spit such that it is in danger of breaching. If this occurred then extensive and ecologically valuable salt marshes behind would be inundated and the whole circulation and configuration of the western Solent could be altered. Some spits appear to have a natural history of breaching though not this one. The recent problems appear to have occurred because of a deficit of sediment supply due to protection up the coast, then a hinge effect on the spit itself due to the presence of structures. It has now been decided, in consultation with English Nature, that a replenishment and partial protection scheme will take place. Thus a naturally mobile feature is now being controlled mainly because of its relation to another valuable conservation area.

The second example is from the River Bollin near Wilmslow in Cheshire. The Bollin Valley has been opened up with a path through it and a Bollin Valley Management scheme to enhance enjoyment and conservation of this attractive valley. One section of the valley is very interesting and valuable as a scientific site because it contains a magnificent set of small active meanders which illustrate processes of meander development, of bank erosion, of floodplain formation and cutoffs (Fig. 1) (Hooke & Redmond 1992). One bend was eroding a 4 m high river cliff in a high river terrace of the valley. The footpath goes along the top of this terrace and the zone between the bank and a hedge was decreasing. Last year the Bollin Management decided that in order to maintain the footpath it must straighten the river. A large loop was cut through, though minimal, soft protection was put in to stabilize the new channel. A valuable site which did not have SSSI status but undoubtedly should have had RIGS status was devalued from a scientific and educational viewpoint.

Conclusions

Active geomorphological sites present particular challenges for conservation because the dynamic processes are an integral part of the characteristics of the site. Their continuation may also be necessary for the sustenance of other sites with continual 'knock-on' effects if one site is destroyed. Change is an essential ingredient of the site/area and this may include natural adjustment to human activities. A division between natural and anthropogenically induced change becomes very difficult, e.g. where runoff has been altered by land-use change but a river channel is free to adjust to this. In most areas we cannot go back and start again, so decisions on where to draw the line on change to be conserved and sustained and change which is unacceptable are very difficult. An essential prerequisite is to have a long-term perspective on the rates and patterns of change and to be able to assess the impacts both locally and regionally of various strategies.

References

Bray, M. J. 1992. *Cliff Erosion Contribution to the Sediment Budgets of Cells Along the South Coast of England.* Department of Geography Working Paper No. 22, University of Portsmouth.

——, Carter, D. J. & Hooke, J. M. 1991. *SCOPAC Coastal Sediment Transport Study.* Report to SCOPAC, Portsmouth Polytechnic, 535 pp.

Gardiner, J. L. (ed.) *River Projects and Conservation: A Manual for Holistic Appraisal.* Wiley, Chichester.

Hemphill, R. W. & Bramley, M. E. 1989. *Protection of River and Canal Banks.* CIRIA, London.

Hooke, J. M., Bayliss, D. J. & Clifford, N. J. 1991. *Bank Erosion on Navigable Waterways.* NRA Project Report 225/1/T.

—— & Redmond, C. E. 1992. Causes and Nature of river planform change. *In*: Billi, P. *et al.* (ed.) *Dynamics of Gravel-bed Rivers.* Wiley, Chichester, 549–563.

Hydraulics Research. 1991. *A Guide to the Selection of Appropriate Coast Protection Works for Geological SSSIs.* Report to NCC.

NCC. 1990. *Earth Science Conservation in Great Britain – A Strategy.* Nature Conservancy Council, Peterborough.

Newson, M. 1992. River conservation and catchment management: a UK perspective. *In*: Boon, P. J., Calow, P. & Petts, G. E. (eds) *River Conservation and Management.* Wiley, Chichester, 385–396.

Geological conservation of Holocene peatlands in the national parks of England and Wales

DANIEL J. CHARMAN

Department of Geographical Sciences, University of Plymouth, Plymouth, Devon PL4 8AA, UK

Abstract: Peat deposits cover large parts of the British Isles and have long been regarded as important for their wildlife conservation interest. However, they are also very significant in geological and archaeological terms and occupy a unique position at the interaction between these three disciplines. The key geological attributes of these deposits are described and their distribution within the national parks of England and Wales is evaluated. More than 70% (2884 km^2) of peatlands in England and Wales are contained within national parks. Most parks hold significant deep peat deposits in terms of percentage of land area covered and absolute extent. The development of attitudes to conservation within the national parks is traced with respect to geological considerations and it is suggested that while recent recommendations indicate a broader outlook for environmental matters in general, there is no specific mention of geological conservation. Some peatlands are likely to be considered important for their biological attributes but it is dangerous to assume that this will be true for all geologically valuable sites. It is argued that geological attributes of peatlands should be more widely and explicitly recognized in the management of national parks and other land uses.

Peatland deposits and landforms are widely distributed throughout the British Isles, especially in the north and west of the UK and Ireland. These areas have long been held to be important in terms of their wildlife conservation interest (e.g. Goode & Ratcliffe 1977) and in recent years there has been much controversy over their future for this reason. For example in the Flow Country of northern Scotland (Stroud *et al.* 1987; Lindsay *et al.* 1988). However, these peat deposits have also played a significant role in geological research on the Devensian Lateglacial and Holocene periods and they also often contain archaeological interest. These three apparently separate groups of attributes are also linked in various ways. The geological record is of interest to neo-ecologists as it incorporates long-term ecological data such as tree migration rates in response to climatic variations and successional change in lake and mire plant communities. Likewise, the peat archive also contains information relating to past human populations and their interaction with the environment. Peatlands therefore occupy a unique position linking recent geological change with ecological and archaeological sciences. Because of their distribution, peatlands in England and Wales are important components of the national parks. This paper enumerates the geological significance of peatlands, reviews the extent of peatlands currently contained within national parks and examines the ways in which national park policies currently cater for geological peatland conservation.

Geological attributes of peatlands

Geological attributes fall into four main categories (Charman 1994):

(1) *Historical*: Late nineteenth and early twentieth century palaeobotanical studies utilized peatlands almost exclusively and as a result a number of sites could be considered as 'classic' locations for this pioneering work.

(2) *Vegetation history*: Clearly this is crucial to an understanding of Holocene environmental change. There have been some important palynological studies on individual sites, but the widespread nature of peat deposits means one of their greatest attributes is the potential for high spatial resolution of the Holocene record. Additionally, there will be many older, especially Lateglacial, deposits which are not evident from surface inspection. Plant macrofossil studies, particularly those utilizing buried tree remains, are also becoming increasingly important.

(3) *Palaeoclimates*: Peat stratigraphy was the basis for early climatostratigraphic subdivision of the Holocene. More recent research using *Sphagnum* mosses, humification and other techniques is producing more detailed

proxy records of Late Holocene palaeoclimates.

(4) *Archaeological*: many Holocene records of vegetation change are related to human impact and peat deposits adjacent to archaeological sites yield records of direct relevance to specific settlements and structures. Additionally, some structures and artefacts may be preserved within peats themselves.

Peatland distribution and national parks

Recent figures compiled by the former Nature Conservancy Council (NCC) and Scottish Natural Heritage (SNH) for Britain suggest that the original extent of deep peatlands (peat depth > 1 m) in the UK was approximately 1.4×10^6 hectares (Lindsay 1993a). This is divided into ombrotrophic mires (often termed 'bogs') which receive all their nutrients and water from atmospheric sources, and minerotrophic mires (fens) which also have a hydrological input from groundwater or surface runoff. Ombrotrophic peatlands cover around 1.3×10^6 ha as blanket or raised mires and the remaining c. 100 000 ha is fen peats, although this latter figure may be underestimated. If the occurrence of peatlands is compared with the distribution of national parks, it is clear that there are large areas which occur within these protected landscapes. Table 1 provides a summary of data relating to deep peat distribution in the national parks as total peat area, percentage of the total peat resource in England and Wales, and percentage of individual park area.

Together the parks contain around 2884 km^2 of deep peatlands, which constitute over 70% of the total in England and Wales. This clearly highlights the importance of the parks in the management of this resource within the region. Mean peatland coverage within parks is 20.7% but ranges between 0% and 47.3%. The most important park in terms of the total peat resource is the Yorkshire Dales National Park with 576 km^2, which equates to 32.7% of its total area. However, there are several other parks which contain in excess of 400 km^2 and the only parks with less than 50 km^2 are Exmoor, the Pembrokeshire Coast and the Broads. The percentage of land area covered is in excess of 10% for most parks and Northumberland National Park is almost 50% peatland.

Despite this regional importance, when compared with the national picture including Scotland, the significance of the parks diminishes as total deep peat area within the parks represents only about 21% of the national total. There is c. 1×10^6 ha of deep peat outside the national parks system in Scotland, which is by far the most important British country in terms of peat deposits.

Peatland protection and the national parks

Protection of peatlands in Britain is mainly the province of a variety of nature conservation organizations, either locally by trusts linked to the Royal Society for Nature Conservation or nationally by designation as Sites of Special Scientific Interest (SSSIs) by the NCC and now SNH, English Nature and Countryside Council

Table 1. *Original extent of deep peats in national parks of England and Wales estimated from overlays of national park boundaries and peat boundaries in the National Peatland Resource Inventory (NPRI) based at SNH, Edinburgh. Figures are approximate. Percentage calculations are based on 4000 km^2 total peatland in England and Wales (Lindsay 1993a)*

National park	Peat area (km^2)	Peat in England and Wales (%)	Park area containing deep peatland (%)
Brecon Beacons	228	5.7	17.0
The Broads	0	0.0	0.0
Dartmoor	264	6.6	27.9
Exmoor	10	0.3	1.5
Lake District	508	12.7	22.2
North York Moors	96	2.4	6.7
Northumberland	488	12.2	47.3
Peak District	412	10.3	29.3
Pembrokeshire Coast	2	0.1	0.3
Snowdonia	300	7.5	13.8
Yorkshire Dales	576	14.4	32.7
Total	2884	72.1	20.7

for Wales. The main criteria in the selection of sites for conservation have been based on biological attributes (Ratcliffe 1977) of area, diversity, naturalness, rarity, fragility and typicalness with additional factors of recorded history, position in an ecological/geographical unit, potential value and intrinsic appeal. More recently, and in the future, greater emphasis is being placed on the geological values of peatlands (Lindsay 1993*b*). However, to date, the general philosophy of biologically orientated conservation has been continued with the values embodied in statements in various national park documents.

National parks in England and Wales were set up by the National Parks and Access to the Countryside Act 1949 which stated their objectives very clearly as 'for the purpose of preserving and enhancing the natural beauty ... and promoting their enjoyment by the public'. The most recent review of the national parks was carried out in 1991 (Edwards 1991) and the above statement was echoed in the recognition of two purposes; environmental conservation and quiet enjoyment and understanding. A third purpose was added to encompass the social and economic well-being of communities within the parks. Clearly the first purpose of preserving and enhancing natural beauty is critical for conservation as a whole (and therefore for geological conservation specifically), but a great deal depends on the definition of 'natural beauty'. Throughout the early years of the parks, despite the definition as 'including flora, fauna, and *geological and physiographical features*' (Section 114(2), quoted in Edwards 1991. My italics.), in practice this was interpreted as meaning scenic quality. However, the review (Edwards 1991) recommended widening the definition of this term and placed greater emphasis on all environmental attributes, redefining the first purpose as 'to protect, maintain and enhance the scenic beauty, natural systems and landforms, wildlife and cultural heritage of the area'. This change of attitude must be of benefit to all conservation interests, but to determine the likely impact on geological conservation and peatlands it is important to examine the detail of this broad approach.

One of the critical parts of the implementation of the new recommendations is an environmental inventory and a list of elements is given in Edwards (1991). This includes landscape diversity, vegetation diversity, flora and fauna and archaeological artefacts. There is no explicit recognition of geological values but it could be argued that this is implicit in 'landscape diversity'. As far as peatlands are concerned, because of their unique position in the interaction of biological, archaeological and geological sciences, many sites will be considered important under a number of the above categories and in this sense may be rather better off than other geological sites. It may also be argued that peatlands are a very important component of scenic landscapes in many national parks. The high altitude blanket mire of the Dark Peak area of the Peak District is central to the wilderness quality of Kinder and Bleaklow. Likewise the blanket mire and treacherous valley mires on Dartmoor are also an essential part of the visual experience of the region. While clearly this is another good argument for peatland conservation in general, it is unrelated to geological values. It is important to emphasize that, to date, any preservation of geologically important peatlands can be considered accidental and is not related to any serious consideration of such values. Furthermore, reliance on such coincidence may be dangerous for several reasons. First, there may be geologically important sites which do not qualify for protection on biological grounds. Secondly, there are sites which are of particular interest to Quaternary science because they have been partially excavated or otherwise altered by some activity which would not be tolerated in biological conservation management. Finally, there are situations where biological conservation management actually damages the geological record, such as recontouring of raised mire sites prior to attempts at rewetting. There is, therefore, a need for geological considerations to be recognized in both the selection and management of peatland conservation sites.

Conclusion

It is clear from the above discussion that peatlands are an important element in the recent geological record of the British Isles. Furthermore, they occupy a unique position in the interaction between ecology, geology and archaeology. Despite, or perhaps because of this, there has been relatively little attention given to their conssservation on geological criteria. The national parks of England and Wales hold a significant proportion of peat deposits and are important for the future preservation of this resource. However, although the stated purposes of national parks are broader than they were, little attention is given to geological conservation in general and therefore, by implication, to geologically significant peatland sites. Peatlands may be protected for ecological reasons but it is still important to acknowledge the geological attributes of these deposits. This paper has concentrated on the role of national parks in England and Wales in conserving Britain's peatlands for their geological attributes.

However, the most important peatlands in terms of extent are concentrated in Scotland, where there is c. 1×10^6 ha (approximately 70% of the total). This is an important consideration in future national planning strategies.

References

CHARMAN, D. J. 1994. Holocene ombrotrophic peats: conservation and value in Quaternary research. *In*: STEVENS, C., GREEN, C. P., GORDON, J. E. & MACKLIN, M. G. (eds) *Conserving our landscape: evolving landforms and ice-age heritage*. English Nature, Peterborough.

EDWARDS, R. 1991. *Fit for the future: Report of the National Parks Review Panel*. Countryside Commission, Cheltenham.

GOODE, D. A. & RATCLIFFE, D. A. 1977. Peatlands. *In*: RATCLIFFE, D. A. (ed.) *A nature conservation review. Volume 2*. Cambridge University Press, 206–244.

LINDSAY, R. A. 1993a Peatland conservation – from cinders to Cinderella. *Biodiversity and Conservation*, 2, 528–540.

—— 1993b. Ombrotrophic mires (bogs). *In*: *SSSI guidelines (Draft document)*. JNCC, Peterborough (unpublished).

——, CHARMAN, D. J., EVERINGHAM, F., O'REILLY, R. M., PALMER, M. A., ROWELL, T. A. & STROUD, D. A. 1988. *The Flow Country: The peatlands of Caithness and Sutherland*. Nature Conservancy Council, Peterborough.

RATCLIFFE, D. A. (ed.) 1977. *A nature conservation review. Volume 1*. Cambridge University Press.

STROUD, D. A., REED, T. M., PIENKOWSKI, M. W. & LINDSAY, R. A. 1987. *Birds, bogs and forestry*. Nature Conservancy Council, Peterborough.

Conserving the Holocene record: a challenge for geomorphology, archaeology and biological conservation

P.C. BUCKLAND[1], B.C. EVERSHAM[2] & M.H. DINNIN[1]

[1] *Department of Archaeology and Prehistory, University of Sheffield, Sheffield, S10 2TN, UK*
[2] *Institute of Terrestrial Ecology, Monks Wood, Abbots Ripton, Huntingdon, Cambridgeshire PE17 2LS, UK*

Abstract: Many Holocene deposits are rich in biological material, and provide a detailed record of Postglacial floristic, faunistic and climatic change. At many sites, especially in the lowlands, a record of human impacts on the landscape is also well preserved. When this constitutes 'archaeology' – human artefacts or other unequivocal evidence of human activity – its documentation and conservation are relatively assured. Where the human impacts are only indirectly evident, the relevance to pure archaeology (and thus to the organizations responsible for archaeological site protection, such as English Heritage) is less clear. At the same time, Postglacial deposits are marginal to the concerns of geological conservation; and human-modified sites are of less interest to geomorphologists than entirely natural ones. The problem is illustrated by recent interdisciplinary studies of commercially-excavated Holocene peat deposits at Thorne Moors, South Yorkshire, a lowland raised mire which is shortly to become a National Nature Reserve. The relevance of the biotic record of the past 5000 years to the future rehabilitation and wildlife management of the site is stressed. The authors suggest that there is a gap in the current legislative and research framework for the protection and study of Holocene deposits, whose value is divided between archaeology, palaeoecology, landscape history and biological conservation.

Many of the papers presented in this volume consider the conservation of geological sites *per se*. This paper examines one group of geological sites – Holocene deposits – in the broader context of biological and archaeological conservation.

Holocene deposits and biological conservation

Many Holocene deposits support habitats of importance for wildlife. Riverine and montane ecosystems often retain relict species of very restricted distributions. For instance, many Scottish fluvioglacial and riverine Sites of Special Scientific Interest (SSSIs), such as those on the River Feshie (Brazier & Werritty, this volume) are nationally important for endangered *Red Data Book* insect species (Shirt 1987).

In the lowlands especially, deposits of Holocene peats often support flora and fauna which are valued highly. The site used as an example throughout this paper is the largest lowland raised mire surviving in Britain, Thorne Moors in Yorkshire. The site covers 1900 ha, and its geomorphology and natural history has been described by Gaunt (1981, 1987), Limbert *et al.* (1986), and Skidmore *et al.* (1987). Designated a SSSI under the Wildlife and Countryside Act, 1981, on grounds of habitat quality and species, it is the richest site for insects in Britain north of the River Thames. The site qualifies for protection under international conventions, specifically the Ramsar Convention on Wetlands of International Importance, and the European Community directive on Special Protection Areas for bird populations (Eversham 1991*a*). The site is also likely to figure under the forthcoming EC Habitats Directive. Part of the site is managed by English Nature as a National Nature Reserve, and the remainder may be acquired by English Nature shortly.

The main current threat to Thorne Moors is peat extraction. This has taken place at Thorne since at least AD 1200, but only since 1965 has it been mechanized, and at an accelerating pace able to threaten the future of mire vegetation, and the bulk of the Holocene record (Everham, 1991*b*). Planning permissions for peat extraction at Thorne pre-date nature conservation legislation. One effect of the industry has been the recent exposure of *c.* 4 km^2 of the lowest peats and the underlying Lake Humber deposits, whose palaeoecological and archaeological significance is currently being assessed.

Biological conservation has been the main motivation for the protection of Thorne Moors.

As long as its objectives for wetlands emphasize whole-site conservation, hydrological integrity, and the maintenance of a high water table, they will probably accord with requirements of other disciplines.

Holocene deposits and archaeology

The archaeological value of the locations of human settlements is clear-cut – if well-preserved, they are of great intrinsic interest, and lend themselves to public interpretation. More isolated human artefacts, such as trackways across wetlands (which, in Britain, usually lie between current and former settlements) may be excavated, or conserved *in situ* or in museums in the region.

More diffuse indications of landscape management are not so clearly the province of the archaeologist. At Thorne, there is evidence for management of the forests which existed before the initiation of peat formation, in the late Neolithic and throughout the Bronze Age (Buckland & Dinnin 1993). Preliminary excavations suggest these may have been managed as an open-canopy pasture–woodland dominated by oaks; the trees would provide green fodder and timber, and the grassland between them would be grazed by livestock. This form of forest management is well-known in Medieval Britain, but its earlier origins are unclear, and the data fragmentary and inconclusive (Harding & Rose 1986). The evidence for pasture–woodland in the Bronze Age at Thorne rests on details of the size and placement of individual trees, and the existence of a grassland flora contemporaneous with mature timber. There is also some evidence of controlled burning as a tool in woodland management, but this is hard to distinguish from natural fires in wildwood. Such features are unlikely to be recognized as 'real' archaeology until the sites of settlements are located, or other unquivocal evidence of human activity is found. It has so far proved difficult to interest the statutory archaeological conservation agency, English Heritage, in the site. The questions which the exposure raises about landscape management and development, are of limited interest to archaeologists, although the potential for enhancing the dendrochronological framework for archaeology in the region, for instance by providing the first long time-series chronology for Scots pine in England, would be valuable to mainstream archaeology. Climate change signals from the peat, and from tree-ring analysis, may have still wider applications (Barber 1985).

Archaeological work on the Bronze Age at Thorne is now being funded by the private landowner, the peat extractors, partly as a result of media and political pressure.

Geomorphological conservation and holocene sediments

Glacial, periglacial, fluvioglacial and fluvial features are well represented in SSSI, from Scottish moraines, river shingle formations, to lowland pingos and kettle-holes. These sites are mainly 'natural', with minimal sign of human impacts. In the lowlands, human impacts are more pervasive. The value of the larger, human-modified lowland sites is less readily recognized, and their future is often less secure.

A problem for the geological conservationist is that the sequence of deposits is often incomplete. A cut-over peatland, from which the last 2–3000 years of deposits have been removed, is less likely to be the focus of geomorphological conservation. Conversely, where the full sequence is still present, even if a mire has lost its surface interest (and is thus unlikely to meet the biological criteria for SSSI notification), the site may merit protection on geomorphological grounds.

A common approach of geologists to site conservation – identifying a 'representative exposure' – may be inappropriate for some organic deposits, especially those in wetlands. Drainage of part of a site usually causes a lowering of the water table throughout. Desiccation of organic sediments leads to rapid decomposition and oxidation. The archaeological and palaeoecological content (plant macrofossils, pollen, and insect remains) may decay beyond recovery within a few months or years of the lowering of the water table, whether they are physically exposed on the surface or not. Thus, conserving part of a mire, while allowing the drainage or removal of the rest, is unlikely to succeed. The usual approach of biological conservation, of attempting to safeguard whole hydrological units, is more appropriate to the protection of wetland archaeology and palaeoecological interest (Buckland & Dinnin in press).

There is further dichotomy of approach relevant to the method of site conservation. Traditionally, most pollen analysis and studies of sub-fossil insects have examined changes through time, at one or a few points on the land surface. In recent years, 'landscape palaeoecology' has begun to examine contemporaneous samples over a wider area, seeking to reconstruct earlier landscapes (such as the Bronze Age forest at Thorne). Only thus can land management in the Holocene be understood.

Damaged sites

Even if a large part of the Holocene record has been damaged or removed, a site may still be of value for the material which remains. Thorne Moors has lost most of the last 2000 years of peat over much of its surface, but the lowest deposits (older than 2000 years BP) are still in place and, in part at least, are waterlogged. The reconstruction of Postglacial history is thus still possible.

Extractive industries also affect the biological conservation value of sites enormously. Bare peat has little current biological value, and is seldom scheduled except to help maintain the integrity of a hydrological unit, essential for the survival of adjacent vegetation. A further criterion in SSSI notification is 'habitat potential', in the case of peatlands, the opportunities for post-extraction restoration. However, to restore peatlands, or other habitats, one must understand the processes involved. Whereas woodland and grassland management use techniques from forestry and agriculture, and achieve results within the timescale of a human life-span, the creation of acid mires may take upwards of 500 years, and during that time, the process could be deflected or halted by quite small environmental changes.

Five models of mire initiation at Thorne have been proposed (Buckland & Smith in press), and these may be falsified only by a careful study of the Holocene record at Thorne. Recent radiocarbon dates for basal peats suggest a polyfocal origin and a large-scale moving front of paludification and subsequent mire genesis. A natural rise in sea-level may have been responsible for changes, but it is equally possible that anthropogenic changes to river channels resulted from woodland clearance further up the catchments (Buckland & Sadler 1985).

Flooding and drying events recorded in the peat may also help restoration planning. A prominent flooding horizon, indicated by dominance of *Scheuchzeria palustris*, a plant which became extinct in England during the nineteenth century, is a feature of lowland peat at several sites. Desiccation events recorded in the peat are even more timely in their message: recovery from a century of drought, whether caused by climatic or sea-level fluctuations, is the closest available analogue to recovery from a century of drainage and peat cutting.

Past changes in species composition on the mire surface are relevant to the future, especially the loss of important mire-building bog mosses: *Sphagnum imbricatum*, disappeared around the fourteenth century at Thorne (Smith 1985), perhaps through early human impacts (Rose & Wallace 1974) or through competition with *S. magellanicum* (Smith 1985), a species which itself has declined in the past 200 years, probably through atmospheric pollution (Smart et al. 1986).

The problem now facing biological conservation at Thorne is to define the mire surface which can grow in the absence of these two mire-building *Sphagnum* species. Peats laid down in the past 500 years would provide the best clues. It may be that biological conservation can hope only to maintain the species composition of the mire surface as a 'stopgap', in hope of a reduction in pollution levels and a resurgence of more appropriate mire *Sphagna*. There is still no clear picture of the state of the mire surface before drainage (c. 1850) but after the losses of the main peat-forming *Sphagna*. These would provide the most useful insights possible for the planning of mire restoration for the future.

Policy implications of the Thorne study

A number of inferences may be drawn from the experience of conservationists in several disciplines at Thorne over the past 20 years.

- Biological conservation legislation cannot be effective against prior planning permissions.
- Biological conservation caters for archaeology and palaeoecological interests only incidentally.
- Conservation of Holocene deposits *per se*, and of soils, is likely to be neglected except at sites which are entirely natural and intact.
- Archaeological and palaeoecological value are sensitive to degradation induced by drainage, even if the organic sediments are not physically removed.
- If a large part of the sediments has been removed from a site, there may still be enough to provide valuable climate-change signals, and information on land management for the periods where sediments are intact.
- Geomorphology, archaeology and palaeoecology may provide information crucial to the future biological conservation planning and management of wetlands.
- The heritage value of signs of prehistoric land management, which are not rich in human artefacts, cannot readily be accommodated in the current legislative and research framework in Britain.

This paper was prepared as a result of deliberations over the future of Thorne and Hatfield Moors in the Thorne and Hatfield Conservation Forum. The comments particularly of J. P. Sadler, P. Skidmore and B. M. Smith are acknowledged.

References

BARBER, K. E. 1985. Peat stratigraphy and climatic change: some speculations. *In*: TOOLEY, M. J. & SHEAIL, G. M. (eds) *The climatic scene*. George Allen and Unwin, London, 175–185.

BRAZIER, V. & WERRITTY, A. This volume. Conservation management of dynamic rivers: the case of the River Feshie, Scotland.

BUCKLAND, P. C. & DINNIN, M. A. 1993. Holocene woodlands: the fossil insect evidence. *In*: KIRBY, K. J. & DRAKE, C. M. (eds) *Dead wood matters: the ecology and conservation of saproxylic invertebrates in Britain*. English Nature Science 7, English Nature, Peterborough, 6–20.

—— & —— (in press) Peatlands and floodplains: the loss of a major palaeontological resource. *In*: STEVENS, C., GREEN, C. P., GORDON, J. E. & MACKLIN, M. G. (eds) *Conserving our landscape: evolving landforms and ice-age heritage*. English Nature, Peterborough.

—— & SADLER, J. 1985. Late Flandrian alluviation in the Humberhead Levels. *East Midland Geographer*, **8**, 239–251.

—— & SMITH, B. M. (in press) Equifinality, conservation and the origins of lowland raised mires. *Thorne and Hatfield Moors Papers*, **5**.

EVERSHAM, B. C. 1991a. Land use change and wetland invertebrates in Britain. *In: Conserving and managing wetlands for invertebrates*. Council of Europe/Ramsar Bureau, Strasbourg, 107–110.

—— 1991b. Thorne and Hatfield Moors: implications of land use change for nature conservation. *Thorne and Hatfield Moors Papers*, **2**, 3–18.

GAUNT, G. D. 1981. Quaternary History of the Southern Part of the Vale of York. *In*: NEALE, J. & FLENLEY, J. (eds) *The Quaternary in Britain*. Pergamon Press, Oxford, 82–97.

—— 1987. The geology and Landscape Development of the Region around Thorne Moors. *Thorne Moors Papers*, **1**, 5–29.

HARDING, P. T. & ROSE, F. 1986. *Pasture-woodlands in lowland Britain*. Institute of Terrestrial Ecology, Huntingdon.

LIMBERT, M., MITCHELL, R. D. & RHODES, R. J. 1986. *Thorne Moors: Birds and Man*. Doncaster and District Ornithological Society, Doncaster.

ROSE, F. & WALLACE, E. C. 1974. Changes in the Bryophyte Flora of Britain. *In*: HAWKSWORTH, D. L. (ed.) *The Changing Flora and Fauna of Britain*. Academic Press, London, 27–46.

SHIRT, D. B. (ed.) 1987. *British Red Data Books, 2: Insects*. Nature Conservancy Council, Peterborough.

SKIDMORE, P., LIMBERT, M. & EVERSHAM, B. C. 1987. The insects of Thorne Moors. *Sorby Record*, **23**, (suppl.), 89–153.

SMART, P. J., WHEELER, B. D. & WILLIS, A. J. 1986. Plants and Peat Cuttings: Historical Ecology of a much exploited peatland – Thorne Waste, Yorkshire, UK. *New Phytologist*, **104**, 731–748.

SMITH, B. M. 1985. *A Palaeoecological Study of Raised Mires in the Humberhead Levels*. PhD thesis, University of Wales, Cardiff.

Keynote address

Malta: a model for the conservation of limestone regions

ANNA SPITERI

Malta Council for Science and Technology, 112 West Street, Valletta, Malta

Abstract: The most frequently addressed issues at the end of this century are undoubtedly the environmental challenges that the Earth now faces. New integrated (i.e. multidisciplinary) management approaches and techniques have been developed to meet the new environmental challenges.
  Conserving our geological and landscape heritage is one of the challenges that has to be met. Every landscape of notable importance has to be considered as a resource in its own right, but especially as a touristic resource. The same applies to landscape conservation strategies which would have a better chance of becoming part of the environmental law process if they were drawn up as part of the overall conservation strategy.
  The integrated resource management approach and conservation strategies are, of course, best worked out within a sustainable development framework which is the most globally relevant. Sustainable development is defined as 'development that meets with needs of the present without compromising the ability of future generations to meet their own needs'. In the case of conservation of our geological and landscape heritage it would mean the preservation of areas that contain an invaluable store of knowledge and information on our geological and geomorphological history and human evolution. Limestone terrains the world over deserve special attention.
  Recent technological revolutions in remote sensing from space and in land information systems can provide the necessary technical means to ensure the physical survival of geological sites of scientific importance.
  This paper discusses new management approaches that are being developed in Malta to meet local environmental challenges; outlines the national policy for promoting cultural tourism, and proposes how the application and integration of these techniques and strategies can contribute towards making Malta a model for geological and landscape conservation for limestone regions in a sustainable development framework.

Our perception of the Earth is at present undergoing a drastic change. We are realizing collectively for the first time that every facet of the environment plays a role in our lives. The international discussion on environmental conservation issues, as covered by the Rio Conference in 1992, focused mainly on biological conservation. The concept of conserving our geological heritage was not on the agenda.

Earth science conservation, because of the perceived static nature of the features requiring conserving, will never become as topical as biological conservation. Indeed the relatively few who understand the equal importance of conserving our geological heritage have the responsibility to ensure the immediate survival of these geological sites.

The Maltese Islands have many such sites. The Dwejra inland sea and Blue Grotto are two examples. To wait for public awareness of their scientific importance to materialize, may take too long to guarantee their preservation. The alternative and parallel approach is to earmark them as tourist attractions, and consider them as a land-based resource, thus attempting to protect them from harmful development.

Sustainable development

In the Bruntland Report on 'Our Common Future', sustainable development is defined as 'development that meets the needs of the present without compromising the ability of future generations to meet their own needs'. This concept applies to every aspect of environmental management. In the context of conservation of our geological and landscape heritage it would mean the preservation of areas that contain an invaluable store of knowledge and information on our geological and evolutionary history for further study and understanding by present and future generations.

This is especially significant in limestone regions. Limestone terrains contain the key to the sea-level history of the Earth for at least the last million years and, therefore, deserve special attention the world over (Ford & Williams 1989). Caves, for example, which are a phenomenon of limestone, are considered nature's vaults, containing an irreplaceable and datable record of biological, climatic and landscape history. The Maltese Islands present some classic examples. (Ford & Williams 1989).

The Ghar Dalam cave is an underground karstic river system in which Quaternary sediments were deposited during the interglacial period about 250 000 years ago. The remains discovered there of Pleistocene fauna, hippopotami, elephants, red deer, brown bear, fox and wolf; and even a fused molar of Neolithic man, render the cave a site of regional scientific significance. This particular site is a valuable touristic resource and its management should, therefore, be extended beyond the mere display of fossils.

Moreover development decisions in the area should reflect the full value of the natural and cultural environment. There is an apparent need in Malta to introduce the idea of sustainable development and shift the focus away from the traditional growth versus development argument. This shift demands a new management approach.

Integrated resource management

Prevailing systems for decision-making in many countries tend to separate economic, social and environmental factors at the policy, planning and management levels. This influences the actions of all groups in society, including governments, industry and individuals, and has important implications for the efficiency and sustainability of development and the environment. (UNCED 1992).

Small-island developing states are a special case both for development and environment. They are ecologically fragile and often, as in the case of the Maltese Islands, geologically vulnerable.

In general, the uniqueness of limestone terrain arises from the fact that limestone is relatively highly soluble in water. This is reflected especially in surface and subterranean solution, both of which give rise to particular features. Limestones provide the most distinctive suite of landforms and applied problems of any rock type and require a special approach for their conservation and management (Cooke & Doornkamp 1990).

Here, the integrated resource management (IRM) approach is being proposed as a way to balance rationally the use and conservation of the island's resources (Martin 1991). The limestone landscape of the Maltese Islands was always considered as offering very limited resources. The IRM approach may prove to be the most appropriate management tool to deal with our particular situation.

To begin with, IRM allows for a new assessment of what constitutes our resources and for the first time makes space for a paradigm shift in the perception of how we look at our islands. For example, a 'bare' landscape can represent an attractive geological landscape and a valuable development resource.

A shift is taking place towards promoting cultural tourism, which includes many of the geological karst features as part of the islands' landscape heritage. Some of these sites are already being considered as resource sites in their own right and should, therefore, now be managed and protected as such.

The planning authority is aiding this objective by incorporating protection of geological and geomorphological sites in its first local plan study of the area of Marsaxlokk. There are also plans to carry out the same kind of exercise for the rest of the islands. The criteria used for ranking each geological site of notable importance in the Marsaxlokk area, were assessments of the various 'human-use' parameters of each site selected. These parameters ranged from educational, research, historical, aesthetic, touristic to recreational. The next step is that some of the areas marked out for protection need to be integrated into the tourism plan. This will become possible under the proposed IRM package, where active inter-sectoral communication and co-operation, i.e. exchange of data and also of expertise, will help to define what should constitute the optimum management of a site.

The tools

The driving force of IRM is a central databank, better known as GIS, Geographic Information System. Some of the data to be stored are readily available, for example, data on groundwater resources, climate, land-use, building development plans, telecommunication infrastructure and energy resources. Data on other resources, such as human, financial, scientific, technological and cultural resources, as well as on tourist resources (whether natural or not), can also be assessed and included in the GIS set-up.

Real-time spatial data acquired from various remote sensing platforms and from geophysical investigations can be transmitted straight into the GIS. These data can provide information such as status, limits and mapping of resources with a degree of accuracy and detail not previously available, which makes these techniques effective instruments for timely decisions.

Thus in an IRM environment, the survival and protection of important scientific geological sites as a valuable (tourist) resource will be ensured.

Cultural tourism

The application of IRM will be even more effective if there is an overall policy of treating each and every resource in a sustainable way. This is especially relevant to the tourist sector. It is often the case, especially in the Mediterranean region that the type of tourism that brings the greatest economic benefits also results in serious environmental degradation, and therefore cannot be justified in the long run.

There is still a need in Malta and also in the region to develop greater awareness and understanding of the significant contributions that tourism can make to the environment and the economy. Sustainable tourism development does involve promoting appropriate uses and activities that draw from and reinforce landscape character and site opportunity.

Tourism development involving any loss of existing natural landscape or sites of geological importance will increasingly indicate how future generations will be compensated. It is obvious that the loss of these natural assets can no longer simply be substituted by capital wealth created by new development (Inskeep 1991).

Tourism can provide the incentive and help pay for the conservation and maintenance of geological, archaeological and historic sites that are attractions for tourists that might otherwise be allowed to deteriorate or disappear. Tourism can be a major stimulus for conservation of geological sites because their conservation can be justified, in part or whole, by their being tourist attractions (Inskeep 1991). Unfortunately, it must be noted that many residents of the Maltese Islands prove to have a limited interest in, and even less concern about, the natural environment and its conservation. The economic factor of these sites for touristic purposes can encourage local awareness in the wise management of these sites.

The potential danger is over-use by tourists of the fragile limestone environment. Conservation of the natural features should take precedence over visitor use although the small size of the Maltese Islands makes their complete control extremely delicate. Land zoning, for example, although attempted, is not very possible, i.e. protection of the area adjacent to a site on such a small island is not considered practical.

The Maltese Islands' geological heritage

Having discussed the concepts of IRM and cultural tourism it is appropriate at this stage to describe the most significant geological sites that are being proposed as representing a valuable touristic resource. The following are classic examples of geological features from limestone and karst terrains. Some of these sites are being presented to UNESCO to be included as World Heritage Sites.

(1) Blue Grotto – natural arch
(2) Dwejra inland sea – circular dolines
(3) Il-Maqluba sinkhole
(4) Il-Hofor – circular dolines
(5) Ghar Dalam cave – underground karst river system
(6) Ghar Hasan cave – underground karst river system
(7) Il-Maghlaq fault
(8) Victoria lines fault
(9) Kullana wave-cut platform

Malta as a model

A number of initiatives are presently already being carried out, aiming to transform the Maltese Islands into a model for the conservation of limestone terrains. A Maltese Earth science working group has just been set up with representatives from the planning authority, the Environment Ministry, the Malta Council for Science and Technology, the Ministry of Tourism, NGOs and the university, under the auspices of the Environment Secretariat.

A co-ordinated regional policy of promoting cultural tourism in the Mediterranean, inspired by the Maltese model, could safeguard the geological heritage not just of the Maltese Islands, but of that of the Mediterranean region as well, for generations to come.

References and further reading

BISWAS, A. K. & GEPING, Q. (eds). 1987. *Environmental impact assessment for developing countries*. United Nations University, Tycooly Publishing, UK.

COOKE, R. U. & DOORNKAMP, J. C. 1990. *Geomorphology in environmental management*. Oxford University Press, New York.

COUNCIL OF EUROPE. 1987. *Management of Europe's*

natural heritage, Twenty-five years of activity. Strasbourg.

ENVIRONMENT SECRETARIAT. 1993. *An 'Agenda 21' for Malta.* Ministry for the Environment, Malta.

ESTES, J. E. et al. 1992. *Advanced data acquisition and analysis technologies for sustainable development.* MAB Digest 12, UNESCO, Paris.

FORD, D. & WILLIAMS, P. 1989. *Karst Geomorphology and Hydrology.* Unwin Hyman Ltd, London.

INSKEEP, E. 1991. *Tourism Planning, an integrated and development approach.* Van Nostrand Rheinhold, New York.

JENNINGS, J. N. 1985. *Karst Geomorphology.* Basil Blackwell Ltd, Oxford.

MALTA COUNCIL FOR SCIENCE AND TECHNOLOGY. 1992. Vision 2000: Malta Regional Hub. Valletta, Malta.

MARTIN, F. 1991. Integrated Resource Management, the answer to a Socio-economic Problem. *GeoJournal,* **25** (1), 109–113.

MITCHELL, B. 1989. *Geography and resource analysis.* Longman Scientific and Technical.

PLANNING SERVICES DIVISION. 1990. *Structure Plan for the Maltese Islands.* Ministry for Development of Infrastructure, Malta.

RAPER, J. (ed.) 1989. *Three Dimensional Applications in Geographic Information Systems.* Taylor and Francis Ltd, London.

SPITERI, A. 1991. *Geological/Geomorphological Survey for Marsaxlokk Bay Development Area. A Conservation Strategy and Guidelines for Local Plan Studies.* Planning Services Division, Malta.

UNCED 1992. *Earth Summit '92.* The Regency Press Corporation, London.

Caves as unique conservation education resources

GEORGE N. HUPPERT

Department of Geography and Earth Sciences, University of Wisconsin–La Crosse, La Crosse, Wisconsin 54601, USA

Abstract: Caves are unique and fragile places harbouring many resources. These resources are extremely varied, which presents many challenges for their management. This report considers three significant caves in the United States that are managed in different ways for the protection of the cave, research and education. The caves, listed in order of the least to the most restrictive management policy, are as follows

1. *Tumbling Creek Cave*, Missouri: the site of the Ozark Underground Laboratory (OUL) which is a privately operated institution devoted to cave conservation education. The OUL offers tours of the cave and the land above it, along with a lesson on cave geology, hydrology and ecosystem dynamics. Most tours are designed for school groups. The cave, which contains several endangered species, is closed during bat hibernation.
2. *Coldwater Cave*, Iowa: a privately owned cave open for research and exploration. Entrance to the cave is strictly controlled by the owners to those individuals and groups that they know. This 22 km undeveloped and nearly pristine cave is designated as a National Natural Landmark by the federal government.
3. *Lechuguilla Cave*, New Mexico: is located within the boundaries of Carlsbad Caverns National Park, and is over 112 km long and over 474 m deep. Speleologists consider this cave as one of the most unique and most fragile in the world. Extremely limited numbers of individuals are allowed into the cave for exploration and research. The Lechuguilla Cave Protection Act was signed by the President on December 2, 1993. This Act (Public Law 103–169) adds another layer of legal protection for the cave, especially for those passages that may be discovered outside of the national pasrk boundaries.

These examples are just a few of the protection plans used to conserve caves in the US. There are numerous caves that have some sort of protection; however protection is only as good as the enforcement of the law, which varies significantly between jurisdictions.

There may be as many as 50 000 caves in the US (Dickey 1974), yet few of these caves are regarded by the general public as valuable natural resources. Some of the values that caves may have are scientific (geological, hydrological, biological, archaeological, and palaeontological), recreational, aesthetic/religious, and historic (Huppert & Wheeler 1986). Obviously, caves can be storehouses of many resources; however, they are almost always very fragile environments.

Most caves are protected by federal or state laws such as the Federal Cave Resources Protection Act of 1988 (PL 100–691) which went into full effect for the Department of Interior Lands (National Park Service and Fish and Wildlife Service) with the release of the implementing rules in April 1994. The Department of Agriculture (Forest Service and the Bureau of Land Management) has yet to release its rules of implementation. This law is discussed in detail by Huppert & Thorne (1989). Laws are only as good as their enforcement which is usually variable, for several reasons. First, the offenders have to be apprehended. Second, the legal system must consider caves important enough to press charges and punish the offenders rather than to dismiss the incident. Dismissals have unfortunately been the most common outcomes until recent years. Educational programmes by the American Cave Conservation Association, the National Speleological Society, and the Nature Conservancy are aimed at law and court officers as well as potential offenders.

In response to the general ineffectiveness of the law to protect specific caves until after the fact of damage, many individuals and conservation organizations have made the effort to physically protect specific significant caves. An exact number is not known, but perhaps several hundred caves in the US have physical protection beyond the law. This protection may range from a simple (or complex) gate to an on-site caretaker. The effectiveness of these barriers is mixed. In general, those that are well-gated with caretakers are best protected from vandalism and trespass. In this paper, three caves that are designated and protected as scientific, educational, and recreational (in a limited sense) reserves are described. Two of the caves are privately owned and the third is in a national park.

Tumbling Creek Cave, Missouri

This cave, located in southwest Missouri, is the site of the Ozark Underground Laboratory (OUL). The cave is owned and operated by Tom and Cathy Aley as a research and educational reserve. The cave was purchased in the late 1960s after a long search for the right cave. The selection criteria included a protected surface drainage and a great diversity of biota. Tumbling Creek Cave has these attributes. Aley (1977) reports that there are about 100 species of flora and fauna (seven or eight previously unreported ones, some indigenous) and one new genus. There is also a large summer colony of endangered grey bats in this >3 km long cave. The cave is so biologically significant that it has been designated as a National Natural Landmark by the federal government.

While research at the cave has produced a number of theses and research reports, the main purposes of the OUL are the preservation of the unique ecosystem and the education of as many people as possible without harming the resource. Numerous university and school students participate in field trips at the laboratory each year. The field trip not only familiarizes students with the cave environment, but also explains its intimate interactions with activities on the surface. Literally, thousands of people have been educated about cave science and protection due to the efforts of the OUL over the years.

Coldwater Cave, Iowa

Coldwater Cave, located on private land in northeastern Iowa, was discovered in 1967. Originally, the cave was entered by diving into a large spring. The State of Iowa leased rights to the cave from the landowners in 1970, with the intention of creating a state park because the cave is perhaps the best decorated cave in the Upper Midwest. The State drilled a 28 m shaft and installed a permanent ladder to make access much easier to non-divers. However, three locked barriers including a fence and a pole shed with two different doors, were built to restrict access to only those individuals with permission. Later, the spring entrance to the cave was permanently gated.

Due to the great potential cost of tourist development, the State of Iowa let the lease lapse a few years later. At that time the owners decided to institute informal management to control access and to protect the cave. Huppert (1987) details the story of this successful cave reservation. The control of access was turned over to a chapter (called a 'Grotto') of the National Speleological Society (in this case the Rock River Speleological Society). Until recently the cave was essentially open to all responsible groups on the third Saturday of each month. Many types of groups would show up at the cave. They were mostly cave explorers, but often they were nature clubs, school and university groups and the like. Researchers generally had free run of the cave for approved projects. Groups had to have a leader who knew the cave and was known by the managing grotto or the owners. Group members would be instructed about safety and the importance of conservation. Close control of these groups allowed any damage to the cave to be assigned to the responsible party. This process worked well until recent years when traffic became quite heavy. A few unfortunate incidents by inconsiderate individuals led to a semi-closing of the cave. The owners assumed complete control of the cave's access. Only a limited number of individuals or groups personally known by the owners are allowed in at this time. This restriction still allows the majority of those who explored, surveyed, or did research on the cave in the past to continue to work there. It does, however, restrict new groups from working in the cave.

The result of this restrictive policy, along with periodic flooding of the cave, is that, with very minor exception, the cave is nearly as pristine as the original explorers found it. The control (thus conservation) of the cave is successful because of concerned and co-operative owners who live on the site and maintain the three locked barriers. The cave is quite secure and should be a resource for teaching science and conservation for many years in the future. Coldwater Cave has also been designated as a National Natural Landmark.

Lechuguilla Cave, New Mexico

Lechuguilla Cave, discovered in 1986, is located within the boundaries of Carlsbad Caverns National Park. Thus the cave is managed by the federal government. The cave is 112 km long and 474 m deep, the deepest cave in the US. Portions of the cave are incredibly decorated with fantastic formations. The beauty of this cave is well illustrated in the recent book by Taylor (1991). The cave is isolated in a wilderness area and has a gate at the bottom of a 27 m entrance drop. While there is some impact along main travelled pathways, strict environmental rules have kept the cave nearly pristine.

The National Park Service strictly controls access to the cave by a permit system open only to bona-fide cave mappers and researchers. The Park Service turned over the logistics of managing access to the cave to a private group formed for the purpose called The Lechuguilla Project. Of course, the group was required to work within the environmental restrictions of the National Park Service permitting system which mandates limits on group size, number of groups, length of stay in the cave, etc. This agreement was recently dissolved (1992) because of misunderstandings of the obligations of the Project and personality differences. A new agreement is under negotiation at this time. However, research, exploration, and surveying continue under the direct supervision of the Park Service.

A significant threat to Lechuguilla Cave is the desire by commercial groups to convert the cave into a public show piece. Most speleological experts who have seen the cave agree that it would be extremely difficult and expensive, if not impossible, to develop the cave for tourism and to protect all of the cave resources at the same time. The cave is not only in a national park but within the boundaries of a designated wilderness area, so there is a double layer of legal protection. To commercialize the cave would require the retraction of wilderness status of the land above it. While this is not unknown to occur, it takes Congressional passage and is extremely difficult. However, a recent effort by conservationists to have the cave itself declared a wilderness on its own merits failed, due to intense lobbying efforts of local business groups.

A greater threat is posed by the fact that passages are being explored and mapped so rapidly that the cave will soon extend beyond the boundaries of the national park and the wilderness area into an area called Dark Canyon. Dark Canyon is managed by another federal agency, the Bureau of Land Management (BLM). Protection by the BLM is much less assured because that agency is allowed to lease public lands under its control to private companies for oil and gas exploration and development. In an effort to protect the cave from damage, a bill (HR 698) was introduced into the US House of Representatives early in 1993 (Krause 1993). Of particular concern is gas leaking from well bores and migrating through the karst into the cave where it would pose a great hazard to the cave and cavers. The bill proposes to withdraw Dark Canyon from commercial exploitation. Hopefully this action will be taken soon as the BLM is actively pursuing the permitting of numerous wells in the canyon potentially close to cave passages. This bill was signed into law by President Clinton on December 2, 1993 as Public Law 103–169. It specifically protects potential cave passages in the Dark Canyon area from mineral exploitation.

An Environmental Impact Statement (EIS) has been completed and enacted by the Bureau of Land Management in January 1994 covering BLM lands in Dark Canyon adjacent to Carlsbad Canyon National Park. This EIS and the Lechuguilla Cave Protection Act are somewhat redundant. However, the Act is more encompassing and the EIS only relates to oil and gas exploration and development.

Cave conservationists successfully worked toward the goal that all potential threats to Lechuguilla Cave, including tourist development, oil and gas development, and all types of degradation from visitation, can be prevented or at least minimized. The plans for Lechuguilla Cave allow little or no direct contact with the cave by the general public. This plan does allow research to continue. It is hoped that the research results will be extremely helpful to further the public's understanding of caves and karst.

Conclusions

This paper has summarized the efforts and some of the problems of landowners, land managers, and conservationists trying to protect three very significant and different caves in the US. Each cave is a unique entity used in various ways by different types of owners/managers to preserve the resources and to educate the public. There are many more caves in the country managed in similar ways but these three are truly outstanding.

References

ALEY, T. 1977. The Ozark Underground Laboratory. In: ALEY, T. & RHODES, D. (eds) *National Cave Management Symposium Proceedings – 1976.* Speleobooks, Albuquerque, New Mexico, 95–96.

DICKEY, F. 1974. *Report of the Caver Proliferation Committee.* National Speleological Society, Huntsville, Alabama.

HUPPERT, G. 1987. A study in landowner relations: Coldwater Cave. *American Caves*, 2(3), 14–17.

—— & THORNE, J. 1989. Federal cave protection in the United States: The Federal Cave Resources Protection Act of 1988. *In*: KOSA, A. (ed) *Proceedings of the 10th International Congress of Speleology*. Hungarian Speleological Society, Budapest, 188–90.

—— & WHEELER, B. 1986. Underground wilderness: can the concept work? *In*: LUCAS, R. (ed.) *Proceedings – National Wilderness Research Conference: Current Research*, United States Department of Agriculture, Forest Service, Intermountain Research Station Technical Report INT-212, Ogden, Utah, 516–22.

KRAUSE, A. 1993. It's time to act. *NSS News*, **51**(5), 122.

TAYLOR, M. (ed.) 1991. *Lechuguilla: Jewel of the Underground*. Speleo Projects, Basel.

Sixty-five years of legislative cave conservation in Austria: experiences and results

HUBERT TRIMMEL

Union Internationale de Spéléologie, Draschestrasse 77, A-1230 Wien, Austria

Abstract: On June 28, 1928, the Austrian Parliament agreed a Federal Law concerning the protection of caves, one of the first laws in the world dedicated especially to the protection of geo-scientific phenomena. This established new dimensions of protection and preservation of Austrian caves. The most important of the criteria for declaration of a protected cave was its value for natural science. This meant that scientific studies and speleological research were vital in establishing the conservation measures. Between 1928 and 1938 and from 1945 to 1974, some 177 caves or cave areas in Austria were declared a 'protected natural monument'.

Since 1975, measures for the protection and preservation of caves have become regionalized. The regions ('Länder') follow their own policies in their legislative decisions but, in general, protection of caves, of the surroundings of the cave entrances and of karstic phenomena connected with caves is now a field of special legislation.

The economic development of alpine regions through tourism now makes the protection of major karst regions more important than the protection of single caves. The existing measures of cave protection form a sound basis for the active development of more extensive protected karst areas.

On June 28, 1928, the Austrian Parliament agreed a Federal Law concerning the protection of caves, supplemented by a series of decrees in the following year. One of these decrees concerned conservation-orientated rules for commercial caves and the education of cave guides; another, a scheme for continuous permanent documentation of protected caves. Scientific research as a basis for all conservation measures was undertaken before 1938 by an Institute of Speleology, and after 1945 by a Speleological Department in the Federal Bureau for the Protection of Monuments.

This was one of the first laws in the world dedicated especially and exclusively to the protection of geo-scientific phenomena. At this time the law established new dimensions of protection and preservation. The most important of the criteria for the declaration of a protected cave was its value for natural science. This meant that it was possible to declare a cave a 'protected natural monument' not only because of its prehistoric or palaeontological importance, but also because of geological structures, important sediment layers or ice formations. In recognition of the relationship between the ecological development of cave chambers and conditions at the surface, the law also made it possible to protect the surroundings of the cave entrance and related karst-features at the surface. So, this law agreed 65 years ago, creates a very modern impression.

Between 1928 and 1938 and from 1945 up to 1974, in Austria, 177 caves and cave areas have been given the status 'protected natural monument'. In the first instance, all important show caves have been protected and the first steps have been taken to resolve the conflict between natural environment and tourism in caves. Since the Second World War, many newly discovered cave systems have been protected in collaboration with cavers and caving societies. Most of the known cave systems have been explored since 1945, and it was very important to limit human influences in these systems before undertaking possible complex scientific documentation. Today, the total number of registered caves in the central documentation system is nearly 11 000 – an important potential for future research. But this number is increasing relatively rapidly. In this situation it is more important that cavers have a proper understanding of the problems of protection than that a sound law exists.

Historically, experience with the Austrian 'cave protection law' has been good. Success has been possible mainly for the following reasons:

(1) The law has been administered by objective scientific institutions – in general well-accepted by the public and led (or regularly advised) by speleologists.
(2) Permanent collaboration with the cave clubs by these institutions and federal authorities has guaranteed good information and documentation as well as educational measures for the cavers.
(3) Caving is not a mass sport in Austria, and

access and descent in caves, especially in the high-alpine regions, are often very difficult.

In practice, the situation regarding the protection of caves has changed for several reasons and in several ways.

First, measures for the preservation of caves by law have been regionalized. Now, the regions ('Länder') follow their own policies in their legislative decisions. In many regions, protection of caves and of karst phenomena connected with caves is now the field of special legislation; in other regions, cave protection is now part of the general legislation for the protection of nature. In many cases, problems arise because the law is administered by local or regional authorities without any knowledge of important geo-scientific factors and often more or less in response to local economic influences.

Second, the economic development of alpine regions, especially through tourism in both summer (mountaineering by funiculars) and winter (skiing) necessitates the protection of major karst regions including all the accessible caves. An important aspect of this need to protect regional karst landscapes is the protection of karst waters: nearly 50% of the Austrian population is supplied with drinking water from karst springs, and it seems likely that 'karst water protection' will complement the planned creation of national parks in the karstic Limestone Alps.

Thus, existing measures for cave protection in Austria form a sound basis for the active development of more extensive protected karst areas.

Protection of limestone pavement in the British Isles

HELEN S. GOLDIE

Department of Geography, University of Durham, Science Laboratories, South Road, Durham DH1 3LE, UK

Abstract: Limestone pavements in the British Isles are in the unusual position of having parliamentary legislation aimed at protecting them: Section 34 of the Wildlife and Countryside Act 1981. This situation arose as it was realized that these landforms, with their great botanical, geological and archaeological interest, were being destroyed astonishingly rapidly for commercial sale for garden rockeries, mainly from the 1960s onwards. There had been earlier use made of such outcrops, but recent decades saw an acceleration. This was coupled with increased awareness of environmental damage, and concern to decrease or prevent this particular example, especially in view of the limited extent of limestone pavement in Britain. The areas where most damage was occurring were in NW England, including the Craven district of Yorkshire and the various smaller outcrops in Cumberland and Westmorland (now Cumbria), and Lancashire.

Legislation was made feasible by the existence of a comprehensive conservation survey of pavements in Great Britain carried out in the early 1970s, which has been the basis for protective work carried out since the Act was passed. Several public and private bodies are involved in the conservation and protection process. The work of these bodies has been essentially small scale and local in nature even though the legislation is national in its overall applicability. It includes the collection of information on the pavement outcrops for the notification process as well as detailed consultation and negotiations with landowners and tenants. There is natural variation in the way the work is pursued between different areas even within northern England.

The protection process is well in-train, but is not without its difficulties. It is most advanced in Cumbria and Lancashire where most effort and resources have been applied to the issue, but there are local difficulties, including the sheer scale of the task in the Yorkshire Dales, whilst elsewhere in Britain there has been less progress, perhaps because the perceived threat in these areas is less than in northern England.

Signs of efficacy of the legislation are hard to judge at a time of slump in demand for rockery stone; however, it appears to be having some effect in Cumbria, Lancashire and elsewhere in northern England, but to have caused a problem in Eire.

Limestone pavements are limestone outcrops which have been stripped of most or all of any pre-existing soil or other cover by some scouring mechanism, generally, but not exclusively, glacial scour. Mechanically strong limestones are also required. Internationally the term is accepted for a landform assemblage, including limestone blocks (clints) separated by joints opened by solution (grikes), and with a wide range of surface solution forms on both clint tops and grike sides (Williams 1966). British limestone pavements have less varied surface solution features compared with some, particularly the Alpine sites; even so, they range from near-perfectly smooth undissected surfaces to shattered outcrops, and have very varied solution sculpture.

The largest areas of limestone pavement in the British Isles are in Burren, Co. Clare and on the Arainn Islands of Co. Galway, both in Eire; and in the Craven district of Yorkshire. In addition there are extensive outcrops elsewhere in northern England, mainly in Cumbria and Lancashire. Numerous small outcrops occur in South Wales around Ystradfellte, and in North Wales between Great Orme and Denbigh and on Anglesey. Derbyshire has some small pavements thought to date from before the Devensian. All these sites are on Carboniferous Limestone. Finally, there are small but interesting pavement outcrops in Scotland, on Dalradian limestone in Perthshire and on Cambrian limestone on Skye and in the Durness area of Sutherland (see inset map, Fig. 1).

Geomorphologically, limestone pavements are very interesting and attractive. Figure 2 illustrates a typical undamaged pavement of moderately well-dissected clints at Newbiggin Crags, Cumbria. Pavements are also fascinating botanically, sustaining a rich species range in the shade and protection of the grikes. Some sites are extremely rich floristically (Ward & Evans 1976), but less so geomorphologically, and vice versa. Other sites are rich in both senses. The

From O'Halloran, D., Green, C., Harley, M., Stanley, M. & Knill, J. (eds), 1994, *Geological and Landscape Conservation.* Geological Society, London, pp. 215–220.

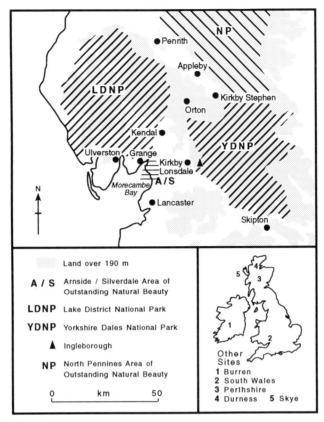

Fig. 1. Landscape protection in northwest England and location map (inset).

reason for protecting limestone pavements is that their features have been much damaged or even destroyed by various human activities over the centuries, and this damage has accelerated in recent decades.

Various aspects of limestone pavement protection have been discussed elsewhere by the author (Goldie 1993), and the ways in which limestone (and other) landscapes are protected in Britain need little description here. However, not all arrangements are effective or legally strong (Waltham 1983). Limestone pavements are protected variously: local naturalists' trusts nature reserves, e.g. Lancelot Clark Storth, Cumbria; National Nature Reserves (NNRs), e.g. Great Asby, Cumbria; Sites of Special Scientific Interest (SSSIs), e.g. Clawthorpe Fell, Cumbria; national parks, e.g. Lake District National Park, Yorkshire Dales National Park; and Areas of Outstanding Natural Beauty (AONB), e.g. Northern Pennines. Of these, reserves provide the strongest protection since, even in the national parks and SSSIs, economic pressures cause problems.

The main pressure on British pavements has been from clint removal to provide garden rockery stone (Goldie 1976, 1987), which has occurred intermittently since at least the late nineteenth century, but increasingly since the 1960s. Clints have been sold at garden centres to purchasers who are probably ignorant of their beauty and interest *in situ*. Such limestone blocks could equally well have been quarried. Pavements thus damaged have a much altered geomorphology (Goldie 1986), a messy, ugly, broken surface with much loose debris and rough remnant clint tops, lacking attractive runnelling, and with a much depleted, or even totally destroyed, flora. A damaged site at Andrew Scar in Cumbria (Fig. 3) illustrates the contrast between undamaged and damaged pavements.

Limestone Pavement Orders

As a result of this damage these landforms became the first to be specifically protected by legislation, Section 34 of the Wildlife and

Fig. 2. Well-runnelled massive undamaged limestone pavement on Newbiggin Crags, Farleton Knott, Cumbria.

Countryside Act 1981 (HMSO 1981), under which Limestone Pavement Orders (LPOs) can be made. An LPO is a legal instruction to the owner or occupier of land with limestone pavement, prohibiting its removal or disturbance. The orders are made by local planning authorities (LPAs), i.e. County Councils or National Park Authorities.

The legislation was passed after the survey carried out by Ward & Evans (1976) identified about 2100 ha of limestone pavement in Great Britain; half had already been damaged and damage was continuing. At this time, the value of these landforms was widely known, there was heightened concern over conservation of natural features and a form of conservation law was

Fig. 3. Partially damaged limestone pavement on Andrew Scar, near Great Asby Nature Reserve, Cumbria.

being discussed, with a Bill on the subject in 1979.

MPs were lobbied by numerous interested parties, including naturalist and conservation bodies, to specify pavements for protection. It was questioned why one landscape feature should be protected and not others, but in the case of limestone pavements the detailed, nationwide information available from the Ward & Evans survey meant that the law could be clear about these features, whereas a general law for a wider range of geomorphological features would have been too loose and vague to be of practical value. In addition, one MP argued strongly for the inclusion of limestone pavements in the Bill. Thus the Wildlife and Countryside Act was passed in 1981 with limestone pavements specified in Section 34.

Section 34 requires the Countryside Commission and English Nature to notify the LPA of the existence of limestone pavement 'of special interest by reason of its flora, fauna or geological or physiographical features'. The LPA then has the power to make an LPO to protect the site.

Furthermore, it is suggested that if the character and appearance of any land notified are likely to be 'adversely affected by the removal of the limestone or by its disturbance in any way whatever', then 'the Secretary of State or that authority may make an order designating the land and prohibiting the removal or disturbance of limestone on or in it'. Interpretation of 'adversely affected' is open, but in NW England it is taken to emphasize the need to make the orders.

Several stages are required before an LPO is in force: site identification; consultation with owners and occupiers; notification to the LPA; order making by the LPA; the informing of district and parish councils, as they will be enforcing and monitoring orders; and publication in the local press. Finally, the Secretary of State confirms the order within 9 months unless there are problems. This work is followed by monitoring of the pavement site and enforcement.

Notification requires much information on location and status (local government area, parish, etc.), the landscape, flora and fauna, geology and physiography. In particular, site features of special interest must be identified. Boundaries and ownership of the site must be detailed, and maps depicting these must accompany the descriptive information. For small sites these boundary and ownership details may be simple, but many quite moderately sized sites involve half-a-dozen or more owners. Orders for large areas with many owners and occupiers, for example in the Yorkshire Dales National Park (YDNP), are particularly time-consuming to prepare, with the added problem that land can change hands surprisingly frequently entailing further clarification of detail.

Progress with LPOs

Lancashire: By early 1993, 11 LPOs had been made, with 2 further notifications in hand. Many small areas in Lancashire are still to be covered, and work on numerous further notifications continues.

Cumbria and LDNP: Orders for 29 pavements had been made, leaving 7 notifications to be made into orders and 31 notifications still being worked on covering many small outcrops.

Many pavements in these two areas are small and wooded. Though the exposed sites are better known (e.g. Hampsfield Fell near Grange), wooded sites are still being found, often showing well-developed, attractive runnels in considerable variety even where disguised by moss (Fig. 4).

YDNP: 12 orders have been confirmed, one awaits confirmation and two are being actively worked on, leaving 18 notifications still needed. The YDNP pavements include about half of Britain's total, including the best known, on Ingleborough and above Malham Cove. It is a difficult area to protect, partly because of the great, but patchy, extent of pavement, and partly because of the area's sensitivity. Making LPOs here is a delicate but still necessary job, though many sites are protected in some ways already, not least by their high public profile and exposure to visitors.

Scotland, Wales and Derbyshire: Very little work applying the 1981 Act has been carried out, partly due to pressure of other conservation work and partly because the pavements are regarded as less threatened by the rockery trade than in NW England. However, these areas should eventually also be put under LPOs.

Problems with making LPOs

Numerous problems have arisen in the interpretation of Section 34, for example over the terms 'limestone pavement', 'special interest', and 'wooded'. Other debatable points concern whether damaged pavements need protection, the interpretation of a threat of damage, the issue of discontinuous patches of pavement,

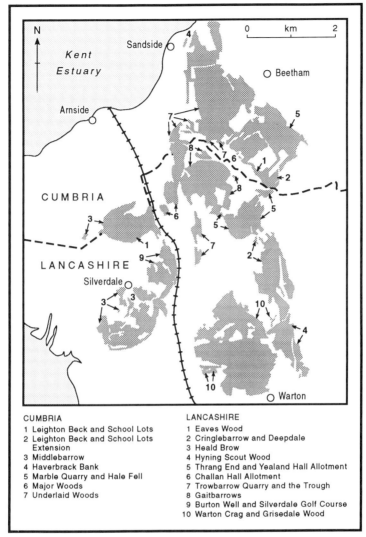

Fig. 4. Sketch map to show location of current LPOs (as at August, 1992) in the Arnside–Silverdale AONB (area A/S on Fig. 1).

whether other interesting limestone features can be included in the LPO, site access and, finally, the question of threats to limestone pavements which do not involve the removal of limestone.

In NW England numerous appeals against LPOs have been made by landowners. However, these have been mainly resolvable with only a handful of solid objections. Inclusion of non-limestone areas at patchy sites has been the main difficulty.

Are LPOs effective?

Although it is too early to say definitely, there are signs that LPOs are effective in Britain. The orders, with publicity in the press and leaflets, have raised public awareness of the value and attraction of limestone pavements and the need not to damage them. There is evidence that contractors have ceased trading in clints from NW England, but that unfortunately their attention has moved to Eire. Carboniferous Limestone blocks for garden rockeries have been imported into Great Britain from Burren, Co. Clare, Eire, where, sadly, there is less control and protection. In addition, a slump in the housebuilding market has meant a decline in landscaping activities demanding the clint blocks. Publicity and education may have had some effect by the time the economy picks up, but there is the obvious threat that Irish pavements will then suffer instead of British.

By early 1993 the enforcement function of councils had been needed on very few occasions. Enforcement depends on monitoring since, when offences are committed, successful prosecution depends on strong evidence. Such evidence could result from regular policing of the sites but this is impracticable through cost, and therefore the evidence of casual witnesses is useful. Regular monitoring would, however, aid enforcement particularly by providing updated information on the state of the pavement sites. Regular site visits by officers of the district councils for instance, with photographs and documented, dated information, are good practice.

Management

Questions arise over pavement site management once LPOs have been made. Should there be active positive management and should LPOs be part of other activities to manage the environment? These would clearly involve labour and cost problems, amongst others. Should public access to limestone pavements be improved so that direct public good can be seen to derive from the LPOs? Or should site access be restricted to protect their relatively delicate landforms and flora? Landscape improvement grants available to farmers in Britain could be appropriate at some pavement sites, given that there is no financial compensation to landowners for income lost due to LPOs. There is encouragement to manage the sites in ways compatible with conservation where management agreements have been entered into, and more work on these lines is one way forward.

British limestone pavements in their present form result from thousands of years of human management, as well as natural processes, and their present environment is not static; natural processes continue, as do human activities. Pavement management should thus be aimed at maintaining the balance between these, in order to conserve the variety of attractive and interesting geomorphological and biological characteristics of these landforms.

Conclusion

National legislation has addressed the problems of British limestone pavements, though the urge to conserve these landforms for the common good derives largely from local and personal appreciation of their scientific and aesthetic worth. The actual legal protection process entails certain difficulties, and much work, perseverance and goodwill are required at the local site level, the real level at which conservation must operate.

The author would like to thank Nick Cox and Martin Green for helpful comments.

References

GOLDIE, H. S. 1976. *Limestone pavements: with special reference to North-West England*. DPhil. thesis, University of Oxford.
—— 1986. Human influence on landforms: the case of limestone pavements. *In*: PATERSON, K. & SWEETING, M. M. (eds) *New Directions in Karst*. Geobooks, Norwich, 515–540.
—— 1987. Human impact on limestone pavements in the British Isles. *In*: KUNAVER, J. (ed.) *Karst and Man*. Department of Geography, University E. Kardelj, Ljubljana, 179–199.
—— 1993. The legal protection of limestone pavement in Great Britain. *Environmental Geology and Water Sciences*, **21**, 160–166.
HMSO 1981. Wildlife and Countryside Act. HMSO, London.
WALTHAM, A. C. 1983. A review of karst conservation sites in Britain. *Studies in Speleology*, **4**, 85–92.
WARD, S. D. & EVANS, D. F. 1976. Conservation assessment of British limestone pavements based on floristic criteria. *Biological Conservation*, **9**, 217–233.
WILLIAMS, P. W. 1966. Limestone pavements with special reference to Western Ireland. *Transactions, Institute of British Geographers*, **40**, 155–172.

Karst and environment: a Romanian approach

EMIL SILVESTRU

Emil Racoviță Speleological Institute, Str. Clinicilor 5, 3400 Cluj, Romania

Abstract: Karst terrains have a high specificity: they produce the only known relief which induces a sort of 'negative', a subterranean replica (endokarst), which in its turn conditions the surficial form (exokarst). Karstological research illustrates that the dynamics of karst landforms can be considered to be a continuous search for balance between exo- and endokarst. This additional dimension (subterranean and subaerial) of the landform makes karst a special active geosystem, extremely sensitive to any external influences.

Romania has 4400 km^2 of limestone terrains and other soluble rock terrains (rock gypsum and rock salt). In some areas, limestones are affected by bauxite mining and quarrying; other areas are fairly densely inhabited, which adds to the environmental stress, since no attention has been paid to karst fragility.

The conservation of karst landforms has been a concern to Romanian karst experts ever since the world's first speleological institute was founded at Cluj, Romania, in 1920. However, an ecological approach to karst areas cannot be seriously considered until Romania stabilizes after recent governmental changes, and until the more inefficient, state-supported, mining activities become less actively supported.

Karst and its distinctive character

To the ordinary three-dimensional relief, karst landforms add a 'fourth dimension', namely the subterranean relief, a sort of a negative replica of the surficial patterns, to which it is closely connected.

There is an intricate relationship between the surficial (exo-) karstic forms and the subterranean (endo-) karstic ones. Briefly, the former give rise to the latter which, in their turn control the evolution of the former, mainly by hydrological means. There are many 'external' factors controlling this relationship. Most important are, however, geological structure and the behaviour of water (in terms of both climate and drainage). Large soluble rock surfaces (karst does not necessarily develop only in limestones; rock gypsum and rock salt are also karstifiable) will gradually have all their surface drainage transferred underground. Subsequently, the karstification process will rapidly obliterate the initial surface drainage network. Smaller, isolated soluble rock surfaces, either have no real surface drainage network (meteoric waters rather infiltrating) or share a part of the regional hydrographic network, which will thus control karst evolution.

Obviously the above-mentioned additional landform dimension (subterranean as well as subaerial) increases the fragility of karst landforms, as compared to other non-karstic ones. Since the karstification process has a strong chemical component (the H_2O–CO_2–$CaCO_3$ system, in the case of limestones) the karstic geosystem has a highly sensitive component affecting its fragility. Changes in water chemistry may cause either the acceleration of corrosion or a slowing or even suppression of it. Such natural (or unnatural) changes are clearly recorded by the best known by-product of karstification – speleothems (stalactites, stalagmites etc.).

If we are to reduce the whole issue to the most basic level, we may consider karstification as a process of returning limestones (or other soluble rocks) to their original environment – the sea – having meteoric waters as the main vector. The process takes place simultaneously at the surface and in the interior of the rock (mainly through corrosion) with a continuous competition between the two, competition that rarely reaches a point of balance.

Let us suggest a model to illustrate this better and to conclude this section: most typically, a surface stream in karst terrain enters a subterranean course via a **swallet** (or swallow-hole). Quite often the swallet is located at the foot of a rock wall. Downstream a cave develops and at its far end the stream issues through a **karstic spring**. When the karstic spring is also located at the foot of a rock wall or a steep slope, a **pocket valley** tends to form, by gradual translation of the spring upstream, because of repeated collapses of the undermined rock. If no important external changes occur, the spring will move continuously towards the swallet, until no cave is left, only a gorge. Thus, the drainage will be subaerial again and endo-karstific features will practically cease to exist.

From O'Halloran, D., Green, C., Harley, M., Stanley, M. & Knill, J. (eds), 1994, *Geological and Landscape Conservation*. Geological Society, London, pp. 221–225.

Karst terrains in Romania

Karst terrains (limestones, dolomites, rock salt and rock gypsum) cover only 4400 km² in Romania (i.e. 1.4%), mostly because of the large extent of Quaternary deposits which cover older, carbonate rocks (Bleahu & Rusu 1965; Fig. 1). Most of these karst terrains are located in the Southern Carpathians and the Western Carpathians, usually called Apuseni, for which reason most of the following examples deal with these two geographic units.

Typical karst landforms include spectacular gorges (Romania's most beautiful ones), surprisingly flat high plateaus dotted with dolines, uvalas and an equivalent of poljes[1] – termed karstic-catchment depressions. There are also more than 12 000 caves and pot-holes.

Unfortunately (for environmentalists at least), several economically attractive deposits are associated with karst terrains, ruling out any recreational interest: bauxite, fire-clays and coal; and the limestone itself is an important resource. On the other hand, water reserves in karstic aquifers also represent valuable resources. Most tragically, urgent economic needs and the poor tourist infrastructure, have quite often resulted in serious environmental damage.

Man and karst

Settlement

Karst landforms on Romanian territory have been an attraction for people, from the Neolithic and up to the present. Apart from the above-mentioned economic objectives (to which metals may be added in the early Bronze Age), settlement in karst terrains (especially in the Apuseni) is well established. This is illustrated by the fact that a famous cave – Ghetarul de la Scàrisoara (the Ice Cave of Scàrisoara) had been marked with this name on Austrian maps, two centuries ago, when the closest village was Scàrisoara, 20 km away from the cave entrance. Nowadays, the closest village is 200 m from the entrance, in the heart of a karstic plateau, sharing a single water source, a well for more than 60 houses, with no running water, no electricity or any other modern facility. As for sewage, its disposal is entirely natural, the

---

[1] There is an ongoing scientific debate as to whether or not there are poljes in Romania. The author considers that there are no poljes on Romanian territory.

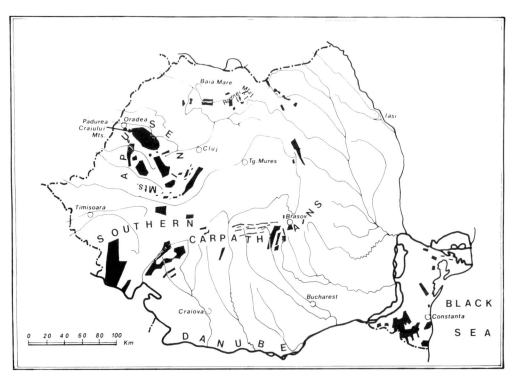

Fig. 1. The distribution of karst terrains on the Romanian territory.

karst absorbing all liquid discharges, those coming from animal farming being quite substantial in volume. Given the scattered distribution of the houses, practically all karstic aquifers in the area are likely to be polluted.

Such a settlement relies heavily on local resources like firewood, pasture and soil (no matter how thin and poor it is). However, for the geographic unit of the Apuseni as a whole, the situation described here, is rather local and does not (yet) seriously affect water resources. Nevertheless, the attraction of this very special landscape, is resulting in an increasing number of visiting tourists. The very poor tourist facilities kept this number, until now, below critical levels but the future will surely see these levels exceeded. Therefore, all new tourist facilities must be created on ecologically-safe principles. All Romanian karst experts are deeply concerned about this issue and are trying to convince the newly established legislature to back-up this concern with effective laws. The general point of view is that it is the entire geosystem that has to be taken into consideration and protected, rather than isolated areas on the surface and/or underground.

Exploitation

There are two sides to this problem. First, there is local exploitation due to basic needs: firewood, timber, various crafts and grazing. Most typical from this point of view are again the Apusenis. Firewood and timber had been exploited in a rather rational manner, for centuries and the local inhabitants understood the value of their wood, before the communist regime took over and nationalized all forests. So, no serious damage was recorded before 1948, except for short clear-cutting episodes under the Austrian Empire in Transylvania.

The natives of the central Apuseni – the **moti** (motz) – are incredible craftsmen in wood working. Their most important products are barrels and wooden kitchen utensils. A good craftsman can produce as much as seven 200 l barrels and 2–3 m^2 of shingle, out of a 30–40-year-old spruce or fir. Given the natural conditions, the motz economy is dual: wood working is done during winter, whilst in summer, the men load their horse-drawn wagons with their products and travel to rich agricultural counties (sometimes hundreds of kilometres away) to barter for wheat. Until their return, the women and children tend the animals (mainly sheep and cattle).

This almost natural balance, has been dramatically broken after the events of December 1989, mainly because of slackening of police and forestry controls. Huge amounts of timber are now cut and sold, while the legislation remains confused and opportunities tempting. The local inhabitants consider that the simple fall of communism has restored their ownership of the forests, while the state pretends to be still the rightful owner of all forests. If the present rate of cutting is maintained, the effects will soon be visible: the soil will rapidly vanish, leaving a stony desert which would eventually exacerbate the karstification process. Soil recovery on karst terrains is a very delicate business and requires major investment.

Within the same context, another serious problem is overgrazing. The karstic plateaus in the Apuseni provide very rich pastures. As local cattle and sheep are not very numerous, the local inhabitats used to lend their pastures to shepherds from southern Transylvania (transhumance is still very active in Romania). Therefore, large areas of pastures are quite often overgrazed. The result is clearly shown by a small protected area in which more than 60 original species of indigenous flora recorded in 1960 are all present, whilst in the whole surrounding area, only two, the most robust, are left nowadays, as a consequence of overgrazing.

The other side of the exploitation problem relates to various industrial activities. Deforestation is the most frequent such activity and affects karst terrains throughout the country. No attention has ever been paid to the fragility of karst soils and consequently, to the need to protect them. There is nothing much to be done for the already deforested areas, except to wait for a providential natural recovery. Even if the communist legislation took into consideration limited deforestation, the desperate need for resources and primitive local management led to uncontrolled cutting. A small karstic area in northern Romania (the Rodnei Mountains) provides a good example: a whole mountain slope had been cut clear (even if in the official papers the cutting appears as selective), the soil immediately starting to move down the 35° slope, exposing bare limestone to weathering. Moreover, since the limestone dip roughly corresponds to the slope angle, rock slides occurred within 4 years (after the cutting), because of intense karstification. At present, Romania's second deepest cave – Izvorul Tàusoarelor (465 m) which lies in the centre of this area, is endangered, as the land slides may bury the entrance.

Mining is another industrial activity which severely affects limestone terrains, in many respects. Quarries of ordinary limestone or marble produce the most important morphological changes. Most of the quarries are located in the slopes of the numerous karstic gorges or in klippen-type massifs. Any quarry represents an anomaly in slope geometry, therefore inducing a chain reaction which tends to re-balance the slope, by collapsing and slides. Solutional patterns are rapidly established and relatively soon, the initial cut regains a natural look. This is, the less serious side of the issue. Most frequently however, the damage does not consist of mere subaerial artificial cuts. Endokarstic features may also be affected. Sometimes even to the extent that underground drainage is diverted or disrupted, which always results in important regional changes and has long-term consequences. Even more destructive, are the auxiliary works like roads, buildings and everyday activities (sewage disposal, machinery, storing of various polluting materials). In most cases, abandoned quarries are littered with rusting equipment (a rather general feature for centralized economies in Eastern Europe).

No less destructive is bauxite mining. As bauxite formed in a Lower Cretaceous palaeokarst, subsequently covered by more than 200 m of Mid-Cretaceous limestones, which are intensely karstified today, mining takes place both in surface quarries and in subterranean galleries.

Surface quarrying usually begins with clear-felling of woodland. Excavations are then dug into the limestones, most of them completely closed, with depth varying between 10 and 60 m and diameter between 100 and 300 m. A closed hollow of this type, takes a very long time to recover, and represents a strikingly unnatural feature in the heart of karstic landforms. In addition, the waste dump left on the slopes is either terra rosa or reddish sediments associated with bauxite, both practically sterile for vegetation. Therefore, the recovery time is even longer.

When surface activities are associated with subterranean ones, the situation becomes even more dramatic as all karst processes are practically obliterated. Underground drainage is forced along artificial path, all voids are filled or sealed and, above all, new voids are created (the mining galleries as well as the ones left by the removal of bauxite). Great amounts of rusting equipment are abandoned underground. Under such circumstances, one can hardly speak of karst, after mining has been discontinued. In addition, secondary activities, especially transport, often by lorry, all aggravate the disaster, adding an important element of pollution, both to the atmosphere and to the karst landscape.

Large areas in the nothern Apuseni – the Pàdurea Craiului Mountains – have been subjected to such activities, and now form a land of desolation. And all of this, to produce aluminium exported in its very first industrial stage – ingots! Another example of insane communist planning.

If bauxite mining yields (as a result of digging) mainly limestone spoil, fire-clay mining brings to the surface insoluble rocks. Waste dumps are subsequently washed-away and the fine, clay-rich sludge can insulate limestones and suppress karstification. The same Pàdurea Craiului Mountains suffer the consequences of such mining, around and above Romania's longest cave – Pestera Vîntului (45 km). The underground stream is heavily polluted, floating foam being a usual feature. Other caves in the area have also suffered, all kinds of strange, non-karstic mineral deposits being discovered recently underground.

Hydrological damage is also present in limestone areas. Apart from those produced by the previously mentioned activities, one must also take into account hydroelectric and water storage works. By definition damming on limestone terrains is an extremely risky and costly enterprise. However, risks and expenses were a stimulating factor for the former Romanian rulers and dams were erected in unsuitable karst areas and were hardly viable from the beginning (because of water escaping through karst conduits). Deep underground, in the heart of limestone terrain of the Pàdurea Craiului Mountains, a shaft was dug in the early 1980s, for a future hydroelectric plant. Despite dense test-drilling, a natural underground drainage channel was intersected and the shaft, 100 m deep and 6 m wide, filled with water within hours and the work was abandoned. It was a rather bizarre way to check karst experts' warnings.

Finally, but no less important, we must emphasize that karst (both caves and the upper layer of limestone, called by biospeleologists mesocavernous surficial substratum (MSS)) shelters extremely rare species of animals like molluscs, arthropods and some vertebrates, most of them troglobionts, i.e. living only underground. Many of these species are strongly endemic, sometimes two neighbouring caves sheltering two different species which cannot be found anywhere else. These animals are extremely sensitive to the slighest changes.

Possible remedies

Before talking about remedies, we must take a rather philosophical approach and try to define what is good and what is not good within this context. There are probably many people who think that economy comes first: 'Forget about aesthetics and romanticism, jobs and raw materials are the only important things', could be their slogan. Others however, will say: 'Karst landforms are among the most beautiful relief of the planet and therefore we must not touch them;. And then, there are those who believe that yet another solution is possible: 'Jobs, raw materials and beauty, all together!'

This is no solution in the author's view. I believe that a decision must be taken at the very beginning: tourism **or** industry. There is no place for both in karst terrains. And if the decision is tourism (it is pointless to explore the other alternative), we must recognize that:

Karst is a fragile geosystem and must therefore be protected as a whole!

Now, having established the principle, let us see if there really are remedies to human insensitivity towards karst. No matter which of the components of the karstic geosystem have been harmed, there are two basic ways to re-establish a natural balance:

- returning to the original state;
- shaping of a new dynamic balance, according to the general principles of the karstification process.

Returning to the original state

First, of course, all non-tourist activities must be discontinued. Then all traces and equipment must be removed. In the case of quarries, all working faces should be shaped as close as possible to natural ones (which is not a difficult task). The vegetation should be helped to take over as soon as, and everywhere, possible. If quarrying produced stepped working levels it is appropriate to create a soil cover on each step, as well as a plant cover.

In the case of bauxite quarrying, the bottom of the excavations should be covered with gravel and sand, and then the quarry should be filled with the original excavated material (when still possible) or with rocks similar to it. The gravel and sand would prevent infilling of the karstic conduits (of all dimensions) by the non-compacted filling, thus allowing water to continue its circulation.

When underground mining is involved, the removal of all equipment should be preceded by works aiming to re-settle drainage in its natural paths. Furthermore, if important cavities have been cut, they should be sealed in order to regain their original climate.

Shaping of a new dynamic balance

One must be aware of the fact that no effective recovery is possible if the geosystem's structure had been harmed. In such cases, the only solution is to create a new structure (sometimes, paradoxically, by destructive methods), choosing of course the natural model which requires the least destruction. Such a choice is a question of high professionalism and demands great responsibilities. Only a very experienced karstological team can honestly get involved in such decisions.

The human side of the problem, i.e. pollution by habitation and/or tourism, is mainly a question of education and civilization. Both are closely related to the overall economic state of the country. As the founder of the world's first speleological institute (in 1920, at Cluj, Romania) E.G. Racovitza put it – no matter how draconian the protection laws, they remain simple written words, if there is no money to reinforce them.

Romania places great hope in its tourist potential, but if money is not found for the infrastructure of tourism and environmental protection as well, there may be no such potential left in years to come. I am deeply convinced that, if there is a real desire in the rich countries of the world to help Romania, one of the most effective ways is to help tourism become a profitable enterprise, and thus to produce money for environmental protection (and karst landforms are among the most important touristic attractions in Romania).

Reference

BLEAHU, M. D. & RUSU, T. 1965. Carstul din România. *In*: *Lucràrile Institutului de Speologie "Emil Racovita"*, t. IV, Bucuresti, 59–74.

Keynote address

Protected volcanoes in Iceland: conservation and threats

GUDRIDUR THORVARDARDOTTIR & THORODDUR F. THORODDSSON

Nature Conservation Council, Hlemmur 3, PO Box 5324, 125 Reykjavík, Iceland

Abstract: In Iceland, natural phenomena such as volcanoes, can be protected via the Nature Conservation Act and are classified as natural monuments. Volcanoes can also be protected in nature reserves, where entire biological and geological features are protected. Another possible means of protection is where large areas have been conserved under separate legislation. Nearly all types of volcano found on the Earth are represented in this country, although the majority here are formed on short fissures and tubular channels or on long fissures.

Many scars are found on volcanic landscapes caused mainly by construction, agriculture and tourism. The most obvious scars are those made by mining pits in lava and pumice formations. Volcanic material is especially desirable for construction work as it is lightweight, porous, resistant to frost and usually has very thin or no vegetation cover. Nowadays, from the point of nature conservation, the emphasis is on fewer, but larger, pits.

The Nature Conservation Council has listed more than 200 areas of special conservation interest. In Iceland, the emphasis on nature conservation is to set aside and protect large wilderness areas, preferably including as many natural phenomena as possible. Three areas are in particular need of protection. These are:

(1) Askja caldera, including its fissure swarms and Kerlingardyngja and Trölladyngja;
(2) Snæfellsjökull;
(3) volcanoes in the southern highlands, including volcanoes like Hekla, Katla, Lakagigar and Eyjafjallajökull.

Active volcanoes in Iceland

Volcanic activity in Iceland is governed by complicated interaction between excessive mantle upwelling at the Iceland hotspot and rifting at the Mid-Atlantic plate boundary.

Active volcanoes in Iceland are a continuation of prehistoric Postglacial activity and mainly confined to the volcanic zone that runs through central Iceland, which divides into two in the southern part of the country. The western part of the zone stretches from the Reykjanes peninsula in the southwest, northeastwards to the Langjökull area, where it has been displaced by transform fault movement about 100 km to the east. Its eastern part stretches from the Westman Islands through Mýrdalsjökull to Vatnajökull and there joins the western part and stretches to the north coast. Other active zones are the Snæfellsnes–Mýrar area in the west and the Öræfajökull area in the southeast (Gudmundsson 1986). The active zones cover approximately 25% of the land area. According to Thorarinsson and Sæmundsson (1979) the activity has been characterized by mass production of predominantly basaltic lava and a greater diversity of volcanic phenomena than could be expected on an oceanic island.

Nearly all the different types of volcanoes found on Earth are represented in this country. Magma type, extrusion rate and extraneous factors like water mixing with magma give basaltic volcanoes various forms. The main types are formed on short fissures and tubular channels or on long fissures. They can be lava rings, lava shields, agglutinate cones, scoria cones, tephra cones, tephra rings, maars, crater rows, tephra cone rows and explosion chasms.

Protected volcanoes

The Nature Conservation Council has the power to declare unique geological formations and natural formations of outstanding beauty or scientific interest protected according to the Nature Conservation Act (no. 47/1971, article 22). These are classified as natural monuments and include waterfalls, volcanoes, hot springs, caves, rock pillars, as well as beds containing fossils or rare minerals. The article stipulates that a sufficiently large area around the monument must be protected as well, so that the preserved formation can be appreciated. As of today and according to the Nature Conservation Act, 29 sites have been protected as natural

From O'Halloran, D., Green, C., Harley, M., Stanley, M. & Knill, J. (eds), 1994, Geological and Landscape Conservation. Geological Society, London, pp. 227–230.

monuments, including nine volcanoes and pseudo-crater areas.

Volcanoes may also be protected as part of a nature reserve (article 24). Nature reserves are areas that are considered important because of their landscape, flora or fauna. An example of this form of conservation is the Surtsey Nature Reserve (Náttúruminjaskrá 1991). The entire island was declared a nature reserve in 1965 and, following the revised Nature Conservation Act in 1971, the protection of Surtsey was revised in 1974. According to the IUCN classification of protected areas (IUCN 1990) the Surtsey Nature Reserve is classified in category I; that is a site of scientific importance, where natural processes are allowed to take place in the absence of any direct human interference and where public access is prohibited.

Although the Mývatn–Laxá region is protected by a special Act (no. 36/1974) it is effectively a nature reserve. The area covers 440 000 ha and has many volcanoes, of which Krafla and Askja are best known. The purpose of the Act is to further the conservation of the river Laxá and Lake Mývatn and the surrounding area. The main emphasis is on biological conservation, but the area is renowned for its bird life. The Act includes the rule that any form of construction or disruption of the land is prohibited, without permission from the Nature Conservation Council.

Threats

When referring to threats facing landscape areas, we can often use the term 'endangered spaces', since if the landscape areas are altered it is not possible to remedy the damage. The greatest threats to volcanic sites in Iceland are due to construction, agriculture and tourism.

Construction

The greatest alteration on volcanic sites is created by construction in the form of gravel mining, road building and building of power plants accompanied by tracks, pipes and pylons.

If an area is not protected by the Nature Conservation Act, landowners are permitted to mine gravel, rock, scoria or pumice on their estate. This has led to the production of many mining pits, especially where lava and pumice are found, as these have thin or no vegetation cover, are lightweight, porous, without frost activity and give good insulation: they therefore make an ideal basis for road construction, for example. Mining for red pumice or ash has also become popular, for use in roads in summer house complexes, for footpaths, in gardens, and most recently, for use in barbecue gas grills.

Many abandoned mining pits are found in Iceland. They look like battlefields in the landscape, where the best material has been taken away. It will take a long time, great effort and cost a fortune to remedy these old abandoned pits, and one must bear in mind that they will never regain their original structure. Many interesting volcanic formations have been lost since the Second World War due to advanced mining technology. Nowadays, the mining is planned with more care and the emphasis is on avoiding craters. However, fewer, larger mining pits create greater scars on the surface, but this situation is more acceptable from the point of view of nature conservation than many smaller scars. The Nature Conservation Council is totally against damaging lava formations unless the relevant area has been set aside as a construction site.

Another form of environmental impact is connected with the use of hydroelectric or geothermal power. Even in the preparation stage of a power plant, off-road driving in connection with scientific work to find the best spot for electricity pylons, leaves tracks in sensitive areas. Roads are constructed for use during the building stage and new gravel pits are opened. These roads and the powerlines create scars in the open wilderness landscape as there are no trees to hide them. There is an increasing effort to minimize the negative environmental impact by selecting the best possible site for the pylons and other constructions from an aesthetic point of view.

A geothermal power plant is located inside the Krafla caldera in the Mývatn district. Accompanying its construction are boreholes and platforms as well as pipelines into the station and pylons to carry the electricity to the users. At the beginning of the construction work the area was mapped in order to protect the most valuable conservation sites. During the design process the aim was to avoid the sites of conservation value and to make as little negative impact on the environment as possible. After the construction work was finished the area was opened up for tourists, and nowadays is a popular tourist site.

Another famous volcano, Askja, has been pointed out as an ideal location from the power company's point of view, for a geothermal power plant in the future.

Agriculture

Among changes made to volcanoes by farming are changes in vegetation, where soil reclamation takes place in order to create pasture for sheep. Land holdings or estates are divided by fences and in order to erect them, for instance in lava fields, scars are made by bulldozers to make the job easier. In autumn, sheep are gathered and in order to get as close as possible to the sheep and their pasture, farmers have made tracks for four-wheelers. If these tracks are not closed they are used by travellers as well and, unfortunately, these tracks increase off-road driving.

The State Soil Conservation Service has used aeroplances since 1958 to disperse seeds and fertilisers. This method has created vegetated strips in volcanic areas as elsewhere. Although this is not a severe damage to a volcanic site, it changes its value as an aesthetic and authentic volcanic site. It must be noted that in some areas, as on the Westman Islands, in the area of the volcano Hekla and in the Lake Mývatn district, it is necessary to bind the loose ash with vegetation cover to discourage the blowing ash from damaging vegetation, buildings and cars.

Tourism

Geological formations are among the most popular tourist attractions in Iceland and volcanoes are no exception. The negative environmental impact on volcanoes and volcanic sites made by tourists is mainly paths, often many and irregular, on lava rings or cones. This is especially striking where mosses have covered the crater's slope, but trampling breaks the moss cover, leaving scars open for erosion. Tephra formations and some lava formations are so fragile that they disintegrate underfoot.

Accompanying tourism is stone sampling which occurs in volcanic sites as well as elsewhere in Iceland. This is a minor impact but irritating nonetheless, as the samples often end up in litter bins at the airport or at home where people realize that the sample they collected is not as beautiful on the shelf as it is in its natural environment.

Nowadays there is an increased demand for building roads to and onto volcanoes in order to make the site more accessible to all and to limit time spent getting there. An example is a request to build a tourist road to the top of Hekla volcano. Off-road driving in Iceland is a severe problem and it seems that tourists, Icelanders and foreigners alike, cannot resist the temptation to try out their four-wheel drive vehicles and their ability to drive, or rather to spin their wheels, on slopes of volcanic craters. The attitude to off-road driving has changed greatly in recent years through co-operation of non-governmental organizations, like nature conservation groups, touring clubs and travel agencies, in repairing the damage already made and working for better outdoor recreation experience.

Necessary conservation

As in many other countries, Iceland has not been able to ensure long-term survival of its natural heritage. This is mainly due to the fact that conservation policy will rely on economics rather than aesthetic and moral values. In spite of this, the Nature Conservation Council has protected 69 sites according to the Nature Conservation Act and listed more than 200 areas as sites of special conservation interest.

The emphasis on protecting volcanic landscape has mainly been on enlarging the natural monument Askja in Dyngjufjöll. The idea is to include the entire caldera and its fissure swarm as well as two shield volcanoes, Trölladyngja and Kerlingardyngja, and as such, to protect unspoiled volcanic wilderness landscape. Another area is the western most part of the Snæfellsnes peninsula, the strato volcano Snæfellsjökull and its surrounding area. Thirdly the Nature Conservation Council is looking at the possibility of establishing a volcanic park in the southern highlands, an area including famous volcanoes like Hekla, Katla, Lakagígar as well as Eyjafjallajökull. Among other special sites are Skjaldbreiður, from which the scientific name 'shield' volcano obtains its name. Even though a mining pit exists in the northern part of Skjaldbreiður, along with electricity pylons, a road and a number of mountain huts, it is of great importance to protect because of its location, being only about 60 km from Reykjavik and of such important educational value.

For the future, the emphasis will be on minimizing negative environmental impact on volcanic formations. As many of the volcanic phenomena in the country are of international importance there is a need for a general plan of conservation. In this matter we may need international cooperation to protect our endangered spaces.

References

GUDMUNDSSON, ARI TRAUSTI. 1986. *Islandseldar, Eldvirkni á Íslandi í 10 000 ár.* Vaka–Helgafell, Reykjavik.

IUCN. 1990. *1990 United Nations List of National Parks and Protected Areas.* IUCN, Gland, Switzerland and Cambridge, UK.

Lög um náttúruvernd, Stjórnartiðindi A, nr. 47/1974.

Lög um verndum Mývatns og Laxár í Suður-Þingeyjarsýslu, Stjórnartiðindi A, nr. 36/1974.

NÁTTÚRUMINJASKRÁ. 1991. Friðlýst svæði og aðrar skráðar náttúruminjar. Reykjavík. 6. útgáfa.

THORARINSSON, SIGURDUR & SAEMUNDSSON, KRISTJÁN. 1979. Volcanic activity in historical time. *Jökull,* **29**, 29–32.

Development and management of geological and geomorphological conservation features within the urban areas of the Western Ukraine

YURI ZINKO

Geography Department, Lviv University, 41 Doroschenko Street, 290002 Lviv, Ukraine

Abstract: The problem of establishing and managing Earth science conservation in urban areas is analysed with reference to the example of the West Ukrainian region. The following aspects of the problem are considered: (1) conceptual and methodological basis of the interaction between Earth science conservation (geological and geomorphological) and the urban environment; (2) development stages and distinctive roles of Earth science conservation in the cities of Western Ukraine; (3) basis and implementation of urban Earth science-based Local Conservation Systems (LCSs).

It is established that the geological and geomorphological nature conservation features of urban areas are the most commonly preserved elements of the natural and historical environment of cities. At present the extension of the network of protected natural features in cities, and their organization on the basis of certain conservation systems is taking place. Effective management of these features calls for the study of different ways of 'blending' them in to the structure of the city environment.

There are three main stages in the development of nature conservation in the West Ukrainian region, and the network of features, their status and role in the environment changed at each stage. Deterioration of the ecological situation in the cities of the region and the need to improve their natural and architectural landscape set the agenda for modernization of Earth science conservation in both structural and operational terms. For cities with a diversity of geological and geomorphological conditions, Earth science-based LCSs are suggested. They are based on a selected set of conservation features (protected areas and protected geological and geomorphological features), on various forms of area amalgamation (zones, belts) and they have certain functions within the city environment (ecological, site protection, recreational). The structural and functional aspects of the Earth science-based LCSs are illustrated in the example of the city of Lviv, Podillya. The main ways of implementing Earth science-based LCSs are via an inventory of the existing and suggested conservation features and via elaboration of special projects of 'linking' and preserving conservation features in the city environment.

It is pointed out that the formation of Earth science-based LCSs in the cities of the main West Ukrainian region must be based on the distinctiveness of geological and geomorphological conditions and on an appreciation of the functional type of the city.

The problem of protecting geological and geomorphological features in cities involves special attention to certain conceptual and methodological issues. From the conceptual point of view it is necessary to distinguish the following aspects of the problem:

(1) elements of the geological substrate and relief forms are the most commonly represented natural features in the area;
(2) anthropogenic activity in cities leads to modificiation of the ground surface and of geological deposits, and at the same time increases geological exposure and the variety of relief forms;
(3) conserved geological and geomorphological features may be either natural or artificial in origin.

In evaluating the conservation interest of these features it is important that the following methodological issues should be recognized:

(1) geological and geomorphological features in cities are the best known and serve as a standard for regional investigations of geological deposits and relief;
(2) the conservation status of geological and geomorphological features in urban areas is of scientific, landscape and architectural importance;
(3) geological and geomorphological conservation has served different purposes during the process of city development.

Modern development of conservation in urban areas involves attention to a number of tasks relating to Earth science features. There are three important groups of tasks: organizational, functional and protective. Organizational tasks are connected with the expansion of the protected site network and with the creation of territorial structures and local Earth science conservation systems as a major element in the reconstruction of urban environments. To fulfil the aims of urban conservation, they should be incorporated in the management strategies or urban areas. The issue of protection involves a search for ways of limiting anthropological impacts upon geological and geomorphological features in the urban environment.

The Western Ukraine is characterized by great geological and geomorphological diversity, including the glacial and alluvial lowlands of Polissya, dissected uplands of Volyn–Podillya and low folded hills of the Carpathian molasse. Demographically, the Western Ukraine is characterized by a considerable density of population (80–100 people per km) and a pattern of settlement dating from the twelfth to the seventeenth centuries.

The distribution of urban areas in relation to geological and geomorphological conditions makes possible the identification of the following city types in the region:

- cities with monotonous geological and geomorphological conditions;
- cities with some diversity of geological and geomorphological conditions;
- cities with contrasting geological and relief conditions, situated on boundaries between regions.

The development of nature conservation in the cities of Western Ukraine covers a period of one hundred years. Geological and sometimes geomorphological features were the first objects of conservation in cities and their environs. Earth science features in urban areas were especially concentrated in cities with contrasting geological and geomorphological conditions (e.g. Lviv, Kremenetz). Several stages can be recognized in the development of Earth science conservation in urban areas, and at each stage Earth science conservation fitted distinctively into the pattern of urban settlement and land-use.

The first objects of urban geological conservation were singled out at the end of the nineteenth and the beginning of the twentieth century, in connection with the geological mapping of the region. Geological exposures within urban areas were the main basis for the development of regional stratigraphic schemes giving them status in terms of their scientific value (protected areas). Landscapes and artificial relief forms (rocks, caves and mines) also attracted conservation status. Within cities, scientific interest and site preservation were the main aims of conservation (Pavlovski et al. 1914).

The second stage in the development of urban Earth science conservation in the Western Ukraine began in the middle 1950s. At the time intensive geological investigations of the regions, changes in nature protection legislation and a rapid rate of urbanization in Western Ukraine took place. Detailed analyses of geological investigations led to an increase in the number of conserved features in cities, and to their systematization according to their scientific interest – i.e. stratigraphic, tectonic, palaeontological, geomorphological (Korotenko 1987). Thus among type stratigraphical sections, eight stratotypes of the Miocene of the Volyn–Podillya area (Vendlinsky 1979), are situated in cities or near them. On the other hand, the considerable rate of urbanization, which was followed by reclamation and afforestation of quarries and the landscaping of relief, led to the loss of some existing and potential Earth science conservation features. Changes in the nature protection legislation meant that geological and geomorphological reserves were included in protected areas, such as city parks and forest-parks (Voinstvensky 1986). During this period (1950s–1970s) geological and geomorphological reserves and valuable Earth science features in forest- and city parks had educational, scientific and recreational roles. Because of their small number and area their site preservation role within the cities of the region was not significant. At that time the problems of preserving Earth science features in cities of the region became urgent. Typical anthropogenic influences were unauthorized exploitation of local pits and dumping of rubbish, recreational degradation of the ground surface and of deposits, and economic activity close to the conservation objects.

The role of conservation in the cities of the Western Ukraine was actively discussed in scientific and economic circles at the beginning of the 1980s. Such attention arose from the need to improve the ecological state of cities, their landscape and architectural appearance and to eliminate imbalances in the functional zoning of cities. To undertake these tasks, special programmes were set up, e.g. in the cities of the Lviv region, the main stress was placed on the renovation of the existing features and on the establishment of new conservation and recreational features (Zinko et al., 1988). In accordance with this programme, principles and a series of schemes for the development and management of urban Earth science conservation were worked out at the Chair of Geomorphology of Lviv University. The conceptual basis of a Local Conservation System (LCS) for Earth science conservation was put forward to those cities with a considerable diversity of geological, geomorphological and hydrological conditions. An LCS was to be characterized by the association of features in the reserve area, their representativeness, and by territorial integrity.

The LCS fulfils specific functions within urban areas (e.g. ecological, site protection). The formation of an Earth science-based LCS involves the consecutive solution of the following problems:

(1) recognition of a specific set of conserved features (elements of the system) and identification of their spatial relationships (territorial structures of the system);
(2) management differentiation in accordance with the type of location in the urban area (e.g. residential, recreational, derelict);
(3) protection of the system as a whole and its elements from anthropogenic influence.

Realization of these strategies is expected to involve several specific tasks which include: (1) making an inventory of the existing and suggested Earth science conservation features; (2) recognition of their spatial distribution and their spatial relationship to the different types of urban environment; (3) working out the systematic architectural, engineering and technical means to decrease anthropogenic influences on the protected features.

During the identification of the component features (element network) forming the basis of the LCS it is necessary to add to those elements that already have conservation status, any newly recognized ones. Expansion of the network of conservation features within urban areas can be achieved by designating natural relief forms, exposures and hydrological features. In the nature protection legislation it is suggested that they should be called 'conservation objects of an inanimate nature' (Earth science conservation features). The creation of such a category of conservation features is important for cases where the object of conservation is protected in conditions which are no longer natural. In establishing the LCS for the city of Lviv (Podillya heights), conservation of geomorphological and geological features became the basis for the expansion of the element network of the system (Fig. 1).

Expansion of the element network within urban areas can also be achieved by separating geomorphological features as objects of conservation with the same status as geological features (Zinko, Serenko, Velykopolska 1988). Previously, geomorphological conservation features were regarded as a separate group of geological conservation features in the Ukraine (Korotenko 1987). According to their significance, geomorphological conservation features can be divided on the basis of various characteristics into groups: genesis, age morphology/ landscape/historic architecture/architecture.

In cities with strong relief in the Western Ukraine (Polissya, The Carpathians) geomorphological features may form the foundation of the element network when forming the Earth science-based LCS. It is necessary to point out the important role of artificial features, which can have a historic interest in cities. Separation of complex conservation features on the basis of geological and geomorphological features is important for urbanized areas. In the cities of the Western Ukraine it is suggested that the most typical hydrological, geomorphological and botanical–geological reserves should be singled out (Zinko et al. 1988).

In forming the element network of the LCS it is important to adhere to the principle of representativeness. Different natural classes should be included among the elements of the LCS within the city. In the city of Lviv (Fig. 1), there are five geological–geomorphological classes (natural regions) which are represented by unique reserve areas and by typical protected features for each natural region.

Establishment of the element network on the Earth science-based LCS is approached by way of a cartographic inventory of existing and suggested conservation features. This allows a more exact definition of the boundaries of features, the recognition of microzones in accordance with their use and the identification of the main types of anthropogenic influence.

Territorial structures which help to unify the system are an important constituent part of an urban Earth science-based LCS. They are formed on the basis of protected Earth science features linked to form spatial belts and zones. Such belts and zones are established on the basis of well-defined hydrological features and include morphological, lithological and hydrological complexes. These belts and zones may include both protected natural features and areas with relatively undisturbed or little-used geological and geomorphological interest. Earth science-based belts and zones fulfil a number of important functions in the city environment: ecological, site protection, recreational. The

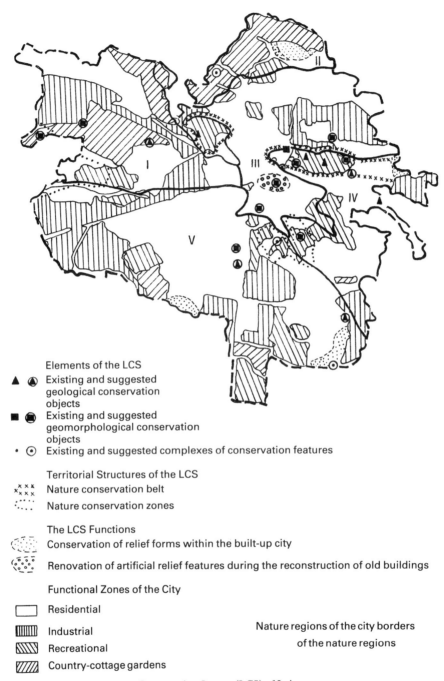

Fig. 1. Structure of the Landscape Conservation System (LCS) of Lviv.

city of Lviv is situated on the main European watershed. A nature conservation belt of the LCS is near the watershed summits, nature conservation zones are connected with ridges and valleys and hydrological complexes (Fig. 1).

When forming an urban Earth science-based LCS it is important to establish a purposive basis for all the elements and structures. For the conservation of geological and geomorphological features it is important to preserve both traditional (education and conservation) functions and specific functions connected with the characteristics of urban land-use. In the residential zone of cities conservation of geological and geomorpholigical features plays an essential role in the formation of the natural and architectrual landscape (landscape function), in the planning zonation of the built-up area (planning function) and in the reconstruction of the natural environment in derelict areas (renovation function). In areas of new building construction it is important to adopt the principle of preserving the original geological and geomorpholigical conditions by 'blending' buildings into the existing natural landscape. For the territory of Lviv, where new buildings are being constructed in various morphological situations (Fig. 1), the choice of technologies with minimum impact on relief and geological deposits is suggested. In recreational areas of the city (parks, forest-parks) the role of Earth science conservation is expected to include specific recreational and landscape functions. Within urban areas polyfunctional roles of elements and structures of the LCS are ensured by establishing specially planned localities to coordinate the demands of nature conservation and urban building requirements. A series of such schemes was realized during the development of the 'General Plan for the City of Lviv in 2010' (Zinko et al. 1989).

It is important to safeguard the continuity and permanence of strategies for dealing with anthropogenic influence relating to urban development to ensure the effectiveness and functioning of an Earth science-based LCS. Legislative, planning and organizational measures supporting the Earth science-based LCS are important. The include: (1) the creation of a managed nature conservation zone (system) with a specifically defined status and territorial structure; (2) elaboration of engineering and technical reclamation for conserved areas affected by anthropogenic influences.

The formation and functioning of the suggested Earth science-based LCS within urban areas in the Western Ukraine will depend on the regional geological and geomorphological distinctiveness and the size and functional type of the city. The diversity of lithological and hydrological complexes in large polyfunctional cities (e.g. Kovel) on the plains of Polissya plays a major role in the formation of the LCS for this area. On the heights of Volyn – Podillya the diversity of morphological and lithological complexes played the main role in the formation of the LCS, as has been shown in the example of Lviv. Both large polyfunctional cities (Rivne, Lutzk, Ternopil) and small medium-sized towns with specific functions are suitable for the formation of LCSs. Cities of the Carpathian areas which have the most diverse geological and geomorphological conditions are very favourable for the development of Earth science-based LCSs, especially in small and medium-sized recreational and health resort centres.

References

KOROTENKO, M. E. 1987. *Geological Reserves: Reference Guide-Book*. Naukova Dumka, Kiev.

PAVLOVSKI, S. et al. 1914. *Nature of Lviv and its Distinctiveness and Reserves*. Scientific Society of N. Kopernick, Lviv.

VENDLINSKY, I. V. 1979. *Stratotypes of Miocene Deposits of the Volyn–Podillya Heights and of the Precarpathian and Transcarpathian Depressions*. Naukova Dumka, Kiev.

VOINSTVENSKY, M.A. (ed.) 1986. *Nature Conservation Fund of the Ukrainian SSR, Register – Guide to Protected Areas*. Urozsay, Kiev.

ZINKO, YU., SIRENKO, I. & RUDOVSKY, L. 1989. Conservation Cartography of the Relief for the Planning of the Region. *In: Ecological–Geographical Cartography*. Irkutzk.

——, ——, VELYKOPOLSKA, L. et al. 1988. Geomorphological and Hydrological Conservation Feature of the City of Lviv. *Bulletin of the Lviv University. Geographical Basis of the Rational Use of Nature*, **16**, 63–67.

Geological protected areas and features in Estonia

R. RAUDSEP

Geological Survey of Estonia, EE0001 Tallinn, Estonia

Abstract: Environmental protection and nature conservation has a long tradition in Estonia. At present there are numerous whole areas and individual features that are protected in Estonia. The main types of geological protected areas fall into the following categories: (1) national parks comprising interesting geological structures, landscapes and numerous erratic boulders; (2) nature reserves of widely different types, designed for the protection of landscape as a whole or of special geological features, e.g. landscape reserves include the beauty spots of the country – valleys, hills, the highest parts of coastal cliffs, etc, while an example of a geological reserve would be Kaali Meteorite Craters (on Saaremaa Island); (3) 'natural monuments' which include isolated features of landscape – banks, hills, waterfalls, caves, karst formations, erratic boulders, stratotypes (outcrops) – which are selected on the basis of their rarity and proneness to destruction. Scientists are currently compiling a monograph *Estonian Ancient Natural Monuments* for registration of all geological and landscape features.

Estonia, one of the three Baltic states is situated in the northeast of Europe, on the east coast of the Baltic Sea (Fig. 1). Its area is 45 200 km² (plus another 2000 km² currently occupied by Russia).

Geologically, Estonia lies on the southern buried slope of the Fennoscandian shield, being a part of the vast East European Plain. Its surface is relatively flat. In general, the Estonian landscape reflects ancient and recent geological

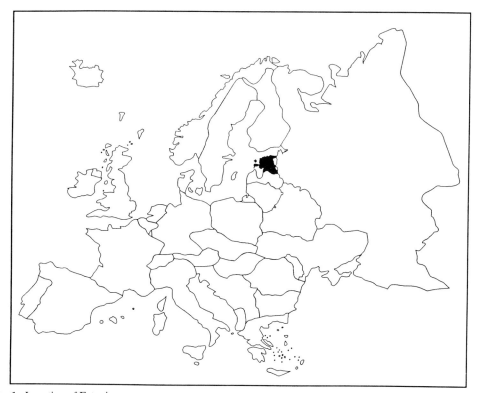

Fig. 1. Location of Estonia.

From O'Halloran, D., Green, C., Harley, M., Stanley, M. & Knill, J. (eds), 1994, *Geological and Landscape Conservation*. Geological Society, London, pp. 237–241.

events. Raised beaches and low cliffs (so-called clint or glint) give the Estonian coastline considerable variety. The clint along the northern coast, is an exhumed cliff of ancient origin, over which unexpected waterfalls flow. It is up to 56 m high in its eastern part and about 30 m in the western part. The dominating bedrock is Ordovician–Silurian limestone in the northern and Devonian sandstone in the southern part.

The variety of landscapes is due to the retreating ice cover which changed the appearance of the country completely after the last period of glaciation some 10 000 years ago. Nowadays, except along the northern coast, the cover of Quaternary deposits largely conceals the underlying rocks. These deposits are unevenly distributed, almost lacking at the northern coast and being up to 100–200 m in the south. Hilly regions with numerous lakes can be found in the southeast (the Haanja and Otepää Uplands), where the highest spot of the Baltic states – Suur-Munamägi (Big Egg Hill), rising to 317.6 m – is located. Some of the rivers in south Estonia have scenic valleys with high sandstone banks. In many parts of the country there are beautiful eskers and kame fields, and drumlins are rather common (Kaasik & Lahtmets 1992).

The variety of natural monuments and landscapes has brought about the need to conserve the most typical ones and to create a basis for the protection of individual features and areas.

Description of areas and features

Environmental protection and nature conservation has a long tradition in Estonia. In 1935 the first nature conservation law was approved and the State Nature Protection Council was founded. Some 47 nature reserves were established before 1940.

After World War II nature conservation activities continued in the Soviet Republic of Estonia.

At present there are numerous protected areas and individual features in Estonia. In total, 12% of the territory is protected via different nature conservation regimes (including the biosphere reserve established in 1990; Fig. 2 and Table 1).

Geological protected areas generally fall into the following categories:

(1) national park
(2) nature reserves
(3) 'natural monuments'

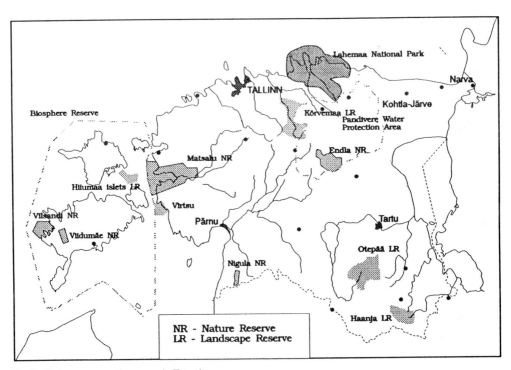

Fig. 2. Nature conservation areas in Estonia.

Table 1. *Protected areas (taken from Kaasik & Lahtmets 1992)*

Protected area and its type	Number	Territory (hectares)
National Park	1	
Lahemaa		64 400
Biosphere reserve	1	
West Estonian islands		1 560 078
State nature reserves	5	
Matsalu		48 000
Viidumäe		1 194
Vilsandi		10 940
Endla		8 162
Nigula		2 771
Hydrogeological reserve	1	
Pandivere		350 900
Landscape reserves	14	
Botanical–zoological reserves	6	
Mire (bog) reserves	30	
Botanical reserves	4	
Ornithological reserves (bird sanctuaries)	20	
Geological reserve	1	
Kaali Meteorite Craters	1	
Nature park	1	
Haanja		
Locally protected areas	53	

National park

Lahemaa National Park was established in 1971. It comprises interesting geological structures, stratotypes, landscapes and numerous erratic boulders. Pre-Quaternary bedrock consists of Upper Proterozoic, Cambrian and Ordovician carbonates and terrigenous rocks. The thickness of the overlying marine, glacial, fluvio- and limnoglacial deposits is usually between 1 and 2 m, sometimes greater.

The relative height of the clint is about 20 m. Its upper part is composed of Middle Ordovician limestone, the lower part of Lower Ordovician limestone and sandstone and Lower Cambrian sandstone. On the limestone plateau occurs a unique terrain type with a specific plant community – alvars. Besides Estonia, alvars occur only in Sweden. The characteristic feature of alvars is a very thin (up to 10–20 cm) but humus-rich soil layer, often directly overlying the Ordovician limestone. Karst phenomenon can also be observed here.

It is impossible to imagine Lahemaa National Park without erratic boulders (Viiding 1981). They are symbolic of implacable struggle throughout thousands of years: at first between the land and glacial ice then between the land and sea.

House-sized granites and giant rapakivi are rather common in Lahemaa, yet quite rare in other regions of northern Europe which were covered by ice in the Pleistocene. In Estonia, about fifty gigantic boulders have been recorded, eight of which occur in Lahemaa. The erratics are derived from Archean and Proterozoic rocks. On the basis of mineral content and texture it is possible to distinguish between more than fifty rock types (Viiding 1981).

The human attitude towards stones comprises curiosity and interest, honour and satisfaction. These boulders reflect the history of the Estonian land and country; indeed some of them are the bearers of legends and folktales, and many have served as sacrificial places. Such a role would have not been related to their size, but to their location, composition and a shape which was suitable for the deepening of a hollow on their surface.

Almost every family had its own 'babystone', usually behind the house or shed. In children's eyes it had a magic power: they had only to get the parents' permission and knock at the stone and, in time, the stone gave them a little brother or sister.

Lahemaa should maintain its position as a land of numerous gigantic boulders and belts of stones for future generations as well.

Nature reserves

There are many nature reserves of very different types designed for the conservation of landscapes. Several relief forms are included in the landscape protection. The highest parts of the north Estonian clint are protected (Saka–Ontika–Toila cliff – 56 m high; Rannamõisa bank – 20–25 m high). The so-called Silurian clint is locally protected in western Estonia and on the islands.

Few of the Pleistocene relief forms have been preserved in their original shape, and as they constitute the principal sand and gravel deposits, there is little hope for their remaining intact unless effective protective measures are taken. At present, very few of them have been placed under protection.

With its clearly defined ancient coastal formations, Estonia is a classical example of the whole Baltic region. The deposits of the Baltic Sea, with subfossil mollusc shells, are of great scientific interest (Viiding 1976). The existence of such formations is threatened by the excavation of sand and gravel pits and by land improvement; protected areas of that kind have, therefore, been established in different regions of Estonia.

A geological reserve (Kaali Meteorite Craters) consisting of eight craters and formed about 3500 years ago is located on Saaremaa Island. Another, the so-called Ilumetsa Craters is situated in southeast Estonia. Both of these geological reserves are under state protection and most of the territory they occupy is maintained in order.

There are, however, some areas in Estonia that need taking under protective legislation as soon as possible. The most famous of these is the Kärdla Impact Structure (on Hiiumaa Island) formed about 400 million years ago. Another one is Osmussaar Island situated in the extreme northwestern part of the East European Plain. The lithological elements of the bedrock (Lower and Middle Ordovician) are interesting from a scientific point of view. The island abounds in classical coastal formations and today it can be considered almost as a laboratory showing its formation. A problem is posed here by a specific type of erratic boulder (gneiss-breccia), the origin of which in Fennoscandia is not known. Until quite recently this island was a sealed area of the Russian (former Soviet) army.

There are also some rather interesting islands in the Finnish Gulf, which belonged to Finland before World War II. For example, Suursaari (Hogland) is partly formed by a rare volcanic rock complex (Hoglandian). This island was investigated by Estonian geologists in the 1970s but it has potential international scientific interest and could become a special geological protected area.

Natural monuments

These isolated features of landscape are as follows: banks, hills, waterfalls, caves, karst formations, erratic boulders, stratotypes (outcrops), etc. They are selected on the basis of their rarity and proneness to destruction.

Estonia is a typical area of the Lower Cambrian, Ordovician and Silurian, with numerous stratotypes which need to be protected. In total, the number of stratotypes amounts to 55 but only 18 of them are protected.

Caves and geological landforms, in total 64 in number, occur in different parts of the country. Each of these features is kept under surveillance according to the system which is considered the best to ensure its preservation. Like nature reserves, many of the natural monuments are widely known on account of their striking appearance and historic or legendary association, thus attracting large numbers of tourists and school excursions.

The distributions of karst formations, their genetic types and conditions of development have been studied for a number of years. In the course of time, especially during the last few decades, human activity has affected the development of karst formations considerably. In Estonia there are about 70 karst areas and only 10% of them are under protection.

Since the nineteenth century, the abundance of large erratic boulders in north Estonia has attracted the attention of many scientists. Until 1940, about 900 large boulders has been registered in the card index in the archives of the Section of Geology and Geography of the Naturalists' Society affiliated to Tartu University. Later, that catalogue has served as a basis for selection of the boulders to be taken under protection. Today about 1900 huge boulders (with a diameter of over 3 m) and groups of boulders are known, 226 of which are protected. Fifty boulders are gigantic (with a perimeter of over 30 m and length of over 10 m). No other country in northern Europe has so many giant boulders. The largest is a block of rapakivi, called Kabelikivi (Chapel Rock) near the capital city, Tallinn, with a perimeter of 56.5 m and an overground bulk amounting to 728 m^3.

Conclusions

Sometimes there is discussion about the necessity and reasonableness for a small country to have so many protected areas and reservations. Yet, it seems clear that damage or destruction of unique geological and other features would be harmful not only to Estonian nature but to the entire northern part of Europe. At present, scientists of the Institute of Geology are engaged on compiling a monograph *Estonian Ancient Natural Monuments* for registration of all geological and landscape features.

References

KASSIK, T. & LAHTMETS, M. (eds). 1992. *National Report of Estonia to UNCED 1992*. Ministry of the Environment Republic of Estonia, Tallinn.

VIIDING, H. 1976. *Eesti NSV maapõue kaitsest*. Valgus, Tallinn.

—— 1981. *Lahemaa kivid* [*The boulders of Lahemaa*]. Valgus, Tallinn.

Conservation of national geological sites in China

PAN JIANG

Geological Museum of China, Beijing 100034, People's Republic of China

Abstract: With such a large territory, China has its best geological (fossil and geomorphic) sites dispersed all over the country. Six Chinese natural heritage sites feature on the World Heritage List – Mt. Taishan in Shandong (in 1987), Mt. Huangshan in Anhui (in 1990), Peking Man Site at Zhoukoudian in Beijing (1987), Wulinyuan in Hunan, and Huanglong and Jiuzhaigou in Sichuan (1992). Amongst these, Mt. Taishan, Mt. Huangshan and the Peking Man Site not only have important geomorphic and palaeobiologic significance, but are also of great importance in global research of cultural sites.

Since 1982, the State Council of the People's Republic of China has also identified 44 national parks and geological sites, such as Guilin in Guangxi (the tropical Karst landform) and Wulingyuan in Hunan Province, (mud and debris flows, Devonian sandstone pinnacles).

Subject to an effective legal implementation system, the natural heritage sites will be under permanent protection. In the case of geological relics, conservation regulation will be formulated in the near future. The State Council has accepted the document submitted by the Chinese Academy of Sciences on protection of vertebrate fossils, initiating important future policy as regards vertebrate fossil protection. Most national parks in China include important geological sites and special organizations have been established for the protection of these sites and the provision of support services.

To date, on-site museums have been established at many famous geological sites, such as the Peking Man Site Museum, Guilin Karst Museum and Shanwang Palaeontological Museum, amongst others. The construction of these on-site museums has an important impact on the public in promoting Earth science through educational programmes and media publicity. As a result, many sites are well protected.

Within the vast territorial area of China, a multitude of important geological and geomorphological sites have been identified. In this paper, some of the key sites are identified, the legal framework for the protection of these sites is described and the role of education is discussed.

National geological sites

Since 1982, the State Council of the People's Republic of China has identified 44 national parks, including national treasures such as Mt. Songshan, Mt. Taishan, Guilin and Wulingyuan. Geological interest ranks high amongst the criteria used to select such sites. Special organizations have been established to protect and promote those sites (Pan Jiang 1991).

On the World Heritage List, six Chinese sites were identified: Mt. Taishan in Shandong (in 1987), Mt. Huangshan in Anhui (in 1990), the Peking Man Site at Zhoukoudian in Beijing (in 1987), Wulingyuan in Huan, and Huanglong and Juizhaigou in Sichuan (1992). Mt. Taishan, Mt. Huangshan and the Peking Man Site not only have important geomorphological and palaeobiological significance, but are also of great cultural significance.

In February 1991, the Task Force, appointed by UNESCO to establish procedures and criteria, and to draw up a provisional Global Indicative List of Geological Sites (GILGES), met in Paris. China recommended seven geological sites to the meeting (Cowie 1991). They are listed below:

Eastern Asia (Category XI)
Priority 1.

(1) *Peking Man Site*, Beijing, China.
(2) *Miocene Shanwang Fossil site*, Shandong Province, N. China. Four hundred species of plants and animals have been identified here, and of particular importance are the 270 species of insects, and many ancestral mammalian faunas. Most are now housed in an on-site museum. The outcrops weather very rapidly and collecting has not been forbidden; however, an extensive fossil resource remains.
(3) *Zigong Middle Jurassic Dinosaurs*, Dashanpu, Zigong City, Sichuan Province, China. Some $2800\,m^2$ of the site is covered by a museum building, which displays 100

From O'Halloran, D., Green, C., Harley, M., Stanley, M. & Knill, J. (eds), 1994, *Geological and Landscape Conservation*. Geological Society, London, pp. 243–245.

individual dinosaurs, reptiles, amphibia and fishes.
(4) *Chengjiang Fauna Reserve*, Mt Maotianshan and Mt. Dapotou, Yunnan Province, China. This site revealed an early Cambrian marine metazoan, with over 60 genera of sponges, worms, medusoids, brachiopods, hyoliths, trilobites and other arthropods represented. The specimens are now housed in Nanjing Museum.
(5) *Songshan National Park*, Henan Province. China. Structures and stratigraphy from the Archean to the Cenozoic can be seen.
(6) *Wuligyuan National Park*, Hunan Province, China. Devonian mud and debris flows are found in this area, along with a forest of over 3000 flat-topped pinnacles. Early Silurian and Middle Silurian aquatic vertebrates have been found.
(7) *Guilin National Park*, Guangxi Zhuangzu Autonomous Region, South China. This site has tropical karst developed in Upper Devonian and Lower Carboniferous limestone.

Legal framework

The People's Republic of China is a member state of the "World Heritage Convention" and plays an active role in the conservation of world culture and natural relics.

At the national level, legislation is in place which protects China's geological heritage and encourages good management practice. For example, the law on 'preservation of cultural relics of the People's Republic of China' passed in 1982, clearly stipulates that palaeovertebrate fossils and human fossils should have the same status in law as cultural relics. This legislation was drawn up after wide consultation.

In 1961, the State Council endorsed a document submitted by the Chinese Academy of Sciences on the protection of vertebrate fossils. Local government officials have worked closely with the security departments to eliminate illegal fossil collecting activities. The problem has been exacerbated somewhat by the fact that some of these vertebrate fossils or 'dragon bones' are used in traditional Chinese medicine and in some south-east Asian countries, they are preserved as 'holy relics'.

The legislation outlined above forms the basis for the preservation of our geological heritage. Agencies have been set up with the objective of conducting scientific and technical studies to give a better appreciation of the resource. This approach is consistent with Article B of the UNESCO Convention concerning protection of Cultural and Natural Heritage.

Management plans are drawn up for each national park and national geological site in accordance with regional regulations.

Role of education

In the last 10 years, China has enhanced its educational programmes, and geological study now assumes a much higher profile. Palaeontology, mineralogy, petrology and mineral resources are all subjects which are covered in the standard textbooks available to primary and middle-school students.

The Chinese government is currently endeavouring to strengthen appreciation of, and respect for, our cultural and natural heritage, mainly through educational and information programmes. Thirty television programmes or films have been produced in recent years with this express purpose in mind. Without doubt, the role of education and media in the promotion of Earth heritage conservation is very important. At present, four regular journals are published in China, namely *The Earth*, *The Fossils*, *Nature* and *Knowledge of Geography*. The purpose of such publications, which are sponsored by government, is to popularize Earth science and to strengthen education activities generally. Achievements in this repect have been remarkable.

Specialized on-site museums have been constructed at some important geological heritage sites, so as to promote conservation and greater awareness of our geological heritage. Some of the finest examples are listed below:

(1) Shanwang Palaeontological Museum near Shanwang, Linqu County, Shandong Province. N. China (Pan Jiang 1991).
   In February 1980, the State Council listed Shanwang as a National Protection Area. After several years, a museum was constructed. Now about 10000 specimens, including diatoms, plants, insects, fishes, amphibians, reptiles, birds and mammals of Middle Miocene age are kept in the museum.
(2) Peking Man Site Museum
   This site is located at the foot of Dragon Bone Hill, near Choukoutien (Zhoukoudian) village of Fangshan County, Beijing (Peking). The museum was constructed in 1950.
(3) Guillin Karst Museum (Jingrong *et al.* 1991)

This museum, established in the 1980s contains displays relating to the formation of karst scenery.
(4) Zigong Dinosaur Museum (Ouyang Hui & Ye Yong 1991).
This museum, located at the Dashanpu Dinosaur burial site, near Zigong city, Sichuan Province, Southwest China was built in the 1980s.
(5) The Natural Protection Area of the Proterozic Sequence in Jixian county of Tianjin city (near Beijing) (Zhang Huimin & Li Huaikun 1991).

In recent years, many Chinese and foreign geologists and scholars have been interested in the study of this Proterozoic section, and finally in 1984, the Jixian section was approved by the State Council as a Natural Protection Area.

Construction of these geological on-site museum facilities has had an important impact and has helped to promote awareness of Earth heritage conservation. As a result, geological sites are now valued by the people of China and will be protected for the benefit of future generations.

Reference

COWIE, J. W. 1991. *Report of Task Force Meeting, Working Group on geological (inc. fossil) Sites*. A co-operative project of UNESCO, IUGS, IGCP & IUCN.

JINGRONG, L., JIANG, P. & DAOXIAN, Y. 1991. Tropical Karst Geological Heritage of Guilin and Protection Measures. Ier Symposium International Sur la protection du patrimione geologique. *Terra abstracts*, **43**, 12.

OUYANG HUI & YE YONG. 1991. On Conservation of Danshampu Dinosaur Buriel Site, Zigong city, Sichuan province, SW China. *Terra abstracts*, **46**, 12.

JIANG, P. 1991. On the conservation of Geological (inc. Fossil) Heritage in China. *Terra abstracts*, **6**, 2.

ZHANG HUIMIN & LI HUAIKUN. 1991. Jixian section – the first national conservation zone for Geological Heritage in China. *Terra abstracts*, **48**, 13.

Earth science conservation in Bulgaria

TODOR A. TODOROV

Geological Institute of Bulgarian Academy of Sciences, Sofia 1113, Bulgaria

Abstract: The territory of Bulgaria is characterized by a varied topography, geology and climate. Over 360 'natural geological movements' are now protected by the Government to safeguard this rich geological diversity. Some of these geological localities are also of international significance and may be adopted as World Heritage sites in due course.

In the literature, the term 'natural geological monument' was first introduced by the German scholar and natural explorer Alexander Humbolt in the last century. Now the term 'Earth science conservation' is widely used, often in connection with features such as geological formations which are unique in their shape, structure or aesthetic appeal. Sites scheduled for conservation may include whole rock massifs, picturesque landscapes, or a wide variety of landforms, fossil horizons, ancient mines or ore deposits. These sites all play an important role in understanding the geological evolution of Bulgaria. No less important is their aesthetic appeal. We must, therefore, protect our irreplaceable Earth heritage for the future, as a record of the geological evolution of our planet and as a source of spiritual inspiration.

Geological sites and nature conservation legislation in Bulgaria

Legislation designed to protect the natural environment was passed in 1967. Protected natural sites or species are grouped into seven categories:

(a) reserves;
(b) national parks;
(c) protected areas;
(d) natural beauty spots;
(e) historical sites;
(f) protected animals;
(g) protected plants.

Geological sites are included in two of these categories, namely natural beauty spots and protected areas.

Using the author's information in conjunction with data from Iliev & Petrov (1989), examples of the types of sites which are protected are given below.
- Stone bridges – protected sites include the Marvellous Bridges, God's bridge, the Saddler, the Rock Window, the Devil bridge;
- Moraines – protected sites include Bristritsa moraine, Golden Bridges;
- Karstic gorges and valleys – protected sites include Vratsata, the River Erma gorge, the Trigrad Gorge, Chernelka, Topchiiska River;
- Caves – protected sites include Ledenika, Snezhanka (Snow-princess), Lepenitsa, Saeva Dupka, Magurata, the Devil Gorge;
- Dunes – protected sites include the Pearl, Kavatsite, the Old Woman, Alepu, the Golden Fish;
- Extinguished volcanoes and associated deposits – protected sites include Fur coat, the Volcano crater, Kozhuh.
- Lakes, swamps and firths – protected sites include Vaya, Atanasovo Lake Srebarna, Kaikusha, Persin Swamps, Smolyan Lakes, Rhila Lakes;
- Waterfalls – protected sites include the Paradise Sprayer, Tufcha, Skakavitsa, Skaklya, the Karlovo Sprayer;
- Mineral deposits – protected sites include Urdin Circus;
- Deposits of fossils and petrified trees – protected sites include areas near the villages of Opanets, Yasen, Tarnene, Oryahovitsa, Stavertsi, the village of Ahmatovo-Pravoslaven, the village of Dorkovo;
- Trace of ancient mining activity – protected sites include the oldest copper mines in Eastern Europe, near the Stara Zagora mineral spas.

Of course, there is still much to be done and work continues to ensure that all the key sites in Bulgaria are conserved for the benefit of future generations.

Over 360 geological sites in Bulgaria are already protected by the Government. In spite of their variety, these sites fall within one of five groups:

(1) picturesque rock formations (*c.* 150 sites);
(2) waterfalls (*c.* 70 sites);
(3) lakes, swamps, firths and karstic springs (*c.* 20 sites);

(4) caves (*c.* 100 sites);
(5) sites within reserves and national parks (*c.* 20 sites);

The distribution of these sites is directly dependent on the geological structure and rock types which exist in each of the different districts of Bulgaria.

The number of protected sites has risen from 55 in 1964 (Spassov 1965) to 224 in 1974, and now stands at around 360 (Todorov 1989). This growing number is testament to the increasing interest in Bulgaria for Earth heritage conservation. The Commission for Protection of Nature at the Bulgarian Geological Society has played an important role in this process.

Monitoring of geological sites

Another area of interest is the restoration and management of geological sites in Bulgaria. Geological sites are subject to constant erosion by water, air, wind and rain, however, no less substantial is the role that humans play, who 'with one hand create and with the other unreasonably and blindly destroy and deaden Nature', as an ancient Greek philosopher stated 2000 years ago. Humans are an important factor, who by their activity are capable of entirely changing the face of certain parts of the Earth's surface. Just as some sites have been created by human activities, for example by quarrying, others have been destroyed by development. Restoration and remedial works have an important role to play in making many of our most important sites safe to visit, and management plans need to be compiled identifying stabilization and other remedial works which are necessary for effective long-term site management. This work should commence immediately, with the objective of fully integrating the results of these studies into the national system which already exists for monitoring ecological sites.

Data under the following headings are required for each site (Todorov 1990):

(1) name, location and type of site;
(2) geological description of the geological site and surrounding area;
(3) level of study of the protected geological site, main factors causing destruction, recommendations for future studies and measures for its protection and restoration;
(4) reference list.

Conclusion

The main tasks for future Earth heritage conservation in Bulgaria are:

(1) more active environment protection, with the integration of safeguards for geological sites into plans for general environmental improvements;
(2) study and protection of geological sites through the creation of additional national and regional parks and reserves;
(3) steps should be taken to include Bulgaria's most famous natural geological sites in UNESCO's list of World Heritage Sites;
(4) all the most important natural features should be preserved for science and future generations. This could be achieved by compiling a catalogue of key sites, as already exists for animals and plants.

References

ILIEV, X. & PETROV, P. 1989. Types of geological–geomorphological sites in Bulgaria and some problems of their protection. *Extended Abstracts of 14th Congress of Carp.* Balkan Geological Association, Sofia, 1563–1565.

SPASSOV, CH. 1965. Protected natural sites in Bulgaria. *Review of the Bulgarian Geological Society*, **26** (2), 229–230.

TODOROV, T. 1989. Beautiful natural geological sites in Bulgaria: status and some problems. *Extended Abstracts of 14th Congress of Carp.* Balkan Geological Association, Sofia, 1574–1577.

—— 1990. Monitoring of status of natural geological sites in Bulgaria. *Review of the Bulgarian Geological Society*, **51** (3), 102–204.

Geological conservation in Hungary

TIBOR CSERNY
Hungarian Geological Survey, Stefánia u. 14, Budapest 1143, Hungary

Abstract: The organization responsible for the protection and conservation of biological and geological sites in Hungary is the National Authority for Nature Conservation. Several professional institutions have been active in geological conservation, with the Hungarian Geological Survey (HGS) in the lead. The activities of HGS in this respect include:

(1) The exploration and study of important geological sites and sections in Hungary. At present, over 400 recognized formations represent an almost complete range of geological units. About 200 of these are accessible as exposures or outcrops, and the others are known from boreholes. All formations are investigated by HGS geologists and other institutions (universities, National Museum, etc) as interdisciplinary projects, and the results are integrated.

(2) The publication of the results of this interdisciplinary study and conservation of the most important geological sections and sites. Leaflets have been published describing over 150 sites and about 200 boreholes.

(3) The establishment and maintenance of geological and ecological study trails, conservation areas and museums. For example, a quarry at Tata has become a nature conservation area, with the first open-air museum of geology, opened in 1992. The 4 ha park includes a display of mineral resources, a silica mine dating from the Palaeolithic, some important geological sections from the Mesozoic, and an arboretum.

(4) The organization of training courses in geological and environmental sciences. Adjacent to the nature conservation area at Sümeg, a field-training base has been operating for several years, providing fieldwork training for students of geology and geophysics. The base also runs postgraduate courses in geology and environmental protection.

Nature conservation in Hungary has a history dating back to the end of the last century. In 1879 the first law regulating forestry, which made some reference to the need to protect the natural environment, was passed. Although the scope of this legislation was limited, it has provided some opportunity for the protection of the environment.

This was followed in 1935 by a second forestry law, which included a specific section relating to nature conservation. As a consequence of this legislation, natural areas, principally forests, were registered as protected sites in Hungary. The Grand Forest in the city of Debrecen, the cave at the village of Abaliget and fossil remains found at the village of Ipolytarnòc, registered in 1939, 1941 and 1943, respectively, were the first areas to be identified primarily for their Earth heritage importance.

In 1961 a third piece of nature conservation legislation was passed. This law made provision for forestry and nature conservation to be considered separately and also established the National Authority for Nature Conservation (Kopasz 1976). This law provided an opportunity for the setting up of national parks as well as the introduction of the term 'site of national value'. Considerable efforts were made during the following years, which eventually led to the creation of the first national park in Hungary. This was established at Hortobágy in 1973.

In 1984, a fourth environmental protection law was passed, which reflected a tendency to confine nature conservation to specific areas. However, new legislation which is currently in preparation, will extend conservation measures to include every living species and their environments in a more integrated fashion. Thus, traditional nature conservation is being replaced by renewed efforts to maintain both the natural environment and wildlife habitats. This legislation also makes provision for the ownership of natural sites of especially high value to be passed to nature conservation bodies (Csepregi 1988).

Present-day nature conservation in Hungary

The current legislative framework makes provision for the protection of living organisms and their natural environment. There is a correspondence between the value of a designated site and the variety and rarity of the natural features which it displays. The factors which are taken into account in any assessment include the geology, hydrology, botany, zoology, landscape and cultural history of the region. Geological assessments are made by the Hungarian Geological Survey.

There are three types of conservation area in Hungary.

(1) Nature conservation areas which comprise unique natural sites (e.g. geological key localities or sites). Their protection is of particular importance both for scientific and educational purposes. These areas have to be exempt from any kind of development.
(2) Landscape protection zones have been established to maintain the most typical features of the country as well as the natural environmental equilibrium. These areas can be subject to forestry and agricultural practices provided the original character of the landscape is preserved.
(3) National parks are the highest grade of nature conservation designation. They serve as a basis for the conservation of the most important natural and semi-natural areas occurring in Hungary. In certain parts, any economic activity is strictly prohibited; in other areas, some indigenous activity, which is compatible with effective nature conservation, is permitted.

Designated and nominated natural conservation areas can be of international, national or of local importance. There are 5 national parks, 147 nature conservation areas and 1 nature memorial of international and national importance. In addition, there are 799 nature conservation areas and 76 nature memorials of local significance. The area under protection exceeds 6600 km^2 representing more than 7% of Hungarian territory (Mtt 1991; Garami 1993).

Of the 2797 caves designated as protected sites, about 100 are nominated under strict conservation provision, including the longest and finest karst cave in Europe, at Aggtelek.

Activity of the Hungarian Geological Survey in geological nature conservation

Activities of the Hungarian Geological Survey include maintaining an open-air geological museum at Tata; an integrated programme of research on designated areas and establishment of geological study-trails throughout the country. Assistance is also given in operating a geological museum and environmental study-trail at Ipolytarnòc, participating in the protection of caves and the organization of training courses in Earth heritage conservation (Fig. 1). Much has already been achieved in improving documentation and increasing public awareness.

Geological museum and geological site of international importance

Activities include the establishment of a museum near the village of Ipolytarnòc. This geological locality figures among the candidates for the World Heritage sites list and is therefore potentially of global significance (Tardy 1990). In the vicinity of the village of Ipolytarnòc, an excellent Early Neogene section, which has now been fully described, is open to the public. A geological trail and material displayed in two adjacent exhibition halls serve to interpret the site for the benefit of the visitor. More than 2000 footprints occurring in Early Miocene terrestrial sandstones, are regarded as the most important fossilized footprints in Europe. They were made by a variety of animals including rhinoceros, four carnivorous species, large and small hoofed mammals as well as four bird species. The footprints were created in the once soft sand of a marshland environment and were subsequently covered by rhyolite tuffs. The flora found in the Miocene beds contains elements which are indicative of a warm subtropical climate (Kordos 1990a).

Similar significance can be attributed to the Hominoidea fossil locality discovered in the wall of the opencast iron pit at the village of Rudabánya, which has been designated as a nature conservation area. The c. 10 million-year-old paludal complex has yielded the skull and bone remains of *Rudapithecus hungaricus* and *Anapithecus hernyaki*, an ancient anthropoid of African origin and an ancient monkey, respectively. This locality is of particular importance not only for these fossil remains, but also as the terrestrial stratotype of the Carpathian Basin which includes an important palaeobotanical assemblage and abundant ostracods and molluscs (Kordos 1990b).

Geological open-air museum and geological key locations and selections

This continuing project has had significant results with the selection of Important Geological (Geomorphological) Sites, which include key geological sections such as stratotypes.

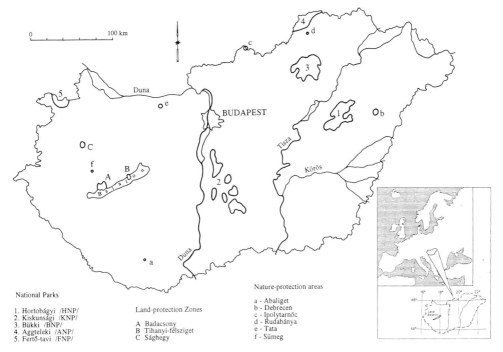

Fig. 1. Conservation areas in Hungary

Some 400 formations representing the geological history of Hungary have been registered. About 200 of them can be observed in exposures, and the rest in borehole cores. Descriptions of around 150 surface exposures along with 200 borehole logs have been published in geological newsletters as well as in short leaflets. Unfortunately, quite a number of these exposures are contained within sites operating as legal or illegal waste disposal sites and this represents a serious obstacle to their effective conservation (Császár 1992). The opening of the first open-air museum for geology and nature conservation in an old quarry and in its surroundings in the city of Tata in 1992 was the result of concerted efforts over the last 30 years. The museum occupies an area of 4 ha. A whole range of activities relating to geological conservation can be experienced at this site including a visit to a 30000-year-old silica mine dating from the Palaeolithic culture; some key sections of the Upper Jurassic–Lower Cretaceous boundary, which is interpreted as a continuous marine sequence; and a display on the mineral resources of Hungary (Fülöp 1984). The geological park is an outstanding example of what can be achieved by dedicated geologists with a desire to tell the geological story to the public – adults and children alike (Reading 1992).

Integrated geological research of nominated and designated areas

Comprehensive geological study of the Hortobágy-, Kiskunság- and Fertő-lake National Parks, the Tihany-peninsula, the Badacsony mountain area and in other protected areas has included mapping and other investigations (Cserny et al. 1981). The results have been published in a map series illustrating the geology, geomorphology and hydrology of the areas concerned.

Geological study-trail on a Site of Special Scientific Interest

Geological and biological study-trails have been established in protected areas, in partnership with the National Authority for Nature Conservation. The first study-trail of this sort was inaugurated by the National Authority for Nature Conservation in 1989 in the Ság-hill at the city of Celldömölk (Tardy & Oravecz 1990). A volcanic cone which rises some 150 m above the surrounding gravel plain of the Marcal river was active about 5 million years ago, but is now reduced to one-third of its original size. An erosional remnant consisting of volcanic tuff thrown out of the crater is overlain by basaltic lava protecting the underlying loose Pliocene beds from denudation. From the study-trail, which is in part an ancient mine, visitors can see the crater of the basaltic volcano and have the opportunity to learn about the geological evolution of the area. This process is assisted by a series of explanatory boards which tell the story of the volcano of the Ság-hill.

Caves

Almost 3000 caves in Hungary have been designated as sites of geological value. More than 100 of them are registered as strictly protected stalactite and crystal caves. Their complex scientific study is carried out by an independent organization, the Speleological Institute (Tardy 1987).

Training courses and competitions

With the active participation of the Hungarian Geological Survey, the National Authority for Nature Conservation organizes graduate and postgraduate training courses in nature conservation (Haas 1987; Edelényi & Haas 1988; Bihari 1993). This programme, which includes a substantial element of fieldwork, involves high school and university students, teachers participating in postgraduate courses and some other specialists. Every year since 1991 the National Authority for Nature Conservation has run a competition entitled 'Our geological heritage' and about 350 high school students have participated. An average of 150 papers are submitted each year, demonstrating a growing interest in the younger generation for the conservation of our physical environment (Bihari 1993).

Improving documentation and increasing public awareness

Many important geological (geomorphological) sites have been documented and details of over 150 are published. Leaflets describing these sites can be obtained from the Hungarian Geological Survey.

There are also booklets written specifically for the public about nature conservation which include descriptions of regionally important geological sites. In addition, some geological guides, which are aimed more at the scientific community, have also been published.

Conclusions

Nature conservation in Hungary has a history dating back to the end of the last century. The current law reflects a tendency for confining nature conservation efforts to special areas, such as national parks, landscape protection zones, nature conservation areas, caves or selected plant and animal species. The organization responsible for the protection and conservation of biological and geological sites in Hungary is the National Authority for Nature Conservation. Several institutions have been active in geological conservation, with the Hungarian Geological Survey (HGS) taking the lead. The responsibility of the HGS in the field of nature conservation includes the assessment, scientific investigation and description of natural or temporary sections, together with co-ordination of their conservation and interpretation. In addition, the Survey provides essential help to the National Authority for Nature Conservation in establishing legal protection for key sites.

References

BIHARI, Gy. 1993. *Földtani örökségünk* (Our geological heritage). Newspaper article, Budapest.

CSÁSZÁR, G. 1992. *Geológiai értékeink sorsa* (State and future of our geological values). Értékmentô, 2993.6., Budapest.

CSEPREGI, I. 1988. *A természetvédelem jogi szabályozásának töténeti áttekintése* (Historical overview of the legislation about nature protection in Hungary). PhD dissertation.

CSERNY, T. & GELEI, G-né & GUÓTH, P. 1981. Badascony környékének építésföldtana (Geological features of the nature conservation area of the Badacsony mountain). *Földtani Int. Évi. Jel.*, **1979-rôl**, 283–292.

EDELÉNYI E. J. & HAAS, J. 1988. A Sümegi földtani kutatási terepgyakorlat 10 éve (Ten years of field training in geological research at the Sümeg base). *Földtani Kutatás* **XXXI**.1., Budapest, 85–92.

FÜLÖP, J. 1984. *Tata. Kálvária domb* (Geological nature protected area in Tata).

GARAMI, L. 1993. *Képes útikalaus, védett természeti értékeink* (Illustrated guide to our Nature Protected Values). Panoráma, Budapest.

HAAS, J. 1987. *Sümeg, Mogyorósdomb, Magyarország geológiai alapszelvényei* (Geological key localities and sections). Hungarian Geological Survey, Budapest.

KOPASZ, M. (ed.) 1976. *Védett természeti értékeink* (Our protected natural values). Mezôgazdasági Kiadó, Budapest.

KORDOS, L. 1990a. *Ipolytarnóc, természetvédelmi terület, Magyarország geológiai alapszelvényei* (Geological key localities and sections). Hungarian Geological Survey, Budapest.

—— 1990b. *Rudabánya, Hominoidea-lelőhely, Magyarország geológiai alapszelvényei* (Geological key localities and sections). Hungarian Geological Survey, Budapest.

MTT. 1991. *Magyarország természetvédelmi térképe* (Map of the Nature Protection areas of Hungary). Kartográfiai vállalat, Budapest.

MÁ, V. (ed.) 1992. *Természetvédelmi adatok* (Nature protection data). Környezetvédelmi és területfejlesztési Minisztérium, Természetvédelmi Hivatal, Budapest.

READING, H. G. 1992. New waves of geology. *Episodes*, **15**, 147–148.

TARDY, J. 1987. *Budapesti termálkarszt hidrotermális barlangjainak és forrásainak védelme* (Project proposal for the protection of hydrothermal caves and springs of the thermalkarst of Budapest) Manuscript, 1987–1933.

—— 1990. *Ipolytarnóc*. Videorecord.

—— & ORAVECZ, J. 1990. A sághegyi tanösvény (The study-trail of mountain Ság). *Földtani Közlöny*, **T.120** (1–2), Budapest, 129–131.

TÁJAK KOROK MÚZEUMOK KISKÖNYVTÁRA SOROZAT (Landscapes, Ages, Museums – series). *Celldömölk-Sághegy* (1980), 110–113; *Badacsony* (1982), 164; *Tata* (1984) 196; *Ipolytarnóc* (1985). Országos Környezet-és Természetvédelmi Hivatal, Budapest.

Geological and landscape conservation in India

K. N. PRASAD

8, Venkataraman Street, R.A. Puram, Madras-28, India

Abstract: The Indian sub-continent is an area of marked contrast in structure, physical features and stratigraphy. There are three distinct regions: the Peninsular Shield, an ancient denuded plateau; the Extra-Peninsula, a region of folded and overthrust mountain chains and fast-flowing Himalayan rivers; and the Indo-Gangetic Alluvial Plains, an important area for the cultural and social history of India.

There are many landscapes and sites of geological interest within these three regions that are in need of conservation. These include the Kurnool Caves, with fossils of Pleistocene fauna; the Deccan Volcanics; Indo-Gangetic Alluvium; Extra-Peninsular region, containing the Siwalik Formation with one of the most important early mammalian faunas; the Mesozoic rocks of the Peninsular region; and Triassic and Jurassic dinosaur sites.

An improving communication system has meant that hitherto inaccessible geological heritage sites have now become easily accessible to a growing population. Legislation and education of the general public is required to ensure the adequate conservation and protection of these sites.

The physiographic map of India shows strikingly that the Indian sub-continent can be divided into three well-marked regions each having distinguishing characters of their own. The first region is the Peninsular Shield (a geologically ancient stable part of the crust) to the south of the plains of the Indus and Ganges river systems. The second division comprises the Indo-Gangetic Alluvial Plains, stretching across northern India from Assam on the east to Punjab on the west. The third is the Extra-Peninsula, the mountainous region formed on the mighty Himalayan ranges with extension to Baluchistan on the one hand and Myanmar (Burma) on the other. These three divisions exhibit marked contrast in structure, physical features and stratigraphy.

General landscape

The Peninsula is an ancient plateau, which has been exposed throughout geological ages to denudation leading to peneplanation. The relict type of mountains of the Peninsula which contain hard rocks have survived weathering during the course of their long history. The topographic expression may not, therefore, be directly attributable to their structure. The rivers, for the most part, traverse relatively flat country with low gradients. They have formed shallow and broad valleys in the plateau region.

The Extra-Peninsula is a region of tectonic, folded and overthrust mountain chains of geologically recent origin. During the Middle Miocene, the third, and perhaps the most violent, episode of mountain building on the northern borders of India took place. This must have been accompanied by raising and folding of the strata laid down in the Tethys to form mountain ranges with large intrusive igneous bodies in the cores of the folds. A narrow depression resulted to the south of the rising mountains. This depression (fore-deep) is the site of deposition of the Siwalik Formations which began to accumulate during the Middle Miocene. The youthful Himalayan rivers are continuously eroding their beds in their precipitous courses, carving out deep and steep-sided gorges. It is here that conservation of geological heritage, in the form of fossils ranging from Cambrian to Tertiary, needs protection as most of the sedimentary sequences containing these remains of ancient marine organisms are being continuously eroded by the fast-flowing Himalayan rivers.

The Indo-Gangetic Alluvial Plains are broad and humanly speaking, of greatest interest and importance, as being the principal theatre of Indian history, and geologically speaking, as the least interesting part of India. In other words, they are of absorbing interest regarding human history, being thickly populated and being the scene of many important historical developments and events in the cultural and social history of India. In the geological history of India, they are only the annals of yester-year with alluvial deposits of the Indo-Gangetic Systems, borne down from the great Himalayas and deposited on the plains below. These deposits have covered a deep mantle of mudstones and valuable records of past ages which may throw additional light on the physical history of the Peninsula and the Himalayas.

The snowline is the lowest limit of perpetual snow on the side of the Himalayas facing the Indo-Gangetic Plains. It varies in altitude from about 14 000 ft on the Assam Himalayas to 19 000 ft on the Kashmir Himalayas. Glaciers of different types are now confined to the highest ranges of the Himalayas. The Himalayan rivers and their tributaries are fed by these perpetual glaciers and these snow-fed rivers in summer, bring copious quantities of water to the plains, occasionally resulting in floods. The present-day glaciers are remnants of extensive glaciation during the Ice Age (Pleistocene), when large areas of the Himalayas were covered with snow and ice. More significant than these, are the fluvio-glacial deposits and moraine-filled glacial lakes which need conserving as part of the landscape preservation.

Conservation of landscape and heritage sites

In the Indian sub-continent, many geological heritage sites can be identified. The nature of the geological sites is controlled by the geology and geomorphology of the respective regions. Landscape and site conservation are closely related. A multidisciplinary approach to these monuments, national parks and biosphere reserves is needed, with long-term planning. Designation of natural parks containing scenic and landscape features has, to some extent, generated public awareness, although in the long run, a legislative framework will be needed to control and relieve visitor pressure, especially in a country where population growth is enormous, bordering on 1000 million people by the year 2000. A conservation policy based on an integrated approach may be desirable to protect landscapes, naturally occurring geological sites and biosphere reserves in order to achieve a sustainable management of the Earth's resources and features like landforms, river systems, and major relief features (e.g. the Plateau of Peninsular India and the orogenic belt of the Himalayas).

Some of the major landscapes and sites of geological interest are briefly outlined here.

Kurnool Caves

There are 121 limestone caves of Pleistocene age in the Kurnool District alone. The Kurnool Caves contain fossils of extinct animals of Pleistocene age. These endogenous caves have been mostly formed by karst activity and the stalagmitic floors when excavated contain relics of prehistoric animals, some of them living but mostly extinct. The Kurnool Cave fauna also helps us in understanding the evolutionary processes and migration patterns of some of the animals. As Lydekker (1886) pointed out, the genera *Cynocephalus* and *Manis* are identical with the living African species, while *Rhinoceros*, *Hystrix* and *Viverra* are distinct from the species now living in India. The occurrence of *Cynocephalus*, *Hyaena*, *Equus* and *Manis*, indicates that many of the existing Ethiopian mammals were derived from India. Interestingly, some of the forms have gradually died out in India to become dominant in Africa. Those which have not totally disappeared are either poorly represented or smaller in size.

The total disappearance from India of *Hippopotamus*, *Giraffa* and antelopes of modern African genera, once prevalent in the Siwaliks, is striking but difficult to explain. Bone implements (Osteodontokeratic culture) comprising awls, barbed arrowheads, spear, scrapers, chisels are also known from the floors of the caves; prehistoric humans probably inhabited these caves since artefacts are abundant in the cave floors. These Pleistocene caves need protection. During the monsoon, many areas are washed away by swiftly flowing currents.

Deccan Volcanics

Among the hard rocks, the Western Ghat hill ranges primarily composed of Deccan Volcanics–Trap rocks (basalt) contain relics of interesting geological structure like the columnar joints of basalts – as in St Mary Island, a few kilometres off the West Coast in the Arabian Sea. Some of these areas, including 'Pillow lava structures' (Mardihalli), need further conservation.

Some of the major relics in the form of terraces occur along the principal rivers in the Peninsula. The famous Kohinoor Diamond was obtained from the Krishna river terraces three to four centuries ago. Can these geomorphological features containing gold and precious stones be protected?

Indo-Gangetic Alluvium

There are many historical monuments and cultural sites which need proper management. However, geologically south of the Indo-Gangetic Plains, the Narmada basin is known to contain Pleistocene mammals. The importance lies in the occurrence of early human remains. One of the recent discoveries is part of a skull of *Homo erectus*. Indiscriminate collection of fossils from this potential hominid site by students and others, has to be ended.

Extra-Peninsular region

In the Extra-Peninsular region there are several geological heritage sites, like the Siwaliks which contain one of the most important early mammalian fauna known in the world, ranging in age from Miocene to Pleistocene. Landscape conservation is essential on the ridges containing these ancient relics which are being continually eroded by agricultural practices, afforestation and constructional activities.

Mesozoics of the Peninsular region

There are several sites known for their geological heritage content like the Frog Beds of Bombay, the Cretaceous Formations of Southern India, the Gondwana Formations, and the Dinosaurian Beds of Jurassic and Cretaceous age.

Triassic and Jurassic dinosaur sites

The sites are located in the Adilabad District, Andhra-Pradesh, South India, their significance being as the only sites containing tetrapod reptile fossils, dinosaurs, fishes, and micromammals of Jurassic age. The sites were discovered by Sir William King, Director of the Geological Survey of India in 1872 during the course of geological mapping. This rare occurrence of so many varieties of fossils belonging to several groups in one area requires declaration as a national monument. So far, no decision has been taken with regard to this. The sites are located in deciduous tropical forest which shelters rare black bucks, sambhars, tigers, leopards, crocodiles and a nature reserve.

Methods of conservation

There are many geologically and geomorphologically interesting sites in India. These need conservation partly by legislation and partly by educating the general public – stressing the importance of these ancient monuments which cannot be replaced once they are destroyed. The significance of these heritage sites can be brought to light by means of popular lectures, puppetry shows, local stage shows, music and through the media by popular handouts. By these means some conservation can be ensured. India has 14 major official languages and several tribal dialects. There is a need to orientate mass communication, by popular methods to suit to various ethnic groups spread all over the country. Such an approach may be successful. The heritage sites in many parts of India were inaccessible in the past and, therefore, escaped destruction, but in recent years, a well-developed communication system has exposed these invaluable heritage sites to easy access and destruction. Students at Higher Secondary School level and volunteers should be encouraged to visit areas of geological and archaeological interest, including landscape conservation areas, in order to appreciate the significance of nature's treasure. This is a major challenge that has to be faced in India.

Reference

LYDEKKER, O. 1886. The fauna of the Karnul caves. *Palaeontologica indica*, **IV** (III), 19–58.

Geological and environmental mapping as an aid for landscape conservation in Lithuania

JONAS SATKUNAS

Geological Survey of Lithuania, Konarskio 35, 2600 Vilnius, Lithuania

Abstract: Large-scale geological maps provide important information for landscape protection, conservation and land-use improvement. In recent times the demand for geological information from landscape architects, land-use planners, environmental authorities and other decision-makers has increased in Lithuania.

Geological mapping has been carried out systematically in Lithuania only since 1958. Geological maps at a scale of 1 : 200 000 are available for the whole country. Maps at a scale of 1 : 50 000, which are more appropriate for applications in land management, are available for only 20% of the country; for these areas the coverage includes maps of the geomorphology, Quaternary geology, groundwater, mineral raw materials and pre-Quaternary geology. The maps serve as a basis for compilation of customized maps or can be used directly for various practical purposes. Until recently the main aims of the 1 : 50 000-scale mapping have been the location of deposits of raw materials and of groundwater resources. The goals have been expanded in recent years and given expression in a national programme for geological mapping which was approved by the government of Lithuania in 1992. Two main directions for mapping at this scale were established in the programme – one for comprehensive integrated geological mapping of industrialized areas with serious environmental problems, and one for areas with less anthropogenic influence and with geological and landscape features to be preserved. The latter include reservation areas and national and regional parks. The type of study envisaged in these areas is aimed at improving protection, management and understanding of the value of the areas concerned.

The relief of Lithuania is mainly of glacial origin. Geological mapping has revealed many sites of scientific importance, particularly relating to Quaternary geology and geomorphology. They include stratotype outcrops, characteristic glacial landscape patterns and various geological features of special interest. All of these must be considered as a part of the national heritage and preserved as such.

Main achievements of geological mapping in Lithuania

Systematic geological mapping at a scale of 1 : 200 000 was initiated in Lithuania in 1958. As a result, maps of the Quaternary deposits, geomorphology, pre-Quaternary, hydrogeology and other maps at a scale of 1 : 200 000 were finished for the whole of Lithuania in 1976. Together with descriptions and additional data they provided general information on the geological structure and origin of relief of Lithuania. This mapping was integrated to produce a set of maps at a scale of 1 : 500 000.

Geological mapping at a scale of 1 : 50 000 has been carried out in Lithuania since 1966. Integrated hydrogeological and engineering-geological mapping was carried out for land-use reclamation purposes in northern Lithuania, while this mapping was more limited in southern and eastern Lithuania and was linked mainly to mineral resources forecasting. A set of geological maps at a scale of 1 : 50 000 (i.e. Quaternary, geomorphology, mineral resources etc.) was compiled for 12 855 km^2 or 20% of Lithuanian territory by the beginning of 1993 (Fig. 1).

Another kind of mapping – aerial photo-geological mapping at a scale of 1 : 50 000 has been carried out by means of interpretation of aerial photographs and fieldwork and covers a greater area of Lithuania, more than 30%. Aerial photogeological and aerial photogeo-morphological maps have been compiled without special drilling, geophysical or laboratory investigations of samples, therefore their geological quality is lower than the maps produced by integrated mapping at the same scale, however they provide very significant information about the origin, geological structure and practical and scientific value of landscape.

National programme of geological and environmental mapping

The economic and political situation in Lithuania after the re-establishment of independence dictates new tasks and requirements for

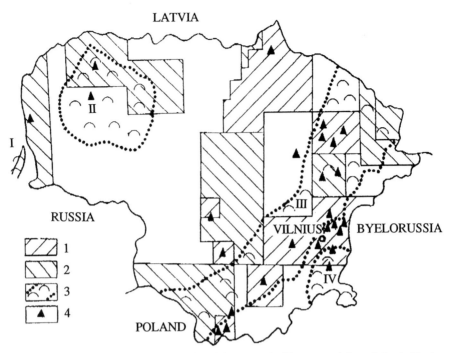

Fig. 1. National geological mapping and distribution of the most valuable areas and sites relating to Quaternary geology and geomorphology in Lithuania. Legend: 1, mapped area at a scale of 1 : 50 000 (at 1 January 1993); 2, area of planned mapping at a scale of 1 : 50 000 in the national programme; 3, areas of particular value of landscape (I, Curonian Spit; II, Zemaitija Insular Height; III, Baltic Marginal Highland; IV, Medininkai Highland); 4, Quaternary geological and geomorphological sites of international importance.

geological mapping. The programme of national geological mapping up to the year 2015 was prepared by the Geological Survey of Lithuania and approved by government in 1992. National geological mapping is regarded as systematic regional geological research carried out with the purpose of obtaining data on the geological structure of the subsurface and providing basic geological information for the substantiation, planning co-ordination and fulfilling of optimal governmental actions in the sphere of land-use and environmental protection. The new aspects of geological mapping that are especially significant for landscape protection and conservation are emphasized in this programme: to investigate the subsurface not only as a source of mineral resources but as a basement for all the ecosystems influencing living standards in society and to pay special attention to the description and forecasting of interaction of geological and technologically based processes (Satkunas 1993a).

Integrated geological mapping at a scale of 1 : 50 000, planned in the national programme, should be carried out first of all in the areas with the greatest technological impact on the landscape, where the most intensive environmental and subsurface pollution exist or may appear, where large factories are operating or being designed, or where undesirable geological processes are anticipated. Data from integrated geological mapping at a scale of 1 : 50 000 form a qualitative basis for various customized maps aimed at landscape protection and other needs.

Special environmental mapping at a scale of 1 : 50 000 is planned in areas with special status, i.e. existing and planned national and regional parks, and in zones with ecologically sound economic activities. Geological–environmental investigations carried out in the course of this kind of mapping should include the main components of landscape: the upper part of the Quaternary strata, relief, mineral resources, groundwater, sediment and soil geochemistry, and should provide the information base for the determination of effective means of environmental protection, regulation and planning of economic activity, and scientific studies of protected areas. The following maps will be compiled within projects of special environmental mapping: Quaternary geological, geomorphological, landscape-geochemical, groundwater level and vulnerability, groundwater chemistry, mineral resources, protection of subsurface, and regulation of economic activities. One of the most important parts of the special environmental mapping is the investigation of the geochemical composition of soil and groundwater in order to establish the natural geochemical background and pollution level. Such information is useful not only for the characterization of the specific area investigated but also for regulation and supervisory actions. It could be used as a standard for the area when mapping and monitoring of industrialized, urbanized territories is carried out.

One project of special environmental mapping has started in 1992 covering southern Lithuania, where the regional parks and other protected areas are more concentrated. This project will be of international significance because the agreement between the Geological Surveys of Lithuania and Poland for cooperative mapping of border territories of both countries was achieved in May 1993. The maps of the joint project entitled 'Belt of Yotvings – Green Lungs' will be designated for the preparation of land-use and landscape conservation projects and will probably serve for the co-ordination of environment protection policies between Lithuania and Poland.

Peculiarities of Quaternary cover and glacial landscape of Lithuania

The relief of Lithuania was formed during Quaternary glaciations and interglacials and is represented mainly by assemblages of landforms of glacial, glaciofluvial and glaciolacustrine genesis (Drobnys *et al.* 1981). Quaternary deposits are distributed everywhere as continuous cover. In the northern part of Lithuania, in the area of prevailing glacial erosion, cover is usually 10–20 m thick. The Quaternary deposits are at their thickest (200–300 m) in the ice-marginal accumulation area in SE Lithuania and the ancient buried valleys. The geological structure and geomorphology of the Quaternary cover are the main mapping objectives of the national programme.

The geological features and patterns of landscape that are characteristic of Lithuania are under protection as geological monuments and reservations, geomorphological reservations, landscape reservations, regional and national parks. By 1993, 47 geomorphological and 62 landscape reservations with a total area under protection of 86 586 ha were in place. Specific geological features are protected in 20 geological reservations and as 164 geological monuments (Lincius 1989). Among them predominate features of Quaternary are – accumulations of erratic boulders and single large boulders, glaciofluvial landforms, outcrops with sediments having stratigraphic significance. All geological features under protection are significant elements of the landscape.

The territory of Lithuania, as well as the whole eastern Baltic area, has served as a key region in solving many topical problems of Quaternary geology, and during recent years has witnessed ever growing interest among western Quaternary researchers (Raukas 1993). Some features of glacial landscape and peculiarities of geological structure in Lithuania are of international scientific significance. Unique landforms like the coastal dunes of the Curonian Spit, accumulations of erratic boulders, the marginal relief of the Baltic Highland and the limit of Weichselian Glaciation can be observed and studied. A very important feature of the territory under consideration is the occurrence of sediments of all glaciations known in eastern Europe (Raukas 1993). Stratotypes of all the interglacials (Merkine (Eemian), Butenai (Holsteinian), Snaigupele (Middle Saalian), Turgeliai (Middle Elsterian)) and of the Lower Pleistocene established in the regional stratigraphic scheme of the East Baltic area are located in Lithuania and serve for recognizing and correlating corresponding stratigraphic units in Latvia, Estonia and other adjacent countries. All stratotypes and key sites comprising Quaternary sediments must be considered as a heritage of international significance and preserved as such. Unfortunately not all of them currently have the status of a geological monument.

Mapping of glacial landscape patterns and stratigraphic type sites

The exploration of all stratigraphic type sites and characteristic patterns of glacial landscape, geological reservations and monuments is one of the obligatory scientific tasks of national geological mapping. It usually gives invaluable data for solving problems of stratigraphy and palaeogeography of specific areas. On the other hand, any geological monument explored with mapping methods increases in importance as a key site of great scientific value. Thus, in the context of mapping at a scale of 1:50 000, in the Utena area, the stratotype site of the Butenai

(Holsteinian) Interglacial was examined in detail using special boreholes and laboratory investigation (Kondratiene & Bitinas 1989). In the same way the Indubakiai geological reservation of erratic boulders was investigated and their origin in connection with flat-top hills was reconstructed. In addition, the prominent geological monuments of Quaternary age and patterns of glacial, aeolian and fluvial relief in the environs of Vilnius were investigated as part of a project of integrated mapping between 1987 and 1990. The limit of maximum advance of Weichselian glaciation, and abundant kettles with lacustrine layers of Eemian Interglacial age distributed in the glacial topography of the Medininkai Highland of Saalian age (Fig. 1) were detected and examined (Satkunas 1993*b*). Recommendations for protection and conservation of the most valuable sites and patterns of landscape of the Vilnius area were presented together with other results of mapping.

Experience available at the Geological Survey of Lithuania demonstrates that data of geological or special environmental mapping at a scale of 1:50 000 are necessary for estimating the scientific value of landscape and subsurface of the area under consideration. The information presented in Quaternary geology, geomorphological and special customized maps should be taken into account by all decision-makers – professional conservationists, environmentalists and land-use managers.

References

DROBNYS, A. *et al.* (eds). 1981. *Atlas of Lithuania.* Cartography, Moscow. [In Lithuanian].

KONDRATIENE, O. & BITINAS, A. 1989. Stratigraphy and palaeogeography of the Middle and Lower Pleistocene of the Southern Baltics in view of palaeobotanical data. *In*: YANSHIN, A. L. (ed.) *Quaternary age: palaeontology and archaeology.* Shtiintsa, Kishinev, 103–110. [In Russian].

LINCIUS, A. 1989. *Geological reservations of Lithuania.* Science, Vilnius. [In Lithuanian].

RAUKAS, A. (ed.) 1993. *Pleistocene stratigraphy, ice marginal formations and deglaciation of the Baltic states.* Estonian Academy of Sciences, Tallinn.

SATKUNAS, J. 1993*a*. Geological Mapping – Basis of Geological Information. *Journal of the Geological Society of Lithuania*, **1**(9), 11–14. [In Lithuanian].

—— 1993*b*. Stratigraphy of the Upper Pleistocene in the marginal zone of last glaciation, Eastern Lithuania. *In*: GRIGELIS, A., JANKAUSKAS, T.-R. & MERTINIENE, R. (eds) *Abstracts of the second Baltic Stratigraphic Conference.* The Geological Society of Lithuania, Vilnius.

Keynote address

Geological conservation in New Zealand: options in a rapidly eroding environment

JOHN S. BUCKERIDGE
Department of Civil and Environmental Engineering, UNITEC, Private Bag 92025, Auckland, New Zealand

Abstract: New Zealand straddles two major, and very active tectonic plates; resultant crustal dynamics have produced a microcosm of Earth geology, encapsulating diverse geological environments. These include active faulting and volcanism, modified by aeolian, fluvial, glacial and coastal processes. The archipelagic nature of New Zealand ensures that no part is far from the sea, with many of the uniquely important geological sections exposed to active erosion. The coastal strip has a relatively dense population and is also the New Zealander's most frequented playground: in many locations conservationists are faced not only with the need to protect against natural erosion, but also against humankind. The high public profile environmental issues enjoyed in New Zealand, coupled with recent parliamentary acts in resource management and conservation, set the stage that led to the development of conservation management strategies. The intent of these strategies is to ensure the conservation of New Zealand's historic and natural resources. In 1986, Earth scientists embarked upon the task of preparing a national inventory of important geological sites and landforms in New Zealand. This was co-ordinated by a group of six Earth scientists, but with input from more than 300 nationally. By March 1993, this inventory included 3681 sites, 320 of which were considered of international significance. Of these, a greater percentage lie in the vulnerable coastal environment, and 27% are volcanic landforms. This paper discusses the legislative and planning framework within which the New Zealand Geopreservation Inventory will contribute, and analyses some of the more at risk, and unique characters, of the New Zealand landscape.

The geological setting of New Zealand

The North and South Islands of New Zealand straddle two converging lithospheric plates: the Pacific and the Australian. Movement between these two plates is significant, with the current shift pole at 62°S 174.3°E, having a rotation of $1.27° \pm 0.05°$ per million years (Walcott 1984). The present geology reflects this situation with orogenesis in both islands and with the North Island further characterized by active volcanism (Fig. 1). A westward-dipping subduction zone extends below the North Island and is manifest, 80 km above, as the Taupo Volcanic Zone (Reyners 1980).

The Taupo Volcanic Zone is the largest active volcanic belt in New Zealand. It extends from northeast of the mainland, passes through the active volcano White Island, and the coast near Whakatane, to terminate some 180 km inland at the Tongariro Volcanic Centre. It is comprised of an active andesitic arc to the southeast (incorporating a total of 19 basalt/medium-K andesite/dacite vents), bounded to the northwest by a marginal basin (characterized by silicic volcanism). The marginal basin includes the spectacular geothermal fields of the Rotorua area, and the geothermal power-generating facilities north of Taupo.

Recent volcanism also occurs to the west (Egmont) and to the north (Auckland, Hauraki and Northland). The Auckland volcanic field extends over $140 \, km^2$, and is made up of monogenetic basaltic volcanoes that have erupted some $7 \, km^3$ of rock (Heming & Barnet 1986). It is however, relatively small when compared to other areas of Cenozoic volcanism in New Zealand, but it is important in that it is vulnerable to extensive quarrying and to a lesser degree, coastal erosion.

Tectonism of the southern part of the South Island is dominated by an eastward-dipping subduction zone. Active earth deformation between the two lithospheric plates occurs along the reverse-dextral Alpine Fault, with concurrent rapid uplift in the mountainous Southern Alps. It is in the uplifted and glaciated southwest of the South Island that some of New Zealand's most spectacular scenery is found. Clearly, in

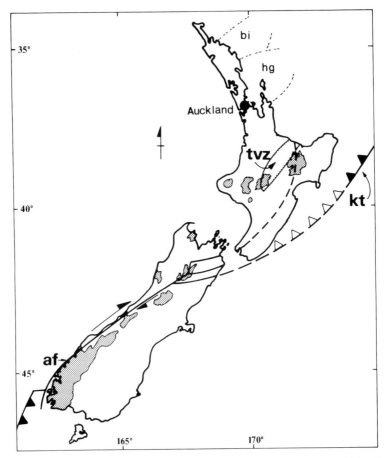

Fig. 1. New Zealand: sketch map showing main structural elements of the geology. ap: Alpine Fault; kt: Kermadec Trench; tvz: Taupo Volcanic Zone; bi: Bay of Islands Maritime Park; hg: Hauraki Gulf Maritime Park. National park areas are stippled.

areas where rainfall exceeds 4000 mm per annum, these glaciated valleys and fjords will be susceptible to very high erosion rates; further, because of their physiography and disposition, they are also excellent sites for hydroelectric schemes.

The tectonic placement of New Zealand ensures that it represents a microcosm of Earth geology, encapsulating active faulting and volcanism, modified by aeolian, fluvial, glacial and coastal processes. The archipelagic nature and high relief of much of the land ensures that both a range of climatic systems will operate, and that no part is distant from the sea. As such, many of the uniquely important geological sections are exposed to active erosion.

The coastal strip also has the greatest population density and is also the most frequented area for recreation. In many locations conservationists are faced not only with the need to protect from natural erosion, but also from humankind.

The New Zealand Geopreservation Inventory

In 1887, New Zealand's first national park was created within the Tongariro Volcanic Centre (now also a World Heritage Site). Although reasons for the creation of this park were perhaps less than altruistic (it was seen as an expedient means of solving a land ownership dispute), its creation paved the way to the present, at which time approximately 20% of the land area (or 5×10^6 ha), are set aside as parks and reserves under Crown management. There are now twelve national parks, generally restricted to the mountainous uplands of both main islands. The primary aim in the setting up of these parks was to preserve the rich biological heritage contained therein, but they have also served to conserve some of the country's best Earth science features, including glaciers, waterfalls, fjords in the South Island and four active volcanoes in North Island.

Until the early 1980s, action to conserve sites having specific geologic interest was uncoordinated, with lobbying for protection of specific locations carried out on an *ad hoc* basis. Even so, a number of very important sites gained protection. These included the raised beaches at Turakirae (near Wellington), the Muriwai pillow lavas, and more recently the Wiri lava cave in Auckland.

In 1983, to maximize the efforts of Earth science conservationists, the Geological Society of New Zealand promulgated the concept of compiling a list of the nation's rich and diverse Earth science heritage. From 1986 to 1993, under the guidance of Dr Bruce Hayward, and with financial support from both the Lotteries Board (Ministry of Internal Affairs) and Department of Conservation, the Joint Earth Sciences Working Group on Geopreservation recorded over 2500 Sites of Special Scientific Interest (SSSIs) (Table 1). The aim of the Geopreservation Inventory is:

> ...to identify and list information about all the internationally, nationally and many of the regionally important earth science sites throughout New Zealand, irrespective of their current protected status. This will allow sites to be confidently listed in management plans and for a programme of action to protect high priority sites to be devised.
>
> (Kenny & Hayward 1993)

In addition to this, the Working Group developed a 'mission statement' emphasizing the educational importance of Earth science sites:

> ...Preservation of these features and processes is essential for the continued prosecution of earth science education and research both now and in the future. The sites should be managed for public presentation and education where this is compatible with their long term preservation.
>
> (Bruce Hayward, pers. comm.)

Although the inventory includes both geological and pedological sites, the selection for each was based on very different criteria: geological sites were compiled initially on the basis of their categories, i.e. fossil sites, landforms, Earth

Table 1. *Number of sites recorded in each of the inventory categories as at March, 1993*

Site categories	Importance rating			Total
	A	B	C	
1. Landforms	23	180	365	568
2. Earth deformation	10	31	187	228
3. Caves and karst	13	34	56	103
4. Geothermal	17	34	95	146
5. Quaternary volcanoes*	4	50	102	156
6. Quaternary volcanoes†	65	138	142	345
7. Igneous geology	8	38	163	209
8. Sedimentary geology	13	111	136	260
9. Fossil	40	208	70	318
10. Metamorphic geology	10	39	36	85
11. Minerals	20	69	47	136
12. Structural geology	10	42	60	112
13. Historic	9	82	121	212
14. South Island soils	48	170	225	443
15. North Island soils	30	75	255	360
Totals	320	1301	2060	3681

* Refers to volcanoes of the Taupo Volcanic Zone.
† Refers to northern volcanoes.

Importance ratings are A: international, B: national, C: regional. The actual number of locations nationally is 2575 (see also Table 2). The 3681 sites listed identifies that some localities are registered in more than one site category (from Kenny & Hayward 1993).

Table 2. *The vulnerability of sites listed on the New Zealand Geopreservation Inventory*

Class categories	Total
1	110
2	566
3	1852
4	37
5	10
Total	2575

Note: 1: Sites highly vulnerable to human modification; 2: sites moderately vulnerable to human modification; 3: sites unlikely to be damaged by humans; 4: sites that could be improved by human activity; 5: sites that have been destroyed.

deformation features, geothermal fields and features, Quaternary volcanoes, caves and karst sites, igneous geology sites, geologically related historic sites, structural geology sites, metamorphic geology sites, sedimentary geology sites and mineral sites. The next phase was to define these on a regional basis, and much of this has now been published, e.g. Kenny & Hayward 1993.

Pedological sites, however, posed a very different problem. Whilst many of the excellent geological sites had been preserved in the upland national parks and reserves, these regions excluded a significant proportion of the regional soils, particularly those of the very modified lowland. Pedological sites then, were selected on the basis of existing reserves, and as such they represent a very selective remnant of New Zealand's unmodified soils.

A vulnerability rating for each site (or location) is given on the inventory. The locations that are of concern are those in categories 1 and 2 although categories 4 and 5 are fully recorded in each of the Regional Inventories. Destroyed sites include those that have been demolished by human activities (e.g. quarrying of the Auckland scoria cones) and those lost by natural activities, such as the 'Pink and White Terraces' of Lake Tarawera, destroyed in the 1886 eruption.

The vulnerability rating and/or value placed on almost all sites will be subjective. In order for the inventory to gain popular support and acceptance, a degree of balance in the selection of these 'protected' sites must be demonstrated. The more than 2000 locations identified to date, and the restriction in human activity these imply, may be seen by some as an encroachment on individual liberty. Earth scientists are now faced with the task of defending why particular sites are worthy of conservation, for ultimately every exposure has a degree of uniqueness. Further, once a site is identified as either valuable, or vulnerable, it immediately becomes a potential target for disenfranchised members of the community – with a resultant increase in vulnerability. A typical case exists on Motutapu Island, near Auckland, where a unique and fascinating faunal assemblage is exposed on the shore platform. The exposed fossils are large and spectacular (therefore desirable!), but they are also easily accessible and therefore vulnerable to the sea, and to humans. The conundrum then, is to determine a management plan that appropriately incorporates a balance between site publicity and anonymity, in the light of the knowledge that inevitably much of the material will be lost through erosion.

The Legislative Framework

The Conservation Act (1987) has stipulated that all Department of Conservation Conservancies, and all Department of Conservation managed land, must develop a Conservation Management Strategy for the next 10 years. Conservation Management Strategies include the 'management for conservation purposes', of all land, and of all other natural and historic resources, and the 'promotion of the benefits' to present and future generations of the conservation of said resources. In addition to this, the Resource Management Act (1991) requires that regional councils prepare Regional Coastal Plans. These mandatory plans will provide policy to guide the regional and local authorities in their day-to-day management of coastal activities. The draft Coastal Policy Statement (Department of Conservation 1992) proposes seven principles for the sustainable management of New Zealand's coastal environment.

These are to ensure that the:

- coastal environment is available for sustainable use and development;
- values of the coastal environment shall be protected;
- management of the coastal environment shall be carried out to provide for the social, economic and cultural wellbeing of people and communities, and for their health and safety;
- use, development and protection of the coastal environment shall sustain the reasonably foreseeable needs of future generations;
- use, development, and protection of the coastal environment shall safeguard its life-supporting capacity;

- people shall avoid, remedy or mitigate the adverse effects of their activities on the coastal environment;
- management of the coastal environment under the Resource Management Act 1991 will be shared between the Minister of Conservation and local authorities.

Although somewhat subjective, the intent of these proposals is appropriate, but the means of 'enforcement' are perhaps less clear. The draft goes some way towards effecting this by devolving power to local authorities. It is clear, however, that in order to effectively manage the coastal environment, it is necessary to educate the public on the value of the environment. The aims and mission of the New Zealand Geopreservation Group certainly go some distance towards achieving this with the geological heritage.

The intent of the Resource Management Act has been to preclude passive balance in environmental management, i.e. to balance human needs and aspirations against the detrimental effects that these actions may have on the environment. The Act, however, in outlining conditions under which enforcement procedures such as abatement notices may be served, infers that it is the duty of individuals 'to avoid, remedy or mitigate any adverse effects...'; this concept of duty is novel in law, as it infers a moral obligation (Buckeridge 1992). It also provides a clear mandate for further educational initiatives in environmental management.

The special character of New Zealand

New Zealand's glacial, volcanic and geothermal inheritance has provided some of its most spectacular scenery, facilitating the capture of essential tourist revenue. The apparent omnipotence of these features, however, masks their real vulnerability. This has been most effectively demonstrated in geothermal areas like Waiotapu and Whakarewarewa, near Rotorua, where unregulated steam extraction has decimated some of the world's most spectacular geothermal activity. In the 1800s, more than 130 geysers erupted regularly in the central part of the North Island. Today, this number has dropped to less than 15 (Hayward 1989).

Whakarewarewa's Geyser Flat included 7 large geysers in what was previously one of the highest concentrations of large geysers in the world. The first domestic bores for steam recovery were drilled in the Rotorua area in the early 1930s. Since this time, more than 800 domestic and commercial wells have been drilled, and into the middle 1980s, 550 of these were still extracting. Further, the average diameter and depth of bores has doubled over the last few decades, as higher water temperatures became more elusive. Of the seven large geysers, the largest, Wairoa, has not erupted since the 1920s, the highest, Waikite, has not erupted since 1965 and Papakura has been inactive since 1979. The remaining geysers, including the famed Pohutu (Fig. 2) erupted less and less frequently, threatening the closure of the entire complex. In 1987 government legislation was passed, not only to halt further drilling, but to seal off all existing bores within 1.5 km of the geysers. Following a slight increase in bore pressures, it is pleasing to note that the decline in activity has stabilized. But the saga is yet unresolved, as those who had enjoyed the luxury of cheap heating for decades are appealing against the new laws.

Auckland, the largest city in New Zealand is built on an isthmus peppered with volcanic

Fig. 2. Whakarewarewa geothermal area, showing an eruptive phase of Pohutu Geyser, with the smaller Prince of Wales Feathers Geyser erupting to the left. Taken in December, 1976 by the author.

cones. These cones provided an ideal source for construction materials, and since the 1800s have been used extensively for buildings, roading and fill. All of the 48 original cones have suffered damage, with more than 50% being covered over, or like those at the Three Kings cones, Maungataketake and Mount Gabriel, quarried out (Fig. 3). Fortunately the survival of many of the remaining cones, such as One Tree Hill, Rangitoto and Brown's Island has been enhanced by their placement within reserves (Fig. 4).

The need to balance conservation and development is heard no more strongly than in Auckland, where in 1989, 22% of New Zealand's total aggregate production was quarried (Kermode 1992). Mount Wellington presently provides about 800 000 tonnes of aggregate per annum, and at this rate, the quarry will be exhausted by 2003. The ultimate fate of the quarry is still uncertain, although there is potential for bioconservation: the Auckland Regional Authority has offered NZ$20 000 000 for the site, to be used initially as a landfill; plans exist for an end use that may include cultural amenities, botanic gardens and reserves. Although quarrying is clearly not compatible with geoconservation, the end use of sites may still have considerable conservation value (Buckeridge 1993).

Dramatically glaciated South Island's west coast attracts many tourists, and helps complement New Zealand's clean, green and natural image. Although sensitive to human interaction, the area is at even greater risk from the environment, where fluvial processes have the ability to not only mask the glacial features, but ultimately to destroy them. There is a need then, to put geological processes into perspective, defining whether humankind should interfere in natural degradation processes in order to preserve what is a unique, but vanishing, landscape.

Discussion

The New Zealand Geopreservation Inventory has now reached an agreeable level of sophistication. It will be instrumental in both preserving parts of our Earth heritage that are at risk, and in providing an educational window to ensure that Earth sciences are more fully appreciated by the general public. The project as a whole, however, is an ongoing one. New Zealand presently lacks an Earth scientist whose appointment is specifically directed at Earth science conservation. We have come a long way, but it has been on the basis of voluntary and part-time labour. It is certainly a matter of concern that no professional geoscientist has yet been appointed within the Department of Conservation.

Fig. 3. Basalt scoria quarry at Maungataketake, south Auckland, showing almost complete loss of the cone. Aerial view, taken in June, 1988 by the author.

Fig. 4. The basaltic scoria cones of Brown's Island, near Auckland. Conservation is partly assured by inclusion of the island within the Hauraki Gulf Maritime Park. The site has been threatened however, by heavy rodent infestation. Taken in June, 1988 by the author.

The author wishes to thank Dr Bruce W. Hayward, New Zealand Geopreservation Society, Auckland Institute and Museum for valued and informative discussion. Dr Jill Kenny, also of the Auckland Institute and Museum provided data on the vulnerability of sites listed on the Geopreservation Inventory. Dr Hamish Campbell, Institute of Geological and Nuclear Sciences, Lower Hutt, critically read the manuscript.

References

BUCKERIDGE, J. S. 1992. Morality and Resource Management: what are the issues? *Proceedings of the New Zealand Institute of Surveyors. 1992 Annual Conference: Conservation and Development, Partnership or Paradox?* Invercargill, New Zealand, 43–49.

—— 1993. Sustainability – an Australian perspective of resource management. *Proceedings of the New Zealand Institute of Professional Engineers*, February, 25–30.

CONSERVATION ACT. 1987. New Zealand Government Printer, Wellington.

DEPARTMENT OF CONSERVATION. 1992. *Draft New Zealand Coastal Policy Statement*. Wellington, New Zealand.

HAYWARD, B. W. 1989. Earth Science Conservation in New Zealand. *Earth Science Conservation*, **26**, 4–6.

HEMING, R. F. & BARNET, P. R. 1986. The Petrology and Petrochemistry of the Auckland Volcanic Field. *In*: Late Cenozoic Volcanism of New Zealand. *Bulletin of the Royal Society of New Zealand*, **23**, 64–75.

KENNY, J. A. & HAYWARD, B. W. (eds) 1993. *Inventory of Important Sites and Landforms in the Auckland Region*. Geological Society of New Zealand Miscellaneous Publication, **68**, 1–64.

KERMODE, L. O. 1992. Geology of the Auckland urban area. Scale 1:50000. *Institute of Geological and Nuclear Sciences geological map*, **2**, 1–63. Institute of Geological and Nuclear Sciences Ltd., Lower Hutt, New Zealand.

RESOURCE MANAGEMENT ACT. 1991. New Zealand Government Printer, Wellington.

REYNERS, M. 1980. A microearthquake study of the plate boundary, North Island, New Zealand. *Geophysical Journal of the Royal Astronomical Society*, **63**, 1–22.

WALCOTT, R. I. 1984. Reconstructions of the New Zealand region for the Neogene. *Palaeogeography, Palaeoclimatology, Palaeoecology*, **46**, 217–231.

Nature conservation and beach management: a case study of Köycegiz–Dalyan Specially Protected Area

ERGÜN ERGANI & PERIHAN KAMIS

The Authority for the Protection of Special Areas (APSA), Koza Sok. No. 32 G.O.P. Ankara, Turkey.

The Authority for the Protection of Special Areas (APSA) was established in 1989 for the environmental management of Specially Protected Areas (SPAs) of Turkey to implement the international protocol concerning Mediterranean Specially Protected Areas of the Barcelona Convention. Pursuant to this convention, 12 SPAs were declared in the southwestern part of Turkey. SPAs are sensitive ecosystems with archaeological, cultural and natural value.

Among the 12 SPAs, the Dalyan area is the most characteristic wetland ecosystem, the current morphology dating back to the Quaternary. Here, a lagoon and canals link the Köycegiz lake with the Mediterranean. This whole area is very important biologically: four different species of marine turtle and 125 different bird species nest here. There are also unique endemic plants and vegetation. Besides this, ancient ruins of the Karia civilization attach additional importance to the area. Köycegiz and Dalyan are the two major settlements, and agriculture, fishing and tourism are the main local economic activities.

Development in the area is controlled and orientated by the rules and regulations contained in the land-use master plan prepared by APSA. Building and population densities, site coverages, floor area ratios and the heights of buildings are defined in the regulations of the plan. Applications violating plan regulations are stopped and legal enforcement is applied. The preservation of traditional architectural patterns is one of the objectives of the plan.

Urgent conservation measures are enforced through this master plan. At the same time, researchers are compiling ecological inventories and mapping habitats. In accordance with the results of the scientific research on flora and fauna, the master plan is checked and revised, if necessary. Having determined ecosystem properties qualitatively and quantitatively, the development needs of all sectors are addressed with a conservation perspective. Rules and regulations are re-set to limit and orientate the activities in the zones of the plan.

Thus, in the example of the Köycegiz–Dalyan SPA, a 5 km sandy beach separates the sea from the lagoon and, as a result of regulations in the APSA plan, a touristic development here, involving the construction of hotels with 3000 bed capacity, was stopped. Furthermore, tourists are allowed to visit the beach in the daytime only, beyond the turtle nesting zone which is marked by wooden poles. The beach is also cleared daily, and portable sanitary facilities with leak-proof septic tanks have been put into service. Wooden landing platforms have been erected in several localities to regulate boat traffic; and information signs, brochures and 24-hour beach guards guide visitors for effective conservation and controlled tourism within the SPA.

Environmental geology maps for national parks and geomorphological reserves in Lithuania

JURGIS VALIŪNAS
Institute of Geology, Sevcenkos 13, Vilnius 2000, Lithuania

Abstract: Geoscience mapping for planning tasks has been undertaken at various scales in Lithuania since 1989 (from 1:200 000 to 1:10 000). The geoscientific set at 1:50 000 includes, as a rule, three maps: a geological–geotechnical map, a geological potential map, and a map of land-use recommendations. The geological–geotechnical map presents information about stratigraphical, genetic and lithological types of sediments, about the character of soil from an engineering point of view, and about geodynamic processes. The geological potential map shows the basic characteristics of mineral and groundwater resources and vulnerability to pollution. The map of land-use recommendations presents propositions for efficient management of territory from a geoscientific point of view. The 1:50 000 scale is used for environmental geology mapping of administrative districts. Because the areas of the national parks are quite large (up to 300 km^2), this scale is used for their mapping too. It is very important that decision-makers in local administrations and planners in national parks are able to use the same information.

Geomorphological reserves, however, occupy quite small areas (average 4.13 km^2), so the best scale for environmental geology mapping here is 1:10 000. Their set includes four maps: a map of actual land-use, a geological map, a geomorphological map, and an eco-geological map. This set of maps is compiled during the proposal stage of a reserve. Environmental geology maps are used in national parks to define their protection zones and reserves boundaries, to create rational management programmes and to select the most valuable features. To date, map sets for two national parks and four geomorphological reserves have been compiled, and a set for another national park is in preparation.

Until 1989, there were only sporadic attempts to use geoscientific information for nature protection and land-use problem solving. Then, the Environmental Geology Department was founded in the Institute of Geology and systematic environmental geology mapping began in Lithuania.

Environmental geology mapping is undertaken at various scales (from 1:200 000 to 1:10 000), but the main experience has been gained in compiling maps for administrative districts at 1:50 000. As a result, sets of three maps have generally been compiled.

The geological–geotechnical map is compiled on the base of a Quaternary map. This map contains information about stratigraphical, genetic and lithological types of sediments, about the character of soil from an engineering point of view, and about geodynamic processes. The second map is the geological potential map. Geological potential is a component of the potential of the whole natural environment. This definition includes objects (e.g. deposits, anomalies), conditions (e.g. soils, structure), phenomena (e.g. earthquakes) and processes (e.g. erosion, karstification) (Baltrūnas *et al.* 1993). Factors which encourage the development of society could be called positive geological potential. Those which limit development, negatively influencing the quality of the human living environment, could be called negative geological potential – these can be natural or human-induced. Such estimation is relative, as mostly it depends on scientific knowledge about the environment, and the technical and economical possibilities of society. The geological potential map presents information about mineral deposits, groundwater resources, structure of the Quaternary cover, zones of neotectonic activity, etc.

The third map is a map of land-use recommendations. On the base of the two previous maps and other geological data, recommendations for the protection and rational utilization of mineral and groundwater resources and efficient land-use management from a geoscientific point of view, are proposed.

Environmental geology maps for national parks

The Aukštaitija National Park was founded in 1974 and, for a long time, it was the only one in Lithuania. However, during the last few years, more have been founded and today Lithuania has five, the areas of which vary from 80–550 km² (Fig. 1). The aim of national parks is to preserve, explore and exhibit the variety of Lithuania's natural and cultural landscape, and to save the most valuable natural and cultural inheritance monuments or their complexes. The parks are founded, as a rule, on the base of existing conservation areas. The pattern of protection combines recreation as the main land-use with, in sustainable forms, forestry, agriculture and other land-uses.

The major management problem in Lithuania's national parks is that it depends on local government policy. Environmental geology maps will be less useful if they have been compiled for the national parks only. It is very important, therefore, that decision-makers in local administrations and planners in national parks have the same information. For these reasons, environmental geology maps for national parks are compiled at the same 1:50 000 scale as those for administrative districts.

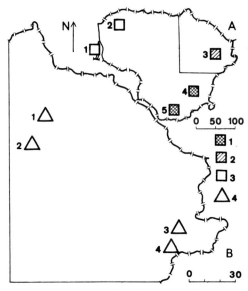

Fig. 1. Location map. (A) National parks: 1 – National park of Courtland Spit; 2 – National park of Žemaitija; 3 – National park of Aukštaitija; 4 – Trakai Historical National Park; 5 – National Park of Dzūkija. 1–3 based on environmental geology maps: 1 – compiled; 2 – in work; 3 – uncompiled.
(B) Investigated geomorphological reserves: 1 – Prusagale; 2 – Lévuo; 3 – Nevaišiai; 4 – Sirveta (based on geomorphological reserves). Scale in kilometres.

Trakai Historical National Park

The Trakai Historical National Park was founded in 1990 in the vicinity of the old Lithuanian capital. This area is worth protecting for its historical past and for its unique scenery – the Trakai Lakes (Fig. 2).

Environmental geology maps were compiled in 1989 for the Trakai administrative district, in which the park is located (Baltrūnas *et al.* 1992). For this reason, geological information was used in the planning stage, especially for recommendation of the protection zone and land use there. The main emphasis was on protection of the Trakai Lakes, for which more detailed special hydrogeological investigations were carried out later (Eitmana /ičius 1992).

The main threats to the natural state of the lakes are related to possible pollution and to lowering of the water level. Water balance is largely maintained by groundwater recharge from the south. The groundwater occurs in fluvioglacial deposits which are permeable and vulnerable to pollution. For this reason, the possibility of pollution from groundwater was investigated through special detailed hydrogeological research. It was found that pollutants are transported mainly with surface water and sewage from Trakai town. Groundwater is polluted only in some local areas near livestock farms and recreation sites which are very close to the lakes. Human activity must be regulated and livestock farms moved further from the shore zone.

More serious is the possibility of the water level in the lakes being lowered because of human impact. A water intake to satisfy the drinking-water needs of Trakai town (5000 m³ per day) is being proposed. Development of a shallow intermorainic aquifer is being planned, and aquifer level lowering will have an impact on water level in the lakes.

There are also large gravel resources to the south of the lakes. When the Serapiniskes I gravel deposit (which had about 15×10^6 m³ of reserves), was reaching exhaustion, the Miskiniai gravel deposit (c. 6×10^6 m³) was planned to be developed next. This deposit, however, is situated in the groundwater-feeding area of the lakes. The forecasting estimate showed that, with full development of the groundwater intake (165 000 m³ per day) and of the Miskiniai gravel deposit, the level of the Trakai Lakes could be lowered by 24–41 cm (Eitmanavičius 1992). To counter this, the Serapiniskes II gravel deposit was recommended for development. This lies outside the groundwater-feeding area and also has

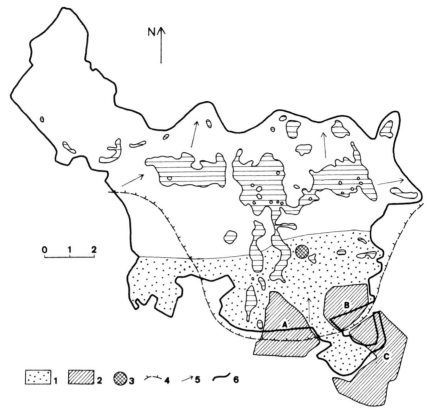

Fig. 2. Trakai Historical National Park. (1) unconfined groundwater layer; (2) gravel deposits: A – Miškiniai, B – Serapiniškès I, C – Serapiniškès II; (3) water intake; (4) boundary of Trakai Lakes groundwater-feeding area; (5) direction of groundwater movement; (6) boundary of national park.

larger reserves than the Miskiniai deposit ($c.\ 23 \times 10^6\ m^3$).

Dzūkija National Park

The Dzūkija National Park displays a landscape of plains, valleys, rivers, forests, and mainland dunes (Fig. 3). The environmental geology maps for the park and the Varena administrative district, in which the park is situated, were compiled at the same time. The mapping was undertaken after the foundation of the park.

The most important landscape element of the park is the mainland dunes, which nowadays are covered by forest. The areas of dunes are shown on the geological–geotechnical map. In these areas, deforestation is recommended only on a small scale, combined with reforestation. Otherwise there is a danger of renewed deflation, soil degradation etc. For more reliable protection of dunes, four geomorphological reserves in the park are proposed. Also in the park, four more geomorphological reserves for protecting erosional and glacial relief complexes are projected (Kavaliauskas *et al.* 1991).

Rivers, fed mainly by groundwater, are another important element needing protection. Fluvioglacial sandy deposits occur over almost the whole area of the park. For this reason, great attention has been paid to the vulnerability of groundwater to pollution and to the directions of its movement. This information is useful in determining areas where the danger of pollution is greatest and where activity leading to the risk of chemical pollution must be prohibited.

Rapid stream flow erodes river banks which, in some places, reach a height of 15–20 m. Some villages and roads are very close to these active precipices. Segments of rivers where active bank erosion takes place are presented on the map. Building, road construction and deforestation must be carried out very carefully in these places.

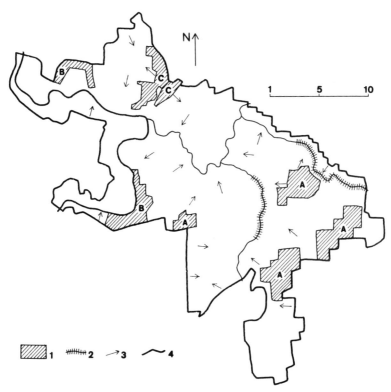

Fig. 3. Dzūkija National Park. (1) Proposed zones for geomorphological areas: A – mainland dunes protection; B – erosion features protection; C – glacial mezoforms protection; (2) active erosion of rivers banks; (3) direction of groundwater movement; (4) boundary of national park.

Environmental geology maps for geomorphological reserves

The geomorphological reserves in Lithuania are founded in order to conserve geomorphological localities, features and complexes for research, education and training. In these areas, mechanically changing or destroying protected relief forms and changing natural watersheds, is prohibited. In hilly terrains, land-users must carry out anti-erosion measures (Sasnauskas and Raščius 1989).

Nowadays, there are 16 geomorphological reserves and 30 more are planned for the near future. Their areas vary from 0.5–10.61 km^2 (an average of 4.13 km^2). The most suitable scale for environmental mapping is 1 : 10 000. The set used includes four maps: a geological map, a geomorphological map, a map of actual land-use, and an eco-geological map. A scheme for defining reserve boundaries is also included. So far, map sets for four geomorphological reserves have been compiled (Valiūnas & Šinkūnas 1991) (Fig. 1).

The aim of eco-geological map compilation is to present cartographically the ecological peculiarities of any interaction between the upper part of the Earth's crust and humankind. Because the map is compiled for geomorphological reserves, more attention is paid to factors which have evident impact on the stability of relief complexes. Geological and geomorphological maps, and archive and field investigations data, are used for eco-geological map compilation.

Evidently, relief is changed during the extraction of raw materials. Quarries and pits, depending on their actual state and damage to the reserve, are divided into active, reclaimed and non-reclaimed, and are shown on the map. Also, the undeveloped mineral deposits are presented on the map; they might be developed in the future and the possible impact on the reserve must be estimated.

Erosion greatly influences the development of relief, especially in hilly terrains. The main factors conditioning the scale and type of erosion are vegetation, relief morphology, geological substrata, soil type, and surface water flow. Steep slopes and geological substrata are shown on the map. The maps make possible land-use recommendations on the most vulnerable areas to avoid erosion activation. The eco-geological map presents necessary information about actual geological processes that influence morphology: landslides, erosion, karstification, deflation, etc.

Besides the factors related to relief complex protection, other environmental geology elements are presented on the map: groundwater, vulnerability to pollution, technogenic load, amongst others.

The concrete recommendations for reserve management and land reclamation, and the scheme of reserve boundary definition, are given on the base set of maps. The main goal of boundary definition is to include the whole geomorphological complex in the protected area, especially those areas where danger of natural or human impact degradation is most serious.

Conclusions

Environmental geology mapping is important in the management of conservation areas. Geoscientific research in the area, in conjunction with other research, makes it possible to determine optimal boundaries of conservation areas and their protection zones.

The second important issue which environmental geology maps help to resolve is the optimization of land-use, combining recreation, scientific research, education and other activities. Recommendations for agriculture on steep slopes, anti-erosion measures, ways of protecting groundwater and minimizing damage from mineral extraction and similar processes are given, and the most interesting objects for research and education are determined.

Environmental geology maps are not the final stage of geological research in conservation areas, especially in those as large as national parks. They stress the need for more specialized research covering the whole area, or in the most interesting or vulnerable localities.

References

BALTRŪNAS, V., ŠLIAUPA, A., VALIŪNAS, J. *et al.* 1992. Compilation of large scale environmental geology maps for rational development and planning in Lithuania. *Engineering Geology*, **4**, 77–86. [In Russian].

——, ——, —— *et al.* 1993. Natural Environmental Potential of Lithuania and the problem of its evaluation (Expert report for Lithuanian Academy of Sciences). [In Lithuanian].

EITMANAVIČIUS, S. 1992. The Trakai National Park: hydrogeology and ecology. *Geologijos akiračiai*, **4**, 46–50. [In Lithuanian].

KAVALIAUSKAS, P., BRUKAS, A., DUMBLIAUSKIENE, M. *et al.* 1991. National Park of Dzūkija. *Mūsu gamta*, **11**, 11–15. [In Lithuanian].

SASNAUSKAS, I. & RAŠČIUS, G. (eds) 1989. *Conservation territories and their management*. Environment Protection Committee, Vilnius. [In Lithuanian].

VALIŪNAS, J. & ŠINKŪNAS, P. 1991. Estimation of ecogeological state of protected areas. *In*: ABOLTINS, O., DANILANS, I. & ZELCS, V. (eds) *Glacial lythomorphogenesis, palaeogeography of Quaternary age, recent egzogenic processes and their geoecological aspects*. Latvian University, Riga, 11–12. [In Russian].

Country sports as an element in landscape conservation

DAVID B. BRAGG

Department of Science Education, University of Warwick, Coventry CV4 7AL, UK

Abstract: In British national parks, private landowners own the largest percentage area of land within the park boundaries. The highest percentage landownership the park authorities (i.e. the State) achieve in any one park is 7%, which is in the Brecon Beacons National Park. This is in contrast to many other countries, where national park land is owned wholly by the State, thereby making the park authorities of other countries more powerful than our own.

Over the last 50 years, many large country estates have been broken up, to the detriment of landscape and wildlife. It has become increasingly difficult to make a living from the countryside, and economics has become the driving force for a change in the way that landowners run their estates. Farmers and landowners are often committed to country sports and to maintaining the habitats of quarry species – fox, deer, hare, grouse – on their land. These country sports, through the conservation of quarry species and their habitats, can be shown to have had a major impact on traditional landscape features. £1.4 billion was spent on country sports in 1990, a large percentage of which was spent on landscape maintenance. Landscape and wildlife in national parks can be nurtured by careful stewardship of farmers and landowners in response to their involvement with traditional country sports.

It may be a truism to say that the national parks of Britain are neither 'national', in the sense of belonging to the nation, nor 'parks', in that they are not enclosed within a pale. But this is understandable in the context of a country with a settled history going back some five thousand years and strong traditions of landownership and tenure which are enshrined in Statute. In our national parks, private landowners are by far the largest area landowners, averaging 73% as their percentage landholding, compared with only 2.3% for the national parks authorities themselves. The largest private landownership is in the Yorkshire Dales (97%), and even where their landownership is the lowest percentage (56%), in Northumberland, it is still more than twice that of the next largest, the Ministry of Defence (with 23%). The highest percentage landownership which the national parks authorities achieve is 7%, in the Brecon Beacons (Anon 1987). This is a quite different situation from that which exists in, for example, the Republic of South Africa, where one enters the Kruger National Park through impressive gates into an area which is partly enclosed by high fences; this is both national and a park in the true sense of both words. Similarly, as the road from Calgary enters the Banff National Park in Alberta, the toll booths, with their accompanying Rangers and ground squirrels, are as clear an indication as one can get that here is a quite different environment that is being entered. No fences as such, but one is without doubt in a national park where *Homo sapiens* is just another animal with as much right as any other animal to exist, but with rather more responsibility. In August 1992, for instance, in Jasper National Park in Alberta, several of the trails around Fort Jasper were closed as a result of a Grizzly Bear (*Ursus horribilis*) having been seen in the area and having attacked a student who surprised it. The young man, descending the trail at some speed on a mountain bike, came upon the bear at the side of the track; it knocked him from the bike and he was lucky enough to 'play dead' successfully until the bear went away and he could remount and get to Jasper Hospital. Some weeks after this, a British couple encountered the same bear within a few hundred yards of their campsite – as is usual in the Canadian National Parks, a highly organized site for several hundred people. The bear attacked and knocked the woman unconscious; the man went to her aid and was killed by the bear. Shortly afterwards the bear was shot by the Park Rangers. Thus, in spite of all the efforts by the national park authorities and their stated policies that the animals have priority over human beings within the Parks, in the final analysis that particular Grizzly became a statistic. It represented the only 'killer' bear for about forty years in Jasper and one of a handful which have been 'culled' for safety reasons by the Rangers. It is largely because of the danger of bears, and the Grizzly in particular, that access within the Canadian national parks is often restricted to the trails which are maintained by the authorities. In many of the more heavily wooded areas, access off the trails is virtually impossible anyway because of the amount of criss-crossing fallen timber. Above the tree line, however, access is very much more like that in British moorland areas, with more numerous animal tracks as well as organized human trails. The end result is that many landscape features are very much more protected than is the case in much of Britain and, even in well-frequented tourist areas such as above the Athabasca Glacier, one can cross snow patches in September which have not felt another human foot.

Whereas in the Canadian national parks the land is owned by the State and people are only tolerated as campers or tourists who are transitory, the case in Britain is quite different. It has already been demonstrated that most of the land in British national parks is privately owned. The State, as represented by the Countryside Commission or English Nature, Scottish Natural Heritage and Countryside Council for Wales, organizes its land tenure through the Park Board or the Park Planning Authority, and has nowhere near the same power as exists in other countries' parks. In many texts on the countryside, for instance Oliver Rackham's *The History of the Countryside*, there may only be one mention of 'national parks' running to a few lines (Rackham 1986). Kenneth Dobell, writing about the Lake District, indicated one of the inherent problems of the British national park when he wrote:

> ...the aim of the National Park shall afford to those who live in it the chance to make a living in a competitive world; and to those who visit it the solace of natural beauty with the challenge of wild mountains...
>
> Dobell 1973

In a country so densely settled as Britain, even in its uplands, it is very difficult to reconcile the economic demands of the inhabitants of a national park with those who visit the park for its 'natural beauty', a beauty which is more often than not the direct result of generations of its inhabitants making a living there. In fact, a leading agronomist is quoted as saying: 'The shooting of grouse is the only economic justification for the maintenance of heather moorland.' (JNRP 1992).

Certainly over the last fifty years the break up of many large country estates has not been to the benefit of either landscape or wildlife. By their very nature, estates integrated many of the essentials of upland landscapes, in particular dry stone walls, hedges and banks, woodlands, and shelterbelts and, of course, the moorlands of which Britain has so much of a limited global resource. These estates also helped to maintain communities within the uplands, communities which are essential, through their labour, to maintain the fabric of landscape. What is now happening is that the State is attempting by other means to do this. The establishment of the Exmoor Environmentally Sensitive Area (ESA) in February 1993 and the proposal for an ESA on Dartmoor in 1994 are examples of the way in which the State is applying the agri-environmental policies which the recent reforms of the Common Agricultural Policy (CAP) have put in place (Booth 1993). As Booth has said:

> Landowners and farmers, who genuinely do not want to see the moorland deteriorate further want to be given the opportunity to explain the economic value of diversification into moorland management for landscape value... the paymasters – the general public through their taxes – must also be persuaded and encouraged to appreciate the benefits of moorland retention.
>
> Booth 1993

The problem is, of course, that making a living from the countryside is becoming more difficult and the break up of estates, or their takeover by institutions, has often meant that the new owners must become more hard-nosed economically. In the uplands it is not that difficult to convert heather moorland into woodland or into pasture; if economics, supported by the CAP will assist farmers to do this, then they are very likely to do it. As a result the situation arises where one arm of State actively opposes what another arm of State supports, something which could not happen when the same person, family or Trust was running an integrated agricultural estate which also had forestry, sheep or deer and grouse interests.

Many of our national parks fall within the broad categories of mountain, fell, moorland and woodland and these are all areas where country sports such as fox and deer and hare hunting, deer stalking, game and rough shooting, game and coarse fishing, and falconry all take place. All of them involve groups of people who are very knowledgeable about the ecology of their quarry species and with a keen interest in the maintenance of the habitats where these species are found. An instance of this is the annual survey of the brown hare *Lepus europaeus* and the mountain hare

Lepus timidus, which is carried out by hare-hunters, showing that after a fall in the population in the 1960s the position is now stable. The 1992 survey also indicates that the introduction of rotational set-aside increases crop diversity on arable farms and this, with more effective control of the hares' predators, could lead to an increase in populations (Stoate 1992). Similarly, the National Trust, in its 1993 *Red Deer Conservation and Management Report* (Savage 1993) discovered that 'the red deer population of Exmoor and the Quantocks was revealed to be close to 7000 and increasing, a figure much larger than previous estimates had indicated'. It was also very clear that hunting was important as an element in the conservation and management of the red deer herds '...in order to achieve the necessary co-operation from landowners and farmers, which is essential to the welfare of the deer, the hunt plays a key part.'

In the past, and to a very large degree at the present, it is very clear that country sports, through the conservation of their quarry species, have also made a major impact on the conservation of unrelated species and in particular on what could be described as the more traditional detailed landscape features: hedges, walls, spinneys, woodlands and lakes, as well as major landscape features such as tracts of woodland and moorland. Outside the national parks, in the Midland Shires, for instance, this is very obvious. Many of the small (less than 20 ha) woodlands scattered across Warwickshire and Leicestershire were planted, and are still owned, by hunts (Hoskins 1955). In many cases, too, there are agreements between particular hunts and the County Wildlife Trust enabling joint use of the woodlands. Within the national parks, access to moorlands where there are no rights of way have frequently been agreed with landowners, the only conditions being the occasional closure on shooting days. Game laws, and the appropriate codes of practice which exist for every country sport are designed 'to ensure the successful breeding of the animals ... hunted for food and sport' (Tivy & O'Hare 1981). They have not only been successful at maintaining viable populations, but have also ensured that the habitats essential for their survival have been maintained.

The countryside, and especially the national parks, are under increasing pressure from the growing demands which the population makes for outdoor recreation. The resource is limited and, in landownership terms, the National Park authorities are a minor partner; however, there is no doubt that there is now a willingness on behalf of private landowners to make some of their land available and accessible (Cobham Resource Consultants 1992). The old established country sports have long been undervalued in terms of their great contribution to maintaining landscapes in such a state that people do want to come and enjoy them. In simple money terms, the amount spent on country sports in 1990 was £1.4 billion. Much of this money was spent on landscape maintenance and in Scotland '75% of the grouse moors operated at a loss, reflecting in large measure the commitment of the landowners to maintaining the Highland habitats' (Cobham Resource Consultants 1992). Similarly, in the valley of Little Langdale in the Lake District National Park the commitment of farmers and landowners to maintaining small larch woods and bogland areas as habitat for the hare is very largely because of its sporting characteristics. Not until regular meets of harehounds were organized could one be sure of seeing a hare in this valley; now they are a regular sight.

There is a wealth of landscape and wildlife in the national parks and this is regarded with considerable envy by visitors from the European Community and elsewhere in Europe. Over the centuries that wealth has been nurtured by the careful stewardship of farmers and landowners, much of it as the direct consequence of their involvement in, and support for, country sports.

References

Anon. 1987. *Chartered Surveyor Weekly*.

Booth, S. 1993. *In*: Joseph Nickerson Reconciliation Project, *Ninth Annual Report*, 30–31. Unpublished.

Cobham Resource Consultants. 1992. Countryside Sports, their Economic and Conservation Significance. *Standing Conference on Countryside Sports*. College of Estate Management, Reading.

Dobell, K. 1973. *In*: Pearsall, W. H. & Pennington, W. (eds) *The Lake District*. Collins, London.

Hoskins, W. G. 1955. *The Making of the English Landscape*. BCA, London.

JNRP. 1992. Joseph Nickerson Reconciliation Project) *Eighth Annual Report*. Unpublished.

Rackham, O. 1986. *The History of the Countryside*. Dent London.

Savage, R. 1993. *Red Deer Conservation and Management Report*. The National Trust.

Stoate, C. 1992. The 1992 BFSS Hare Survey. *In: The Game Conservancy Review of 1992*, British Field Sports Society, London, 100–101.

Tivy, J. & O'Hare, G. 1981. *Human Impact on the Ecosystem*. Oliver & Boyd, Edinburgh.

Scale problems related to the use of the criterion 'naturalness' in the context of landscape protection and road construction in an apparently pristine Arctic environment

SYLVIA SMITH-MEYER & LARS ERIKSTAD

Norwegian Institute for Nature Research (NINA), Box 1037, Blindern, N-0315 Oslo, Norway

Abstract: 'Naturalness' is an important criterion for nature conservation, including landscape protection, but the criterion is vulnerable to scale problems. This may be illustrated by the plans for a new road between the coal mines in Longyearbyen and Svea on Svalbard. The landscape is apparently pristine, but has some encroachments such as tracks from different vehicles. This has been used as an argument for the lack of naturalness and therefore that the consequences in terms of landscape protection will be moderate. Such a conclusion is rejected based on a discussion of the scale problems involved. The criterion naturalness should always be used with care relative to the scale in each individual landscape analysis.

The concept of 'naturalness' is an important criterion for nature conservation. Pristine landscapes are considered valuable both because they are attractive for human adventures, and are important reference areas for scientific studies. There may, however, be considerable difference of opinion as to what a pristine landscape really is. This conceptual divergence can be illustrated by an example taken from Svalbard in an area where the only slightly affected and vulnerable Arctic tundra merges with true wilderness with no detectable human influence.

Present situation

The area in question is part of the 40% (Erikstad & Hardeng 1992) of Svalbard which is not included in existing national parks or natural reserves. Longyearbyen (Fig. 1) is the main

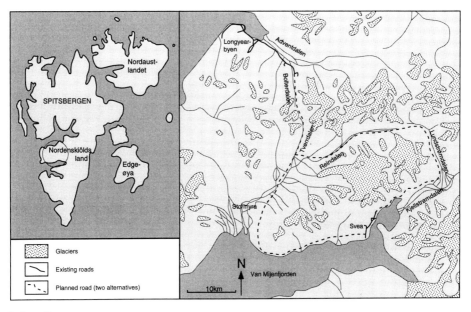

Fig. 1. Location map.

From O'Halloran, D., Green, C., Harley, M., Stanley, M. & Knill, J. (eds), 1994, *Geological and Landscape Conservation*. Geological Society, London, pp. 283–286.

centre on Svalbard with administrative offices, schools, shops, a central airport etc. Svea is a coal mine some 45 km to the southeast. The roads reach some 10 km to each side of Longyearbyen and are built mainly for easy access to various coal mines in the Longyearbyen district. A few huts and a water supply reservoir are found outside the town centre, and aggregates are extracted from a moraine nearby. In Svea, a small settlement has been built, with several kilometres of road, a harbour and a major coal storage site.

Reindalen is the main valley between Svea and Longyearbyen (Fig. 1). The valley has no roads. There are some remnants of oil drilling where the valley starts at the watershed between Reindalen and Lundstrømdalen, and a handful of huts is located in the valley. None of these affect the discussion of naturalness significantly, and in reality they do not play a major part in the visual landscape.

In Reindalen, as well as along the coast of the van Mijenfjorden, there are some tracks made by snow scooters and other vehicles (Fig. 2). These tracks can be easily spotted from the air as thin parallel stripes in the terrain. They may be difficult to see from the ground although they may be locally quite distinct. Some of the tracks date back to about 1918 and are caused by former use of heavier vehicles (Nordang & Hansson 1991).

Planned development

In connection with planned expansion in the coal fields of Svea, a new road and a 66 kV power line were planned over a distance of 60–70 km between Longyearbyen and Svea. The road was planned to be constructed on a 1 m high embankment. Two alternatives, one following the inland valleys and one following the coast of van Mijenfjorden, were presented (Fig. 1).

Landscape and geology

The area is situated within the central Spitsbergen basin of gently dipping sedimentary rocks from Jurassic and up to late Tertiary age (Major & Nagy 1972; Hjelle et al. 1986; Steel et al. 1989). These flat sedimentary rocks result in plateaux at various altitudes even if the high mountains generally have alpine forms (Klemsdal 1989). The valleys and fjords are wide with flat floors (Figs 2 & 3) and plateaux often form shoulders in the sides of the valley. The tributary valleys are narrow and sharp and make a considerable contrast to the main valleys (Rudberg 1988). Large fluvial deposits are found in the valleys as active alluvial plains in the valley floors and large alluvial fans where the tributary valleys meet the main valleys (Sørbel et al. 1991).

As for Svalbard in general, this area has a

Fig. 2. Outer parts of Reindalen towards van Mijenfjorden. A remnant of a Weichselian moraine is central in the picture. Tracks of vehicles are just visible in the lower right-hand corner.

Fig. 3. Tundra flat in central Reindalen viewed towards the southeast.

fairly low glacier coverage, i.e. small- to medium-sized glaciers occupy cirques in the mountains without growing together to form larger ice fields (Fig. 1). Several of the glacier outlets reach the valley floors and have formed ice-cored moraines which are important landscape elements. Most of the glaciers are surging (Liestøl 1976; Hagen 1988) and the ice-cored moraines vary in shape relative to the time elapsed since the surge formed them, i.e. how much of the ice core has melted. A series of pingos (Liestøl 1976) in the northern parts of Reindalen are also important landscape elements within the valley.

Discussion

We have argued that among the large-scale landscapes and landscape elements of Svalbard, Reindalen has natural interests of national importance (Erikstad & Smith-Meyer 1992). This is partly because the valley contains medium-scale landscape elements such as the large Stormyra tundra flat, as well as small-scale geotopes such as the pingos, which are both characteristic of natural interests of that level. Additionally these values, together with the glaciers, the ice-cored moraines, the river and the alluvial fans, form a landscape unit representative of central parts of Spitsbergen. This valley also has an overall pristine character, despite the driving tracks discussed above. These tracks have been of some significance when alternative road routes have been discussed. The planned road partly follows these tracks and this has been used as an argument for a current lack of naturalness, implying that the consequences of road building, in terms of landscape protection, will be moderate. We find that such a conclusion must be rejected because of the scale problems involved. The magnificent character of the broad and diverse tundra landscape, compared to the existing narrow tracks, shows there is a need for discussion of scale when assessing what type of human activities affect the naturalness of a landscape. The existing tracks, regrettable though they may be, must be classified as encroachments on a very small scale. Both at a large scale and at a meso-scale and even for much of the detail, naturalness characterizes the general impression of the landscape. The road, however, is planned on an embankment that will be visible over large areas. It will represent a new landscape element at least on a meso-scale.

A new road, as planned, will undoubtedly have major negative impacts for this landscape. In this case this argument merges with ecological arguments (Nordang & Hansson 1991).

Conclusion

A direct comparison between existing and planned encroachment does not work due to the difference in scale. The same will probably be the case in inhabited areas with a more complex pattern of land-use. The criterion 'naturalness' should therefore be used with care, always relative to the appropriate scale in individual cases of landscape analysis.

References

ERIKSTAD, L. & HARDENG, G. 1992. *Nature Conservation areas*. National atlas of Norway. Statens kartverk. [In Norwegian with English summary].

—— & SMITH-MEYER, S. 1992. A landscape analysis. Consequences of a planned road between Longyearbyen and the Svea mines, Svalbard. *NINA Oppdragsmelding*, **158**, 1–31. [In Norwegian with English summary].

HAGEN, J. O. 1988. Glacier surge in Svalbard with examples from Usherbreen. – *Norsk geografisk Tidsskrift*, **42**, 203–213.

HJELLE, A., LAURITZEN, Ø., SALVIGSEN & WINSNES, T. S. 1986. Geological map of Svalbard 1 : 100 000. Sheet B10G Van Mijenfjorden. *Norsk Polarinstitutt Temakart*, **2**, 1–37.

KLEMSDAL, T. 1989. *Landforms – Svalbard and Jan Mayen, Map in the scale 1 : 1 000 000*. National atlas of Norway (Map 2.1.3).

LIESTØL, O. 1969. Glacial surges in West Spitsbergen. *Canadian Journal of Earth Sciences*, **6**, 895–897.

—— 1976. Pingos, springs, and permafrost in Spitsbergen. *Norsk Polarinstitutt Årbok*, **1975**, 7–29.

MAJOR, H. & NAGY, J. 1972. Geology of the Adventdalen map area. *Norsk Polarinstitutt skrifter*, **138**, 1–58.

NORDANG, I. & HANSSON, R. (eds). 1991. Sentralfeltprosjektet: Miljøkonsekvenser av en veiutbygging mellom Longyearbyen og Svea. *Norsk Polarinstitutt Meddelelse*, **117**, 1–65. [In Norwegian with English summary].

RUDBERG, S. 1988. High arctic landscapes: comparison and reflections. *Norsk geografisk Tidsskrift*, **42**, 255–264.

STEEL, R., WINSNES, T. S. & SALVIGSEN, O. 1989. Geologicap map of Svalbard 1 : 100 000. Sheet C10G Braganzavågen. *Norsk polarinstitutt Temakart*, **4**, 1–22.

SØRBEL, L., SOLLID, J. L. & ETZELMÜLLER, B. 1991. *Reindalen, Quaternary geology and geomorphology. Map in the scale 1 : 100 000*. Geografisk institutt, Universitetet i Oslo.

Quarry slope stability and landscape preservation in the Malvern Hills, UK

JOHN KNILL

Highwood Farm, Shaw-cum-Donnington, Newbury RG16 9LB, UK

Abstract: The Malvern Hills form a spectacular narrow, north–south ridge composed of a horst-like structure of Precambrian metamorphic rocks rising up above younger Lower Palaeozoic and Triassic rocks to the west and east, respectively. The Malverns are the major source of hard rock material within the region and have, traditionally, supplied building stone and other rock products for some centuries. As a consequence, a number of quarries have been excavated into the flanks and core of the ridge. Many of these quarries had been abandoned by the 1950s and active extraction was limited to four sites. At that time there was a considerable increase in the demand for aggregate in connection with the construction of the M5 motorway some kilometres to the east. Enhanced extraction occurred and, in view of the risk of large-scale rock slides breaking through the ridge crest profile, the landscape quality of the Malverns was endangered.

The metamorphic rocks of the Malverns are composed of highly deformed and sheared gneisses and schists, and granites intruded by a few basic dykes. As a consequence of inherent rock mass defects, slope failures have occurred in the quarries on all scales from small rock falls to large-scale rock slides involving tens of thousands of cubic metres of rock. This ease of fragmentation of the rock mass assisted the quarrying practice which relied on undercutting the faces leading to progressive slope failure which was enhanced through wedging and barring by quarrymen working on the faces. Such oversteepening resulted in a number of large-scale rock slides whose edge approached the ridge crest. In addition, it was recognized that the quarry faces continued to fail after quarrying ended and this situation prompted a study of the likely long-term stable slope angle of the quarries, after abandonment. A long-term relationship between maximum slope angle to face height was established for the quarry faces and this was used to encourage adjustments to quarrying practice in order to minimize landscape damage to the ridge profile, and thereby to restrict quarrying. Quarrying has now ceased.

The Malvern Hills are composed of a north–south ridge, some 12 km in length and little more than 1 km across at the widest, located close to the margin between England and Wales. The hills form a spectacular feature rising sharply from the Severn valley to the east, which is underlain by more softly weathering Triassic rocks. To the west, the ridge is in contact with the rolling ridge-and-vale topography of the Lower Palaeozoic comprising alternating limestones and softer shales. The hills are predominantly composed of Precambrian metamorphic rocks including gneisses with metasediments, migmatites, pegmatites and occasional basic dykes. The east and west margins of the hills are fault-controlled so that, effectively, they form a horst-like structure in relation to the surrounding younger rocks.

The Malvern Hills are the main source of hard rock, suitable for building stone or aggregate, in the region so that quarrying has been a traditional practice. There are many small, abandoned quarries which were used for the extraction of rock used for the construction of local buildings. By the end of the 1950s, there were four quarries still active in the Malvern Hills. At about that time the M5 motorway came into construction some kilometres to the east and this led to enhanced rates of extraction of aggregates for road building. The Malvern Hills Conservators, who are obliged to maintain these hills, became concerned that there was a risk that slips would result in land outside the quarry area being lost, and that the vulnerable and visually attractive ridge crest would be breached. The studies which were carried out, on behalf of the Conservators, to examine the effect of slope stability are described in this paper.

Malvern Hills

The Malvern Hills rise from the relatively flat Severn Vale at 15–60 m elevation through the gently sloping 'commons' to a maximum elevation of 425 m at Worcestershire Beacon.

To the west, the mean elevation is about 150 m. The crest ridge undulates southwards along the length of the Malvern Hills ridge from North Hill at 397 m, to Worcestershire Beacon, to Herefordshire Beacon at 338 m and eventually falling away to lesser elevations at Midsummer Hill at 284 m, and Raggedstone and Chase End Hills at the southern edge of the Malvern Hills. The view of these hills is, therefore, quite spectacular from almost every direction, whether as a long ridge rising dramatically from lower-lying country or as a sharp, conical hill complex when seen in profile.

The sides of the hills slope relatively evenly at about 1 vertical to 2 horizontal (just under 30°) for heights of 100 m or more, providing much of their visual attraction. Slopes of lesser height than 100 m may be somewhat steeper. The hills are, therefore, symmetrical, although the easterly-facing slope can be somewhat shallower possibly as a result of enhanced exposure to the sun during freeze-and-thaw cycles in the past. The form of the hills is such that most potential quarry sites would be highly visible, cutting directly into the side of the ridge, well above the elevation of the adjacent country. Only where valleys cut through the line of the hills is there an opportunity for quarrying to take place parallel to the axis of the ridge. At the end of the 1950s there were four active quarries; two of these were at the extreme north end of the hills above Great Malvern. Quarrying in the North Quarry ended following the development of a major rock slide in the 1950s, but extraction from the adjacent Tank Quarry continued into the 1970s. The other two quarries, at Gullet and Hollybush, are in the southern part of the hills. The Gullet Quarry was excavated northwards into the hills from a small east–west valley, and the Hollybush Quarry was driven northwards along a fault-controlled valley between Hollybush and Midsummer Hills. Quarrying also ended at these two quarries within the 1970s.

Rock characteristics

The rocks of the Malvernian complex which have been quarried are Precambrian in age being comprised of igneous and metamorphic rocks of various ages. The oldest rocks consist of a variety of metamorphic rocks including meta-sedimentary schist, gneiss, quartzite and marble with some amphibolite. The bulk of the complex consists of intrusive calc-alkaline acidic and intermediate rocks; there are later intrusive basic and pegmatitic rocks. The rock mass has been extensively deformed on all scales, so that rock within the quarry faces or in individual boulders can disintegrate relatively easily. This ease of fragmentation was of considerable advantage in crushing the rock for aggregate production. There are major chlorite-covered, penetrative listric shears, commonly tens of metres in length, cutting through the rock mass which are locally referred to as 'elephants' backs'.

The quarrying process took advantage of the ease of fragmentation of the rock. Traditionally, the base of a quarry face was drilled and blasted thereby undercutting the face above causing oversteepening, leading to instability bringing down more rock which broke up as it fell to the quarry floor. Instability was encouraged by quarry workers suspended on ropes operating on the face wedging and barring rock, thereby further enhancing instability. As a consequence, the quarrying practice worked the face at about its maximum angle of stability, maintaining that angle for the full height of the quarry face up to a recorded maximum of about 100 m. Such a method of quarrying in the 1950s and 1960s, although more common in the past and appropriate to small quarries, was very different in approach to that adopted in then-current practice, where the rock would be extracted by means of individual benches of about 10 m height.

This method of quarrying resulted in oversteepening of the rock face and, as the unstable rock cover in the face could not be removed fast enough by quarrying, mass failure took place in the form of rock slides. Such slides ranged in volume from a few tens to thousands of cubic metres. Several of these slides cut back through the working crest of the quarry and, as mentioned above, working of the North Quarry ceased for this reason. At the Hollybush Quarry the lip of quarry cut into an Iron Age fort. These slides were primarily controlled by displacement on the listric shears. Movement has continued on the North Quarry slide after working ceased for several decades and, as a result, the displacements were monitored. A regular movement of 0.2–0.3 m per year was observed during the 1960s, but acceleration in the next decade resulted in particularly active movements (Fig. 1). Typically, accelerations of several times that displacement rate over a period of a few months was followed by occasional rock falls, and then a reduction in movement rate. This stick–slip process was repeated on a number of occasions until major movements and active rock falls occurred in the mid-1970s. The slide was observed closely over this period. After a few weeks the rate of movement slowed down, and the slide rock has moved relatively little since.

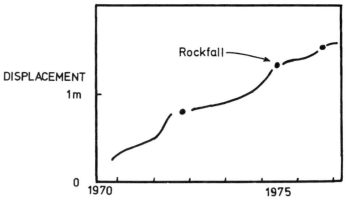

Fig. 1. Displacement of rock slide in North Quarry.

Once abandoned, the quarry faces continue to disintegrate to create a scree slope immediately below the face so that, as disintegration continues, the original face is progressively buried by rock debris. In some of the older quarries over half the height of the original quarry is now obscured below scree. During this process the inclination of the rock face becomes progressively flatter.

Although the structure of the rock mass is inherently heterogeneous, the rock mass behaves in terms of its engineering behaviour as though it were homogeneous. This situation is illustrated on a large scale by the uniformity of the slope angle of the Malvern ridge. *In situ* seismic velocity measurements taken in different orientations in the floors of a number of quarries are very consistent, as are equivalent measurements made on hand specimens of rock.

Quarry stability

The slope angle to slope height relationship for all the quarry rock faces in the Malvern Hills in 1962 is presented in Fig. 2, separated out on the basis of quarries which were abandoned and working at that time. It is immediately apparent that, for both groups, the maximum measured slope angle becomes less with increase in height of the rock face. Such a situation is consistent with a rock material which exhibits both frictional and cohesive strength. The maximum

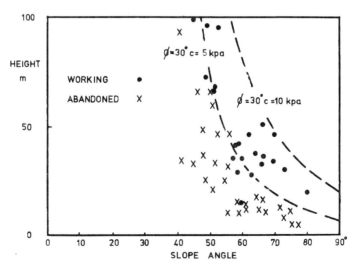

Fig. 2. Slope height–slope angle relationship for the Malvern quarries in 1962.

theoretical slope angle for two different materials defined by the same angle of shearing resistance (30°) and cohesion intercepts of 10 and 5 kpa respectively are traced out on Fig. 2 and these lines define, relatively precisely, the maximum *in situ* shear strength of the Malvern rocks on a short-term (working quarry) and long-term (abandoned quarry) basis. This diagram adds, also, further confirmation to the observed homogeneity of the complex lithological and structural characteristics of the rock mass.

The reduction in the calculated cohesion between the maximum *in situ* strength for the rock mass in the working and abandoned quarries is a reflection of the weathering and age-softening of the chloritic infill to the weaker shear zones.

Application of the slope chart in Fig. 2 established the long-term slope angle of the Malvernian rocks, and this was then used as a basis of negotiation with the quarry companies regarding limitations of working practice within the quarry area. This approach contributed to quarries recognizing the importance of leaving a protective marginal strip at the quarry crest. This situation ensured that face disintegration and rock slides would not cross the planning boundaries, and so the preservation of the ridge crest was assured.

Conclusion

Detailed studies of the mass behaviour of rock in the Malvern quarries provided a numerical basis for the prediction of rock behaviour and, thereby, assisted in the minimization of derogation to the landscape of the Malvern Hills.

Geological and landscape conservation in Hong Kong

D. R. WORKMAN

Department of Geography & Geology, University of Hong Kong, Pokfulam Road, Hong Kong

Abstract: Hong Kong has a particular interest in conserving its countryside and natural history, given the intense pressure of urban development resulting from the combination of a large population (6 million) and small area (1100 km^2). In this respect it is fortunate that about three-quarters of the land area has, up to now at least, been regarded as unsuitable for urban expansion because of its rugged nature.

All land in Hong Kong is held by the government, which sells or grants leasehold interests. The government is thus able to exercise a high degree of control over land-use and changes of use. The government agency with overall responsibility for environmental matters is the relatively new (1986) Environmental Protection Department, but most conservation provisions pre-date this department and remain under the management of other branches of the government. The Agriculture and Fisheries Department, for example, oversees the territory's 21 country parks and various 'special areas', while responsibility for archaeological and palaeontological conservation lies with the Antiquities and Monuments Office within the central Government Secretariat. There are provisions for designating Sites of Special Scientific Interest (SSSIs), but no statutory powers to enforce preservation of SSSIs as such.

In general, the system has worked reasonably well up to now from the point of view of landscape conservation, but areas of special biological and geological importance are vulnerable and in need of better protection than they receive at present, especially outside the country parks. Coastal sites particularly are at risk, from pollution, dumping and reclamation for example, but inland sites can be threatened too, by public projects that may be deemed to have overriding claims for space and by private developments such as housing and golf courses, if they are considered to be in the public interest. In the absence of comprehensive enforcing legislation, much depends on the willingness of government bodies responsible for planning and development to pay due attention to environmental issues where there are conflicts of interest. Therefore, there is every incentive for private groups and individuals to make sure these conflicts of interest do not pass unnoticed.

Hong Kong is a small, crowded place, with nearly 6 million people and a total area of only 1076 km^2, including 41 km^2 reclaimed from the sea. It consists of a mainland area of 783 km^2 and 235 islands, of which Lantau (142 km^2) and Hong Kong Island (79 km^2) are the largest (Fig. 1).

When Hong Kong Island was ceded to Britain by China in 1841 it was inhabited by no more than a few hundred villagers and fishing people. Many of the latter lived on their boats. Lord Palmerston, the then Foreign Secretary, described the place dismissively as 'a barren rock, with hardly a house upon it'. Later, other territory was added, mainly by virtue of a 99-year lease, signed in 1898.

Hong Kong has, of course, seen phenomenal growth. By 1991 the population had reached 5.8 million, making it overall one of the most densely populated administrative entities on Earth – more densely populated than London. For all that, about three-quarters of the colony is, to this day, very thinly populated, and large areas, more or less uninhabited.

Most of the territory is very rugged. Sizeable areas stand at more than 300 m above sea-level and the highest peaks reach more than 900 m; hillsides are typically steep and rocky. There is little land suitable for farming. At present about 6% of the total area is classed as agricultural land, and of that less than half is in use. Rice cultivation has been totally abandoned. More than two-thirds of Hong Kong is open country – woodland, grassland or scrub – and only about one-fifth is classified as built-up.

Legislative framework for environmental protection

Hong Kong's emergence as a major modern metropolis has been essentially a post-war phenomenon, and the legislative framework to control it has evolved as the colony has grown. Given the avowedly *laissez faire* nature of the administration, environmental concerns were some way down in the list of priorities for a long time and in some respects still lag behind as the

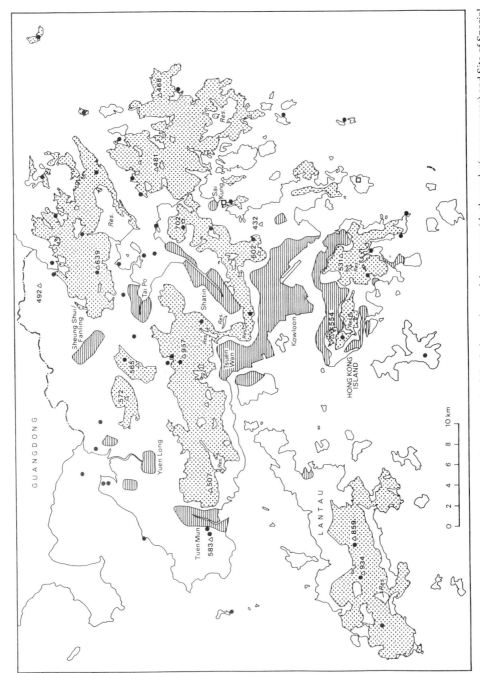

Fig. 1. Map of Hong Kong showing location of country parks (stippled areas), special areas outside the parks (open squares) and Site of Special Scientific Interest (black circles). The main built-up areas are shown by line shading. Spot heights in metres.

economy forges ahead. However, much has been achieved in the past 20 years.

The government agency with overall responsibility for environmental matters is the relatively new (1986) Environmental Protection Department, but most conservation provisions pre-date this department and remain under the management of other branches of the government. The Agriculture and Fisheries Department, for example, oversees the territory's country parks, while responsibility for archaeological and palaeontological conservation lies with the Antiquities and Monuments Office within the central Government Secretariat (HKG 1986b). Another department involved is Planning which, among other things, is responsible for the operation of a set of planning standards and guidelines covering all aspects of development including landscape and conservation. Public participation in all this is mainly by means of representation, by invitation and up to stipulated limits, on advisory boards and committees that make recommendations to the Authority on legislation, proposed actions, disputes, appeals and so on.

Landscape conservation

Much of Hong Kong remains wild and beautiful (Fig. 2) and legislation and planning procedures now in place should ensure that it remains so. Protection most certainly is needed, bearing in mind ever-present population pressures and the fact that, although most of the territory has hitherto been regarded as too steep and rocky for development, there is very little of Hong Kong that is technically or economically undevelopable today.

Fortunately, in matters of land-use, the government has always been in a very strong position. All land is Crown land. Original village lands were leased back to their occupants and users at no or nominal rent, generally with minimal conditions, and land for community purposes is granted by private treaty, but leases of land for development are sold by auction or tender. This enables the government to impose strict controls on what is built, and where.

The government, of course, has complete control over all land it chooses not to let, and can zone it for whatever use it sees fit. One of its most important tasks has been to ensure an adequate water supply, in the absence of any large rivers or aquifers. This has meant large areas being reserved for catchments and reservoirs, and kept free from any kind of development not aimed at maximizing the amount of water retained and stored. When a series of country parks was established in the 1970s, they consisted essentially of these water-gathering grounds, with the addition of various peripheral

Fig. 2. Lantau Peak (934 m), a special area within a country park.

areas, some other uplands, stretches of coastline and nearby islands which had remained unleased. Settled and cultivated areas were largely excluded, even to the extent of creating enclaves within parks, but even so most of the villages enclosed or otherwise isolated by the parks have withered away as a result of migration into the towns.

The country parks

The Country Parks Ordinance (HKG 1986a) was enacted in 1976. The first three of the parks came into existence in 1977; by the end of 1979 there were eighteen more and the total area designated was 40 833 ha – about 40% of Hong Kong's land area (Fig. 1). That remains the situation at present. The biggest park is 5640 ha and the smallest 100 ha. The Country Parks Ordinance provides for control of all kinds of development within the parks, as well as changes of use of leased land.

The Ordinance makes no direct reference to either geological or landscape conservation among its primary objectives, which are (paraphrased):

(i) to encourage their use for recreation and tourism;
(ii) to protect the vegetation and wildlife;
(iii) to preserve sites of historical and cultural significance;
(iv) to provide facilities for public enjoyment of the parks.

As a part of Hong Kong law it does, however, make development and activities such as quarrying more or less impossible without re-defining a park's boundaries, and this has not so far happened to any significant degree.

Even though the Ordinance does not make it explicit, the essence of the country park concept in Hong Kong is undoubtedly seen as countryside conservation, not merely provision of recreational space, although this is important. There are few roads in the parks and none have been built for public use since the parks were created; all service roads such as reservoir access are strictly off limits to the private motorist. There are very few car parks and no plans for any more. Bus routes stop at, or short of, country park entrances, except in a very few places where they serve existing village enclaves or ferry piers within the parks. The parks are for walkers, picnickers and campers. Several long-distance trails have been made from existing paths, and various short nature and family trails have also been laid out with flagged points of interest. Several hundred kilometres of paths are maintained. At the same time, however, many historic paths to now-abandoned villages have practically disappeared, and large areas of some of the parks are thereby rendered inaccessible because of thick bush. Whether or not it is the specific intention is hard to determine, but the effect of the government's management of the parks is to create something for everyone, and indeed for all living things. Near the entrances there are the barbecue sites and the crowds, further away maintained paths and groups of walkers, further away still, tracks kept open by the few hikers who use them, and between those tracks large areas with nobody at all.

The Country Parks Ordinance does not, of course, constitute a guarantee of the parks' future. The government is free to bend or change its own rules. Up to now, the problems that have arisen have been limited to a few attempts to nibble away at the edges of the parks for various reasons. The Ordinance lays down procedures providing for public consultation and for lodging objections to any changes proposed by the Authority, but these procedures have not always been followed. There are currently issues concerning conversion of country park land to use as a golf course (not so far approved) and encroachment of a waste disposal landfill site into a country park (already approved) that have sparked an angry response from conservation activists. However, the country parks have remained intact and almost entirely untampered with so far, and public awareness of the need for conservation (as well as membership of organizations such as Friends of the Earth, established locally in 1983) continues to grow.

Special areas

The Country Parks Ordinance includes provision for the designation of 'special areas' (SAs), both within and outside the parks. The intention is to safeguard areas of historical, natural historical or landscape value. It is not altogether clear what additional protection is provided by designation of a special area within a park, or why it is felt necessary, but to date there are 14, ranging in size from 3–460 ha and with a total area of 1639 ha. Several of the SAs are of particular scenic value and geological interest. Twelve of the 14 SAs are within country parks.

Sites of Special Scientific Interest

SSSIs can be adopted anywhere, and can be terrestrial or aquatic. They are selected for their 'flora, fauna, geographical, geological or physiographical features'. SSSIs are registered by the Planning Department on the advice of the Director of Agriculture and Fisheries. Unlike country parks or special areas, the status of SSSI neither confers statutory power on the government to enforce preservation nor implies any restriction upon owners. SSSIs are not identified on the ground by any signs or boundary marks. Designation as an SSSI is intended to ensure that due consideration is given to protecting the site when considering development proposals that may affect it. Measures to minimize any adverse effects are achieved mainly through co-operation of government departments.

Many of the 49 existing SSSIs have been designated primarily for their geological interest or have valuable geological features, such as outcrops of particular scientific or educational significance. The system does work; in a recent case an officially proposed marine dumping operation was cancelled after objections that it would affect a shoreline SSSI of geological interest. It gives private individuals and organizations such as geological, archaeological and natural history societies a means of contributing to the cause of conservation, by proposing new sites and keeping a watchful eye on existing ones.

Planning standards and guidelines

The Hong Kong Planning Standards and Guidelines (PSGs) are a wide-ranging set of ground rules covering all aspects of development (HKG 1991). They are not statutory, but they are intended to be used by all government and private agencies involved in any kind of development, and they can be included in lease conditions or stipulated as a condition of planning permission.

The establishment of the PSGs has been a lengthy affair, with different aspects dealt with piecemeal, and a decision to publish them was only taken in 1991. What their impact on environmental aspects of planning will be in future is hard to say, but on paper they contribute an impressive governmental commitment to environmental protection.

One of the specified aims of the PSGs is to provide guidelines for conserving landscape and heritage. To this end, a chapter devoted to landscape and conservation (the tenth of eleven chapters issued to date) was published in a preliminary version (not yet officially approved) in August 1992 (HKG 1992).

The PSGs, although not binding, do address important aspects of landscape protection not covered by the country parks, special areas and

Fig. 3. A new residential complex on the border of one of the country parks on Hong Kong Island, seen from a hilltop within the park.

SSSIs. For example, they provide a framework for control of development in areas not designated as country parks, and especially on the fringes of the parks. They contain statements like 'development on hill tops, scenic ridges and prominent positions should be avoided wherever possible'. Presumably this means that it will at least be harder in future for developments like the one shown in Fig. 3 to get the go-ahead.

Conclusion

In general, the system has worked reasonably well up to now from the point of view of landscape conservation, but areas of special biological and geological importance are vulnerable and in need of better protection than they receive at present, especially outside the country parks. Coastal sites especially are at risk, from pollution, dumping and reclamation for example, but inland sites can be threatened too, by public projects that may be deemed to have overriding claims for space and by private developments such as housing and golf courses, if they are considered to be in the public interest. In the absence of comprehensive enforcing legislation, much depends on the willingness of government bodies responsible for planning and development to pay due attention to environmental issues where there are conflicts of interest. There is therefore every incentive for private groups and individuals to make sure these conflicts of interest do not pass unnoticed.

References

HKG. 1986a. Country Parks Ordinance (revised). Hong Kong Government.
—— 1986b. Antiquities and Monuments Ordinance (revised). Hong Kong Government.
—— 1991. Hong Kong Planning Standards and Guidelines, Chapter 1: Introduction. Planning Department, Hong Kong Government.
—— 1992. Hong Kong Planning Standards and Guidelines, Chapter 10: Landscape and conservation (preliminary version). Planning Department, Hong Kong Government.

Some problems of geological heritage in the Russian Commonwealth

S. A. VISHNEVSKY

Institute of Mineralogy and Petrology, Novosibirsk, Russia

Abstract: The geological heritage of the Russian Commonwealth is huge and undoubtedly unique. Although, public opinion is not yet activated in this area, people working within the fields of geological science, natural history and education are partially involved. The reaction to the dissemination of information from the Digne Symposium (1991) and to an invitation to co-operate yielded only 10 responses from 50 potentially interested organizations. Nevertheless, the number of organizations and people involved in geological heritage is sufficient for initial conservation activity. Information about 24 new sites has been received, and the need for a working group has been established.

The involvement of groups, however, is still not maximized. The Il'men Mineralogical Reserve exists only due to State support. In several regions (Murman, Armenia, etc.), initial conservation is carried out with State support also. Local initiatives involve the investigation of a site, recognition of its uniqueness, and attempts at its conservation. These attempts are hampered by an absence of support, by vegetation and insufficient protection. The real threats are from the State, business, and uncontrolled tourist activities (El'gygytgyn, etc). Successful development of local initiatives is limited by economic problems and lack of attention from the State.

Measures seem necessary to activate public opinion and community goodwill. These might include:

- State investment with special support for local needs;
- setting up a working group;
- concentration of effort on the successful and complete development of a few specific sites;
- heritage popularization and improvements in education;
- support from UNESCO.

Spreading over approximately a sixth of the Earth's land surface, the Russian Commonwealth is enriched not only in climatic zones and in the diversity of life forms, but there is, perhaps, no other area of the world which exhibits such a variety of geological structures. These structures record the history of evolution of the Earth over the last 4.5 billion years – a history of dramatic acts of creation and destruction. Numerous geological heritage sites of various types and rank are present here. For example, out of approximately 150 meteorite craters known on Earth, 35 are found in the Commonwealth (Fig. 1). Among them, the greatest and most interesting from a scientific and cognitive point of view is the Popigai crater (100 km in diameter), which was recently included in UNESCO's provisional geological World Heritage List (Cowie 1991).

A simple attempt to attract attention to Russia's geological heritage at the Digne Symposium (1991) yielded 24 new sites, some of which are considered below. However, the geological heritage of the Russian Commonwealth is still little appreciated or understood, although the academician I. Borodin wrote as long ago as 1914 that: 'Spreading over huge territories, we are the possessors of a unique natural heritage; like Rafael's pictures, it is easy to destroy but impossible to recreate' (Borodin 1914).

Historical background

The politicized mass social conscience of the Commonwealth is not yet addressing the problems of conserving and utilizing geological heritage. The stereotypes and political clichés which called for the subjugation of nature for the last 70 years were practically opposed to ideas for conserving geological heritage. The rare and timid voices which called for conservation of natural monuments were drowned by the strong chorus of the 'nature explorers'. The situation was also complicated by one-sided development of the national nature conservation system, such that areas under protection very often became inaccessible to visitors, except for hunting and entertainment of the government officials. Under these conditions,

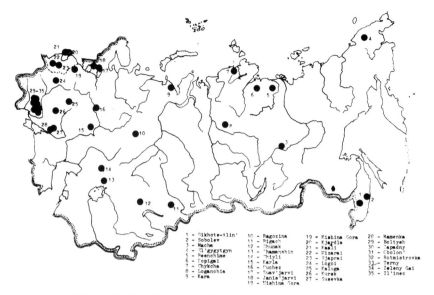

Fig. 1. Astroblems within the Russian Commonwealth.

a division between people and conservation occurred. Nature reserves never were 'sacred cows', and their areas were changed and reduced many times. None of these circumstances helped the dissemination and survival of ideas about the priceless importance of nature conservation as a whole, and the conservation of geological heritage in particular, in the minds of the people.

Only in the 1970s and 1980s were the first government decisions to improve the conservation of nature and its monuments taken, but the

Fig. 2. Non-impact sites within the Russian Commonwealth.

neglect and mistreatment of nature still continued as a whole. The situation began to improve slightly, and only really in recent years when various specialized departments began to be organized; for example, the Russian Ministry for Ecology and Use of Nature. However, numerous ecological problems, together with economic problems, still did not allow these departments to start the real and essential work on geological heritage sites.

Current situation

Some scientific, natural history and educational organizations possess the creative forces to enable them to value the geological heritage of the Commonwealth, and to work out the measures for its conservation. But what is the present response to the needs of geological heritage against such a background?

The reaction to the 'International Declaration of the Rights of the Memory of the Earth' from the Digne Symposium (1991) and to an invitation to co-operate, distributed by Novosibirsk Institute of Mineralogy and Petrology on the author's initiative in 1991, yielded only 10 responses out of 50 potentially interested organizations: Institute of Geochemistry (Irkutsk), Geological Institute (Yakutsk), Palaeontological Institute (Moscow), Institute of Geological Sciences and Armenian Academy of Sciences (Yerevan), North-Eastern Complex Institute (Magadan), Geological Institute (Ashkhabad), Fersman Mineralogical Museum (Moscow), Institute of Geology (Apatites), Karpinsky Geological Institute (Sankt-Peterburg), and Regional Geological Department 'ZapKazgeologia' (Aktubinsk). The fact that the remainder of the organizations contacted failed to respond to our invitation may be due, in part, to a lack of interest in geological heritage. It was especially dissappointing that the magazines *Priroda* and *Nauka i Zhiz'n* did not publish our information for their readers. However, thanks to our Bulgarian colleague Dr T. Todorov (Todorov 1991), information about the Digne Symposium was published in the journal *Razvedka i Okhrana Nedr*. Undoubtedly, many enthusiasts have still to come forward. Nevertheless, the number of organizations and people involved with geological heritage is sufficient for initial activity to begin. The suggestion to organize a National Geological Heritage Working Group has been repeated more than once in the responses from our correspondents.

By means of a questionnaire, various new geological heritage sites have been revealed. Among these are the meteorite craters at El'gygytgyn (V. Bely) and Zhamanshin (V. Fjodorov), the unique Precambrian cross-section on the Circular Baikal Railway (V. Levitsky), astrophyllite and other rare mineral occurrences in Khybiny and Lovozero Massifs (V. Krjuchkov, I. Kondratovitch, G. Andreev), the 'Mamontova Gora' cross-section of fauna-enriched Cenozoic rocks (P. Lazarev), The Kugitang-Tau Mountain Jurassic dinosaurs footprints and Kara-Ges Mountain Pliocene camel footprints (O. Odekov), the Shuldak river valley cross-section of palaeovolcanic rocks (V. Fjodorov), key stratigraphic cross-sections of Cretaceous–Palaeogene rocks in Kazakhstan (V. Fjodorov) and Turkmenistan (O. Odekov), and the caves of Kugitang-Tau (O. Odekov).

Contact with colleagues shows that group involvement is not maximized, although there is a certain amount of success in some cases. Among these is the Il'men Mineralogical Reserve, which exists due to State support. The initial work on conserving geological heritage in several regions is also carried out with State support. Examples include the Murman Region, where the I. Kondratovitch Group operates (Krjuchkov & Kondratovitch 1988), and Armenia, where the List of Natural Monuments if being prepared by request and support of the Armenian Nature Conservation Department. The A. Lapo Group, Karpinsky Geological Institute (Sankt-Peterburg) is starting work on an inventory of Russia's geological heritage, with support from the Geological Committee of the Russian Ministry for Ecology and Use of Nature.

In many cases, local initiatives involve the investigation of a site, recognition of its uniqueness, and attempts at its conservation. But these initiatives are hampered by an absence of support, by vegetation and by insufficient protection, amongst other things. However, the real threats to a site are unfavourable conditions, the attitudes of the State or business, and uncontrolled tourist activity. Examples include the management problems of the Popigai National Park (Vishnevsky 1991), and of the meteorite crater at Lake El'gygytgyn (NCI 1993).

Future prospects

One of the paradoxes of today is that, due to a lack of real protection from the State and poor public opinion, a threat to geological heritage comes through publicizing sites to a wider audience; this may attract the attention of unsupervised amateurs, and amateurs wishing to exploit the site. So it is desirable to pose some points relevant to an international convention on geological heritage conservation (like the ageements already made in relation to ecology and conservation of fauna and flora). The background to such a convention may wish to address the Russian experience, particularly:

(a) the predicted explosive growth in the number of unique geological sites;
(b) the low level of public participation in the needs of geological heritage;
(c) economic problems.

Both items (b) and (c) show that it is impossible to organize quickly the proper conservation and use of geological heritage. As a result, a lot of new unique sites may experience problems, like the problems encountered in the case of ivory and crocodile skins, and be subjected to plundering and illegal trade abroad. An international convention may stop or limit the development of such activities.

Finally, measures are necessary to activate public opinion and community goodwill throughout the Russian Commonwealth. This might be achieved through:

- State investment, with special support for local needs;
- establishment of a National Geological Heritage Working Group, with the broadest participation of academic science;
- initial concentration of efforts on the successful and complete development of a few specific sites;
- popularization, as broadly as possible, of geological heritage and improvement in programmes of education at all levels;
- support from UNESCO (Kurochkin 1991).

References

BORODIN, I. P. 1914. *The conservation of Natural monuments*. Sankt-Peterburg.

COWIE, J. W. 1991. *Report of Task Force Meeting on Working Group on geological sites. Paris, France, February 1991*. Trondheim, Norway.

KRJUCHKOV, V. V. & KONDRATOVITCH, I. I. 1988. *The Natural monuments of the Northern Kola*. Murmansk.

KUROCHKIN, V. 1991. One of the channels of loss of the Russian Money, or the Paris mysteries of UNESCO. *The 'Izvestisia' Newspaper*, November 16, N 249, p. 1.

NCI. 1993. *The Nature of the El'gygytgyn Hollow: the problems of study and conservation*. North-Eastern Complex Institute, Magadan.

TODOROV, T. 1991. The International Symposium on the conservation of the geological heritage of the Earth. *Razvedka i Okhrana Nedr*, N 11, 36–37.

VISHNEVSKY, S. A. 1991. The Popigai astroblema, a possible site of geological heritage: properties and sights, problems and ideas of conservation. Terra Abstracts (Suppl. 2). *Terra Nova*, **3**, 6.

Theme 3: Local Conservation and Community Initiatives

Legislation and attitudes to geological conservation in Queensland, Australia: past, present and future

PHILLIP G. L. HARLOW

Ipswich Grammar, Ipswich, Australia

Abstract: The attitudes and legislation that governed geological conservation in Queensland in the past, the changes that are occurring in the present (with the consequential changes in action), and the future, are discussed, with examples.

Past geological conservation at a state government level was extremely limited. Conservation efforts concerned with non-geological tourism and particular projects, for example, those initiated by universities, were exceptions to the rule of non-interest in geological conservation.

Local government was sporadically active in conservation of geological sites for geological tourism, resulting from the activities of individuals within the organizations who had a personal interest in geological conservation, rather than the deliberate policy of the local government involved. Conservation by universities and individuals was limited in general to conservation of sites economically important to those involved.

However, in the past five years, a programme of sponsoring increased interest in geological conservation in the community and in government by the Geological Society of Australia has resulted in the formation of deliberate policies, backed up by legislation and action, at both state and local government level. The results of these changes are discussed in detail, with examples. In summary, they have enabled the conservation of sites that would have previously been destroyed. Given current activity, the future of geological conservation in Queensland would appear to be much rosier than the past.

This paper discusses the changes in legislation and attitudes to geological conservation in Queensland, Australia, during the past five years and the consequences for the future. To establish a basis for comparison, conservation activities of the past are examined and are compared with current activities and legislation.

Discussion

Geological conservation in Queensland can be separated into six distinct areas. These include the conservation of:

- landforms for the purposes of non-geological tourism;
- sites of scientific interest;
- sites for purposes primarily concerned with biology;
- sites for educational purposes;
- resources for the purposes of geological tourism;
- heritage sites;

by four distinct groups:

- the state government, often acting in concert with local government, universities, and/or private individuals;
- local government;
- educational institutions (primarily universities);
- individuals.

Queensland had 32 years of Conservative coalition rule before the election of the current Labour government in December 1989. There was no defined policy towards geological conservation during this time, indeed the then Premier, Sir Bjelke Peterson, displayed considerable antipathy towards the conservation movement in general, but his government did support individual cases. In particular cases, such as the conservation of Fraser Island, the government actively pursued avenues which were perceived by conservationists to be counter productive and launched personal attacks on leading conservationists; e.g. Mr John Sinclair, the leader of the Fraser Island Defence Organisation, had his integrity and honesty questioned in Parliament in relation to his conservation activities. In other cases, however, the government aided the establishment of particular reserves and measures to conserve geological landforms in conjunction with individuals, educational institutions and local government bodies. These measures were, however, primarily concerned with sites that were perceived to be of direct economic value to the group looking to conserve them. Historical, educational and scientific sites with no direct economic benefit were ignored.

Thus conservation in Queensland was concerned with:

(1) Conservation of geological structures for the purposes of non-geological tourism. The extension of the Tweed River groynes in 1962–1964 to ensure boating safety led to sand depletion of the Gold Coast beaches and coastal erosion through interference with longshore drift. The state government, under pressure from the local councils and tourist operators, constructed further groynes at individual beaches between 1971 and 1976. When this was not successful, the state government funded dredging operations to restore the beaches after heavy storms. These efforts have met with short-term success and have been primarily directed to conserve tourism, rather than the beaches, *per se*.

(2) In some cases in this era, geological landforms were successfully preserved coincidentally with the conservation of areas primarily concerned with biology. Thus, the Numinbah Valley and Fraser Island have been established as national parks and the geological landforms, the Natural Arch in Numinbah (Fig. 1) and the sand morphology of Fraser Island, have become tourist attractions in national parks established to preserve biodiversity.

This coincidental preservation has occurred in more recent times with the extension of the Lawn Hill National Park. One of the premier palaeontological sites in Australia (UNESCO 1991) at Riversleigh, Queensland, is included in the Lawn Hill National Park extension. This park was established primarily for purposes of conservation of biodiversity and the current regulations governing national parks, primarily concerned with the preservation of biodiversity, seriously inhibit geoscientific work. No scientific reserve, which would better enable the study of fossils, has been declared to this point.

(3) Sites for educational purposes were sparse within the state prior to 1989, conservation being limited to where an educational institution could see a direct economic benefit, or where sites were preserved by individuals under various titles in a spirit of altruism.

Thus, the establishment of The University Experimental Mine, now housing the Julius Kruttschnitt Mineral Research Centre (Fig. 2), by the University of Queensland on an old silver, lead, zinc mine in 1951 was followed up by the vesting of five hectares in freehold title in 1967, and the gazettal of a University Reserve in 1976.

These measures were of significant economic benefit to the university in providing an experimental mine in close

Fig. 1. Natural Arch, Numinbah Valley, Queensland.

Fig. 2. Julius Kruttschnitt Mineral Research Centre, University of Queensland.

proximity, allowing the development of a nationally important mining and metallurgy department.

Reserves of this type were few and driven by the educational institutions, or the personnel within them. Some cases of personal altruism have occurred, one example being the preservation of an old gold mining site near Charters Towers by Dr Eric Heidecker of the University of Queensland under a personal mining lease for educational purposes.

(4) The documentation of sites of scientific interest by concerned individuals was excellent during this period, with numerous publications appearing; e.g. de Jersey et al. (1976), Willmott et al. (1981), Rienks et al. (1984), Willmott, W.F. (1984). However, preservation of geological sites of scientific interest was rare, and often became more concerned with concealing sites from the public to prevent their destruction by the merely curious than establishing them as permanent reserves, e.g. the burial of the insect locality at Denmark Hill, Ipswich under $52\,000\,m^3$ of soil.

(5) Preservation of geological resources for the purposes of geological tourism was fairly widespread throughout the state. This was undertaken by companies and individuals for profit and led to the development of many small sites in the 1970s, e.g. Mount Hay in 1974, Thunderbird Park in 1974 and The Jasper Farm in 1976. Government action in this area was lacking until 1985, when the success of these ventures and the increasing popularity of fossicking, partly as a result of the publicity given to these ventures, led to the establishment of the first public fossicking reserves at Anakie and the 1985 Fossicking Act. In some cases, anonymous individuals signposted sites of geological interest within their sphere of influence; e.g. the Brisbane City Council. Unfortunately, if these individuals ceased employment with the council, this work was often not kept up.

(6) Preservation of our geological heritage for its own sake was extremely limited. In the few instances where it occurred, it has led to the development of small museums, often with co-operation between concerned individuals, mining companies and local government, and in most cases associated again with tourist developments; e.g. the Gold Museum in Gympie. No effort was made towards the conservation of larger or less accessible sites, such as the Palmer River Goldfields, which were largely left to revert to natural bush.

As can be seen from the above, the government of the day was not particularly interested in geological conservation. Thus, there was no provision within the national parks and wildlife legislation specifically aimed at geological conservation (geological activities within parks, such as gold panning or fossicking, were illegal). Conservation was generally left to individuals and organizations with an economic interest, usually tourism.

Most geological sites visited by both school and university students were essentially private visits to operating mines or sites on privately owned land. It was becoming apparent to secondary school and university staff by the mid 1980s that many of these sites were becoming unavailable, particularly through land development, and that some efforts should be directed towards conserving sites as educational reserves. These concerns led to the push by the Historical Monuments Subcommittee of the Geological Society of Australia to increase the community awareness, and particularly government awareness, of the need for a concerted effort towards geological conservation.

However, during the last five years, there has been a concerted effort by the geological fraternity to make the government and the community aware of the geological sites and heritage of Queensland. These activities have made use of the media to bring greater public attention to the geological heritage of Queensland for both geological recreation and scientific study, as well as a simultaneous lobby campaign aimed at both state and local government representatives portraying geological conservation as politically positive. This strategy has been very successful in bringing about change.

Examples of these activities include:

(1) During the reconstruction of the Cunningham Highway, at Mutadapilly, in 1988, a large deposit of Triassic plant fossils was uncovered in the road bed and in the adjacent cutting. This deposit was used to gain national coverage in the print and electronic media. A joint dig was undertaken between Ipswich Grammar and the University of Queensland and some of the resulting fossils were displayed at Expo 1988, leading to further publicity. Subsequently, a temporary educational reserve was declared on a portion of future main road reserve. This established a basis for further action.

(2) Mr Mervin Feeney, a private landowner of Ipswich, was approached to donate freehold land, which had been previously identified in the literature (Willmott et al. 1981). This discussion was followed up with consultations with council and Department of Minerals and Energy staff concerning the mechanics of managing possible future reserves within the city. The educational benefits to students of the city (and the electoral benefits to the Mayor) were also discussed. It was pointed out to the council representatives that land subdivisions in Queensland required that a portion of the land to be subdivided should be vested in the government for public use as a condition of subdivision. That this portion could be invested in an educational or scientific reserve was proposed and accepted in principle. In this way, the land owner/council could provide the land for the reserve at no cost. This mechanism has since been used successfully in another local government area.

(3) A private landowner at Thane, with a family history associated with gold mining in the area, Mr Tim Costello, was approached and donated land for the establishment of a historical reserve in addition to a fossicking area out of a sense of altruism, and the local council was persuaded to manage the reserve. The establishment of these reserves, and the public spiritedness of this man, not only led to good publicity for conservation, but also helped to convince the state government of the need for these reserves.

(4) Public lobbying of the government, particularly the campaigns led by the educational institutions and the lapidary club members, into the difficulties associated with the Mineral Resources Act gazetted in 1991, also led to increased awareness by the government for the need to properly address conservation issues in geology.

As a result of these, and other, activities, several desirable outcomes have been achieved.

(1) The National Parks and Wildlife Service are drawing up joint management plans for a proposed tourism lease in conjunction with the Undara National Park, which contains the most extensive lava tube system in the world (Figs 3 & 4). The Minister for Heritage and Environment sees developments like these as supporting both conservation of the environment and tourism. In the case of Undara, the larger portion of the lava tubes will continue to lie within the national park thus preserving them as is for scientific study, while the tourist resort will use the remainder; part of the money from this then supports the national park.

(2) The new Nature Conservation Act (1992) includes provision for consideration of

Fig. 3. Surface expression of Undara lava tubes, Undara National Park.

Fig. 4. Lava tube, Undara National Park.

geological aspects in future national parks, e.g. Palmer River.

(3) The Department of Minerals and Energy is now pursuing an active role in the preservation of geological sites of scientific, historical, educational, and geologic recreational interest around the state. These initiatives include:

(a) working with the National Parks and Wildlife Service to manage reserves under joint trusteeship as multi-use reserves under the new Nature Conservation Act (1992). Management will be via management guidelines which will prioritize objectives for each reserve. One example is the proposed Palmer River Reserve, where it is suggested that the first priority is preservation of key historic relics (mainly associated with geological history), second is mining, third recreation, fourth nature conservation, etc.
(b) preservation of geological sites of local significance for educational purposes, such as the Thane Creek gold reserve;
(c) preservation of small geological sites of national significance, such as the Triassic/Tertiary fossil site at Dinmore;
(d) preservation of small geological sites of international significance, such as the mammal site at Murgon;
(e) preservation of large geological sites of international significance, such as the World Heritage listing of Riversleigh and Fraser Island.

(4) There is a new fossicking act presently being drafted to manage more successfully fossicking in the state. The Department of Minerals and Energy is actively establishing new fossicking areas, with a gold reserve at Nanango awaiting final approval by the shire council, an opal reserve under consideration at Quilpie, and other sites under discussion. The department is producing an opal brochure which will soon be in print.

(5) Local authorities are becoming more active in preserving the geological heritage they own. The drawing of the mining museum plans, at the site where coal was first discovered in Queensland, by the Ipswich City Council in 1988, and its continuing growth, is a symbol of this new awareness.

Conclusion

As a result of the changes in attitudes and legislation that have occurred at both state and local government levels, sites which would have been previously ignored or destroyed are being preserved for the benefit of both the local and international community, now and for future generations. Given that the change in attitudes and legislation has occurred at both the local and state government level, and the establishment of secure management procedures for the reserves, the future of geological conservation in Queensland appears assured.

References

DE JERSEY, N. J., STEVENS, N. C. & WILLMOTT, W. F. (eds). 1976. *Geological elements of the National Estate in Queensland*. Geological Society of Australia Incorporated, Queensland Division.

RIENKS, I. P., WILLMOTT, W. F. & STEPHENSON, P. J. 1984. *Geological sites in central and northern Queensland, Geological elements of the Nation :l Estate in Queensland, Report 3*. Geological Society of Australia Incorporated, Queensland Division.

UNESCO. 1991. *World Heritage Working Group Task Force Meeting Report, February 1991*. UNESCO

WILLMOTT, W. F. 1984. *Geological Sites in Brisbane. Nominations for the Brisbane Conservation Atlas*. Geological Society of Australia Incorporated, Queensland Division.

——, WEBB, J. A. & WADE, M. (eds). 1981. *Geological sites in southeast Queensland, Geological elements of the National Estate in Queensland*, Report 2. Geological Society of Australia Incorporated, Queensland Division.

The role of voluntary organizations in Earth science conservation in the UK

CHRISTOPHER P. GREEN
Department of Geography, Royal Holloway University of London, Egham, Surrey TW20 0EX, UK

Abstract: Questionnaire responses from 36 voluntary organizations in the UK with Earth science interests indicate that the number of people involved on a non-professional voluntary basis is probably fewer than 6000, and those involved in Earth science conservation probably very many fewer. Voluntary organizations are shown to have four main roles in Earth science conservation: monitoring the condition and accessibility of geological sites; management of sites; dialogue with a wider public through publication and interpretative provision in the field; and representing the case for Earth science conservation to those with statutory environmental responsibilities. The most commonly perceived threats to Earth science sites are identified as problems of access, natural degradation, permitted landfill, illegal dumping, and insensitive collecting. The scale of voluntary sector funding for Earth science conservation is shown to be small.

There is a general awareness in the Earth science community in Britain that voluntary organizations play a significant part in the conservation of geological and geomorphological sites, but the scale and nature of that involvement have never been investigated in detail. This paper reports the results of a questionnaire investigation conducted under the aegis of the Geologists' Association and addressed to some 70 organizations with known Earth science interests.

There is a long tradition in Britain of voluntary involvement with the natural sciences, and there are now many voluntary organizations in Britain with specific interests in Earth science. These organizations fall into six broad classes:

- national general interest geological organizations;
- national and regional specialist geological groups;
- regional general interest geological groups;
- wider interest groups, often natural history and archaeological societies and often county based;
- county wildlife trusts; and
- volunteer geological groups associated with museums.

The questionnaire was addressed to groups which are known to have a large proportion of non-professional members. Responses were received from 36 organizations, including representatives of all the classes noted above, as follows:

- 23 regional general interest geological groups;
- six natural history societies or wildlife trusts with geological sub-groups;
- two regionally based specialist geology groups;
- two museums with active geological volunteer groups; and
- two national geological organizations.

The data are unfortunately by no means comprehensive, but they can reasonably be assumed to be representative of active organizations in which the significance of Earth science conservation is recognized. However, the response certainly does not include all such organizations in Britain.

The results of the questionnaire provide insights into the scale of involvement in voluntary Earth science conservation, and into the nature of that involvement. In addition, from the results, it is possible to recognize some of the limitations to the effectiveness of voluntary organizations in Earth science conservation.

Involvement in voluntary Earth science groups

The numbers of people involved on a voluntary basis in individual Earth science groups, and in the voluntary sector overall, is relatively small. Numbers range from four or five individuals forming volunteer groups associated with museums, to a few hundred in the larger regional societies, and just over 2000 in the largest national groups with significant non-professional memberships – the Geologists' Association and the Open University Geological

Society. Overall, perhaps five or six thousand people, other than professional geologists, are involved in voluntary Earth science organizations. The numbers active in Earth science conservation are probably very much smaller. There is a significant contrast here with some other groups representing environmental concerns. For example, the Royal Society for the Protection of Birds (RSPB) has a membership of about 750 000, the Wildfowl Trust of some 30 000, and the Woodland Trust of some 20 000.

Most Earth science interest groups are not primarily conservation groups. So what conservation role do they play? Of those responding to the questionnaire, a significant minority of 13 reported some sort of formal or recognized responsibility for the management and maintenance of sites of geological interest. Nine of these 13 were regional geological groups and the remainder were distributed, one each, among a national geological group, a wider interest group, a wildlife trust and a museum. The wider context of this site management role is examined more fully below. The most common role of voluntary groups is one of monitoring what is happening to geological sites.

Approximately 70% of the groups responding reported that the present or prospective condition of specific geological sites had been a matter of concern to the group during the last five years. In some 130 separate issues identified, six broad areas of concern are evident.

(1) Access to sites, including both physical obstruction and objections by owners.
(2) Loss of sites to permitted development, including building, landfill and agriculture.
(3) Damage to sites by people with geology-related interests, including the effects of excessive collecting, rock coring and the use of power tools.
(4) The effects of leisure activities, including caving, motorcycles, mountaineering and walking.
(5) Natural degradation of sites.
(6) Illegal dumping of rubbish.

Table 1 shows the more commonly perceived issues and the proportion of respondents referring to each issue. Those causes for concern enumerated in the questionnaire but not appearing in Table 1 (caving, walking, mountaineering, use of power tools) were all cited at least once as actual or threatened issues. An area of concern which was not identified in the questionnaire itself, but which was highlighted by several respondents in the 'other' category relating to 'loss of sections as a consequence of permitted developments', was the loss of sites resulting from the construction of coastal defences and the reclamation of land at the coast.

Table 1. *Respondents (%) reporting concern over Actual/Threatened issues*

	Actual	Threatened
Objection to access	48	12
Natural degradation	48	8
Landfill	44	36
Deteriorating access	40	48
Illegal dumping	32	4
Excess collecting	32	8
Rock coring	28	0
Commercial collecting	20	16
Building developments	16	12
Motorcycles	12	4

Note: Regional Geological Societies only, $n = 25$.

The figures in Table 1 in the 'threatened' column may give an insight into the scope for encouraging foresight in the protection of vulnerable sites from particular threats. Thus, for example, the threats of deteriorating access and of permitted landfill are apparently quite often recognized, whereas the threats of illegal dumping or, more surprisingly, of natural degradation, are more rarely perceived.

Conservation work by voluntary groups

In all the areas of concern identified in Table 1, voluntary organizations have a role to play, which can be summarized as: improving access, either through physical measures or by negotiation with site owners; representing the geological interest to those with statutory responsibilities for management of the environment; and, encouraging good practice within the community in the use of the environment for leisure and recreation. In fact, nearly 70% of respondents reported some form of involvement in conservation activities. These activities fall into two groups (Table 2). Firstly, as noted

Table 2. *Conservation involvement in the voluntary sector (% of groups active)*

	%
Publicizing issues	39
Interpretative guides	33
Interpretative boards	28
Site clearance	25
Site maintenance	22
Site enclosure	5

above, voluntary groups are looking after geological sites in the field, by clearing, maintenance and enclosure. Secondly, and apparently more commonly undertaken, is a role of communication, relating to public awareness and community involvement in care for the landscape. This is achieved through the provision of interpretative boards and leaflets, and through raising local issues with site owners, with local planning authorities, and with national conservation agencies.

The questionnaire provides information for some 49 specific geological sites where issues of concern had been identified. At 32 of these sites, some positive steps had been taken to safeguard the geological interest, and in 21 cases voluntary organizations had played an active role in raising the issue with site owners, local authorities or national conservation agencies.

The effectiveness of the voluntary sector

Communication

Respondents were invited to assess how sympathetically they had been treated in raising Earth science conservation issues. In most cases, respondents recognized a sympathetic reaction. Only one case in a total of 72 responses recorded was placed in the category 'unsympathetic and obstructive'. Otherwise, a 'sympathetic' reaction was recorded, as follows: site owners – 50% sympathetic; local authorities – 60% sympathetic; national conservation agencies – 88% sympathetic; other voluntary groups – 93% sympathetic. The balance in all cases (apart from the one noted above) was perceived as 'indifferent but not obstructive'.

Respondents were also invited to assess the outcome of their representations. The response shows that something was achieved in 70% of the cases, and a reasonably satisfactory outcome ('ideal' or 'adequate – long term') in 40% of cases. This is not a record of 100% success, but it is an important indication that voluntary organizations have the potential to, and do, play a valuable role at site level.

The effectiveness of voluntary involvement in Earth science conservation depends to a significant extent on the effectiveness of communication among the separate, but overlapping, networks of Earth science conservation interest and activity. The questionnaire reveals a strong response by voluntary groups to the relatively new, county-based, RIGS scheme (Regionally Important Geological/geomorphological Sites), initiated by the Nature Conservancy Council (now the national English, Scottish and Welsh conservation agencies) and supported by the Royal Society for Nature Conservation and the Geologists' Association. Twenty groups out of the total of 36 reported 'regular contact' with a RIGS group. Smaller numbers reported regular contact with a museum (14%), a wildlife trust (11%), or the National Scheme for Geological Site Documentation (7%). Only two organizations reported regular contact with all four networks, and a further 8 with three of them.

Involvement

What are the principal limitations to the effectiveness of voluntary organizations in Earth science conservation? First, and perhaps most fundamental, is the small size of the interest groups, especially (as noted above) by comparison with some other environmental interest groups. The questionnaire responses show that general interest geology groups of less than 100 members are often not active in conservation. This is illustrated by the fact that all nine of the larger regional general interest geological groups (membership >100) reported having recognized conservation issues during the last five years, and six reported active involvement in conservation. Whereas for the 14 smaller groups (membership <100), only seven reported concern about conservation issues, and only four of these reported active involvement in conservation. Similar situations are apparent in the development of links with Earth science conservation networks, and in the conduct of negotiations with other landscape interests – it is the larger groups that are more active. This pattern is not unexpected, as in smaller groups there is less likelihood of a conservation interest sub-group emerging. The important point to recognize is that in some geographical regions, concern for the geological dimension of the environment is weakly expressed within the community, and the conservation of Earth science interest correspondingly uncertain.

Important conservation work is being done by some very small volunteer groups, especially where they are associated with strongly conservation-orientated organizations, such as wildlife trusts and museums. However, the issue of involvement highlights the wider problem for Earth science conservation – that few people recognize the welfare of the geological and geomorphological basis of the landscape as a cause for concern. The Earth science interest is, therefore, a relatively weak interest within the broader conservation movement, which itself is a relatively weak interest in the wider social,

Table 3. *Geologists' Association, Curry Fund grants 1989–93*

Purpose of grants	Total (£)	% of total	Voluntary groups (£)
Conservation			
Sites	24 704	14	11 658
Specimens	67 348	38	1 020
Publications	31 192	18	17 672
Personal research	25 333	14	—
Conference organization	15 107	8	3 492
Field centres	12 414	7	4 414
Total	176 098		38 256

economic and political arena. There is a great need, therefore, to articulate this interest to a wider public and ultimately to convince those with responsibilities for the natural landscape – both landowners and local and national government – that rocks and the processes that shape them are as important to our environmental welfare as animals and plants.

Funding

Finally, and closely related to the issue of involvement, is the issue of funding. Of 36 groups responding to the questionnaire, only eight had been able to devote funds to conservation activities, either their own funds or grants attracted from elsewhere. A total expenditure in five years among 36 groups, representing over 2000 people, was just over £2000.

This poor funding record in the voluntary sector is not quite as weak as the questionnaire response suggests. The Geologists' Association, which has always had a strong interest in conservation, has been able to play an increasingly active role in conservation since the establishment of the Curry Fund in 1986. As Table 3 shows, nearly £180 000 has been disbursed in grants from the fund since 1986. Just over half that sum has been devoted to the conservation of sites and specimens. Voluntary organizations have received a total of £38 000 (though evidently not those responding to the questionnaire!). Of this sum, about a third has been towards the cost of site conservation, but more than half has been directed towards communication, through publications and through the development of field centres.

Conclusion

The voluntary sector is playing an important role in the management and monitoring of Earth science sites in Britain. It has, however, another role which is fully as important, and that is the articulation of a more general awareness in communities of the interest and importance of the geological dimension of the natural and cultural environment. Arguably at the present time, this is the most important priority, whether in the professional or the voluntary sectors, because without a strongly articulated public interest in the geological dimension of landscape, it will be difficult to attract the resources that make effective conservation possible.

The RIGS (Regionally Important Geological/geomorphological Sites) challenge – involving local volunteers in conserving England's geological heritage

MIKE HARLEY

English Nature, Northminster House, Peterborough PE1 1UA, UK

Abstract: Historically, geological conservation has been the 'poor relation' in nature conservation circles in England. The conservation of wildlife has, on the other hand, benefited from a strong and influential public following. County-based voluntary groups dedicated to wildlife conservation work closely with land-use managers and planners to select and conserve important sites in their own local areas.

Outside English Nature – the Government's nature conservation agency in England (which has legal powers to protect the nation's best wildlife and geological sites) – geological conservation was, until the end of 1990, only recognized by six local groups. The public's knowledge of, and interest in, their geological heritage, the threats posed to it, and the need for its conservation, has never been great.

Happily, over the last two years, things have changed considerably. One of English Nature's strategic objectives has been to redress this imbalance – this being part of a corporate drive to facilitate greater public awareness of, and involvement in, conserving the natural environment. By assuming the role of catalyst, English Nature has provided the encouragement and support needed to enable local volunteers to set up county-based geological conservation groups. These groups, known as RIGS (short for Regionally Important Geological/geomorphological Sites) groups, operate like their counterparts in wildlife conservation, harnessing the energy, enthusiasm and expertise of local people in identifying and promoting the conservation of key local sites.

The network of RIGS groups now extends to the vast majority of England's 45 counties, with development still ongoing in the very few areas that remain. The RIGS initiative has provided the much needed, and long awaited, stimulus to get local volunteers involved in conserving their geological heritage. In consequence, geological conservation is becoming an increasingly important consideration in local land-use planning, and efforts to raise the public's awareness of this hitherto neglected component of their natural environment are being made in parallel.

English Nature, the Government's nature conservancy agency, has legal powers to protect the best of England's wildlife and natural features. The latter comprise a network of about 1300 geological and geomorphological sites selected for their research interest. These sites are safeguarded by statute as Sites of Special Scientific Interest (SSSIs), and depend upon the local authority planning system and the co-operation and support of landowners for protection.

English Nature is also committed to facilitating greater public awareness of, and involvement in, nature conservation. In England, wildlife conservation has traditionally had a strong local base, with many local people involved and many local sites protected. Geological conservation, however, has lacked popular support. Until the end of 1990, only six locally based geological conservation schemes existed and few local sites were protected.

The challenge

Over the last two years, English Nature has sought to rectify this imbalance with a major project aimed at involving local people in conserving their geological heritage and at giving protection to a nationwide network of geological and geomorphological sites of local importance.

With this in mind, detailed consultations with the groups running existing schemes (geological and wildlife) were carried out, and a consensus of ideas emerged which formed the basis for the development of new schemes. As a result, the RIGS (Regionally Important Geological/geomorphological Sites) initiative was launched, with the vision of local site networks being run entirely by locally based, largely voluntary groups of enthusiastic conservationists and geologists – the role of English Nature being catalytic and supportive.

Getting people together

The concept of sites being selected and conserved at county level by informally constituted groups of local people was promoted extensively. Within each county, members were typically sought from the local wildlife trust, local museum and geological societies. The involvement of local authority planners, teachers, professional geologists and others with interests in geology, geomorphology and conservation was also actively encouraged.

Interest in RIGS grew rapidly and now extends to all of England's 45 counties (shire and metropolitan). Most have an established RIGS group – the few that remain are still recruiting members and defining their organizational base. Each group comprises members with a wide range of knowledge and expertise, with administrative support commonly being provided by a member organization, for example a wildlife trust or museum.

A geological audit

Selecting sites for conservation is a primary task of a RIGS group. But first, the group has to organize a geological audit to find out what sites are there. Site records held by the local museum or local wildlife trust (over 16 000 sites are listed in England as a whole) provide a strong starting point. These are then supplemented with information gathered in local field surveys carried out

---

GUIDELINES FOR EVALUATING REGIONALLY IMPORTANT GEOLOGICAL/GEOMORPHOLOGICAL SITES (RIGS)

PARAMETERS:

Scientific Importance
Educational Value
Historic Associations
Aesthetic Characteristics

ASSESSMENT PROCEDURE:

1. SCIENTIFIC IMPORTANCE

Petrology

a)   Does the site expose igneous/metamorphic/sedimentary rocks representative of ....?

b)   Does the site demonstrate important igneous/metamorphic/sedimentary features?

Stratigraphy

a)   Does the site show stratigraphic features representative of ....?

b)   Is the site important for stratigraphic correlation?

Palaeontology

a)   Is the site important for a particular fossil species or assemblage?

b)   Is the site important palaeo-ecologically?

Mineralogy

a)   Is the site important for a particular mineral or assemblage of minerals?

Structure

a)   Does the site demonstrate any important structural features?

Geomorphology

a)   Is the site an important landscape feature in ....?

b)   Does the site demonstrate geomorphological features or processes characteristic of ....?

by group members (particularly those from local geological societies), and a complete geological record for the county is compiled.

Selecting sites for conservation

The process of selecting sites for conservation is based on clearly defined but locally determined criteria, against which each site in the county geological record is tested. These criteria commonly fall into four broad themes.

- The value of a site for educational fieldwork in primary and secondary schools, at undergraduate level and in adult education courses.
- The value of a site for scientific study by both professional and amateur geologists. Such sites might demonstrate, alone or as part of a network, the geology or geomorphology of an area.
- The historic value of a site in terms of important advances in geological or geomorphological knowledge.
- The aesthetic and cultural value of a site in the landscape, particularly in relation to promoting public awareness and appreciation of geology and geomorphology, its links with society and the need for conservation.

The balance of importance between these themes varies from county to county, although emphasis tends to be placed on selecting sites of scientific and educational value. The number of sites selected also varies, this being a function of

Other Scientific Interests

a) Does the site have any wildlife significance?

b) Is the site protected for its wildlife significance (eg. nature reserve)?

2. EDUCATIONAL VALUE

a) Is the site suitable for teaching the earth science components of the National Curriculum?

b) Is the site suitable for teaching the earth sciences at 'A' level or undergraduate level?

c) Is the site suitable for other educational users (eg. adult classes)?

d) Is the site physically accessible?

e) Is the site safe?

f) Is access to the site permitted for educational visits?

3. HISTORIC ASSOCIATIONS

a) Is the site historically important in terms of advances in geological/geomorphological knowledge?

b) Has the site any associations with culture, folklore or religion?

c) Has the site any archaeological significance?

4. AESTHETIC CHARACTERISTICS

a) Is the site an essential component of an attractive or evocative local landscape?

b) Could the site be used to promote public awareness and appreciation of geology/geomorphology?

NOTES:

Groups will need to compile lists of the salient features of their county for use in the assessment procedure (eg. particular rock types, fossils, minerals etc).

Groups may wish to consider developing a scoring system, with thresholds, to assist in site selection.

Fig. 1. Guidelines for evaluating RIGS.

geological diversity and the amount of outcrop in the county.

To assist RIGS groups in selecting appropriate sites, English Nature worked with a small group of experienced practitioners in this field to formulate a set of site evaluation guidelines (Fig. 1). Whilst not being exhaustive or put forward as a standard for universal application nationwide, these guidelines do, however, provide a checklist against which groups can develop their own criteria to meet specific local needs.

Conserving sites

The local authority planning system is the most effective means of conserving RIGS. As with SSSIs, the system provides the first line of defence in the protection of a site. But the protection given to RIGS is entirely discretionary and not statutory, as in the case of SSSIs. Planners are encouraged to recognize RIGS in land-use planning policies and in nature conservation strategies. RIGS groups supply site information to these authorities and ensure that it is periodically reviewed and updated. RIGS groups also comment on planning applications affecting 'their' sites. By providing a reasoned reaction to development proposals, site damage can be countered and sensitive modifications brought about through negotiations with both planners and developers. To date, 22 RIGS groups have submitted site lists to their local authorities, amounting to a current total of some 1700 RIGS having planning protection.

Once a network of local sites has been safeguarded through the planning system, RIGS groups can look at ways of managing these sites for conservation. Certain sites may benefit from small changes in land management practice to improve their protection. Many of these changes, such as disposing of farm waste away from rock faces, can be carried out by sympathetic landowners. At some sites, particularly those of educational value, further enhancements may be required. With the owners' consent, RIGS group members can undertake a wide range of conservation activities. These might include:

- practical management – creating pathways, clearing and stabilizing rock faces, fencing dangerous areas and establishing stockpiles for collecting;

- interpretive work – producing fact sheets, guide books and display boards, both for educational groups and the casual visitor.

RIGS groups can also encourage the involvement of those who use sites, particularly educational groups, in conservation projects. Besides assisting site conservation, this helps to spread knowledge and acceptance of the need to conserve geological and geomorphological features more widely. In this context, several groups have sought publicity for their activities through television, radio and local newspapers.

Conclusions

With the network of active RIGS groups now extending to most English counties, and with development work ongoing in the very few that remain, English Nature is justifiably proud of having achieved its stated aims in establishing the RIGS movement. The enthusiasm with which the RIGS concept has been met reflects escalating support for geological conservation in England. RIGS are proving to be an effective and widely recognized method of conserving hitherto unprotected locally significant geological and geomorphological sites. RIGS are also a highly successful means of involving local people in this one-time neglected area of nature conservation. An unexpected bonus has been the positive reception of RIGS by local authority planners, their frequent proactive involvement having led to geological considerations becoming increasingly more important in local land-use planning in England.

It has always been English Nature's intention to act as a catalyst in the RIGS initiative, and not to own or run the project in the long term. This still hold true and, whilst continuing to support the development of the project by providing a strategic input to sustaining and enhancing the national RIGS movement, English Nature are gradually passing operational responsibility to the voluntary sector – the rightful owners – under the umbrella of RSNC (Royal Society for Nature Conservation), now renamed The Wildlife Trusts. In producing a forward plan for RIGS, together English Nature and The Wildlife Trusts, in consultation with many of those involved, will be looking at ways of maintaining the momentum and commitment of England's network of voluntary RIGS groups.

Further reading

NATURE CONSERVANCY COUNCIL. 1990. *Earth science conservation in Great Britain – a strategy.* Peterborough.

HARLEY, M. & ROBINSON, E. 1991. RIGS – a local earth science conservation initiative. *Geology Today*, **7**(2), 47–50.

KNELL, S. 1991. The local geologist 2: Making rock records. *Geology Today*, **7**(2), 62–66.

RSNC, THE WILDLIFE TRUSTS PARTNERSHIP. 1993. *Starting RIGS.* Royal Society for Nature Conservation, Lincoln.

Conservation and management of geological monuments in South Australia

ROSEMARY SWART

Mawson Graduate Centre for Environmental Studies, University of Adelaide, Adelaide 5005, Australia

Abstract: This paper presents results from a study of the protection of geological sites in South Australia. It is shown that current protection strategies are only offered in terms of the protection of natural areas or areas of scientific interest, particularly in relation to their inclusion in national parks, World Heritage listing, National Estate listing and similar schemes. These schemes are not sufficiently directed towards geological issues for the sites to fulfil their purpose for scientific research and teaching.

The inclusion of geological sites in the general national parks protection model can work if the site has minimal maintenance requirements and is not one where collection forms part of the research uses of the site. It is inappropriate for small, highly significant areas, especially those in remote areas, where inadequate funding and staffing levels mean that such sites are rarely visited. Lack of understanding of the importance of geological sites within such organizations reduces the effectiveness of any protection available under legislation.

This paper concludes that, although geological sites in South Australia have varying status in terms of international, national, and South Australian legislation, it is apparent from this research that these levels of protection are often outweighed by local controls and public perception.

Natural heritage areas are generally perceived as being of predominantly biological importance. In Australia, Hunwick (1992) writes that national parks are of little use if they are not actively saving our endangered wildlife; but in many Australian national parks, it is geology that is the 'natural heritage centrepiece' (Uluru, Flinders Ranges for example), yet major administrative bodies which monitor such places do not employ geologists (Osborn 1991).

Protection of geological sites in Australia occurs at a number of different levels. Sites of world significance are in some cases protected by default; they have been included on the World Heritage List for their cultural attributes as much as for their natural features (Slayter 1989). In other cases, as will be shown, world class sites are offered little or no protection.

In Australia, geological sites that are of national rather than international importance may be protected by the Australian Heritage Commission, through inclusion on the Register of the National Estate. The Register of the National Estate lists items and areas worthy of protection for future generations. It is the responsibility of the Commonwealth Government to ensure sites are not adversely affected by activities such as mining or development. This, therefore, places restrictions on what Commonwealth activities may be permitted in areas listed in the Register. Most of the areas, though, are under direct State jurisdiction (only the Northern Territory and the Australian Capital Territory are under direct Commonwealth legislation) which take precedence. This means that should a State Government decide to allow mining or other potentially destructive activities in them, they may proceed (Washington 1991). The Australian Heritage Commission has no power over local government or private landholders, and the listing of a place does not change its status with respect to public access. There are no provisions for financial penalties or penal clauses in the legislation, hence it merely provides moral protection (Hardy 1987).

Geological heritage protection in South Australia

In this study, five specific sites in South Australia were selected for detailed investigation. They represent different levels of geological importance and are under different management regimes. This approach was adopted to provide an indication of effectiveness of site protection in South Australia, within a limited time frame. A summary of sites chosen and the criteria used in their selection is given in Table 1.

The Ediacara Fossil Reserve

This is a remote site, located about 40 km south-west of the township of Leigh Creek (about

Table 1. *Selection criteria used in choosing case studies*

Site	Summary of selection criteria
Ediacara Fossil Reserve — Precambrian fossil locality	World Heritage significance as geological monument; on Register of the National Estate, State Heritage listed; threatened by over-collection; exemplifies problems of protection of remote sites.
Sturt Gorge — Precambrian glacial deposit	State importance as geological monument; State Heritage listed; in a conservation park and close to the city; should be protected by National Parks and Wildlife Service.
Reeves Point — Jurassic basalt	State importance as geological monument; State Heritage listed; in historic reserve – relatively well protected.
Christmas Cove — Permian glaciation	Regional importance as geological monument; threatened by development; conflict of land uses.
Horse Gully — Cambrian fossils	Regional importance as geological monument; on private land; owner is responsible.

700 km north of the city of Adelaide), in northern South Australia. The site occurs within the Pound Quartzite, the youngest Precambrian rock formation in the area. Fossil jellyfish were first located in the rock in 1946 (AHC 1985), and when in 1958 a range of fossil soft-bodied animals was discovered at Ediacara, the locality became world famous. The fossils represent the earliest known animal assemblage in the world (Hasenohr 1978). It is a geological site of global significance.

A 'fossil reserve' is a site that is under the control of the South Australian Museum Board, who regulate collection and access. Specific permission is granted by the Museum Board for people to visit the area occasionally, and groups have been taken to visit the site, with people from the museum checking the material found to ensure pieces of scientific value are retained by the museum.

In the last five years, there has been debate about the transfer of control of the site to the South Australian National Parks and Wildlife Service (SANPWS), a State Government body who would administer it as a conservation park. The SANPWS claimed that it was able to provide the strict regulatory framework to protect the site, a claim which has since proved to be untrue. The SANPWS does not have a strong understanding of the 'geological' and monument management issues involved in such site management. The lack of geological expertise and interest shown by the SANPWS with respect to similar fossil sites within the Flinders Ranges indicates their lack of ability to provide the type of care that ideally is required by such a vulnerable area. Fossils from sites within the Flinders Ranges have been given to the SANPWS – only to be lost in a garage!

Sturt Gorge

Sturt Gorge is in the foothills of the Mount Lofty Ranges, 13 km south of the centre of Adelaide. It contains a Precambrian glacial deposit, the Sturt Tillite. Sir Douglas Mawson described Sturt Gorge as 'one of those few areas in any country that should never be alienated from the State, but preserved as a National Reserve' (NPWS 1990). The immediate area containing Sturt Gorge is now a recreation park managed by the SANPWS. Both its geological and historical significance make it an important geological site.

There is, at present, no provision of maps or trail markers in the park, although the management plan notes that this is a medium-term priority, along with the preparation of a self-guided walk to incorporate the geological features. A lack of resources has hindered any development of interpretation material within the park.

The other three sites from this study are 'protected' by a variety of measures, with more or less success.

Reeves Point

In geological terms, Reeves Point is a site of State significance. It occurs within a historic reserve that is State Heritage listed and as such is relatively well protected in legislative terms. There has to be some provision of interpretative material at the site, but little if any maintenance is carried out and, as a result, the geology is largely obscured by vegetation and rubbish dumped in the locality.

Christmas Cove

Christmas Cove has been adversely affected by insensitive development whereby the geology has been obscured and is in danger of destruction. The site is of regional geological importance, but the local council's ignorance of its geological significance, and their desire for economic gain by developing a ferry terminal within the site, overruled the calls by geologists and concerned locals for protection of the site.

Horse Gully

This is a small fossil site, also of regional geological importance, on private land, on the Yorke Peninsula. The owner of the site is interested and committed to its protection. With advice from the South Australian branch of the Geological Monuments Sub-committee of the Geological Society of Australia, he has instituted a permission form system, and those who enter the land without permission are asked to leave. Permission for access is granted to most who apply (20–30 groups per year), but this does not entitle those people to collect material. Specific requests to collect material must be made, and with reason or backing from a research institution, before such permission is granted.

Discussion

Examples such as Ediacara, South Australia, indicate just how difficult it is to monitor sites against over-collection or poor collecting techniques. A permit system applies in national parks in South Australia, where the number of specimens to be collected is specified on a collection permit, but few people are aware of it. This may be effective if there are people at the site to ensure that the system is adhered to, but is very difficult to enforce at sites that are isolated or not constantly attended. One of the major reasons for the over-collection at Ediacara, that has led to the virtual destruction of the site, is its remoteness. Collecting is only allowed at the site with permission, but there is no way of monitoring visitors or activities. The nearest police are at Leigh Creek, 42 km away, with responsibility for an area of thousands of kilometres.

Even sites within the metropolitan area are not necessarily adequately catered for. Sturt Gorge falls within metropolitan Adelaide, but the SANPWS's lack of resources has prevented adequate clearance of the area. Although they acknowledge the need for interpretative material at the site, they are unable to provide it.

The contrast between Christmas Cove and Horse Gully indicates the protection that can be made available with a little community understanding. Not only is the Horse Gully site protected, but the local council in that area aims to actively promote geology and geological heritage within the district as a tourist attraction and education resource. At Christmas Cove, the council wants to increase visitors to the area by installing a ferry terminal at the expense of a picturesque and important geological site and potential tourist attraction (there is an existing ferry terminal nearby, operated by an independent organization).

Large government organizations may not necessarily be the best means of providing adequate protection to geological sites. The SANPWS have responsibility for large areas of South Australia on very limited funding, which restricts the activities that can be undertaken. The most successful methods encountered within the scope of this work were those at the local community level, where interest and local pride encouraged the protection of the local area. Ignorance of the significance of sites may result in their deliberate or inadvertent destruction through the actions of those in direct control, most often the local council. Public perception is most important at this level because it is they who vote councils in and out, and hence it is the voters that councils seek to please when making planning and development decisions. Education of the public must, therefore, be the first priority in better conservation.

References

AHC 1989. *Geological Sites and Monuments*. Background Notes No. 46, Australian Heritage Commission, Canberra.

HARDY, B. 1987. Conservation of Geological Monuments as Heritage Items. *ICCM Bulletin*, 13(1–2), 107–115.

HASENOHR, P. 1978. *Precambrian Metazoan Fossils, Ediacara Fossil reserve, Central South Australia*. Register Assessment Report, S.A. Heritage Act.

HUNWICK, J. 1992. National Parks after the Centenary. *Environment South Australia*, 27, 75–90.

NPWS. 1990. *Sturt Gorge Recreation Park Management Plan*. National Parks and Wildlife Service. Department of Environment and Planning, South Australia.

OSBORN, R. A. L. 1991. Geological Sites and Monuments. *The Australian Geologist*, 81, 6–7.

SLATYER, R. O. 1989. The World Heritage Convention. *Heritage Australia*, **8**(2), 3–5.

WASHINGTON, H. 1991. *Ecosolutions: Environmental Solutions for the World and Australia.* Boobook Publications, New South Wales.

Two geological monuments in the Netherlands: De Zândkoele and Wolterholten

GERARD P. GONGGRIJP

Institute for Forestry and Nature Research, P.O. Box 23, NL-6700 AA Wageningen, The Netherlands

Abstract: Geological monuments have been established in two drumlinized ice-pushed moraines. The first site is situated in a former sand pit, later used as a communal storage yard and cycle-cross terrain. A good relationship with the local authorities favoured the development of an educationally and scientifically important site. The site combines an illustrative exposure of a Saalian till and cover sands from the Weichselian period, with an instructive map of Scandinavia and 'guide' boulders.

The second site is the result of the construction of the A 32 motorway through the till covered drumlin. Big erratic boulders collected from the road cut decorate a till exposure. Connections with the Department of Public Works authorities and a local section of a nature organization led to the establishment of this monument.

In both cases, the strategy – a personal approach combined with involvement and flexibility – turned out to be successful. The educational value of the sites was increased considerably by the publication of booklets giving a popular geological history of the area combined with a cycle trail that introduced the geology to the general public.

The Netherlands is a country of low relief in comparison to many other countries, and accordingly has few natural geological exposures. In consequence, Earth scientists and interested amateurs who want to examine geological sequences have to rely to a large extent on artificial exposures, such as pits and quarries, to satisfy their curiosity. Their reliance on human-made exposures of this sort, however, is not without problems, for such exposures mostly tend to be rather short lived and disappear through the dumping of rubbish, replantation or just plain neglect. To prevent the loss of such sites and the unique information they provide, Earth science conservationists are succeeding in their efforts to have more and more sites conserved and maintained in a 'geologist friendly' condition, for example the geological monument 'De Zândkoele' in the province of Overijssel. Another way to provide more geological sites is to create exposures by digging pits especially for this purpose, as in the case of the geological monument 'Wolterholten' in the same province.

Geological setting

Both geological sites are situated in the Pleistocene half of the Netherlands, in the northern part of the province of Overijssel, and both occur in the same type of geomorphological feature (Fig. 1). The northern part of the country is represented by several low ridges of ice-push moraines up to 20 m high, bulldozed by the advancing Scandinavian ice sheet during the Saalian glaciation. According to the thick till layer upon these removed pre-glacial materials, represented by fluvial, fluvioglacial and aeolian sediments, they must have been overridden by

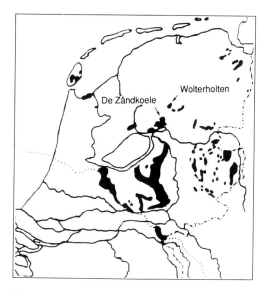

Fig. 1. Location of the geological monuments 'De Zândkoele' near Heetveld and 'Wolterholten' near Steenwijk on a map showing the ice-pushed moraines in the Netherlands.

the ice sheet after first having been pushed. This process led to the streamlining of the small hills into drumlinized forms, with rather thick till deposition on the downstream side. Both sites are located on such drumlinized ice-pushed moraines.

During and after the Saalian glaciation, erosion and solifluction lowered the relief. Arctic conditions in the Weichselian period caused considerable sand transportation and sedimentation. The result was an extensive layer of cover sands deposited on the interglacial Eemian landscape, sometimes flat, sometimes hilly with dune formation, depending on the environmental circumstances. Only the higher parts in the landscape, like the 'Wolterholten Hill', were free of cover sands.

Geological monument 'De Zândkoele'

The history of the monument began in the spring of 1981 (Gonggrijp 1990). On a field survey, the author examined a neglected sand pit, at that time used as a communal storage yard. This pit had been excavated in a drumlinized ice-pushed moraine known as the 'High Land of Vollenhove' – after a small town lying three miles to the west. Although the face of the pit was badly exposed, enough could be seen to understand its importance for education. The exposure showed a Saalian till overlain by a Weichselian cover sand layer, very representative of the geological stratigraphy of the northern part of the Netherlands, but very rarely exposed.

First, the owner, the Municipality of Brederwiede, did not reply to the letter in which safeguarding was asked; however, a year later a new letter established contact. Then it transpired that the local youth had officially been promised the use of the pit for cycle-cross racing and that the pit would be arranged as a crosstrack. However, the physical properties of the till proved unsuited to the needs of this sport and conservation was given another chance. Even the apparent set-back had its silverlining: for, in the preparation of the racing track, fresh faces had been opened to supplement the original exposures. These showed much more geological

Fig. 2. The geological monument 'De Zândkoele' founded in 1984. The exposure in the background shows Saalian till covered with Weichselian cover sands containing several periglacial phenomena. For educational purposes, displays and a 'boulder map' of Scandinavia were constructed in the former pit to inform visitors about ice ages and the origins of erratic boulders.

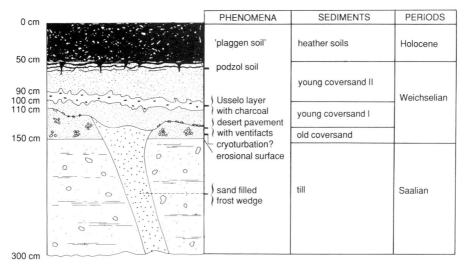

Fig. 3. Outline of the exposed geological section. This sequence is representative of a large part of the northern Netherlands. However, at the moment, this exposure is the only one which shows all these phenomena together.

detail than before. The site turned out to be also of scientific importance, especially the features in the till.

In 1983, after renewed contact with the municipality, the author was asked to report on the geological importance of the site. The report, including information on the geology of the municipality, was talked over thoroughly with the Mayor and Aldermen in a pleasant atmosphere. The author stressed the importance of the site as a nucleus for geological education in the area – the geological history could be reconstructed on the basis of the site's profile – and for the promotion of popular Earth science tourism by using the site as a centre for a geological bicycle route. The varied geomorphology in the area made it a very attractive place to design such a route.

In a second report ordered by the local authorities, some alternative designs for the layout of the monument, varying in complexity and costs, were presented. After a presentation of the plans in a special council meeting, the authorities decided upon the most advanced and also the most expensive alternative. This included the conservation of the sand pit, with till and cover sand sediments, and features like frost wedges, an erosional surface, oxidized next to reduced till, a desert pavement with ventifacts, a remnant of a Late Weichselian podzol profile, and a recent podzol profile partly influenced by long-term addition of sods as a fertilizer (Fig. 3). To make the site more informative to the general public, an outline map of Scandinavia, measuring 30 m by 30 m, was marked out on the pit floor using small boulders. On this were placed the larger ice-carried boulders found in several parts of the Netherlands, each at the location on the map representing its original source, so that visitors could have an idea of what had happened 150 000 years ago (Fig. 2). For convenience, a viewing mound was provided, and there are two interpretative displays related to the map and to the exposure to be seen in the pit.

The money for the construction was provided by the municipality itself and by a general fund for the development of the northern provinces. To keep the costs as low as possible, the work was mainly carried out by two unemployed volunteers.

Several times during the realization of the project, the workers were confronted with vandalism. In consultation with the authorities, the author organized a session with the teachers and the pupils of the schools in the village. After this instruction meeting, vandalism of the site almost completely disappeared. From this, it was learned that involvement of the locals is also a main condition for the successful establishment of a site.

Because of the personal involvement of the Mayor and the efforts of the volunteers, the work was finished in a relatively short time. On September 6, 1984, the State Secretary of the Ministry responsible for Nature Conservation officially opened the site. However, after a while, despite regular police supervision, some vandalism returned. Again the authorities made an appeal to the locals and a woman living opposite was asked to look after the site in return for a small compensation. Since then, destruction has been less.

In general, the management of soft rock exposures is not very easy because of the denudational processes affecting the wall. However, in this case the height is only 2.5 m and the type of sediments are rather favourable, so that regular cleaning is sufficient.

Geological monument 'Wolterholten'

This monument has a quite different history. In 1987, the Department of Public Works started the construction of a new motorway (A 32) cutting the Woldberg, a drumlinized ice-pushed moraine (Fig. 4). The works carried out resulted in beautifully exposed layers of three different types of till. At that time, the department received two requests on behalf of nature conservationists. The first request, by a local section of a national natural history association, suggested not covering the slopes with rich soil to make natural vegetation development possible. The second request, by the author, asked for the realization of a steep bank on a geologically interesting point to use the site for education and scientific research, and for safeguarding the many big Scandinavian erratic boulders that came available by digging through the till covered drumlin. The authorities could not grant the first request because of the erosion problems. The author's request was accepted on the understanding that, instead of the exposure in the bank, a special pit should be dug in a farmyard pulled down because of the road construction (see Fig. 6). Besides, the erratic boulders could be erected in the yard and, with the till pit, form a geological monument (Fig. 5).

This plan was executed in close co-operation with the Department of Public Works and the geological section of the natural history association, which provided the know-how for the determination of the erratics. The geological monument was established in 1989 and comprises a drained pit with an exposure in till and a selection of the most important erratics. This selection was based on size, mainly big boulders to prevent illegal transport; on composition, with a preference for 'guide' boulders; and on the beauty of the stones, well-structured or coloured ones. All the boulders were complemented with name plates. An information display completed the site. With respect to management, it was decided that the Municipality of Steenwijk (named after the very stone-rich area – *steen* means stone) was responsible for the site after conveyance by the Department of Public Works. The geological section of the association who adopted the site watch it regularly and con-

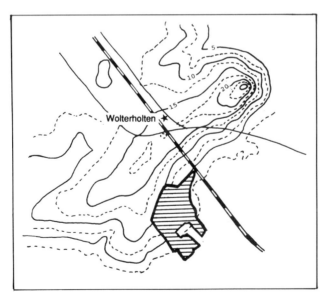

Fig. 4. The geological monument 'Wolterholten' situated on the till-rich tail of the drumlinized ice-pushed moraine.

Fig. 5. Entrance of the geological monument 'Wolterholten', near Steenwijk, situated in a former farmyard between a rail road and the new motorway. The site includes an erratic boulder collection and a specially constructed pit with till exposures.

tact the municipality in case of problems, and also carry out some minor maintenance work.

This site was just like 'De Zândkoele' as it, too, was affected by vandalism. For example, several of the name plates disappeared, but they were replaced as soon as possible by new ones. Some big boulders placed outside the fence were stolen. In this case, it is a disadvantage that the

Fig. 6. The specially constructed pit showing till. Originally the site should have been situated in the wall of the road cut.

site is situated away (two miles) from the town of Steenwijk. Social control does not work here, as it does in Heetveld, where the site is situated just outside the village. The site is now included in a police surveillance route.

Sometimes there are different opinions on the responsibilities concerning the site but, in general, the appointments are being kept rather well, although the municipality is less involved than in the case of the Brederwiede municipality.

Remarks and conclusions

Finally, it is important to list some conclusions and remarks based on several years of experience in the protection of geological sites.

(1) A sophisticated nature conservation law, that of course includes geological conservation, is the basis for an efficient protection policy. However, that is not enough to be successful in the establishment of geological monuments. You need an active group of popularizing Earth scientists to establish a base for the public's interest in the geological history of the landscape.
(2) The preservation of geologically important exposures and relatively small geomorphological features should be preferred at low organization levels, like a municipality or regional section of a (nature) conservation organization, because of the short bureaucratic distance between initiator and landowner.
(3) Probably the most important factor in creating geological monuments is to have involved Earth scientists who can put over their meaning and enthusiasm to the (local) authorities or private landowners. Bureaucratic pressure from higher authorities on unwilling lower ones or private owners seldom really works in nature conservation, especially with vulnerable exposures in pits and quarries where preservation can easily be frustrated.
(4) Adoption of a site by a geological society or another conservation or nature organization on a local or regional level can be a good guarantee for preservation and management. These organizations often have a much better chance of involving the local people and they make the work of non-expert owners easier.
(5) All actions should be focused on arousing feelings of local pride, so that the locals feel responsible for their site(s) and preservation is guaranteed in the future.
(6) Give the site(s) a nucleus function in a regional geological history trail. Geological walks and cycle routes are very instructive ways to demonstrate geological landscapes. Both the above-mentioned sites are nuclei in cycle routes. Of course, in relief-rich areas, car routes can be preferable.
(7) Sites do not always have to be of a very high educational or scientific standard. Even simple sites just showing a rather common, but representative, feature in the area can be developed as a monument for educational purposes.

Reference

GONGGRIJP, G. P. 1990. De Zândkoele. *Naturopa*, **65**, 18–19.

The legal framework and scientific procedure for the protection of palaeontological sites in Spain: recovery of some special sites affected by human activity in Aragón (eastern Spain)

G. MELENDEZ & M. SORIA

Lab. Paleontología, Dpto Geología, Universidad de Zaragoza, 50009-Zaragoza, Spain

Abstract: The legal framework for the protection and preservation of palaeontological sites in Spain is still poorly developed. A practical case is presented for two classical Jurassic sections in Aragón, NE Spain, at the localities of Ricla and Aguilón (province of Zaragoza). These outcrops have been proposed as reference sections for the Callovian and Oxfordian of western Europe and are seriously threatened by numerous rubbish pits. The legal procedure is usually inefficient and slow, so the most effective action against these types of threat seems to be in the hands of palaeontologists themselves. This involves, on the one hand, official claims to the local authorities and, on the other hand, labour-intensive work educating the local population about the great importance of palaeontological heritage in Aragón. At the same time, an effort is being made by the Spanish Palaeontological Society to create a new category of legal status for some palaeontological sites of special interest.

Palaeontological sites in Spain are legally protected by two main laws: the Law of Protection of Historical Heritage and the Law of Protection of Natural Areas. However, these two laws appear to be little related to palaeontological heritage and quite inefficient when it comes to protecting threatened palaeontological sites. The purpose of this paper is to illustrate the current situation and the administrative procedure used to protect and recover any palaeontological site which may have been under any kind of threat. This situation makes it advisable to propose the creation of a new category of legal status, specifically directed at the protection of palaeontological sites, based on criteria related to their scientific interest.

The legal framework

The existing laws

According to the *Law of Historical Heritage*, palaeontological heritage sites are protected. This legal protection covers any palaeontological site or outcrop which has been the subject of scientific publication, thus forming part of the palaeontological heritage. However, the nomination of a palaeontological site in Spain as a specially protected area, within the framework of the Historical Heritage Law, is carried out by the declaration of that area as 'Bien de interés Cultural' (BIC – of Good Cultural Interest) (Fig. 1). This declaration comprises:

'Sitio Histórico' (SH – Historical Site). This is 'a natural place or landscape maintaining some kind of relationship with events of the past, popular traditions, natural creations and works of humankind having some sort of historical, ethnological, palaeontological or anthropological value'.

'Zona Arqueológica' (ZA – Archaeological Zone). This is 'a natural place or landscape where heritage features suitable for scientific study by means of archaeological research can be found; these features may have been excavated or may still be *in situ* and can be located either above or below the Earth's surface, or under territorial waters'.

Within the framework of the *Law of Protection of Natural Areas*, the legal status which allows for the declaration of a palaeontological site as a specially protected place is the so-called 'Monumento Natural' (MN – Natural Monument). A natural monument comprises either 'geological formations, palaeontological sites or other components of the Earth having a special interest because of their inherent importance as a part of the landscape or by their scientific and cultural value'.

The bureaucratic procedure

The nomination of a particular area or site as a specially protected place depends upon the government of each autonomous region into which the Spanish territory is subdivided. Any particular person or official institution is allowed to submit an application form. In the case of a BIC declaration, this has to be submitted to the

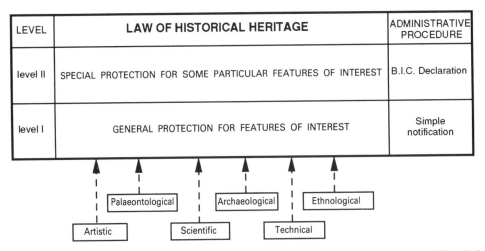

Fig. 1. Diagrammatical sketch of the different levels of protection or legal status which exist in the Historical Heritage Law for palaeontological sites.

Department of Culture, to be evaluated by an expert commission. In the case of the MN category, the application has to be sent directly to the Agriculture Department, no further report by a particular commission being required here.

The long and winding administrative road

A long bureaucratic process has to be endured before a palaeontological site can be declared BIC and adequately protected. According to the text of the law, an administrative response must be given to the application before a maximum time period of twenty months. In the case of a positive result, the place would then be protected by law from further potential threats. The future care of the site will be undertaken by the regional government who, as a rule, will set up an official agreement with the local authorities (the town hall and the municipal institutions), in order to establish a special protection programme. The main drawbacks of this administrative process are related to the status of BIC itself. From the two possibilities that the declaration of BIC allows, that of 'Historical Site' appears rather more appropriate for historical and artistic monuments of special relevance. In fact, only one proposal of a palaeontological site as BIC has been submitted, and rejected, in the region of Aragón. The second possibility of BIC declaration status as an 'Archaeological Zone', clearly subordinates the palaeontological value of a site to archaeological concepts and criteria. Furthermore, once a particular place has been declared BIC, it is possible for it to become overprotected, such that no further exploitation or palaeontological collection is allowed.

The second legal status, MN (natural monument), appears, at first sight, somewhat more suitable for palaeontological site protection. The administrative process takes a shorter period of time (90 days) before resolution and this, combined with a closer link to nature protection problems, make this status apparently suitable for palaeontology. However, the lack of a further detailed programme of protection and the fact that the main criterion of a candidate for designation of a MN has to be either its wealth of landscape types or its wealth of fauna or flora, makes the declaration of a palaeontological site as MN difficult and unlikely. Therefore, in Spain, it can be seen that the inadequacy in the protection of natural features of value is not a consequence of the non-existence of a legal framework, but rather the opposite – the result of a rigid and bureaucratized set of regulations, and a lack of precision in the programmes for further protection.

A case of practical action

Simple notification

Any declared feature of historical interest will be protected by the Law of Historical Heritage by simple notification to the local authorities (Fig. 1). A simple notification is an action that is usually taken when a palaeontological site has suffered from any kind of threat. It involves the production of a short report that should be submitted to the Department of Culture in the regional government. The report should emphasize the scientific value of the site, the technical aspects of the potential threat and the possible solutions.

The case of the palaeontological sites of Ricla and Aguilón (NE Iberian Chain, southern Zaragoza)

Middle to Upper Jurassic sediments crop out widely in numerous outcrops in the environs of Ricla and Aguilón, two classic localities located in the Mesozoic band ranging along the southern margin of the Ebro valley (Fig. 2). Jurassic outcrops at these points provide great stratigraphical and palaeontological interest, either as type sections for particular stratigraphic units (Aurell 1990), or as type localities for certain ammonite species (Meléndez 1989). Some outcrops have been proposed as reference sections: Cariou *et al.* (1988) for the Callovian of Ricla; Goy & Martinez (1990) for the Toarcian of La Almunia and Ricla; Sequeiros & Meléndez (1981), Sequeiros *et al.* (1984, 1986), Lardiés (1989), Meléndez (1989), Fontana (1990), Fontana & Meléndez (1990) and Meléndez &

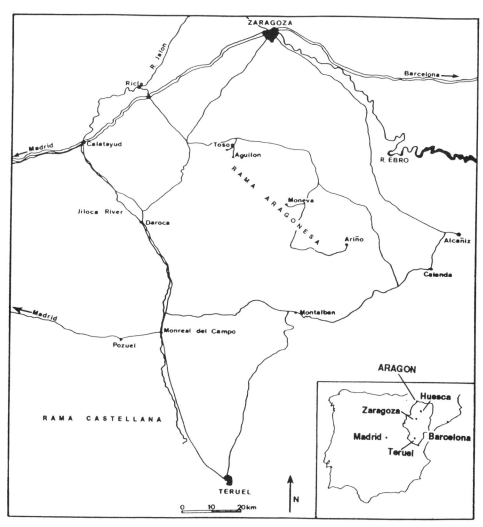

Fig. 2. Geographical location of the region of Aragón (eastern Spain) and the main Jurassic outcrops (Ricla, Aguilón, Tosos and others), along the northeastern (Aragonese) branch of the Iberian Chain.

Fontana (1991) for the Bathonian to Upper Oxfordian of Aguilón and Tosos.

Some of these classic outcrops are suffering serious or uncontrolled threats. The most aggressive action, to date, has been the recent uncontrolled tipping of extensive and highly polluting rubbish next to the classic outcrops of Ricla-2 (Callovian–Oxfordian), Aguilón-1 (Bathonian–Callovian) and Aguilón-2 (Bathonian to Middle Oxfordian) (Fig. 3). A further tip, mainly formed by the scree and rubble coming from the renewal of some village houses, also affects the base of the Jurassic section here, and the Oxfordian outcrop at Aguilón-4. These rubbish tips, particularly that of Ricla, constitute a dangerous environmental threat due to their constant expansion, the concentration of organic matter and the wind dispersal of the rubbish.

Since the affected outcrops do not have a special protective status, the only way to correct such potential and actual threats is the simple notification '*a posteriori*' to the regional government (see Fig. 1) and slow negotiation with the local authorities (town hall). This process was initiated here and should, theoretically, have resulted in a 'quick' protective action. However, it was followed by the production of many scientific reports, and by inactivity of the regional government and the local authorities. The legal status existing in the Heritage Law appears therefore inadequate for palaeontological sites. It is also clear that a simple notification to the regional government can only be presented when the damage, which in many cases is irreparable, has already been done.

Intermediate legal status

It seems, therefore, appropriate to propose the creation of a new, intermediate legal status specifically for palaeontological features, of lower range than the BIC or MN status, but more efficient in the application of protective and preventative actions. This new legal status, the *Punto de Especial Interés Paleontológico* (PEIP – Feature of Special Palaeontological Interest), is regarded as similar in status to those already existing in other European countries, such as the *Site of Special Scientific Interest* (SSSI) in Great Britain. However, it is intended to have a specific palaeontological bias, allowing a clear separation from other sites of purely geological and archaeological interest.

The legal status and scope of the PEIP

This new legal status is proposed to fill in the gap that exists between the ordinary notification *a posteriori* to the authorities – a procedure which appears slow and inefficient when it comes to threats – and the declaration of a palaeontological site as BIC. Any palaeontological site which has been published or referred to in the literature for its palaeontological interest would be suitable for declaration as a PEIP. However, it is difficult to assess in which order of priority these sites should be proposed as candidates for PEIP. This new legal status is mainly intended for those particular sites which are under serious threat or have already been the subject of any form of aggression. In the region of Aragón, a general catalogue of the most important features of interest is being prepared; this is the 'Palaeontological Map of Aragón', which should provide the initial list of sites needing protection. It could serve as a first step for producing a future list of PEIP candidates.

The bureaucratic process leading to the declaration of a palaeontological site as a PEIP

Fig. 3. (**a**) Rubbish tip at Ricla (outcrop Ricla-2; Callovian); (**b**) rubbish tip at Aguilón (outcrop Aguilón-1–Aguilón-2; Bathonian to Middle Oxfordian).

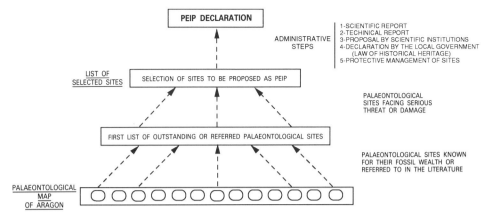

Fig. 4. Pyramidal sketch showing the intended scientific and legal procedures needed to propose a particular palaeontological site as a PEIP (Feature of Special Palaeontological Interest).

is intended to be fast and simple (Fig. 4). A list of selected sites would be presented by palaeontologists at the local universities and research institutions from databases which are currently being prepared (e.g. the Palaeontological Map of Aragón). It would include those sites known for their special importance and those facing serious threat or damage. The administrative steps needed to fulfil the formal declaration would involve (a) a scientific report prepared by the palaeontologists, (b) a technical report also prepared by the palaeontologists but supervised by technical specialists, which would reflect the impact of the threat on the site, and (c) a detailed programme for recovery and/or protection. The scientific advice of the reports, and the formal proposal to local government, would be carried out by the scientific institutions. A final step would include the official declaration of this particular site as a PEIP by the local government, within the framework of the Law of Historical Heritage, and the detailed programme of protective management (penalties for damage, entrance restrictions, fencing, cleaning of the area, panel distribution, etc.).

Conclusions

The current situation concerning palaeontological site protection in Spain appears clearly inadequate due both to the traditional dependency of palaeontological heritage on archaeology, and to a wide and inefficient legal framework. It seems appropriate to propose a new, intermediate legal status, specifically palaeontological, directed towards the protection and/or recovery of threatened and damaged sites. This new status, PEIP (Feature of Special Palaeontological Interest), would involve the palaeontologists and technical specialists in preparing site reports, the scientific institutions in advising on and formally proposing sites, and the local government in the final declaration and the protective management of the sites.

This paper is a part of the research project on palaeontological site protection financially supported by means of a research grant by the Instituto de Estudios Turolenses (CSIC, Spain). Thanks are due to Dr Kevin Page (English Nature, Peterborough) for his comments and advice, and the critical reading of the manuscript, and to the organizers of the Malvern conference for encouraging the presentation of the paper.

References

AURELL, M. 1990. *El Jurásico superior de la Cordillera Ibérica Central (provincias de Zaragoza y Teruel). Análisis de Cuenca.* Tesis Doctoral, Departamento de Geología, Universidad de Zaragoza.

CARIOU, E., MELÉNDEZ, G., SEQUEIROS, L. & THIERRY, J. 1988. Biochronologie du callovien de la Province d'Ammonites submediterranéenne: Reconnaissance dans les Chaînes Ibériques de subdivisions fines distinguées dans le centre-ouest de la France. *Proceedings of the 2nd International Symposium on Jurassic Stratigraphy*, **1**, 395–406.

FONTANA, B. 1990. *El Oxfordiense Medio, Biozona Transversarium (Jurásico Superior) en el borde sur de la Cuenca del Ebro (Cordillera Ibérica). Estudio paleontológico del género Larcheria (Ammonoidea, Perisphictidae).* Tesis de licenciatura. Geología, Universidad de Zaragoza.

—— & MELÉNDEZ, G. 1990. Caracterización bioestratigráfica de la Biozona Transversarium (Oxfordiense Medio) en el Sector Oriental de la Cordillera Ibérica. *Geogaceta*, **8**, 3–5.

GOY, A. & MARTINEZ, G. 1990. Biozonación del Toarciense en el área de la Almunia de Doña Godina-Ricla (Sector central de la Cordillera Ibérica). *Cuadernos de Geología Ibérica*, **14**, 11–54.

LARDIÉS, M. D. 1989. *Estudio Bioestratigráfico y Paleontológico del Calloviense (Jurásico Medio) en el sector Central de la Cordillera Ibérica*. Tesis de Licenciatura. Departamento de Geologia, Universidad de Zaragoza.

MELÉNDEZ, G. 1989. *El Oxfordiense en el sector central de la Cordillera Ibérica. (Provincias de Zaragoza y Teruel)*. Instituto de Estudios Turolenses (CSIC), Teruel.

—— & FONTANA, B. 1991. Sobre la posición estratigráfica de *Perisphinctes wartae* Bukowski y el desarrollo de la Subbiozona Rotoides (Biozona Transversarium) en la Cordillera Ibérica Nororiental. *Geogaceta*, **10**, 38–42.

SEQUEIROS, L., CARIOU, E. & MELÉNDEZ, G. 1984. Algunos ammonoideos del Calloviense Superior de Aguilón (Zaragoza, Cordillera Ibérica). *Estudios Geológicos*, **40**, 293–298.

——, —— & THIERRY, J. 1986. Posición bioestratigráfica y datos paleobiogeográficos de los Reineckeiinae (Ammonitina) del Calloviense de Aragón (Cordillera Ibérica). *Revista Española de Paleontología*, **I**, 73–84.

—— & MELÉNDEZ, G. 1981. Nuevos datos bioestratigráficos del Calloviense y Oxfordiense de Aguilón (Cordillera Ibérica, Zaragoza). *Cuadernos de Geología*, Universidad de Granada, **10**, 167–177.

Naked rock and the fear of exposure

W. J. BAIRD

Department of Geology, Royal Museum of Scotland,
Chambers Street, Edinburgh EH1 1JF

Abstract: The work of newly formed voluntary regional groups in assessing and designating non-statutory Earth science conservation sites in Great Britain is highlighted. Recent work carried out by one such group in Scotland is examined in detail. The task of such groups is discussed in relation to the many geological exposures recently revealed during road building. Evidence for seemingly traditional and irrational opposition to the preservation of exposed rock is considered.

Great Britain gives statutory protection to over 2000 Earth science Sites of Special Scientific Interest (SSSIs). However, voluntary groups working for Earth science conservation around the country are becoming increasingly involved in informal schemes to conserve Regionally Important Geological/geomorphological Sites (RIGS). These RIGS groups within counties or regions are made up of people from a combination of bodies such as wildlife trusts, museums, universities, local councils and geological societies. They include teachers, planners, site owners and geologists, both professional and amateur.

RIGS can be selected for a range of reasons, from educational to historic, but all should provide an opportunity to see and study rocks. They will be of less than SSSI status, but will illustrate certain geological features of regional importance. There are no fixed rules for selecting such sites, but they should be accessible, interesting and reasonably safe.

In Scotland, the Lothian and Borders RIGS group was set up in September 1992 by a group of experienced geologists to assess the regional importance of geological features within their area. Sites presently being considered are the road cutting in the Dreghorn link to the Edinburgh city bypass, areas of the Pentland Hills near the North Esk Inlier, and screes on the north face of Caerketton Hill. A rock face in an Edinburgh city rubbish dump near Craigmillar Castle, and an area of glacial channels near the village of Carlops are also on the list. However, the first site to be designated by the Lothian and Borders RIGS group is Torphin Quarry, on the southwest boundary of Edinburgh. The quarry shows two Lower Devonian basalt lava flows separated by a tuff.

Although education is important in the selection of these sites, it is not the only factor, nor indeed the most important one. Sites that are attractive in themselves, and perhaps linked with wildlife, will probably have the widest public appeal. In the RIGS projects, we may be at last seeing a genuine movement towards preserving geological sites for the general interest and benefit of the community.

Every rock exposure tells its own story, whether it be historical, industrial, palaeontological or mineralogical. Sometimes sites tell several different stories at once. Such sites may be easily preserved, but equally may be easily lost – either through deliberate means such as grading and covering with soil and nets, or simply through neglect and overgrowth with weeds, shrubs and trees. For example, the famous Lower Carboniferous quarry at East Kirkton, near Bathgate in West Lothian, has a diverse fossil fauna and flora including *Westlothiana lizziae*, known to the public as Lizzie the lizard. This site was almost totally overgrown and could have very easily become a landfill site rather than a world famous fossil lagerstatten (a body of rock unusually rich in fossils). Indeed, the nearby West Kirkton quarry was infilled on the instructions of the West Lothian District Recreation Department, despite it being recognized by the same District's Planning Department as an important site in its own right and complementary to the East Kirkton site.

There is often a presumption that, to be worth preserving, a site must be big. Nothing could be further from the truth; often a small site will show one aspect of geology in an excellent fashion. By allowing the visitor to concentrate on this aspect, the site does not overwhelm with information and enables the observer to store away a small snippet which may help later in the interpretation of more complex sites. Features such as contacts, bedding structures and weathering lend themselves to being displayed in such mini sites. In some cases, a string of small interesting exposures can be linked together to form a geological trail. Just such a range of features occurs within the Bathgate Hills in West Lothian and it is likely that these will become a series of RIGS in the future.

A lot has already been learned by the early RIGS groups about the methods required to make sites reasonably safe. Such measures as stepping faces, putting sand or gravel run-outs and bevels at the bottoms of steep slopes, and removing dangerous overhangs all help to alleviate the risk of injury. We are not, however, trying to develop playgrounds, but sites where geology can be experienced in a reasonably free, but responsible manner.

The improvement of roads in Scotland with modern earth-moving equipment and explosives has, over the past decade, produced many spectacular rock exposures. This was wonderfully displayed in the massive rock faces initially exposed during improvements to the A82, west of Loch Lomond. Sadly, many of these magnificent faces were later 'landscaped' and hidden from view with soil and nets. These normally hidden features exposed in road cuttings have sometimes revealed structures and stratigraphic correlations that have added to geological knowledge. So why is it that the reaction of civil engineers to the exposure of bare rock is to cover it at the first opportunity?

Only in Scotland's Highland Region has there been a positive reaction to requests that fine rock cuttings be left exposed. Even here, this has only been true of some more recent and massive rock cuts, such as those north of Loch Laxford along the A838.

The northern Highlands of Scotland are one of the finest areas of geological scenery in Europe and one of the most varied in the world. Yet there is a strange reluctance on the part of those who guide our multi-million dollor tourist industry to take advantage of its unique range of exposures, probably due to a complete lack of knowledge of the treasures they have beneath their feet.

Immediately one crosses the Highland Boundary Fault, just north of Perth, onto the rocks of the Dalradian and subsequently the Moine, there are miles of rock exposures along the sides of the new improved highway to the north, the A9. Clean and crisp between Calvine and Inverness, we see bedding and intrusions, folds, shears and faults in a range of colours from silver-grey to orange-pink. Even short sections amount to full day outings for geological societies and chapter headings in the texts of geological handbooks. But where are the indicators by the roadside, the laybys and information notices? Nowhere!

This is perhaps best illustrated, in the negative sense, by two sites in the famous Inchnadamph area of the northwest Highlands. Here, by the side of the A837 Lochinver road, on the north shore of Loch Assynt, Torridonian rocks lie unconformably on the ancient landscape surface of the Lewisian Gneiss. A place of pilgrimage for geologists, this has never been recognized as an important site by the regional authorities, and has gradually become degraded and obscured over the years. Further east at Skiag Bridge, within a mile of the statue honouring the famous geological duo, Peach and Horne, the spectacular pipe rock of the Lower Cambrian is exposed in a small cliff by the road junction of the A837 and A894. Although the site cries out for a plaque, a notice board and some place to park, there is, as usual, nothing. It is true that Scottish Natural Heritage have carried out recent improvements to the famous Moine Thrust site near Elphin. Here at Knockan Crag, where the Moine lies unconformably on the rocks of the Cambrian, displays, notice boards and guidebooks help to explain this classic site.

It seems strange that in Britain we seek to hide naked rock beneath a covering of soil. In Europe, such cuttings, when of geological interest, are often left exposed. Indeed, special features, such as the Silurian/Devonian stratotype boundary near Koneprusy in Bohemia, are highlighted and signposted. Is there some peculiar British phobia towards naked rocks? Alternatively, have British geologists failed to communicate to civil engineers and the public, the beauty and wonder of such rocks?

Conclusion

The work of Scotland's first RIGS group has identified several potential sites in Lothian and Borders. Many more potential sites throughout Scotland are revealed by new road works. Traditional reluctance to leave rock exposures uncovered suggests that Earth science conservationists have much work to do in informing and educating people to achieve acceptance of the importance of such newly made features.

The proposed Cuilcagh Natural History Park, County Fermanagh: a locally based conservation initiative

JOHN GUNN[1], CHRISTINE HUNTING[1], SARAH CORNELIUS[2] & RICHARD WATSON[3]

[1]*Limestone Research Group, Department of Geographical & Environmental Sciences, The University of Huddersfield, Queensgate, Huddersfield HD1 3DH, UK*
[2]*Department of Environmental & Geographical Sciences, Manchester Metropolitan University, Chester Street, Manchester M1 5GD, UK*
[3]*Fermanagh District Council, Town Hall, Enniskillen, Co. Fermanagh BT74 7BA, Northern Ireland, UK*

Abstract: The middle slopes of Cuilcagh Mountain in County Fermanagh are covered by one of the best examples of a mountain blanket bog ecosystem in Northern Ireland. The same slopes form the catchment of several major cave systems including Marble Arch, Northern Ireland's only show cave, part of which was opened to the public by Fermanagh District Council in 1985. The lower slopes of Cuilcagh exhibit an excellent array of karst landforms, the mid-slopes are characterized by a complex peat pseudokarst assemblage of pipes, pools and dolines, and the upper slopes have notable boulder fields and mass movement features. Since the early 1980s, there has been a significant increase in human activity on the mountain, particularly increased stocking densitites and large-scale commercial peat cutting with accompanying land drainage works and access tracks. These are threatening both the Earth science and the ecological interest of the area, and there is particular concern that the hydrological regime of the show caves may be adversely affected. In 1991, the Limestone Research Group were commissioned to investigate the threat and to prepare a land management strategy for the show cave catchment.

When the project commenced, there was no statutory protection for the area and it appeared that planners were unable to control the, technically illegal, commercial peat extraction, much of which was being carried out by operators from the Irish Republic. Faced with a lack of control at national level, it became clear that a local initiative was necessary and that the only way in which effective control of land use could be exercised was through landownership. Following consultations with the Limestone Research Group, Fermanagh District Council identified a key portion of the caves catchment which they intend to purchase and designate as the Cuilcagh Natural History Park. As the council does not have sufficient funds for outright purchase, additional funding has been sought from European and UK government sources.

Cuilcagh Mountain lies some 20 km southwest of Enniskillen in County Fermanagh, Northern Ireland (Fig. 1). The mountain forms a distinctive ridge profile against the Fermanagh skyline, and is a prominent backdrop to much of the county's lakeland scenery. Cuilcagh summit (667 m) is the highest point in the county, and the summit ridge forms the border with County Cavan in the Irish Republic.

Earth science interest

The summit ridge of Cuilcagh Mountain is composed of gritstones and sandstones which form dramatic cliffs up to 30 m in height. There has been extensive mass movement, many large blocks having slid forward forming extensive and deep gulls. The slopes are littered with boulders 1–10 m in long axis, which extend down to the mid-slopes. These are underlain by sandstones and shales which are covered with a thick (up to 3 m) layer of peat, and form one of the best examples of a blanket bog ecosystem in Northern Ireland. The bog is characterized by an extensive suite of pseudokarst landforms, including pipes and accessible caves, dolines, blind valleys and sinks, and pocket valleys with risings. Below the sandstones and shales are limestones, and the Marlbank area supports one of the finest upland karsts in the British Isles, with impressive blind valleys and sinks, large collapse dolines, karst windows, limestone pavement and a large pocket valley/rising (Fig. 2). The area's extensive caves include the Marble Arch Caves, with a total length of over 6500 m, of which almost 500 m were opened to the public

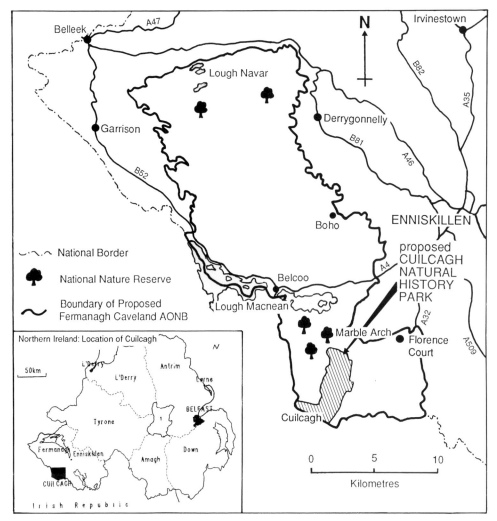

Fig. 1. Location of the proposed Cuilcagh Natural History Park.

in 1985 as the only show cave in the province. The caves contain sedimentary sequences of considerable scientific importance which preserve evidence of past environments stretching back several hundred thousand years. In addition to the geomorphological value of the area, there are important outcrops of Tertiary basaltic lavas in the Cuilcagh dyke and a series of mud reef knolls.

Human impacts

The border location of the Cuilcagh/Marlbank area (Fig. 1) makes it peripheral for economic investment, whilst civil unrest has restricted development of the tourism potential of its scenically beautiful countryside. Poor soils associated with impermeable boulder clay and a high annual rainfall make agriculture in the uplands unprofitable, and rough grazing is the major land use. Such marginal activity has preserved a relatively unspoilt environment, and until the late 1970s, there were few human impacts upon the area in general and the Earth science interest in particular. A stimulus to development was provided by designation as a European Community 'Less Favoured Area'. Subsequent agricultural grants and subsidies have enabled farmers to build access tracks high up onto the mountain, to undertake drainage works, and to operate with higher stocking densities than would have otherwise been the

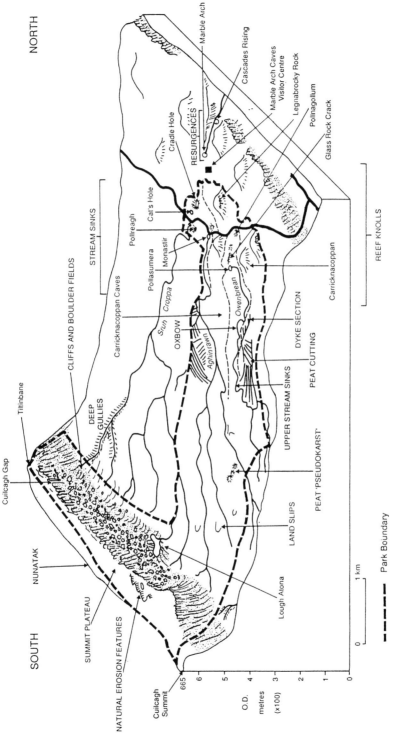

Fig. 2. Landforms of the Cuilcagh Natural History Park.

case. The overall result of these activities has been an increase in the amounts and timing of runoff, leading to accelerated erosion and downstream deposition. Even more significant impacts have resulted from the extension of mechanized peat cutting, using large all-wheel-drive tractors, onto the bog. The most direct impact of this is total destruction of the peat pseudokarst landscape; but peat extraction on the lower slopes of Cuilcagh Mountain falls within the catchment of Marble Arch Caves, and it has been suggested that the flood risk in the show cave has increased as a secondary, indirect result of this activity. There has also been an increase in sediment deposition which is affecting operation of the caves and which appears to be linked to peat extraction. Between 1982 and 1989, commercial peat cutting affected over 60 ha of the caves catchment, and involved digging over 8700 m of new drainage channel and construction of some 6000 m of access track. Concern over flooding problems at Marble Arch Caves, and a more general concern to protect the area's environmental resources, prompted Fermanagh District Council, as show cave owners, to commission a hydrological and environmental investigation which will include a computer-based Geographical Information System for the whole area. This work is being carried out by the Limestone Research Group at the University of Huddersfield.

Statutory protection

It would appear that the commercial peat cutting is being carried out illegally, as no planning permission has been applied for. However, perhaps because small-scale peat cutting for domestic supply has a long history in the area, the existing planning controls have proved ineffective in preventing much larger-scale destruction. Stop notices have been posted but have been ignored by operators who enter the area from the Irish Republic. The whole of the Cuilcagh/Marlbank area has been assessed for designation as an Area of Special Scientific Interest (Department of the Environment for Northern Ireland, 1989). However, the assessment only took account of the area's ecological interest and did not consider its, arguably greater, Earth science interest. This reflects the general lack of a coherent policy for Earth science conservation in Northern Ireland. The authors are in no doubt that had the Geological Conservation Review undertaken in Great Britain been extended to Northern Ireland, then the area would have been designated for its mass movement features, its peat pseudokarst, its karst landforms and its caves.

More recently, the area has been included in the proposed Fermanagh Caveland Area of Outstanding Natural Beauty (Department of the Environment for Northern Ireland, 1992) and in April 1993, the Department of Agriculture for Northern Ireland launched the West Fermanagh and Erne Lakeland Environmentally Sensitive Area Scheme, which covers most of the northern flank of Cuilcagh Mountain. These initiatives, although welcome, are aimed primarily at conserving the landscape and ecological features of the area and do not directly address Earth science interests.

The Cuilcagh Natural History Park initiative

As there appeared to be no possibility of the area receiving statutory protection, at least in the short term, it was necessary to take the initiative at the local level. The most pressing need was to effect land use control and the only way in which this could be unequivocally exercised was through landownership. Hence, following consultations with the Limestone Research Group, Fermanagh District Council identified a key portion of the caves catchment (c. 1170 ha) and proposed that it be purchased and designated as the Cuilcagh Natural History Park (Fig. 1). The proposed park covers a large part of the catchment of the Marble Arch Caves and includes the three major sinking rivers, as well as the two most extensive peat-cutting areas. Clearly, an initiative of this nature cannot be entirely locally funded and hence support has been sought from the Department of the Environment for Northern Ireland, from the National Heritage Memorial Fund and from other UK government and European Community sources. The initiative has received support from Northern Ireland's Council for Nature Conservation and the Countryside, and at the time of writing (May 1993) negotiations are continuing.

Although the proposed Cuilcagh Natural History Park does not cover the whole mountain, or even the whole of the area draining to the Marble Arch Caves, it does include many of the most important sites for Earth science conservation. Moreover, it abuts onto the Florence Court Forest Park, where a further 1220 ha of land owned by the Forestry Commission is managed by the Royal Society for the Protection of Birds. It is hoped that when the park is established, common policies can be adopted with respect to both Earth science and biological conservation interests.

Studying the Cuilcagh Natural History Park and assessing its value for conservation reveals the merits of its natural, physical and ecological systems. However, preservation of this environment as an end in itself is too simple an aim for such an area, where human impact and land use are significant. The very name 'Natural History Park' firmly establishes the importance of nature conservation, yet the past, present and future role of humans in the landscape cannot be overlooked. The welfare of those who live and work in the area forms an integral part of management of the park, whilst public access for purposes of enjoyment of the countryside should also be a priority. In view of this, three primary aims have been put forward for the park:

(1) to conserve and enhance the diversity and value of natural and semi-natural environmental systems and the cultural landscape;
(2) to protect the socio-economic well-being of the local community who earn a living from the land, and to generate their appreciation of the diverse values of the area;
(3) to provide for public recreation and enjoyment of the countryside, which is in sympathy with the aims of nature conservation, and the interests of other users, and to instill something of the significance of the area in visitors.

The initial investigations which led to the Cuilcagh Natural History Park initiative were primarily concerned with the threat to public safety as a result of flooding of the Marble Arch Caves. Hence, an important objective is to control those land uses which are most likely to have contributed to the flooding problem, most notably mechanized peat cutting, land drainage, overgrazing and track construction. However, these land uses are also the ones which are most detrimental to the wider environmental value of Cuilcagh Mountain, and Fermanagh District Council has demonstrated an awareness of the many other issues which require management by commissioning an investigation of the whole catchment. This includes more detailed research into the hydrological and geomorphological impacts of mechanized peat cutting which is being undertaken by the University of Huddersfield and is also sponsored by the Environment Service–Countryside and Wildlife–of the Department of the Environment for Northern Ireland.

Conclusions

The proposed Cuilcagh Natural History Park represents a local initiative with the aim of purchasing and managing a substantial area primarily for its Earth science interest. As such, it is thought to be unique in Northern Ireland and possibly in the British Isles. However, it does not remove the need for statutory protection of the Earth science interest of the wider Cuilcagh Mountain. Moreover, Cuilcagh also has a very high ecological value and this has prompted David Bellamy (pers. comm.), amongst others, to suggest that it should be considered for nomination as a UNESCO World Heritage Site.

The research on which this paper is based was initially funded by Fermanagh District Council with subsequent, and ongoing, support from the Environment Service–Countryside and Wildlife–of the Department of the Environment (Northern Ireland). However, the paper is solely the work of the authors and does not necessarily reflect the views of either organization. An initial Management Plan for the Cuilcagh Natural History Park produced by Michelle Gray as an honours project at Manchester Polytechnic has provided a valuable basis for subsequent work. Thanks are also due to Paul Hardwick, Robert Hyland and Christine Robertson for help with field work.

References

DEPARTMENT OF THE ENVIRONMENT FOR NORTHERN IRELAND. 1989. *Ecological survey report for Marlbank and Cuilcagh* (unpublished), Belfast.

—— 1992. *Fermanagh Caveland Area of Outstanding Natural Beauty Proposal*, Belfast.

The work of the Lothian and Borders RIGS Group in Scotland

NORMAN E. BUTCHER

Lothian and Borders RIGS Group, 22 Drum Brae Walk, Edinburgh EH4 8DQ, UK

Abstract: Set up towards the end of 1992, the Lothian and Borders RIGS Group in Scotland has met on a monthly basis at different venues in and around Edinburgh. Composed of a number of experienced geologists drawn from different backgrounds, together with planners and conservation workers, the group has adopted a broad approach. Much discussion has been devoted to guidelines and criteria for site selection, while work on actual sites has focused on the core of the area formed by the Pentland Hills.

The early years of the 1990s have seen the setting up in most English counties of distinct groups of people dedicated to establishing Regionally Important Geological/geomorphological Sites (RIGS). These sites, chosen for their broad educational, research, historical or aesthetic importance, are recognized as non-statutory in nature, in contrast to the more formal, well-established Sites of Special Scientific Interest (SSSIs).

In Scotland, although the Grampian Region based in Aberdeen has long been noted for its efforts in geological conservation, the latter part of 1992 saw the setting up of its first RIGS Group – in Lothian Region but with an increasing interest in the Borders Region also. There is no doubt that the newly created Scottish Natural Heritage was a significant factor in this development. With its Edinburgh-based headquarters staff, Scottish Natural Heritage has provided a welcome boost to the existing mix of geological and other expertise already provided by the British Geological Survey, the National Museums of Scotland, the Grant Institute of Geology, Scottish Wildlife Trust and the Edinburgh Geological Society.

An initial meeting to establish this RIGS Group took place in the headquarters of Lothian Regional Council (LRC), in Edinburgh, on 1 September 1992. Convened by Dr John Sheldon, Principal of the Natural Resources Unit in the Department of Planning of LRC, there was unanimous agreement amongst those present. Starting on 10 November 1992, monthly meetings of the Lothian RIGS Group have been held in a different member organization's premises in and around Edinburgh. To date, eleven such meetings have been held and the group is currently making a series of site visits to the nearby Pentland Hills. At the time of writing, the Group has just designated its first RIGS, that of Torphin Quarry on the southwest edge of the City of Edinburgh.

Composition of the group

The initial meeting was attended by representatives of the Planning Departments of all four District Councils within Lothian Region, together with those individuals known to be interested in the topic of geological conservation. At subsequent meetings of the established RIGS Group, members attended as individuals in their own right and not as formal representatives of the particular organization to which they belonged. This arrangement has given the group an added strength, being able to develop its own identity and yet able to draw on the expertise residing in disparate organizations. The membership of the Lothian and Borders RIGS Group is as follows:

- (Chairman) Norman E. Butcher, Past-President, Edinburgh Geological Society and former Staff Tutor in Earth Sciences, the Open University in Scotland
- (Secretary) Michael C. Smith, Senior Geologist, British Geological Survey
- Alan McKirdy, Head of Earth Sciences, Scottish Natural Heritage
- Stuart K. Monro, Principal Geologist, British Geological Survey
- Andrew A. MacMillan, Principal Geologist, British Geological Survey
- Bill J. Baird, Department of Geology, National Museums of Scotland
- Alastair Sommerville, Science and Conservation Officer, Scottish Wildlife Trust
- Euan N.K. Clarkson, Reader, Grant Institute of Geology, University of Edinburgh
- Bob Cheeney, Hon. Secretary, Edinburgh Geological Society
- John C. Sheldon, Principal, Natural Resources Unit, Department of Planning, Lothian Regional Council
- George Bruce, Chief Ranger, Pentland Hills Regional Park

- Victor Partridge, Ranger, Pentland Hills Regional Park

The group has benefited from the involvement of other people from time to time. At its meeting on 18 February 1993 at the headquarters of the Scottish Wildlife Trust, the National RIGS Officer for the Royal Society for Nature Conservation (the Wildlife Trusts Partnership) came north to give a valuable presentation to the group on the nationwide activities of RIGS groups. This resulted in a stimulating discussion of the problems posed in the establishment of such sites.

As its work proceeds, the group looks forward to establishing contacts with other individuals and organizations in the future. The group is indebted to the British Geological Survey, Scottish Natural Heritage, Scottish Wildlife Trust, the National Museums of Scotland, Lothian Regional Council and the Grant Institute of Geology for the willing provision of facilities.

Aims and objectives

The prime purpose of the Lothian and Borders RIGS Group, as with all RIGS initiatives, is the establishment of a network of sites throughout the present Lothian and Borders Regions. The extension into the Borders Region has come

---

1. What is geology?

Geology is concerned with the study of the Earth, its origin, structure, composition, history and with the nature of the processes which give rise to its present state.

2. Why is geology important?

Geology has a great influence on everyday life affecting soil type and, therefore, agriculture. The underlying rocks and minerals provide the essential requirements of modern life; water supplies, minerals and rocks for the manufacturing and construction industries, and coal, gas and oil for energy.

3. Do geological features require conservation?

With the increasing pressure on land for all kinds of development and the problems of disposal of waste, particularly near urban areas, there is a danger that this might eliminate or reduce the number of geological features available for educational and research use. Sites may be destroyed, and old quarries and pits infilled.

4. What is the Lothian and Borders RIGS Group?

A group of individuals working on a voluntary basis to locate regionally important geological features.

5. What are its objectives?

To identify important geological sites to ensure that a sufficient number of localities are available for educational interpretation and scientific research use, and sometimes for their significance in the history of geology.

6. What is its expertise?

The site selection is by a group of experienced geologists.

7. What is a RIGS and how can it be part of a flexible system?

A RIGS is a geological feature with local importance for education etc. There may be more than one site of each type, so that alternatives exist if a site is lost to infill or development, for example.

8. What are the criteria used in identifying a RIGS?

The geological and landform content of the site and its value for educational interpretation, research and the history of the science of geology.

9. What are the planning implications of RIGS?

With the permission of the owner/occupier, a RIGS would be notified to the respective local authority. This would mean that there would be no statutory restrictions on the site, but that when considering planning proposals the councils would be aware of the site.

FIG. 1. An information note to owners of RIGS

only recently, with the recognition that the largely Edinburgh-based group is able equally well to work on the more rural parts of southeast Scotland covered by the present Borders Region.

The monthly meetings of the group have been largely concerned with discussion and formulation of guidelines and criteria for selection of sites for proposal and designation as RIGS. This has resulted in an agreed summary (Fig. 1).

In this work, the group acknowledges the help received by the activities of other RIGS groups in England. Since it was felt desirable at an early stage to involve other sectors of the community, principally landowners and farmers, informal contact was made at the outset with other bodies such as the Scottish Landowners Federation, National Farmers Union of Scotland, the Green Belt Trust and the mineral extractive industry as represented by the Sand and Gravel Association in Scotland.

The longer-term objective of the group is to increase public awareness of geology. There is no doubt that, even in Edinburgh of all places, much still remains to be done in this respect. Specifically, by association with the names of James Hutton (1726–97) and Charles Lyell (1797–1875) in particular, it is planned to mark the bicentenary of the death and birth respectively of these two Scottish founders of the science, in 1997, by an appropriate series of events in the Lothian and Borders Regions. Since geology is both science and history, public awareness of geology can often best be achieved in a historical context.

Work to date

With regard to assessing particular sites for proposal and designation as RIGS, the group determined at the outset to concentrate its efforts in the first six months to one year on the core of the Lothian Region as represented in the Pentland Hills. This spectacular range of hills has a generally well-defined entity extending almost into Edinburgh itself and corresponds to the oldest rocks to be found in this part of the Midland Valley of Scotland. There is the advantage that most of the area in question is already a designated Regional Park. Geologically, the rocks to be found in the Pentland Hills belong to the Upper and Lower Old Red Sandstone, with a series of volcanics in the latter being especially prominent, together with three separate and generally well-defined inliers of older Silurian rocks, often very fossiliferous in nature. Also, the Pentland Hills exhibit some spectacular glacial and post-glacial features. Altogether, it is felt that some five or six RIGS may result from present work. The group has selected one particular site, on the northwest edge of the Regional Park, as its first RIGS. This site, Torphin Quarry on the southwest edge of Edinburgh, is currently in a state of dereliction, having been last worked as an active quarry some decades ago. It is felt that its designation as a RIGS would be a positive improvement to the immediate area if a properly managed scheme for the quarry can be brought about. Geologically, Torphin Quarry exhibits good examples of lava flows with bedded ash deposits, together with several faults and mineralization. It forms part of an intriguing area on the edge of the Pentland Hills where the relationships of the Lower Carboniferous, Upper Old Red Sandstone and Lower Old Red Sandstone can be demonstrated by mapping.

Other areas being currently assessed by the group are the Silurian inlier above Bavelaw Castle, extending through the Pentland Hills to the head of the Loganlee reservoir, where the contrast with the so-called felsite of Black Hill is most marked. This is an area of spectacular meltwater channels through the hills.

In addition to examining sites in the Pentland Hills, a number of other nearby and related sites are being assessed as possible RIGS. These include the Dreghorn Spur road-cutting which, although it has all the hazards of road traffic, nevertheless demonstrates an inclined sequence of strata showing many interesting sedimentary features. Another, spectacularly beautiful small upland area is that surrounding Craigmillar Castle on the southeast edge of Edinburgh. Bounded by the Pentland Fault on one side and with the Craigmillar housing scheme on the other, the area was extensively quarried centuries ago, providing stone not only for Craigmillar Castle, but for some of the earliest construction in the Palace of Holyroodhouse and Edinburgh Castle.

Finally, the Lothian and Borders RIGS Group is commenting, where appropriate, to bodies such as Historic Scotland on, for example, the Draft Management Plan for Holyrood Royal Park, Edinburgh. An important exercise for the group, currently in hand, is contributing to the Lothian Region Structure Plan. With the likely reorganization of local government in Scotland in the coming years, it will be a matter of some importance to ensure that due recognition is made of the role of geology in underpinning so many aspects of life in all its activities. The RIGS scheme can contribute much.

Urban site conservation – an area to build on?

COLIN D. PROSSER & JONATHAN G. LARWOOD
English Nature, Northminster House, Peterborough PE1 1UA, UK

Abstract: Towns and cities do not readily spring to mind when considering Earth science conservation, but in England, where many important geological sites lie within urban areas, experience has led to the realization that active site conservation within the urban environment has a vital role to play in selling Earth science conservation to a wider audience. With the bulk of the population located in towns and cities, how better to encourage local involvement in conservation than to promote sites which lie in the centre of communities? Geological sites in urban areas, which tend to consist of disused quarries, river bank exposures and road and railway cuttings, are subject to considerable development and population pressure. However, this pressure on sites can act as a stimulus for local communities, local authorities, industry and schools to become actively involved in conserving and managing a part of their local heritage. Urban sites also tend to be a focus for media attention, with sites either being lost to development or saved for future generations through innovative conservation initiatives. There are few urban sites where nothing happens! This attention serves to raise public awareness of the issues surrounding Earth science conservation and helps to gain social acceptance for the need to conserve our geological and geomorphological heritage.

In Dudley, West Midlands, a town built up during the industrial revolution, geological sites, many of which fuelled the industrialization of the area, are now managed and successfully promoted by the local authority as part of the town's industrial heritage. Managing sites in an urban setting is not easy, but the approach taken in Dudley goes some way to demonstrating the immense potential which exists for promoting and involving local people in geological conservation.

For its size, England has a rich and varied geology which, from prehistoric times, has been subject to a long history of exploitation. Since the earliest economic extraction of flints during the Neolithic (e.g. Grimes Graves, Norfolk), the scale of exploitation of the country's mineral resource has continually increased, with much of England's wealth being founded on the Industrial Revolution of the nineteenth century. Other than urban development around major ports, it is the presence of mineral resources, through influencing the location of industrialization, that has shaped much of England's present-day urban geography. Given this association between mineral resources and urban settlement, it is not surprising that urban environments often provide some of England's key geological localities, either as relics of industrialization (e.g. disused quarries), or as the result of human activities associated with large and expanding population centres (e.g. road cuttings).

The aim of this paper is to raise the profile of urban geology, and to highlight the issues and opportunities connected with the conservation of geological features in an urban environment.

The nature and location of urban geological sites

The distribution of England's main urban areas is shown in Fig. 1. Geological sites within urban areas can be broadly divided into two categories: natural and artificial. Natural exposures may include river sections, natural crags and outcrops, or even coastal cliffs. For example, the Precambrian crags of Charnwood Forest, Leicestershire, and the Tertiary coastal cliffs of Bournemouth, Dorset, both lie within an urban setting. Artificial exposures can include active and disused quarries, mines, mine dumps, and road cuttings. Gilbert's Pit Site of Special Scientific Interest (SSSI), for example, is a disused Tertiary sand/gravel pit in London, whilst Triassic red sandstones form road and railway cuttings in the centre of the town of Frodsham, Cheshire. An important additional resource, often with great educational potential, is the very fabric of a city itself, the stone from which it is built! Whether a locally obtained stone, or a more exotic ornamental stone, much can be gained from an examination of a town or city's buildings and memorials (Robinson 1984, 1985; Robinson & Worssam 1989).

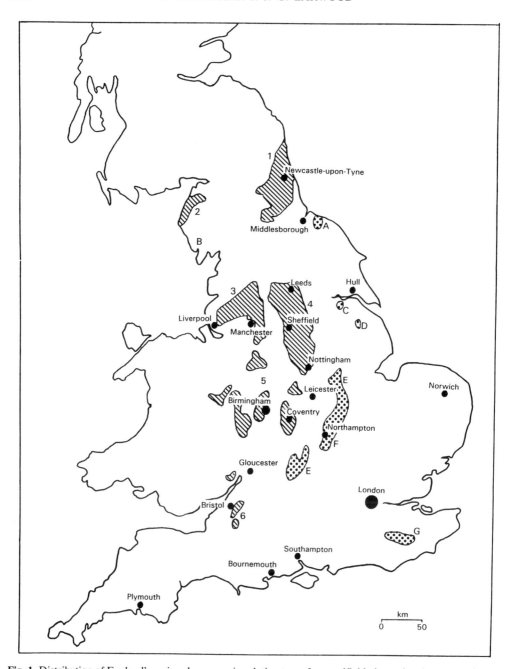

Fig. 1. Distribution of England's main urban areas in relation to surface coalfields (cross-hatch ornament) and major iron resources (dot ornament). Principal south coast urban areas (major ports and resorts) also indicated (from Duff 1992). 1: Northumberland & Durham Coalfield; 2: Cumberland Coalfield; 3: Lancashire Coalfield; 4: East Pennine Coalfield; 5: Midlands Coalfield; 6: Bristol Coalfield; A: Cleveland Ironstone; B: Cumbrian Iron District; C: Frodingham Ironstone; D: Claxby Ironstone; E: Marlstone Ironstone; F: Northampton Ironstone; G: Wealden Ironstone.

In considering the scope of the geological resource in an urban setting, it is important to take into account the immediate countryside surrounding an urbanized area, as well as the urban area itself. The surrounding countryside is often influenced by various aspects of urbanization, and geological features here form an equally valuable part of the urban geological resource.

It is also important to recognize that in urban areas geological sites are often intrinsically linked with wildlife habitats. Disused quarries soon become overgrown and provide a rare wildlife haven in an otherwise inhospitable environment. For example, the natural exposures, cuttings and disused quarries of Permian Magnesian Limestone sites, in the heavily urbanized northeast of England, are nationally important for both their geology and their Magnesian Limestone grassland.

The link between geological resources, industrialization and urbanization has led to a concentration of urban geological sites on or close to England's most economically valuable rocks (Fig. 1). This is clearly reflected in the nature, age and range of geological exposures commonly encountered in urban areas. For example, the coal-bearing Westphalian strata of northern England and the Midlands, and the Jurassic Ironstones of Yorkshire and the East Midlands, provide a good proportion of our urban geological sites. Similarly, exploitation of other associated mineral resources has left many exposures in the Permian Magnesian Limestones of the northeast and the Silurian limestones of the West Midlands.

Pressures and threats

By their very nature, geological sites in urban areas are under continual and increasing threat, with development, tipping and vandalism being common problems. Undeveloped urban land is at a premium, and often represents a favoured location for development, whether through industrial growth or as a result of an expanding population. Similarly, geological sites near to large populations are threatened by the side-effects of urbanization: quarries, for example, are potentially valuable as landfill sites (e.g. Webster's Clay Pit, Coventry, Anon (1992*a*)). Although development represents the main threat to the geological resource in urban areas, other pressures include fly tipping, vandalism and heavy or inappropriate recreational use.

Clearly, there are many problems associated with the conservation and effective utilization of geological sites in urban settings, and a great deal of effort, skill and experience is needed to counter the pressures placed on this resource. However, this continual activity, in a highly populated area, does have the benefit of thrusting urban sites into the media limelight. This media attention, whether highlighting conservation of, or damage to, sites, serves to raise public awareness of the value of our geological heritage and the need to conserve it.

A need for conservation

Having established that there is a resource to conserve, of primary importance is the recognition of the likely site user. Beyond the developmental potential of urban sites, their use can be broadly divided into scientific, educational and recreational. The scientific and educational value of sites can be seen at various levels. Many urban geological sites are of international scientific importance and have, and will continue to be, the subject of much research. Similarly, urban geology forms a vital resource in higher and further education and, as Earth science now forms an integral part of the school National Curriculum, the availability of geological sites, particularly within easy reach of schools, is becoming increasingly important.

The recreational use of urban sites is equally critical, and there are many opportunities for the general public to learn about and enhance their urban geology. A good example is Micklefield Quarry SSSI, a Magnesian Limestone site in West Yorkshire, where the local authority, conservation volunteers and a nearby school have come together to clear the site of rubbish and vegetation. The site has now been adopted by the school and an interpretative information board is being prepared (Smith 1993).

If left unchecked, the pressure placed on our urban geology would result in progressive reduction of the resource. There is, thus, a clear need to set about safeguarding our urban geology for present and future generations.

The Dudley experience

Wildlife conservation in urban areas is generally well supported by both local communities and local authorities, and in many cases is highly innovative. The same opportunities exist for urban Earth science conservation, and it is important that Earth science conservationists develop and market the urban resource. Urban areas provide a large population to influence and involve in conserving our geological heritage, and geological sites are ideal for involving

Table 1. *SSSIs in the West Midlands designated for their geological interest; illustrating the typical range of urban geological site*

Site	Location	Site type	Geology
Daw End Railway Cutting SSSI	Walsall District	Railway cutting and disused quarry	Wenlock–Coalbrookdale & Much Wenlock Limestone formations
Ketley Clay Pit SSSI	Dudley District	Working brick pit	Westphalian–Etruria Formation
Brewin's Canal Section SSSI	Dudley District	Canal-side exposures	Ludlow–Ludlow & Pridoli Series (also lowermost Westphalian)
Bromsgrove Road Cutting, Tenterfields SSSI and LNR	Dudley District	Road cuttings	Westphalian–Halesowen Formation
Doultons Claypit SSSI (part of Saltwells LNR)	Dudley District	Disused clay pit	Westphalian–Productive Coal Measures (Westphalian B)
Turner's Hill SSSI & LNR	Dudley District	Disused quarry	Ludlow–Aymestry Limestone Group
Wren's Nest SSSI & NNR	Dudley District	Disused quarries and natural outcrop	Wenlock & Ludlow–Coalbrookdale, Much Wenlock Limestone & Lower Elton formations
Hay Head Quarry SSSI	Walsall District	Disused quarry	Wenlock–Coalbrookdale Formation
Webster's Claypit SSSI (lost to landfill)	Coventry District	Disused clay pit now infilled	Westphalian–Enville Formation

people and raising public awareness. In many cases, they are relatively common and robust, they are historically and culturally tied in with the origins of the urban area, and are interesting in themselves.

Although there are many urban geological sites identified in England, urban Earth science conservation is still in its infancy. Little has been done to involve the large urban populations or to fully utilize the resource in promoting our science. The few examples where urban Earth science conservation has been practised have generally been successful and much can be learnt from these. One of these more successful areas has been in and around Dudley in the West Milands (Table 1).

The town of Dudley, situated on the western outskirts of Birmingham, lies within an area known as the 'Black Country' – an urban area founded on the former coal and iron industries of the region. More recently, as industry has closed, Dudley has successfully promoted its industrial and geological heritage to attract tourists, and has developed a culture where geology is valued as an integral part of the local heritage and where people are actively involved in conserving this resource.

The geological resource of Dudley (Silurian Wenlock limestone and Westphalian Coal Measures), and its potential for conservation, has long been recognized (Box & Cutler 1988). The international importance of the geology in the Dudley area has led to the designation of the Wren's Nest National Nature Reserve (NNR) (Fig. 2), a number of geological SSSIs and other locally important sites. There are many active local geologists: the Black Country Geological Society, together with local museum staff and with the support of the local authority, have worked to promote and advance local geology through excursions, exhibitions and local events (e.g. the Dudley Rock and Fossil Fair, held in 1992, which saw several thousand people through its doors). The local authority clearly recognizes the potential of its geological heritage and wardens the NNR and some of the SSSIs, which it has declared as Local Nature Reserves (e.g. Saltwells LNR, Tenterfields LNR (Anon 1992b)). Additionally, the relevance of geology to the area is brought home by the Black Country Museum, an open-air museum dedicated to local industrial heritage. The Black Country Museum, by successfully linking the limestone and coal mining history of the area with the history of the local community, has drawn an important link with the geological heritage of the region.

The way ahead

The key to Dudley's geological success has been the recognition of its potential, and the willingness of the local authority and community to then value and promote this Earth science resource. Much can be gained from Dudley's experience in conserving and promoting its geological heritage, and similar lessons can be learnt from other isolated projects around the

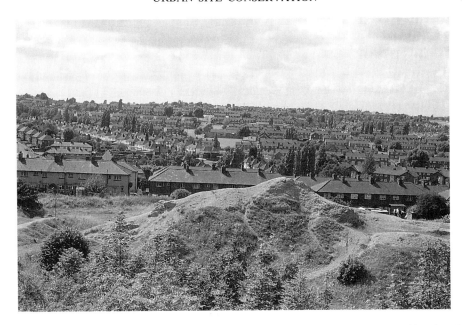

Fig. 2. View looking west across Wren's Nest NNR; Silurian reef-mass in the foreground overlooking a housing estate in Dudley.

country. A number of key factors which appear to underpin successful Earth science conservation in urban areas are highlighted below.

(1) Recognition of the Earth science resource and its potential for conservation and community involvement.
(2) Identification of potential users – recognition of potential users and their perception of urban geology allows the approach to conservation to be tailored to their needs.
(3) Local involvement at all levels – as with urban wildlife conservation, local involvement and community action is the key to success. This is demonstrated in Dudley where community interest and support has been fostered through locally active geological groups and a positive and innovative approach to the representation of geology through exhibitions and events. The involvement of the local authority, as well as providing key funding, has maintained the high profile of geology in this area.
(4) Collaboration – the approach to conservation will vary between urban areas, reflecting both the available resource and its perceived use. Most towns and cities will have well-established groups potentially interested in geology and its conservation, and along with universities, museums, schools and local geological societies, there is clearly a wealth of local knowledge and human resources. It is important to work with established conservation groups whose experience and approach can often be adapted to the needs of Earth science conservation. In most cases conservation of wildlife and Earth science features can go hand-in-hand.

Collaboration with developers and planners is also important. By working with local authorities and developers, the need for Earth science conservation can be recognized at an early stage and appropriate measures can be taken to ensure that development and urban geology does co-exist. The establishment of Regionally Important Geological/geomorphological Sites (RIGS) groups, drawing on a wide range of expertise, is a clear example of how collaboration can work towards successful conservation (Harley 1992).

(5) Develop a strategy – a successful future for urban Earth science conservation lies in the recognition of its importance by local authorities. Development of a strategy considering the resource, alongside the legislation and planning policies as they relate to the conservation management of that resource, will go some way to ensuring the future of geology in our cities and towns.

Conclusion

Urban Earth science conservation in England is an issue which has received little consideration and is certainly an area to build on. It has immense potential for achieving local involvement, both of the community and the local authority, and for integrating Earth science conservation with wildlife conservation. The implications of successful urban Earth science conservation reach far beyond the urban environment. If the importance and the need for Earth science conservation is recognized within the populated urban areas, support for Earth science conservation nationwide can only benefit.

References

ANON 1992a. Webster's Claypit lost. *Earth science conservation,* **31**, 29.

—— 1992b. Imaginative use of school grounds. *Urban wildlife news,* **9**(4), 1–2.

BOX, J. & CUTLER, A. 1988. Geological Conservation in the West Midlands. *Earth science conservation,* **25**, 29–35.

DUFF, P., McL. D. 1992. Economic Geology. *In*: DUFF, P. McL.D. & SMITH, A. J. (eds), *Geology of England and Wales.* The Geological Society, London, 589–627.

HARLEY, M. 1992. RIGS update. *Earth science conservation,* **30**, 22–23.

HARLEY, M. 1992. RIGS update. *Earth science conservation,* **30**, 22–23.

ROBINSON, E. 1984. *London – illustrated geological walks. Book 1.* Scottish Academic Press, Edinburgh.

—— 1985. *London – illustrated geological walks. Book 2.* Scottish Academic Press, Edinburgh.

—— & WORSSAM, B. 1989. The geology of some Middlesex churches. *Proceedings of the Geologists' Association,* **100**, 595–603.

SMITH, D. 1993. Micklefield Quarry: remedial treatment by general consent. *Geology Today,* **9**(1), 5–6.

Local conservation and the role of the regional geological society

A. CUTLER

21 Primrose Hill, Wordsley, Stourbridge, West Midlands DY8 5AG, UK

Abstract: Local geological societies are essentially concerned with providing a programme of Earth science-related entertainment. In the case of the Black Country Geological Society (BCGS), these activities are its foundation, giving the society stability, an essential framework, and a public face which raises the profile of geology in the region and underpins the conservation effort.

Until recently, BCGS has done little in the way of site maintenance or interpretation, having concentrated on site identification and recording, promoting geological heritage and the need for conservation within the local political arena. This has involved developing links with local authorities, planners, leisure services and elected councillors. There have been varying degrees of success, but in Dudley Metropolitan Borough, these links are particularly well developed. In 1992, this was exemplified by a joint project with the Public Works department in cleaning up the important Hayes Cutting Site of Importance for Nature Conservation. Sites of Importance for Nature Conservation (SINCs) are non-statutory sites selected for their local biological and/or geological importance, and are enshrined in the Unitary Development Plans produced by the Black Country local authorities and Birmingham City Council.

The society is a member of the Dudley Nature Conservation Consultative Group, an advisory body comprising officers, elected councillors and representatives of local conservation organizations.

The long-term success of non-statutory site protection needs public support for its success. There is a groundswell of public opinion in favour of wildlife conservation to which politicians respond. Linking geology with other branches of natural history is working well in the West Midlands, without diluting the integrity of geology's special character. Indeed, it promotes a greater understanding of nature conservation in its widest sense and makes for a more complete appreciation of the natural environment.

The Black Country Geological Society (BCGS) was formed in 1975, with geological conservation being one of its two principal aims. Indeed, it was concern for the threat posed to local sites that acted as the catalyst to turn the idea of the society into reality. The society's other principal aim is to provide a programme of activities in the form of lectures, field meetings, excursions and social events to cater for the continuing interest of members' leisure time.

Although amateur based, the BCGS is and always has been a very cosmopolitan society, with professional and academic geologists as members, teachers from primary schools to tertiary colleges, professionals from other disciplines, as well as other interested amateurs. This diverse cross-section of backgrounds and abilities has ensured that the society is well organized in a professional manner. The activities described above give the society a stable foundation, an essential framework and a public face which raises the profile of geology in the region and without which, in our opinion, the conservation effort would not be sustainable.

The society's meetings and other activities are reported in local newspapers, and members have given interviews on both local and national radio. Active contact or dialogue is maintained with English Nature, local wildlife organizations, local museums and, most importantly, with local authorities.

All of these activities promote the society, of course, but most importantly, they promote geology in the region, which is now significant for tourism.

Early conservation work

The BCGS has concentrated primarily on site identification and recording, promoting geological heritage and the need for conservation within the local political arena.

The society embarked on its site conservation programme in 1975 in the face of growing pressure for waste disposal sites and redevelopment. Initially, activities were reactive, these met with little success, usually because of pre-existing planning permission, and many locally well-known sites changed or disappeared for ever. One early success, however, was a compromise worked out in 1976 with the former

West Midlands County Council to preserve a small, but significant, exposure at the southern end of Pouk Hill Quarry, a former dolerite source being used for landfill at Bentley, near Walsall (Ixer 1981).

It became increasingly obvious that the society needed to be pro-active and flag up the conservation value of sites to the planning authorities before threats materialized. It was resolved to draw up a list of significant geological sites throughout as much of the county as possible. This was accomplished in 1977, and by the end of 1978 a list of over 90 sites had been compiled. The information was sketchy and basic; nevertheless it was an important start and the list was supplied to the Planning and Leisure Services Departments at both County and District Council levels. These departments took an early interest in this work, and the society was, and still is, consulted on a variety of planning applications, if on a somewhat ad hoc basis.

Ultimately, the information was transferred to the standard record forms of the National Scheme for Geological Site Documentation (NSGSD), but it was not until 1982/3 that the records were adopted by Stoke City Museum's Geological Records Centre. Since 1987, following the appointment of a geological curator, these records are now held at Dudley Museum.

Although the society was clearly enjoying some success with its conservation activity, two issues gave rise for concern. Firstly, site protection was informal and reliant on the attitudes and goodwill of council officers or landowners. Secondly, despite the increased profile, geological conservation always appeared secondary to wildlife conservation matters.

Local site protection

In the late 1970s, the then West Midlands County Council requested the Nature Conservancy Council (now English Nature) to draw up a list of sites of particular conservation value within the county, but below the standards of statutory Sites of Special Scientific Interest (SSSIs). The list constituted the best of the non-statutory sites – designated Sites of Importance for Nature Conservation (SINCs) and recognized in the County Structure plan. The sites were selected exclusively on biological criteria. A footnote to the schedule referred to a (separate) list of geological sites held by the Black Country Geological Society. At least geology was not entirely forgotten, but in practical terms geological sites were not scheduled and, in effect, marginalized. The omission of geological sites from these lists at that time was symptomatic of a fundamental problem which the author believes still affects geological conservation – that it is not widely perceived as an integral part of nature conservation (Box & Cutler 1988).

The significance and importance of SINC designation may be judged by the County Nature Conservation Strategy, and subsequent local plans, such as Dudley's Recreation & Open Space Plan, which states *inter alia* 'there will be a presumption against development which would reasonably prejudice the nature conservation value of designated Sites of Importance for Nature Conservation'.

The task of obtaining formal recognition of geological sites was achieved partly by changing attitudes and by an odd twist of fate. Denotification of two SSSIs, one in Dudley (Castle Mill Basin) and one in Birmingham (Rubery Cutting and Leach Green Quarries), gave rise to concern that the respective local authorities would assume the sites were no longer of any value and allow damaging development, which actually became the case in Birmingham. Following discussions with the Nature Conservancy Council (now English Nature), the two denotified SSSIs were designated as SINCs and the relevant local authorities informed; a very simple and obvious solution, but quite radical in many respects or so it seemed. The precedent having been established, geological sites were added to the SINC schedules at the next revision in 1988. Each selected site is supported by descriptive material utilizing the record forms of the NSGSD.

In selecting geological sites for SINC status, a number of criteria or factors are taken into account, including intrinsic scientific interest, educational value, historial associations, relationships with other sites, and uniqueness or rarity of exposure in the local context. Formal notification of SINCs to the local authorities is performed by English Nature, following recommendation and justification by BCGS (in the case of geological sites) and the Urban Wildlife Trust (for biological sites). This arrangement does not prejudice the society's direct dealings with local government. The involvement of English Nature gives added weight and credibility to the selected sites.

Geology and wildlife

The BCGS firmly advocates that geology is part of natural history, that nature conservation is not just about birds, butterflies and flowers, and that nature conservation embraces both biological and geological conservation. This view has been strengthened by the society's membership of the Nature Conservation Consultative Group (NCCG), an initiative of Dudley Metropolitan Borough Council (MBC) set up in 1985.

The NCCG is an advisory body comprising officers from the Planning and Leisure Services Department of the council, elected councillors, and representatives of local conservation organizations such as the Urban Wildlife Trust. English Nature is included and, significantly, Dudley's Science Advisory Teacher.

NCCG serves as a forum where conservation topics and planning applications can be discussed in a spirit of mutual co-operation. NCCG is, of course, only an advisory body, but the council committees do take its views seriously and follow its advice. It has also been significant in raising the profile of geological conservation and establishing contact with kindred organizations so essential for protection of non-statutory sites.

When the draft Black Country Nature Conservation Strategy (BCNCS) came before the group, geology had been omitted. However, this was rectified with the overwhelming support of our wildlife counterparts. The BCNCS was commissioned by the Black Country Development Corporation and embraces not only Dudley but Sandwell, Walsall and Wolverhampton Metropolitan Boroughs, so its strategic importance will be appreciated. SINCs do not have statutory protection, but they are recognized by the Metropolitan Boroughs, with their incorporation into the published Unitary Development Plans (UDP) which now acknowledge both wildlife and geological criteria. In the forseeable future, third tier sites will be recognized within the four Black Country Metropolitan Boroughs as a consequence of BCNCS. Broadly speaking, such sites will possess more amenity than scientific value, or have very localized interest. They will be known as SLINCs (Sites of Local Importance for Nature Conservation).

The BCGS experience shows that developing links with its wildlife counterparts is not only productive, but essential for geological conservation in the long term. It also highlights the need to adopt a common nomenclature, on a county basis at the very least, for both biological and geological second tier sites. The society strongly advocates the need for a national second tier designation, as is the case with the top tier (SSSI). Recent experience in helping with a new RIGS (Regionally Important Geological/geomorphological Sites) group for the County of Hereford and Worcester confirms the benefits of co-operating with wildlife organizations and the willingness to adopt common terminology. Unfortunately in Hereford and Worcester, second tier sites are known as 'Special Wildlife Sites'. To alter the

Fig. 1. Brewin's Canal Section SSSI, Netherton, Dudley. Photograph by G. Worton.

published county structure plan is untenable at this stage, but the County Council is willing to accept the addition of geological sites to the schedules and, therefore, acknowledge their importance.

The benefits of linking with wildlife conservationists are best illustrated in the development of Local Nature Reserves (LNRs). Dudley Metropolitan Borough now has four LNRs, three of which have predominantly or substantial geological interest. The latest, Tenterfields LNR, is the first in the UK to be in the grounds of a school. Clearly LNRs will normally only capture public and local authority imagination where there is a biological interest too, as is the case in Dudley (Barker 1992).

Site maintenance

Site maintenance is one area where BCGS has not been particularly active. However, on occasions when there has been a demonstrable need, the society can mobilize volunteers. In the Spring of 1991, Brewin's Canal Section SSSI was cleaned up in a joint exercise between the society and Dudley Canal Trust – prompted by a national waterways festival. The trust provided a narrow boat which served as the means of removing the cleared material (Fig. 1).

In 1992, the Hayes Cutting SINC was cleaned in another joint exercise, this time involving the Public Works Department of Dudley MBC. A brick retaining wall adjoining the Hayes site was to be demolished and the ground behind regraded, improving visibility at a busy road junction. Contractors carried out the demolition and regrading work, and the society was allowed to record the newly exposed section. As the work neared completion, the degraded state of the Hayes Cutting alongside became increasingly apparent. It was very gratifying when the Public Works Department asked if they could help reinstate the exposure. It was subsequently agreed that the contractors would cut and remove all the large vegetation and a disused sewer pipe, prior to BCGS members carrying out conservation of the rock face (Fig. 2). The local authority also provided a skip for the soil and other vegetation cleaned off the outcrop, and arranged for its eventual removal (Cutler & Worton 1992).

Links with Dudley Museum

In parallel with its site conservation activities, the society has enjoyed a long and fruitful relationship with Dudley Museum. In 1977, the society embarked on a campaign for the appointment of a geological curator. The campaign led to contact, not only with local councillors,

Fig. 2. The Hayes Cutting SINC, Lye, Stourbridge. Photograph by G. Worton.

but with Members of Parliament, and even a Government Minister. It was during this period that close contact was established with the Geological Curators Group, culminating in the Group holding its Annual General Meeting in Dudley in 1985. The BCGS acted as hosts, and arranged a well received programme, including a civic reception. Two years later, in 1987, the present geological curator was appointed, and geology is now actively promoted for leisure, tourism and education. Far from being eclipsed, the society's role has been enhanced and members provide a useful support role in many different ways, including site investigation and as an emergency collecting/recording task force.

Conclusion

This paper has attempted to illustrate the many different ways in which a regional geological society can help raise the profile and pursue the cause of geological conservation.

A society as an independent local community group has many more opportunities for publicity than any informal group. And perhaps most importantly, a society should remain stable despite the movement of individual members.

The long-term success of non-statutory site protection needs public support. There is a groundswell of public opinion in favour of wildlife conservation which politicians recognize and to which they respond particularly at local levels. Linking geology with other branches of natural history is working well in the West Midlands, without diluting the integrity of geology's special character. It promotes a greater understanding of nature conservation in its widest sense and makes for a more complete appreciation of the natural environment.

References

IXER, R. A. 1981. The Petrography of the Igneous Rocks from Pouk Hill. *The Black Country Geologist*, **1**, 23–31.

BARKER, G. (ed.). 1992. Imaginative Use of school grounds. *Urban Wildlife News*, **9** (4), 1–2.

BOX, J. & CUTLER, A. 1988. Geological Conservation in the West Midlands. *Earth science conservation*, **25**, 29–35.

CUTLER, A. & WORTON, G. 1992. The Hayes Cutting gets a facelift. *Earth science conservation*, **31**, 27–28.

A person on the inside – opportunities for geological conservation in local engineering projects

GRAHAM WORTON

38 Vale Road, Netherton, Dudley, West Midlands DY2 9HZ, UK

Abstract: New development, or reclamation of derelict land, frequently involves excavation which creates temporary or permanent exposure of geological materials. Locally recognized conservation groups and educational parties are sometimes allowed limited access to record and collect. Such activities are principally carried out under formal agreements with site owners, or controlling bodies such as local authorities or central government departments. However, a potential conservation resource which is often missed is the full-time supervising geologist or engineer on the site, who is in a unique position to liaise between interested parties, while having free access to site excavations. Such site staff often have a detailed knowledge of local geology, are trained in recording information and are usually sympathetic to conservation needs.

The paper will discuss the benefits that may be gained by local conservation groups in developing good relations with such site staff. Case studies will show how such relationships have worked in the Borough of Dudley, in the West Midlands, and will explain some of the perceptions and pressures involved, benefits to be gained and difficulties to be overcome.

Land uses in the local area will change with time. Existing developments will be extended, modified or cleared away for new development to occur. Almost all of these changes will involve some sort of disturbance to the soils and rocks on which they are sited.

The majority of these developments will have existing planning permissions in place which make no consideration of conservation (particularly absent in terms of geology). In some instances, knowledge about ground conditions at the site may be considered to be confidential for commercial reasons.

Even so, the development will create exposure as work proceeds, and it is likely that much exposure will go unrecorded due only to factors of ignorance and working time-scales. The questions are, therefore, what can be done to be better informed about temporary exposures? How can such exposures be accessed, logged and sampled to mutual agreement of all parties concerned? And who, in the local area could best and most efficiently carry out such work?

Small developments

Probably the most abundant newly created exposures will be in very small developments occurring sporadically over an area. As such, it is a very difficult task to maintain an awareness of all potential sites and to be present at the precise moment that exposure is at its optimum. A particular problem is that, on small sites, exposure is likely to be created within small ditches for foundations and services which will be concreted or backfilled very soon after they are excavated.

Members of a local geological society have an important role to play, as they will be best suited to keeping a watchful eye on their local area. It is likely, however, that the society's members may come up against barriers of time, permission, indemnity and attitude towards conservationists. It is important in such instances to remember that developers or contractors have no obligation to such groups to permit access to their site, particularly in respect of their public liabilities, and will probably do so only as an act of goodwill. It is critical, therefore, that the amateur is mindful of the responsibilities of the site staff, and that this is reflected in the approach and conduct of that person when accessing such works.

The professional engineering geologist will usually have a very small part to play in smaller projects. Involvement will be limited to small-scale site investigation to check on ground conditions and, possibly, a brief inspection of excavations at key points in the works. Boreholes and trial pits may be excavated by the geologists, but it is highly unlikely that scientifically relevant data, such as sedimentology, palaeontology, lithology and structure, will be recorded in any detail. In engineering terms, such information would be of little concern and the geologist in question may be criticized for wasting time or resources if such work was

From O'Halloran, D., Green, C., Harley, M., Stanley, M. & Knill, J. (eds), 1994, *Geological and Landscape Conservation.* Geological Society, London, pp. 359–363.

undertaken as part of a contract (however beneficial this might turn out to be). However, an enthusiastic geologist in this position may be able to help the conservation effort in other ways. For example, most geotechnical consultancies will retain samples from a site, sometimes including rock cores which, at regular intervals, must be disposed of. This presents a potential opportunity for interested third parties to log or sample core for scientific purposes, or local schools and the local museum to benefit from donations.

Large engineering projects

Large engineering projects offer many more opportunities for Earth science conservation (NCC 1990). For example, large construction sites, road or tunnel schemes, quarries or large land reclamation projects are likely to have time-scales for planning and execution of works in the order of several years. It is more common now for large engineering projects to require a thorough assessment of their impacts on the local environment before they are given permission to proceed. Such environmental statements commonly consider implications for archaeological and wildlife areas, but seldom, if ever, make reference to geological heritage.

These projects will involve detailed and lengthy site investigation, and will often create large exposures which remain open for considerable periods of time; however, due to their scale and complexity, they are often more hazardous.

Conservation groups or members of the local geological community will usually be given limited access to site excavations under formal agreements, with specific safety instruction and often accompanied by a member of the site staff. Opportunities for such groups will be limited by the extent of goodwill and availability of non-essential time of site technical staff. If visitors to sites become a burden, or do not conduct themselves in a manner conducive to safety, then doors of opportunity will close.

Large engineering projects do, however, have one major, though often unrecognized, potential resource permanently on site, and that is the resident geological/geotechnical staff.

The resident engineering geologist

The geotechnical staff on a site will have a range of professional duties which may include advising on materials for different uses, sampling, testing, supervision of works, and liaison with interested parties on the engineering design and progress of works. It is the author's experience, however, that while most geotechnical staff are very busy, they will usually take a personal interest in the more unusual aspects of the job, and many of their duties are similar to, and often compatible with, practical conservation activities.

Such a qualified person (if aware of Earth science conservation issues) on a large development site can be a very positive force in maximizing the conservation potential of an engineering project, and in minimizing conflict. The advantage of having a 'person on the inside' are manifold, but major advantages would appear to be that the resident geotechnical staff are people who:

- have a personal, if somewhat latent, interest in the subject and may be happy to help, where possible, in interesting projects;
- enjoy free access to all excavations as they happen and may be in a position to alert external parties to temporary features as they are unearthed or, if encouraged, carry out such works on behalf of a conservation group in their spare time (e.g. lunchtimes);
- are trained to record and observe geological detail, and may have a very good working knowledge of the local geology;
- are held in a position of respect by all parties concerned and in the ideal situation to liaise between the parties;
- may be required to show visitors around the site as part of their duties in order to gain good relations with the local population;
- are legitimately on-site and have no problems of liability or indemnity cover.

Clearly, there are many practical advantages to be gained by fostering good relationships with site geotechnical staff; however, the extent of their involvement will be down to the individual and his or her employer. Experience in the West Midlands has indicated to the author that very few practising professional geotechnical staff have any real knowledge of geological conservation, its implications, consequences or benefits for their projects, and are cautious about the subject, having neither the time nor the inclination to find out more. They may perceive conservationists to be interfering amateurs who may cause delays or unaccountable costs, and may also be a liability (Headworth 1982). Where this is the case, it is in their interest to do as little as possible to encourage such visitors.

Where a client and contractor have knowledge of a professional who can advise and inform on such issues, however, barriers can be removed and things can be achieved. One such example is that of a large land reclamation project which was recently completed in the Metropolitan Borough of Dudley, in the West Midlands, known as Dibdale/Burton Road.

Dibdale/Burton Road land reclamation project

Background

Large urban areas, like the Black Country, are subject to planning controls which prevent their spreading into the surrounding countryside. Grants are available to encourage the re-use of land within the urban area. The Dibdale/Burton Road reclamation project was such a site.

The site had widespread and complex problems of waste disposal, chemical contamination and past mining. Over several years, Dudley's planners and engineers, in co-operation with external organizations, designed and implemented a complex land reclamation scheme to restore the site to productive use. This project, like many modern large reclamation schemes, employed the novel use of opencast coal mining on a part of the site to realize the potential of remnant coal, offset the costs of remedial works, and produce a void into which an engineered landfill could be placed. The works created a very large temporary exposure of Upper Carboniferous, Coal Measure rocks, and exposed spectacular mine workings as shown in Fig. 1.

Liaison with site staff allowed small parties to visit the site in a limited way which did not obstruct the works, create unacceptable risks or excessively involve site staff. Due to the author's professional involvement with the site, however, these works were greatly extended.

Project achievements

A very detailed sedimentary log was able to be produced, describing lithology, thickness and sedimentary features of all units in the sequence, (including engineering characteristics, which would be useful to the project and for correlation of the sequence with nearby projects and boreholes).

The work was able to identify and precisely stratigraphically relate several fossil horizons, one of which contained fragmented vertebrate remains, including long curved fish teeth, scales, spines and small jawbones. These teeth have subsequently been identified as a species of the Carboniferous fish *Megalicthys*, however, many of the smaller jawbones and other remains are still unidentified. Dudley Museum's collection previously held a few samples collected from the same horizon – but these were not labelled,

Fig. 1. Pillar and Stall Mine workings exposed in the Stinking Coal seam at Dibdale/Burton Road land reclamation project, Dudley.

other than with reference to their place of collection and donor. This work donated 200 of the best finds to the museum collection; greatly adding to the scientific value of the Coal Measures part of the collection.

Conservation work was also able to examine several worked mineral horizons, including the famous South Staffordshire 'Thick Coal', which was almost 10 m thick at this site. Detailed seam sections were produced and mining details recorded, which have also subsequently been used by Dudley Museum to produce a spectacular mining display – using timbers taken from the mines, artefacts found at the site and a detailed reconstruction of the Stinking Coal face as measured and logged at the site, as shown in Fig. 2 (Reid 1993). The site has subsequently produced one of the most comprehensive site records held by the Dudley Museum Geological Record Centre and has gone some way towards generating good relationships between conservation groups, amateurs and the local geotechnical community.

Conclusions and suggestions

Local construction/engineering projects are continually producing new exposures of rock; however, these are often very short-lived, and usual contractual considerations and attitudes may restrict access and limit conservation potential of external conservation groups and amateur geological societies.

The resident geotechnical staff on such sites can be a very valuable asset in the conservation effort due to their training and position of trust; however, there is a general lack of understanding by these staff or their clients about

Fig. 2. Reconstruction of underground mine in Stinking Coal of Dudley in Dudley Museum and Art Gallery based on the conservation work at Dibdale/Burton Road.

the benefits and consequences of becoming involved with geological conservation, which sometimes generates a reluctance to get involved or help.

The author's experience in Dudley has shown that good relationships between interested parties can benefit both the project and greatly enhance the conservation of an area's common heritage. In the author's opinion, there is much more that could be done, in particular:

- local conservationists could take steps to get to know the local engineering fraternity so that they become a familiar force in the local area;
- local conservation groups could learn to take a more professional approach when dealing with private sector projects to remove concerns of contractors and clients about their presence on site and its consequences;
- steps could be taken to make the local geotechnical community aware of the consequences and benefits of geological conservation for their work, to encourage them to take a more active role and apply their skills positively.

One further opportunity that is still overlooked is the potential to get involved at the planning stage of engineering projects; often an environmental assessment is carried out in which archaeological and wildlife heritage are assessed, and conditions are imposed to protect and conserve important features, even before works are allowed to begin. This is not the case for geological heritage. As geologists, we have much to learn from the success of these other disciplines.

It is the author's hope that the experience in Dudley, which has led to a closer understanding developing between the engineering fraternity, scientists and amateurs, will be helpful to others attempting to carry out such works in other areas, and that the future will see many more examples of co-operative geological conservation and goodwill.

The author wishes to thank the following individuals and organizations for their assistance in the Dibdale/Burton Road projects: Dudley Metropolitan Borough Council, Public Works Deparment (client on the project); R. J. Budge (Mining) Limited (contractors on the land reclamation project); Mr Peter Mills and Mr Graham Bartlett of Johnson Poole and Bloomer (engineers on the project); Mr Stephen Weston for his assistance in logging and recording the succession; Mr Ian Wilkins who made the initial discovery of fish remains; Black Country Geological Society and Dudley Museum and Art Gallery for their support and encouragement for this paper.

References

HEADWORTH, H. G. (1982). Geological Conservation – a conflict of loyalities. *British Geologist*, **8**, 94–96.

NATURE CONSERVANCY COUNCIL 1990. *Earth Science Conservation in Great Britain: A Strategy*. NCC, Peterborough.

REID, C. G. R. 1993. No Stone Unturned. *Museum Journal*, **3**, 3.

Conservation, communication and the GIS: an urban case study

COLIN REID

Dudley Museum and Art Gallery, St James's Road, Dudley, West Midlands DY1 1HU, UK

Abstract: In urbanized areas, new development continually creates temporary exposures whilst threatening existing ones. The ability to monitor this development is vital to the success of urban conservation. In its role as a Geological Recording Centre (GRC), the local authority museum is in an ideal position to carry out this monitoring function.

Dudley Museum is the Recording Centre for Birmingham and the four Black Country metropolitan boroughs of Dudley, Sandwell, Walsall and Wolverhampton. It holds records on over 340 local sites, including Sites of Special Scientific Interest (SSSIs) and Sites of Importance for Nature Conservation (SINCs), the majority of which occur within Dudley borough. By liaising with Dudley's Planning and Public Works Departments, the GRC attempts to keep abreast of all operations that may endanger designated conservation sites or produce fresh exposures. This is not always successful, however, as it depends heavily on the co-operation and goodwill of individual officers.

A Geographical Information System (GIS) being pioneered by Dudley is proving to be a valuable new tool. This system enables the user to build up a digitized picture of council services or activities relating to planning, grounds maintenance, utilities and conservation.

The paper describes how the GIS is being used within the authority to monitor and protect Dudley's 25 SINCs and 6 SSSIs. It also illustrates the importance of good communication between the Recording Centre and other council departments, and examines how the GRC can act upon information gained from GIS and elsewhere to limit or prevent damaging activity and to gain access to mine workings, building sites and other sensitive areas for the purposes of recording and sampling.

In heavily urbanized areas, geological sites are localized and often transient. For this reason, every exposure assumes an importance it might not otherwise have. The urban environment is one of constant flux in which human activity, whether through development, land reclamation or the exploitation of resources, continually creates temporary exposures whilst threatening existing sites. The ability to monitor these activities is, therefore, a priceless tool to the urban conservationist.

Representation on the local authority's planning or conservation committees by a member of the RIGS (Regionally Important Geological/geomorphological Sites) group or geological society is certainly beneficial (see Cutler in this volume). However, the real power lies in being able to monitor potential development at source. The conservationist, therefore, needs to have an ally inside the authority, with ready access to information on new planning applications and existing planning permissions, together with any relevant works being carried out by the authority itself.

In its role as a Geological Recording Centre (GRC), the local authority museum is in an ideal position both to monitor potential development and to provide an element of pastoral care. The benefits of this proactive role and the problems of carrying it out effectively are well illustrated in the case of Dudley Museum and Art Gallery, the Recording Centre for the Black Country.

In 1988, Dudley Museum became the 49th GRC under the National Scheme for Geological Site Documentation (NSGSD), with a remit to record sites within Birmingham and the four Black Country metropolitan boroughs of Dudley, Sandwell, Walsall and Wolverhampton. On its inception, the new GRC inherited records on 340 sites compiled by the Black Country Geological Society (BCGS) during the 1980s. The majority of these are within Dudley itself and include the borough's 6 Sites of Special Scientific Interest (SSSIs), 25 second tier sites, known in the West Midlands as Sites of Importance for Nature Conservation (SINCs), and a National Nature Reserve (NNR) – Wren's Nest, the West Midlands' most important geological locality (Box & Cutler 1988). All the original BCGS records have been entered onto the national sites database and the GRC retains a copy of the database for its own area.

Dudley Museum regularly receives information, logs and specimens relating to new, mainly temporary, sites from members of the BCGS and from other sources. However, demands on the curator's time have made it impossible to collate this information into comprehensive files or update many of the existing records, particu-

From O'Halloran, D., Green, C., Harley, M., Stanley, M. & Knill, J. (eds), 1994, *Geological and Landscape Conservation*. Geological Society, London, pp. 365–369.

larly for sites outside Dudley borough. This, coupled with concern that many temporary sites in the area are going unrecorded, has led to a reappraisal of the Recording Centre's role. It has been decided that, as the GRC is not currently in a position to provide an overt public information service, it would better serve the community by adopting a policing role, using its links within Dudley authority to monitor all activities that might either produce new exposures or threaten existing sites, particularly those with recognized conservation status.

Fundamental to the success of this task is the establishment of good lines of communication both internally, between the GRC and the authority's Planning and Public Works Departments in order to glean information, and externally with the BCGS to act upon this. The latter has never been a problem. The BCGS has played a leading role in geological conservation for many years, and readily provides volunteers and working parties for recording, collecting, site clearance and similar activities. Its relationship with the GRC is a symbiotic one, strengthened by strong personl links, with each organization relying upon the other in order to carry out its own conservational role effectively.

Efficient communication within the local authority has proven much more difficult to establish, particularly where it involves the creation of new geological sites, as no formal arrangement currently exists between departments for informing the GRC when rocks are newly exposed. Consequently, the GRC relies largely on the goodwill and co-operation of individual officers. This is, at best, a tenuous state of affairs, and the GRC has come close to missing the opportunity to record several important temporary exposures.

In 1990, a BCGS member drew the GRC's attention to a new exposure of Ludlow Series (Lower Elton Formation) shales at a building site on Dudley's Castle Hill, apparently under the control of the borough's Planning Department. Fortunately the situation was remedied very quickly. After consultation with the Chief Planner and the site contractors, access was readily gained for recording and sampling. Subsequently, the 25 m section yielded the most comprehensive data on the Ludlow rocks of the area since the nineteenth century. Logging revealed no less than 16 volcanic horizons, while systematic sampling produced an assemblage of invertebrate fossils including brachiopods, trilobites and rare graptolites.

More recently, a 35 m section of Wenlock Series (Coalbrookdale Formation) shales was revealed during the building of a college extension in the middle of Wren's Nest National Nature Reserve, which is ironically the area's most important conservation site. Again, the centre learned of the exposure only by default.

Like the Ludlow Series rocks described earlier, the Wren's Nest strata produced invaluable stratigraphic data and an abundance of fossil fauna. On this occasion, however, the contractors permitted access to the site only after considerable pressure from the borough's Planning Department and production of an indemnity form freeing them from responsibility in the event of an accident.

The recent merger of Dudley's Planning and Leisure Departments (the latter with responsibility for Dudley Museum) promises to improve the situation, with the GRC endeavouring to have put in place a formal policy in which:

(1) planners and public works officers inform the GRC if new exposures are created on sites under their jurisdication;
(2) where deemed appropriate, it is written into contracts that contractors must allow access for recording and sampling, providing, of course, it is carried out with due regard for safety regulations and on mutually agreeable occasions;
(3) the GRC receives regular updates on planning permissions that may produce new exposures.

In contrast to the difficulties in detecting and gaining access to temporary exposures, the GRC has had considerable success in policing Dudley's existing geological sites. This is because a reliable communications network has been in place for some time. Dudley's Planning Department employs an officer whose responsibilities include monitoring operations which might endanger any SSSI or SINC, informing the GRC and BCGS about the work to be carried out and, if necessary, acting as a liaison officer on site.

The efficiency of this system has been further increased by the inclusion of Dudley's recognized conservation sites on the borough's Geographical Information System (GIS). Dudley has been at the leading edge of corporate GIS development for almost a decade. Consequently, the authority holds large sets of digitized information relating to disciplines such as planning, grounds maintenance and utilities. These data can be overlaid on an Ordnance Survey base, enabling a multi-layered picture of the borough to be built up and, more importantly, to be available to many users across the corporate network. Currently the GIS holds plotted records of the borough's SSSIs and SINCs, though not third-tier sites, designated in the West Midlands as Sites of Local Importance for Nature Conservation (SLINCs). These will be added in due course, however.

Dudley's GIS is proving a highly effective method of protecting the borough's conservation areas, as it ensures that any threatening action is immediately drawn to the GRC's attention. Upon receipt of planning applications or similar data, the boundaries are digitized into the GIS by officers from Dudley's Information Technology Department, and automatically referred against existing areas of interest, including conservation sites. In the event of overlap, the officer plotting the data then receives a prompt to call up further information on the area of interest.

In the case of geological sites, this information will include conservation status, a list of activities that may pose a threat to the site, and a contact name and telephone number at the GRC (Fig. 1). At the same time, the system automatically sends off a standard 'consultee' memo through the authority's internal post to inform the GRC of the enquiry, thus giving an early opportunity to comment.

At present the GRC is only a 'passive' user of the GIS system, although the benefits accruing from this in terms of site protection are considerable. In due course, however, the centre will become an 'active' user with its own networked terminal, and will be able to run 'integrity checks', such as periodic audits of planning applications affecting any geological site within the borough. It will also be able to monitor any operations which are likely to produce fresh exposures, although this will still require feedback from the officers on the ground.

The effectiveness of the GIS was demonstrated in 1991 when it flagged up a planning application affecting Coseley canal cutting, a valuable Coal Measure (Westphalian B) section of SINC status (Fig. 2). The application was to put in drainage services on the west side of the cutting and required the removal of an important outcrop. At the GRC's request, a site meeting was arranged to discuss the threat. Subsequently, a new section was created by the contractor a short distance away and no harm was done.

When practised, the nurturing of good communications within the local authority can reap priceless rewards. This is borne out by two recent cases in Dudley. The first concerns a major stabilization project in Stores Cavern, a vast limestone mine under Dudley's Castle Hill.

Extensive limestone mining in Dudley has left a warren of such caverns under the town. Many of these are unstable and must be

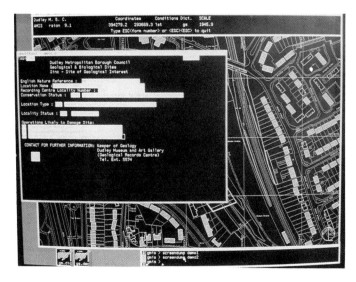

Fig. 1. A prompt button brings up an information box, which includes data on the site's conservation status, a list of activities that may pose a threat to the site, and a contact name and telephone number at the GRC. Once it detects an overlap, the system automatically informs the interested parties within the local authority via an internal mailing facility.

Fig. 2. GIS image showing Coseley canal cutting geological SINC (diagonally shaded area). The boundary of an invasive planning application can also be seen overlapping the NW margin of the SINC.

monitored regularly by a team of inspectors, under the guidance of a mines manager employed by the authority's Public Works Department. The museum retains a healthy interest in the old workings as they yielded many fine Wenlock fossils at the peak of mining activity in the nineteenth century, notably the beautifully preserved trilobites and crinoids for which Dudley is world renowned.

Stores Cavern reputedly yielded superb examples of the rare Wenlock trilobite *Trimerus delphinocephalus* during mining operations in the 1860s. Having been made aware of this, the

Fig. 3. The fruits of good communication. The rare Wenlock trilobite *Trimerus delphinocephalus* was rediscovered in Dudley's limestone caverns after 130 years as a result of close liaison between the authority's GRC and Public Works Department.

mines manager invited the museum's geology curator to inspect the cavern, and as a result the trilobite-bearing horizon was located. Close liaison between the GRC, mines manager and site contractors then enabled 20 tons of material to be removed to a safe site outside the cavern. This is yielding the first *Trimerus* specimens found in the area for 130 years (Fig. 3).

In the second case, officers from the Leisure Department's Parks Section approached the GRC for acvice on the landscaping of a land reclamation site in the Gornal area of Dudley. Once an active sandstone quarry, the site is scheduled for conversion from waste ground into a large open recreation area. It is currently a geological SINC with some reasonable outcrops of Upper Silurian (Pridolian) strata belonging to the Downton Castle Sandstone Formation.

At the suggestion of the GRC (in consultation with the Black Country Geological Society), a new permanent section will be cut during landscaping to reveal the Platyschisma Shale Member which underlies the sandstone. A thin bone bed within this sequence at Ludford Lane in Shropshire has yielded the earliest known terrestial fauna, including trigonotarbid arachnids, kampecarid myriapods, centipedes and mites, together with eurypterids and *Cooksonia*, the earliest vascular plant (Jeram *et al.* 1990). When created in 1994, the new Dudley section should be larger than that in Shropshire and, if as productive, would become a site of international importance.

The Dudley experience proves that the most successful strategy for monitoring and protecting urban sites is one which involves a working partnership between conservationists and the managing authority. It also demonstrates how important it is to keep a finger on the pulse of land development in the urban environment – and the perils of failing to do so effectively. The assumption of a policing role by GRCs, and similar facilities elsewhere, is an area that needs to be examined for the future, particularly in the light of new technology such as GIS, which is now becoming a standard management tool in local government.

References

Box, J. & Cutler, A. 1988. Geological Conservation in the West Midlands. *Earth science conservation*, 25, 29–35.

Cutler, A. (this volume) Local conservation and the role of the regional geological society.

Jeram, A. J., Seldon, P. A. & Edwards, D. 1990. Land animals in the Silurian: Arachnids and Myriapods from Shropshire, England. *Science*, 250, 658–661.

Developing geological site recording software for local conservation groups

C. J. T. COPP

Environmental Information Management, 8, The Paddock, Clevedon, Avon, BS21 6JU, UK

Abstract: The documentation of geological sites by local groups has a long tradition in the UK. This has been given impetus by the National Scheme for Geological Site Documentation (NSGSD) and more recently by the Regionally Important Geological/geomorphological Sites (RIGS) intiative.

These projects have attempted to standardize the data recorded by the use of specially designed forms and guidelines on terminology. A prototype database was developed under contract from the then Nature Conservancy Council and the British Geological Survey in 1991, to hold about 12 000 site records submitted to the NSGSD scheme. The author has subsequently extended and developed the database, in a voluntary capacity, into a version called 'GD2', which has now been installed with 14 local geological records centres, RIGS groups and statutory conservation agencies in the UK. GD2 is written using version 2 of the Advanced Revelation database management system and runs on DOS-based personal computers.

Despite the use of standardized recording forms, users record sites according to their personal interests and experience, which gives rise to highly varied records. The database needs, therefore, to be much more than a simple copy of the data forms. It needs to ensure that certain basic data are always entered and that terminology is controlled so that indexes and retrieval are reliable. The major problems in implementing a standardized geological database are the control of stratigraphic terminology and the lack of a standard palaeontological taxonomic thesaurus. Ideally, data entry routines should offer guidance on valid and non-valid terminology or cope with situations where the two cannot be readily resolved.

Development of software such as GD2 is not a one-off event. There is a continued need for improvement and adaptation to new requirements. Already there has been a change in emphasis of recording from basic site documentation to conservation and education under the RIGS initiative. Technologically, the database will need to respond to the move away from text-based databases to graphical databases, and from simple dot map output to integration with geographical information systems.

All these things can be implemented, but a growing user-base and increased demands for documentation and support will soon outstrip the voluntary resources available. The development of the software on a national scale will eventually require an administrative and funding framework to ensure its longer-term value.

In 1977, the National Scheme for Geological Site Documentation (NSGSD) was launched to encourage and co-ordinate geological recording, with particular reference to sites of interest that did not have statutory protection (Museum & Galleries Commission 1993). It is estimated that in the period up to 1991, county-based groups recorded details of about 19 000 geological sites, of which some 12 000 site records were prepared in machine-readable form by the British Geological Survey (BGS) under contract from the Nature Conservancy Council (NCC). As part of this contract, the author was asked to develop a database to hold the data with the intention of providing copies to contributing groups.

The first version of the database (GD1) was delivered at the end of 1991. Unfortunately, funding was not available to complete the development work under contract and, during 1992 and 1993, development of the distribution version (GD2) proceeded at a much slower pace in a voluntary capacity.

The recording of geological sites on a local basis was given impetus by the launch of the Regionally Important Geological/geomorphological Sites (RIGS) initiative (Nature Conservancy Council 1990). RIGS groups work on a county or regional basis, and their main emphasis is on the conservation and educational potential of non-statutory sites. The emphasis is somewhat different from that of the earlier NSGSD surveys and the geological database (GD2) is being developed further to meet these needs. Other extensions to the original database have been added in response to user requests, notably a specimen cataloguing module and a bibliographic module.

In the process of developing the software, the author has prepared extensive information files on stratigraphy, mineralogy, palaeontology and term lists for other keywords. These are regarded as provisional and will require revision when, and if, the database becomes more widely used and better supported. The long-term objective is to develop a geological information system suitable for use by a wide range of organizations or groups needing access to geological site data.

The original NSGSD project had envisaged that the data from different centres could be brought together to form a single national geological sites database. This database would be of value to the statutory conservation agencies, government and other users wanting a national overview. It was hoped that the national database would be part of a national geological records centre based in the British Geological Survey. This ambition now seems unlikely to come to fruition, at least in the near future, but there is still the possibility that at least summary databases could be maintained in the four national conservation agencies (English Nature, Countryside Council for Wales, Scottish Natural Heritage and Department of the Environment, Northern Ireland) to augment their own database of sites with statutory protection.

This paper describes the factors which have controlled the development of GD2 and discusses some of the implications for the potential for a national network of geological site databases.

Factors influencing the design of GD2

Computer hardware

RIGS groups may be associated with local records centres, educational organizations or local wildlife trusts, most of whom have at least one personal computer, whilst others are highly computerized. There is also a growing number of individuals possessing their own personal computers. The majority of organizations and individuals involved use DOS-based PCs and GD2 was, therefore, written to run on these machines. There has not been any call for copies to run on Macintosh or UNIX-based computers.

Personal computers have become highly sophisticated and most new software is being developed for the Windows graphical environment. Databases running under Windows put significant demands on the hardware, particularly in the amount of memory (RAM and hard disk) required and processing speed. The majority of users in the GD2 target market do not yet possess machines capable of running such demanding software and this was one of the controlling factors in the choice of development software.

Price

The cost of acquiring the equipment and the database is a major problem for voluntary groups. Most do not have any income other than grant aid for site visits from their country conservation agencies. Some will get a degree of support from their local wildlife trust or records centre, who may actually purchase the database software on behalf of the group. All of these organizations are underfunded and there is great pressure to keep the cost of software to a minimum.

Developing sophisticated, user-friendly databases is a labour-intensive activity and in the small, voluntary market can never recoup its costs. In the case of GD2, much of the development work has been voluntary and done on behalf of the NSGSD steering group. The software is given freely to RIGS groups and records centres, although, as there are no funds to support GD2, a charge is made to cover time and travel for installation, training and support. At present, this is averaged out to £200 per installation site, but groups may struggle even to raise this amount.

Simplicity of use

Users can become alienated very quickly from any product that is difficult or tedious to use. Modern database software is very powerful and includes numerous complex tools (e.g. query languages and query by example) which may be difficult to learn and require a detailed knowledge of the structure of the database before they can be used. The majority of general users do not wish to be exposed to these complexities, at least not in the early stages of database use. As confidence increases and requirements change, individual users may need access to these tools and they should be available; but for the novice and less committed user, simple menu guided routes through data input and reporting are essential.

One of the most important aspects of designing GD2 was to make it as self-contained as possible, so that new users could work with it with the minimum of training and without

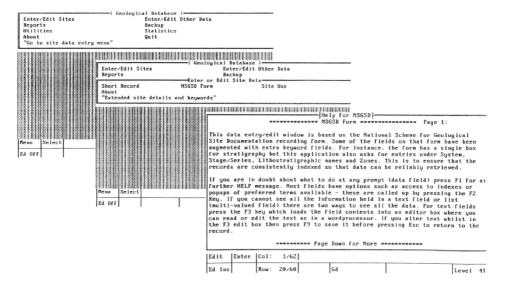

Fig. 1. GD2 is a fully menu-driven geological site recording application. Context-sensitive help is available throughout both as a menu option or by pressing the F1 key.

recourse to bulky manuals. This is achieved through the use of a structured menu system and context-sensitive help notes and termlists (Fig. 1). The overview and guidance notes can be added to easily and edited by the user (e.g. to add organization-specific notes) where desired. Data entry is also made easier by providing 'popup' selection lists of valid terms for database fields wherever possible (Fig. 2).

Data checking and validation

Site record forms present numerous problems for data input. They may be difficult to read,

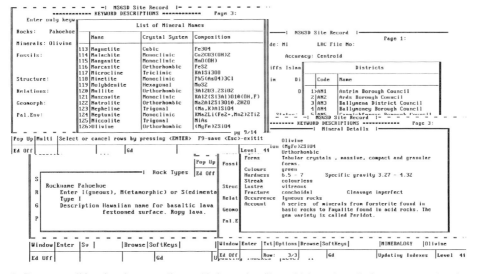

Fig. 2. Data are validated against term-lists and information files, which can be called up as 'popups' and windows during data entry. The information files are dynamically linked to entries in many fields (e.g. stratigraphic keywords, minerals, rocks, fossils, references) and can explain unfamiliar terms to database users.

data may be entered in individualistic ways and the recorder may use terms and syntax differing from those within the database. Data checking and validation routines are, therefore, essential on all data input screens to ensure that data are analysed and entered in a way which will ensure that they are consistently retrievable and potentially transferable. The data entry software should provide 'expert' help (e.g. in recognizing that 'Keuper Marl' belongs to the Triassic system). This is important as even where a centre may have access to skilled geologists for data input, data entry is greatly enhanced by on-line access to term lists, authority files and automatic checking.

The field-by-field validation in GD2 checks the data for syntax (e.g. format of grid references and dates) and for controlled terminology. Some fields are linked so that selection popups are sensitive to preceding information. For instance, selection of Avon in the county field would then give a popup of only the districts in Avon for the district field, and selection of Woodspring as the district would give only a list of Woodspring parishes in the parish field.

Stratigraphy fields are similarly linked between System, Stage and Zone (Fig. 3).

Easy report writing

If information is structured and indexed, it can be searched and sorted in many ways using a computer database. Most database programs offer a range of facilities for creating such reports, but they are often difficult to use and require a detailed knowledge of the database to use successfully. In practice, users often produce variants on a limited number of specific reports and do not always need the complete flexibility and complexity of the system's generic report writers. In these cases, use of the database can be made much easier by providing these reports direct from a menu of choices. This approach was adopted in GD2, where common queries such as finding all the sites in a district or all the sites for a given stratigraphic interval can be easily and rapidly achieved with no prior knowledge of the system. Reporting programs take the user through a series of options, each

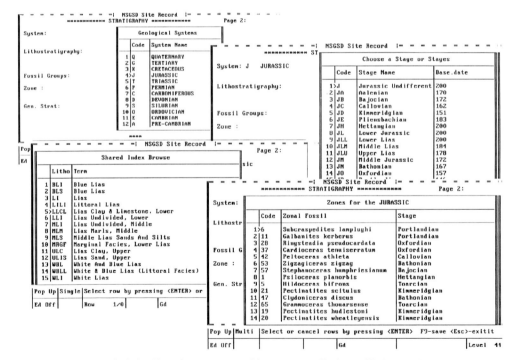

Fig. 3. Data entry can be helped by using context-sensitive popups and indexes. Choices made from the Systems popup control the contents of the Stages popup and Zones popup. The lithostratigraphic index contains old and new terms which can be looked up by entering part of the name, in this example 'Lias'.

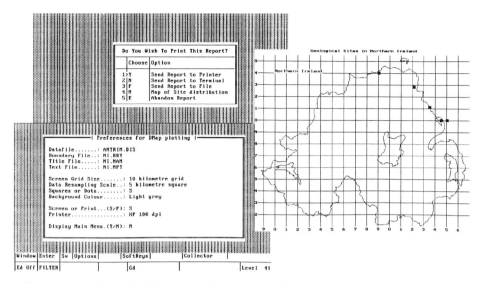

Fig. 4. Reports are easily produced from menu-driven routines. There is a choice of output, including plots of site locations using the DMAP mapping program for which GD2 has its own configuration screen.

selected from a popup menu and ending in a choice of output format (e.g. screen, printer, file or map (Fig. 4)).

The choice of database development software

The rapid developments in database programming tools and the power of computers introduces difficult choices for the developer – how to invest enough time in acquiring the skills to use the new development products as they emerge, how to anticipate the real market leading trends, and how to balance this against the ability of the proposed market to pay for and use the products you develop? For instance, even the decision to develop a Windows-based database may put a hardware requirement (speed of machine and amount of RAM etc.) on the application that cannot be met by the majority of individuals and small organizations. Because of the need for developers to gain a return on their existing expertise and the cost restraints on clients, actual working products normally lag a considerable way behind the leading edge of possibility.

The database software chosen to develop GD2 was Advanced Revelation version 2. The choice was governed mainly by the wide use of this database in conservation organizations in the UK and the United States. It is the standard PC database package in both the UK statutory conservation agencies and the United States Nature Conservancy. Advanced Revelation was used to develop the RECORDER biological records database (Ball 1992), now registered to over 120 users, including many local wildlife trusts and local records centres involved in RIGS. The other controlling factors were several years' experience with the product and that Advanced Revelation has modest system requirements, especially when compared to Windows databases.

Advanced Revelation has the unusual feature that data can be of variable length, such that any field could contain from nothing up to several pages of information. This is ideal for the very variable amounts of data recorded in site surveys. A further feature of Advanced Revelation is that it supports multi-value fields, so that a field can hold a variable length list of items. Variable length and multi-value fields are very useful for managing the information in site-recording applications, but they do not conform to the relational model used by most other commonly used databases which makes data transfer to other databases difficult.

Advanced Revelation is a text-based database and, although version 3 has a number of improvements to its interface including better mouse support and scrolling data boxes, it does not use a graphical user interface and has a distinctly old-fashioned look. The worst feature of Advanced Revelation is the arcane manner in

A standardized national database?

An attempt was made to standardize the data collected during the NSGSD project by use of common recording forms by local groups. The data collection forms were based on the so-called 'MDA' card format used in museums. In this format, information is hierarchically organized within boxes, for instance locality or stratigraphic information is structured from the most general to the most detailed using separators between each term. Concepts may be hierarchically structured, as in the elements of a bibliographic reference.

Recorders were asked to use agreed syntax rules (e.g. for the way references are cited) and controlled terminology (e.g. agreed stratigraphic terms) to ensure compatibility of data. This approach puts a responsibility on the recorder to be rigorous in the use of the form, which is rarely achievable in field recording projects using volunteers. As an example, the NSGSD form had a single box to record the stratigraphy of a site. In this situation, reliable retrieval of information requires that the data be entered in a constant and structured way, e.g. Triassic & Norian & Mercia Mudstone Group. In practice, recorders are just as likely to simply write Keuper Marl. This proved to be the case with the NSGSD data, and none of the fields holding structured information (e.g. stratigraphy and bibliography) could be analysed or manipulated in a meaningful way without further detailed processing.

A further problem with the existing data was that, although recorders were using a common record card and had guidelines to work to, the use of terminology and syntax and the quality and quantity of information recorded about sites was highly varied between recorders and record centres. The reasons for this reflect the policies of the recording centres and the range of interest and experience of recorders. Some record centres recorded all sites that came to their notice, including literature records; other recorded only a selection of prime sites for which there may have been more extensive information.

Implementing controlled terminology on a computer database

One of the major constraints to developing a widely distributed national database is the need to develop reliable termlists. The purpose of controlled terminology is to ensure that information can be reliably retrieved. This is much harder to achieve than it seems. It presumes that recorders are consistent in what they record and that they all have the same level of knowledge. It also presupposes the existence of a rigid hierarchy of unique terms which is agreed and observed by all. This supposition underlines a basic difference between computers and people.

Computers (or computer database designers) like information to be exact and in discrete boxes, preferably of fixed length! The human brain works by generalizing and making predictions about the relationships between things. Knowledge is a fuzzy concept in which things fit into a generalized framework which changes dynamically as our experience changes.

An example of this problem is given by the way stratigraphic terminology was used in the NSGSD project. Despite written guidelines on the recording of stratigraphic terms, the forms submitted to the project are highly inconsistent. Recorders indiscriminately mixed up lithostratigraphic, biostratigraphic and chronostratigraphic terms, and also old and new terms, within each classification. To the experienced geologist, this is not necessarily confusing, but the use of partial equivalents and loose relationships between terms are immensely complex to model using the rigid hierarchies of computer databases.

The introduction of controlled terminology requires a large investment of time and expertise, both in building the original thesaurus and in dealing with records that do not comply. It would be possible to use controlled terminology for virtually all fields in a geological database, but the results would have to be worth the effort. This depends on the context of the application, how the thesaurus will be understood by the recorder and the end-user, and on the compromises required to deal with uncertainty. In practice, only a few fields are worth the effort, stratigraphy being the principal.

GD2 has numerous controlled terminology fields, ranging from type of site and stratigraphy, to various keyword fields, including palaeoenvironment, structure and relationships. Provisional termlists are available with most fields, but users are free to build their own as desired. These fields are also offered on a provisional basis, so that if users do not want to retrieve sites by geomorphological, structural or paleoenvironmental keywords, these features will be dropped from the database.

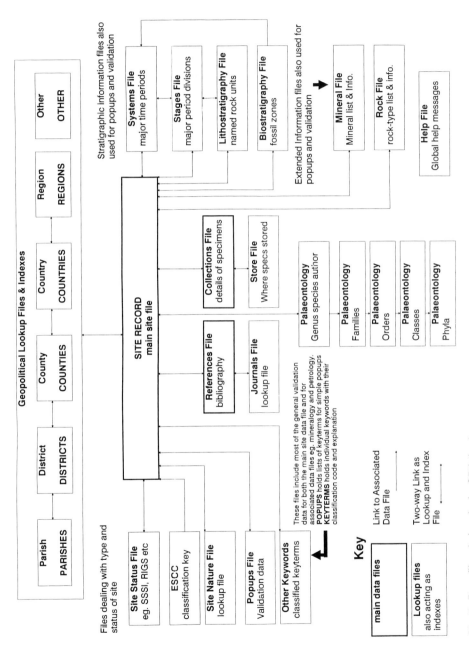

Fig. 5. Data file relations in the GD2 database.

At present, there are not the resources to establish fixed termlists for all the possible keyword fields. As long as users are consistent within their own closed applications, this is not a problem, but problems could arise if data were to be exchanged or if a national network were established.

Stratigraphic terminology

The basics of stratigraphic terminology and procedure are well documented and agreed (Holland *et al.* 1978; Whittaker *et al.* 1991). There are many committees and published reports seeking to stabilize stratigraphic terms and establish the basis of correlation. Unfortunately, most geologists can only keep up with a small area of expertise, and adoption of new changes in terminology lags far behind their introduction. Favourite names also stick with recorders, regardless of current validity. Stratigraphy continues to develop and not all problems of relationship have been resolved or terms revised.

GD2 adopts a pragmatic approach to stratigraphic nomenclature. Within a site record, stratigraphic terms are keyworded into separate fields for System, Stage, Lithostratigraphy and Zone. The fields can hold multiple values and are linked dynamically to one another. The System field has a popup menu of systems from which the user can select as appropriate. The system or systems selected will then affect the popup which appears in the Stage field and also the Zones field. The terms used in the Stage field include both currently valid terms and other, commonly-used, non-valid or fuzzy terms (e.g. Lower Lias). This has to be available because even professional geologists will continue to use 'popular' terms, and these terms may not be directly translatable into other more specific terms (e.g. Hettangian). The Stage dictionary includes base dates, so that selection and sorting of records can ensure, to a degree, that terms come out in the right order or are grouped together as well as possible. The Lithostratigraphy file contains nearly 6000 rock unit names, each with reference to its system and stage(s).

Entering stratigraphic data in GD2 may not, therefore, be a simple matter of copying the data entered on the record card. In the situation where the recorder has given only the rock unit name (e.g. Butcombe Sandstone), the person entering the data will need to add its system and stage, which can be checked automatically from the lithostratigraphy field.

The stratigraphy termlists and lookup files in GD2 are extensive, but far from complete. Users do have the option of adding their own terms, but this will eventually lead to problems of comparing and merging data between centres. The alternative is to centralize update of the stratigraphy tables and issue frequent updates, but unfortunately there is no administrative mechanism to manage this.

Palaeontological terminology

In theory, it should be possible to use a simple thesaurus to check any names of fossils entered into the database. Organizing and sorting records of fossils should also be straightforward because the taxonomy comes with a readily available hierarchical classification. In practice, there are many problems to overcome. The thesaurus needs to cope with multiple synonyms at any level of the hierarchy (from species to phylum), and be able to cope with partial identifications caused by lack of skill, poorly preserved specimens or because the group in palaeontological checklist of this sort using both classical taxonomic indexing or a numerical coding system, but this would require input from a large number of specialists and a steering group to oversee development.

As an interim solution, GD2 offers two ways of entering fossil names. The first is a simple group field which lists a popup of the most commonly recorded fossil groups (e.g. corals, brachiopods, ammonites), regardless of specific taxonomic level. Used in conjunction with other fields, such as Stage, this will readily allow retrieval of, for instance, all Hettangian sites which have yielded ammonites or Llanvirn sites with graptolites. There is also a detail field for listing fossil species, which is backed by a hierarchically linked set of files covering genus/species, family, order, class and phylum. These files are available for users to build their own in-house thesaurus, should they wish. There is little 'intelligence' in the system for trapping synonyms, but selecting names from a popup can, at least, speed entry of long names and ensure consistency of spelling within the database.

Data import and exchange

Users need to import site data for two purposes: the first is to load the database with any pre-existing computerized site records, and the second is to allow for distributed data entry. In the case of new users of GD2, the pre-existing data are generally those from the NSGSD project and they can normally be loaded directly from the NSGSD files (GD1) in a form which can then be edited. This is done for the user prior to delivery of the database.

Distributed data entry is an altogether more difficult problem. Centres may wish to run more than one copy of the database or allow recorders to enter records onto their own machines for later import into the master database. Great care must be taken to maintain the integrity of the database, which can be compromised by record duplication, conflicting record updates and overlapping site definitions.

Despite the long-held aspirations to create a national geological database, most of the groups and organizations involved in recording geological sites for conservation purposes do not yet have a requirement for greater connectivity with other systems. Most can function as stand-alone single user applications, although even here there may be a desire to use several copies of the database for distributed data entry. This will become more of an issue if groups wish to enter into partnerships with local authorities, perhaps sharing GIS facilities, or if initiatives such as the Co-ordinating Commission for Biological Recording (CCBR) programme for the establishment of a national network of data centres come to fruition (Harding & Ely 1993).

Future development of GD2

GD2 is still in the early stages of development (see Fig. 5 for current file structure), with a low number of users, whose feedback will help shape its future format. Many users would be happy to be supplied with a 'complete' application which would remain stable for a long time. Unfortunately, this can never be achieved. Software and hardware are continually developing and 'state-of-the-art' very soon becomes 'yesterday's model'. The availability of new techniques, such as reasonably-priced computer-mapping, will change attitudes to the database. Internally, thesauri need updating, correcting or adapting to match changes in knowledge (e.g. changes in taxonomy or new radiometric dates). The right balance is software that changes often enough to keep pace with major developments, but not so often that users are constantly chasing versions.

Adding pictures

Site descriptions are greatly enhanced by access to photographs and site plans. At present, the main bar to this development is the difficulty of getting pictures into the system and the amount of disk space needed to store them. Advances in technology, such as the availability of cheaper colour scanners and the copying of slides and photographs to Kodak CD format, file compression and cheaper mass storage, will ensure that photographs will be a normal part of database records in the near future. GD2 does not support graphics or any form of multi-media (sound or video) extension, but these are features under consideration for a future, Windows-based version of the database (GD3).

Spatial access to data

Spatial access to site data is becoming increasingly important. A planning enquiry involving a road development may require information on sites in an area up to half a kilometre either side of a sinuous line. This kind of search can only be achieved using a map and the integration of databases with maps lies at the heart of geographical information systems (GIS).

Few of the groups or organizations involved in geological site recording are yet in a position to acquire geographical information systems. The cost of the programs is dropping rapidly, but unfortunately the royalty costs on maps are still prohibitively high. The successful management of a GIS project still requires considerable expertise and this is not available in most records centres or voluntary groups. At present, GD2 can send grid references out to DMAP, a widely-used 'dot mapping' program, and in DBF (Dbase) format to MAPBASE (produced by NextBase) for more sophisticated map display and spatial access. Other future development will need to be tailored to the individual requirements of users.

Support and documentation

All computer applications need to be backed up with support and documentation. Whilst the number of users is low, an enthusiastic author can provide the individual attention needed to answer enquiries and fix problems, but this can very soon grow beyond voluntary capacity. Experience shows that writing the technical and user documentation for an application can take as long as writing the actual software. This is rarely properly accounted for in the costing of projects, and in voluntary projects, the result is usually the neglect of proper documentation. This is the case with GD2 and, without a strategy to fund support and documentation in the future, will lead to a growing number of problems for users.

There is a precedent to this situation in the natural history world with the biological recording package RECORDER. This software developed from the work of its author, Dr Stewart Ball, and was distributed to users in an *ad hoc* way until a point was reached when it became necessary to provide a greater degree of standardization, support and development. Support and funding has been given by English Nature, in conjunction with the Royal Society for Nature Conservation (RSNC), and there is a user group organized through the National Federation for Biological Recording (NFBR).

The development costs of RECORDER are not available, but it is certain that the full economic cost is well beyond what could be recouped from the user-base in charges. The writing of the RECORDER manual, for instance, was a long and difficult task which required signficiant expenditure on contract posts and printing costs. There is a continued need for funding to cover development of the database and user support which goes beyond anything which could be managed by purely voluntary means.

The situation will be identical for GD2 if it becomes widely used. At present, the number of users is low enough for support to be given freely as the need arises; but with 50 users or more, this would not be the case. Even now, development is slow because it has to be fitted in as and when it can be, and users may wait a long time for upgrades to be completed and issued. Ideally, GD2 needs an administrative and funding framework so that its development can be properly planned and supported.

Conclusions

One of the objectives of the NSGSD project was the establishment of a national geological records centre which would be the focus of a national network of local geological records centres. The local records centres exist in a number of forms (e.g. in local environmental records centres, wildlife trusts, museums and RIGS groups), but sadly the national centre has never materialized. In practical terms, however, the local centres are the most important aspect of the network because the practical work of conserving geological sites is essentially a local issue. This work has received a considerable boost under the RIGS initiative and the work of assessing and recording geological sites started by the NSGSD is at a new high level of activity.

The objective of the NSGSD to create a national geological sites database has now been modified to create a user-friendly geological database application to support the recording effort of local RIGS groups and record centres. The development of this product, under the name GD2, has progressed slowly, but is now in use in 14 centres, with several more waiting for installation.

GD2 is a practical, working database that has the potential for development into a geological information system containing information files and thesauri that not only give 'added value' to site records, but can be used to train users and enhance standards. For this to be successful, GD2 will need a framework for development which includes funding for documentation and user-support. Projects such as GD2 can never recoup their full economic costs from users, but are an ideal vehicle for partnership between the statutory and voluntary sectors.

References

BALL, S. G. 1992. *RECORDER User manual (Version 3.1)*. English Nature, Peterborough.

HARDING, P. T. & ELY, W. A. 1993. *A co-ordinated approach to biological recording in the UK*. In D. A. ROBERTS, *European museum documentation strategies and standards*. Museum Documentation Association, Cambridge.

HOLLAND, C. H., AUDLEY CHARLES, M. G., BASSETT, M. G. *et al.* 1978. *A guide to stratigraphical procedure*. Geological Society of London, Special Report, **11**.

MUSEUMS & GALLERIES COMMISSION. 1993. *Standards for Geological Site Recording*. Standards in the Care of Geological Collections, **3**, Museum & Galleries Commission, London.

NATURE CONSERVANCY COUNCIL. 1990. *Earth science conservation in Great Britain – a strategy*. Nature Conservancy Council, Peterborough.

WHITTAKER, A. *et al.* 1991. *A guide to stratigraphical procedure*. Geological Society of London, Special Report, **20**.

Theme 4: Site Conservation and Public Awareness

Keynote address

The protection of geological heritage and economic development: the saga of the Digne ammonite slab in Japan

GUY MARTINI

*Reserve Géologique de Haute Provence, Centre de Géologie,
Quartier Saint Benoit, 04000 Digne, France*

Abstract: The Isnard ammonite slab (Digne, France) is one of the National Geological Reserve Sites on the 150 000 ha of protected land in the Geological Reserve of Haute-Provence. This site was filmed in 1985 by NHK, the national Japanese television network, for inclusion in a series on the formation of the Earth. Years later, the organizers of a very large exhibition on the sea, to be held in 1992 in northern Japan, wished to present the ammonite slab as the emblem of the exhibition.

Thus in 1991, a very special association with Japan began for the Geological Reserve of Haute-Provence, based on techniques developed here for the protection of geological heritage. A cast of the entire site was taken and, in addition to its technological value, this operation helped set up a series of programmes for cultural, economic and touristic development and exchange between the Alpes de Haute-Provence departement and Japan's northern region.

The site at the centre of what is really something of a saga is one of the foremost sites in the Haute-Provence Geological Reserve. This National Nature Reserve, which is in the care of the Ministry of the Environment, covers an area of 150 000 hectares of protected geology, where the last 300 million years of Earth history can be seen in the open air. The reserve, the largest in Europe to be protected because of its geology, includes 18 sites classified as 'Prime National Reserves'. One of these, the Isnards ammonite slab, was the start of this Franco-Japanese adventure.

The site: interest and history

This particular ammonite slab is a piece of sea floor of Sinemurian age (195 ma old) which contains more than 500 ammonites and nautiluses on an exposed subvertical surface having an area of $200\,m^2$ (Fig. 1). The most characteristic fossil is *Corноniceras multicostatum*. The fossils vary in size from 10 cm to 1 m in diameter. The concentration and quality of the specimens present make this outcrop an exceptional site, probably unique in the world. The site has been known for a hundred years. It was discovered when the main road which runs at its foot was built.

In 1960, the Basses-Alpes Scientific and Literacy Society provided very rudimentary protection for the sites fencing and descriptive panels. Although very well-intentioned, this attempt at protection had the opposite effect. The site, with an exposed surface which at the time was less than half that now exposed, was extensively pillaged.

The protection and management which were absolutely necessary for a site of this importance finally became possible through the creation of the Haute-Provence Geological Reserve. At first the protection chosen for the site was in line with the general policy for applying the regulations to all the protected area of the geological reserve. This policy was based on the rejection of conventional protection techniques (essentially coercive measures), instead placing emphasis of greatly heightening awareness among the local population. Thus during the 4 years of investigation prior to the setting up of the geological reserve the term responsible for this investigation engaged in sustained work aimed at the local population.

In the course of this work the value of this geological inheritance which was therefore their property was explained unremittingly. The main objective was that the local inhabitants should acquire a pride in this geological heritage, which was until then unknown or unappreciated. Once this objective had been acheived, it was not desirable that the reserve should then itself be responsible alone for supervising and protecting their heritage. In fact, it could only be promoted

Fig. 1. The ammonite slab site.

with the active involvement of the inhabitants. And this is what happened.

Thanks to the efforts of the local population this highly exposed slab suffered hardly any deterioration after 1980. What is even more remarkable is that work scheduled for that year doubled the surface area of the fossil-bearing outcrop.

A site which is unique in the world, but ...

The ease with which the slab can be seen, with all the intrinsic qualities already mention, and its location immediately alongside a major road a few minutes from the centre of Digne, make this one of the 'star' sites in the reserve. Despite this, not as much was made of this exceptional site as one might desire. A number of factors in the vicinity of the site (which essentially arose from inadequate local authority policies) interfered with its optimum management. The main road, which was too near, the immediate proximity of the municipal rubbish tip (less than 50 m away), and the constant presence of vagrants on the tip, made it impossible to make the most of the site.

Despite the value of the classified site, it was never possible to get an overall environmental management plan from the Digne municipal authority.

Despite everything ...

Despite everything, thanks to the policy of widely publicizing the Haute-Provence Geological Reserve, the site very quickly became popular and became nationally and internationally famous. It was this fame which led the Japanese national television company, NHK, to come and film the ammonite slab in 1989 as part of a series of programmes on Earth history. This series was broadcast in France under the title 'The Miracle Planet'. It was quite successful, but much less successful than in Japan, where the series of broadcasts went out in prime time.

At about this time Iwate Prefecture, in the northeast of the island of Honshu, the largest island in Japan, was undergoing an economic upheaval. The steel works (the oldest in Japan), which provided a living for most of the area, were closing down, and the prefecture had to implement an economic reconversion programme largely based on the development of tourism, which was made possible by the island's very attractive coastline. This gave birth to an enormous exhibition project put in hand by Iwate Prefecture as a joint venture with the Japanese government. The exhibition, costing several hundred million Francs, was to take place in 3 towns in the north of Iwate Prefecture over a 3-month period in the summer of 1992. During this time it was expected that the exhibition would attract more than a million and a half visitors.

Symbolic object

It transpired that the Japanese organizers responsible for the project needed a huge object which could become a symbol for this enormous and unusual exhibition on the topic of the sea. After examining many possibilities, the Japanese designers decided on the one and only thing which was considered worthy of being the symbol of the future exhibition: something which had been seen on television – the Digne ammonite slab.

Thereby hangs a tale

Once the choice had been made, the site had to be found. The Japanese operators of the exhibition recruited a translation company in Paris. Thus in 1990 the Management of the reserve was called by a Japanese interpreter who was acting as an intermediary for his principals. They asked for nothing less than to buy the site in order to dismantle it and reassemble it in Japan. Faced with such a suggestion, written confirmation was essential, and was not slow in forthcoming. It had to be pointed out that the status of the site meant that not even the smallest pebble could be removed.

A refusal, pure and simple, would have been singularly discourteous, and for this reason it was suggested that they could reproduce the entire fossil-bearing site as a cast. The Japanese were sufficiently hooked onto the idea of this symbolic object to accept this proposal. With much toing and froing between France and Japan, many discussions and technical negotiations were needed to bring the project into being, with the signing of a property contract in due form. The contract was signed under the aegis of the Alpes de Haute-Provence Chamber of Commerce and Industry.

The technical and commercial contract was coupled with the launch of a cultural and tourist economic co-operation programme between Alpes de Haute-Provence and Iwate Prefecture, led by the Geological Reserve, which was given the task of seeing through these 'diplomatic' operations, which were far removed from the field of geology.

A highly individual technological operation

Taking a cast from a site of this type in a sub-vertical position over an area of $200\,m^2$ and re-erecting it in Japan is, in any event, an operation which requires fairly unusual moulding technology. During the summer of 1991 a team of specialists which included moulding technicians from the Geological Reserve and the Digne Museum produced an elastomer mould of the whole site. The mould was made up of 30 parts.

In Autumn 1991 the 30 parts of the mould were cast in polyester resin filled with self-coloured chalk powder, weighing a total of 24 tonnes. These casts were sent to Japan by ship, in containers (because of their size, two each, 12 feet long were required) during the winter of 1991. It took until May 1992 to transport the casts from the port of Tokyo to the town of Kamaishi, where they were to be re-erected, and then a team consisting of 8 technicians from the reserve left for Japan to reassemble the 30 pieces of this giant puzzle to produce a perfect facsimile of the famous site at Digne in 2 weeks (Fig. 2).

The operation was completely successful, and being presented in 'museum style', within a pyramid which had been specially designed for it, the cast appeared more lifelike and more beautiful than the original.

In parallel with this work the Franco-Japanese negotiations made progress and, in order to give these development agreements physical and geographical reality, it became essential that this copy of a site 'which was unique in the world' should be displayed in a permanent fashion in this part of Japan, which would

Fig. 2. Fitting the last pieces of the ammonite cast in Kamaishi.

henceforth be a friend to our part of France. This display would have to be in the open air, in order to strengthen identification with the original site.

Display of the copy of the slab as part of Japan Expo in Iwate 92 was clearly a great success. Given extensive media coverage, the resin twin of the Digne site was viewed by more than 2 million visitors. It was this success which then led our Japanese partners to modify their initial ideas. The copy of the ammonite slab will be a permanent feature in Japan, no longer in the open air as intended, but instead within a building specially designed for the purpose, which will form a large extension to the existing Museum of Iron already at Kamaishi.

A prophet is not recognized in his own country

The immense interest shown by the Japanese in this site, and then in its resin copy, in fact produced major spin-offs for the original itself. Not long ago the author emphasized the unfortunate environment of this ammonite slab, which according to the local authorities could not be altered because of the particularly high cost of the work (relocation of the municipal rubbish tip, the vagrant population, etc.). However, because of the high regard in which the Japanese held the site, we were able to emphasize the urgent need for an overall programme for management of the site and finally for tackling the problem of the permanent protection of this ammonite slab. This protection could take the form of the construction of a site museum or at least a structure to protect it from the weather, for example.

It was only after another Japanese delegation had paid a visit, in February 1993, that an overall programme for the management and long-term (century-long) protection of the site was unanimously adopted by the institutions responsible for the reserve.

A prophet is not recognized in his own country!

Epilogue

This experience seems to be an object lesson in many respects, and encapsulates many of the necessary interactions between the protection of geological heritage and the various economic components of our society. Thus a site which was protected because of its geological heritage, and modern technology applied by the teams operating the site, brought about an export operation which, in 1991, was the largest ever (in financial terms) from Alpes de Haute-Provence to the Far East.

Economic, cultural and tourism development programmes linking our region in the south of France to the northern part of Japan have been built up around this know-how.

Finally the success of this operation has helped to protect the original site, through an overall management programme which would never have otherwise been brought into being. Not only that, but it is conceivable that if the site had not been protected it might have been sold to be reassembled in Japan.

Thus a long journey through space and time, from a 195 million-year-old piece of sea bed, from Digne through the Land of the Rising Sun and back to Digne, has achieved final protection for this site. Is it not fitting that international action has made it possible to protect something which is unique in the world and therefore the common property of all humankind?

The American Cave and Karst Museum and the work of the American Cave Conservation Association

GEORGE N. HUPPERT

Department of Geography and Earth Sciences, University of Wisconsin–La Crosse, La Crosse, Wisconsin 54601, USA

Abstract: The American Cave Conservation Association (ACCA) was founded in 1979 specifically to aid in the preservation of caves and karstlands. The organization moved its offices from Richmond, Virginia to Horse Cave, Kentucky in 1986. This move enabled the ACCA to take advantage of the significant karst resources in and surrounding Mammoth Cave National Park.

The Board of Directors decided that a national karst museum would be an appropriate way to educate the public on the uniqueness of karst and caves. Nearly one million tourists visit the show caves of the immediate area each year. To build a museum, loans and grants were sought and soon buildings were purchased and restored in the centre of Horse Cave. This was done at a cost of nearly $1 000 000. Professional museum designers were hired to design and install exhibits at a cost of another $700 000. Preliminary exhibits were finished so that the museum could open in the summer of 1993. The final exhibits of Phase I were completed in time for the Grand Opening in the summer of 1993.

The exhibits include information on all aspects of the cave environment and the impact of human activities. Several of the exhibits are portable so that they can be put on loan.

Located on the museum property is the entrance to Hidden River Cave, a large former show cave that was closed in the early 1940s because of sewage pollution. A tour of the entrance sinkhole is included in the visit to the museum. It is hoped that the cave will be opened again for public tours in the future.

Programmes on caves and karst and their environmental problems have been designed for use for all levels of schools. Special programmes can be arranged for groups at the museum.

The opening of the museum has been made possible by many individuals and groups. Not the least of these are the administrators and people of Horse Cave who saw in the museum the possibility of revitalizing their community with tourism and responded enthusiastically.

The beginning

The American Cave Conservation Association (ACCA) was founded in 1979 as a response to the perception by cave conservationists that speleological organizations were generally ignoring karst and caves as natural resources worthy of protection because their primary concern was with caves as a recreational resource. Most mainline conservation societies generally ignored caves completely. An exception has been the Nature Conservancy which has purchased and protected many caves over the past thirty years.

The major goal of ACCA is to educate the public to the many values of caves (Huppert & Wheeler 1986). Following some consideration by the Board of Directors and the membership it was decided that the most efficient method toward the education of great numbers of people would be through a museum and associated activities. Obviously this institution would need to be in a tourist area with a focus on caves. A number of sites were considered but studies revealed that a location near Mammoth Cave National Park in Kentucky would be best. About one million visitors visit the local show caves and other attractions each year. A major interstate highway passes through the region and provides ready access from eastern metropolitan areas.

The city of Horse Cave, Kentucky, about eight miles from Mammoth Cave National Park, was selected for the museum in 1988. It was decided to restore existing buildings in the city and to maintain the architectural integrity of the structures. Reconstruction of the old buildings was begun in December of 1989 and largely completed by 1991. In the meantime Chase Studios, a professional museum design firm from Cedar Creek, Missouri, was contracted to create the Phase I exhibits for the museum. Installation of the exhibits started in 1991 and much of the first phase was in place by mid-1992 (Figs 1 and 2 show the floor plan).

Financing was needed and this was done in a variety of ways. The largest grants were from the federal block grant programme, the US

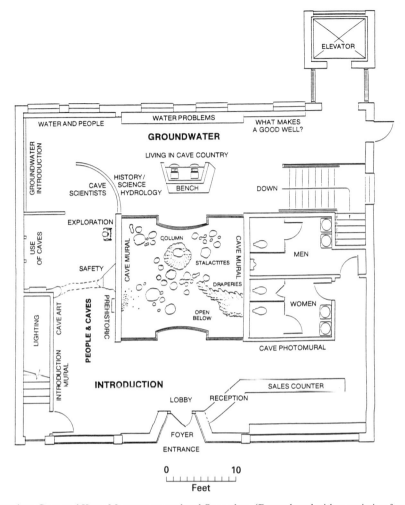

Fig. 1. American Cave and Karst Museum: street level floor plan. (Reproduced with permission from Chase Studios (1990).)

Environmental Protection Agency, the State of Kentucky, the Brown Foundation, the Bingham Fund, and the Kentucky Bicentennial Commission. Many others gave lesser amounts. In addition to the grants, the City of Horse Cave guaranteed a loan of $250 000. To date, the cost of the project has been over $1 700 000. The grand opening date was July 10, 1993.

The future

Phase I has to be finished by constructing and installing the cave biology exhibit in the very near future. Once the museum is operationally stable, fund raising for Phase II will start. This will entail more renovation in order to provide a lecture room/theatre, a gift shop and completion of office space.

The museum grounds contain the large sinkhole entrance to Hidden River Cave, a former show cave closed in the early 1940s due to heavy pollution of the cave stream from domestic and industrial sewage. The waste input problem was corrected in late 1989 with the completion of a new sewage treatment plant. The treated effluent is no longer drained into the cave. In the last four years the cave has made a remarkable recovery and is usable by the public again. Stairs have been constructed several hundred feet into the cave to near the level of the stream for viewing. A limited number of experienced

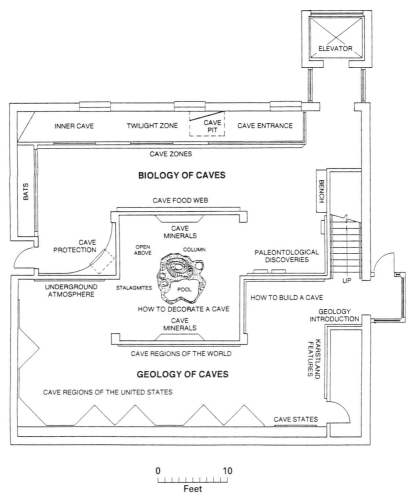

Fig. 2. American Cave and Karst Museum: lower level floor plan. (Reproduced with permission from Chase Studios (1990).)

speleologists are allowed beyond this point for research and exploration. It is hoped that the former tourist trail can be restored in order to show off this environmental success story as an example of how change can be brought about even in a cave that was once known as the most polluted in the United States.

Future plans also include the development of a complete outreach programme to be made available to schools and other interested parties at minimal cost. At present there is a limited stock of commercial videotapes, slide shows and literature available for use in the local area. This stock will be expanded and loaned on a national basis. The museum will be involved in producing similar products. The museum staff had a significant role in the production of a recent televised 'National Geographic Special' on caves.

Staff members, board members, and others in ACCA are active in conservation projects throughout the country. One board member owns Fantastic Caverns, a show cave in Missouri, which is nationally recognized for its conservation message and educational materials which are widely distributed. Another board member, an engineer, is a cave gate designer. He has constructed gates for numerous environmentally sensitive caves, especially those containing endangered bats. He is often under short-term contract with various federal agencies for their cave gate needs. Other board members are collage professors, geologists, bat biologists, a lawyer and local business persons. It is a cross-section of expertise needed to ensure the success of the museum and ACCA.

The museum staff are also in the process of collecting an extensive speleological library. The intent is to provide a clearing house of information for researchers and others. It is hoped that the museum and ACCA will be able to support research with grants and to disseminate research results through various media.

This is all a few years away. However, from the initial response of the public and educators during the past year this will all become reality.

References

CHASE STUDIOS 1990. Exhibit Floor Plans. *In: American Cave and Karst Centre: Schematic Designs.* Cedar Creek, Missouri, p. 2.

HUPPERT, G. N. & WHEELER, B. J. 1986. Underground wilderness: can the concept work. *In:* LUCAS, R. (ed.) *Proceedings – National Wilderness Research Conference: Current Research.* United States Department of Agriculture, Forest Service, Intermountain Research Station General Technical Report INT-212, Ogden, Utah, pp. 516–22.

The private sector – threat or opportunity?

RICHARD DE BASTION

*Historic Touring, The Mansley Centre, Timothy's Bridge Road,
Stratford-upon-Avon, Warwickshire CV37 9NG, UK*

Abstract: The statutory conservators of nature and heritage share the playing field with many agencies, including national and local government, industry, education, the voluntary sector, cultural, sporting and other user groups. The conservators have tended, traditionally, to be in opposition to those influencing or effecting change.

There is an ever-worsening fiscal environment within which all of these agencies seek to fulfil their individual objectives.

This paper examines the ways in which the private sector in general, and the tourism industry in particular, may help in bridging the widening gap between the operational resources and the full requirements of statutory and voluntary conservation.

Against the background of a changing ethos within industry, the move towards a commitment to the principles of sustainability by some sectors of the tourist industry heralds a new era of dialogue. The tourism industry is well placed to act as mediator in bringing together the activities of both conservators and developers.

There is a strengthening body of opinion, within many of the major corporate entities, giving testimony to the fact that the days of the introspective, all-consuming industrial conglomerate, are well and truly numbered. Management consultants report the growing tendency towards the re-structuring of entrepreneurial activity based upon ecological principles. There is also evidence of a shift from the competitive to the co-operative amongst the current generation of management, worldwide.

It is the growth of conscience, of responsibility towards fellow beings and common home, that underlies this movement. It is global, irreversible, a part of the physical and metaphysical evolutionary process. Some have chosen to brand this period in our history the 'new age', others have recognized that humanity is re-establishing the values and belief systems that pre-date those of our so-called civilized world.

Society is in the process of re-defining its relationship with the 'natural' world. This is nothing more than the re-discovery of the total interdependence of human life upon the physical home – Earth.

The evolving status of the tourism industry

There is an urgent need to promote a better understanding of the tourism industry, particularly, the concept of sustainable tourism. Tourism is still widely considered as purely intrusive. This is hardly surprising when considering the many examples of exploitation and environmental rape that characterized the mass-tourism industry of the 1960s and 1970s.

Great Britain competes on the global stage. The travelling public is highly susceptible to any shortcomings, be they actual, or perceived, in the tourist product. Government statistics recently published by the National Economic Development Council (NEDC 1991) demonstrate that, with particular reference to the decline in visitor numbers from Germany over the past decade and the sensitivity of the North American travel market, any negative publicity will have severe implications on the competitiveness of the UK as an international tourist destination.

It is a widely publicized fact that, both globally and in the UK, tourism is replacing manufacturing as the leading industry. The contributions that tourism makes to the Gross National Product, to invisible exports and as an employer, are confirmation of this. And yet, the image of the industry, within education, where future career opportunities are currently decided, the popular media, central and local government and even within some sectors of the industry itself, is still largely misplaced.

Tourism's poor status as an employer is central to many of the industry's shortcomings and is, itself, a stumbling-block to any significant change. It is the quality of product at the point of delivery that ensures customer satisfaction. Delivery, in the British tourist industry, remains largely in the hands of unskilled, temporary employees, often still at school.

This is a far cry from Britain's continental partners, where the status of a career in tourism is highly regarded. Many of their 'front-line' employees share a sense of vocation paralleled

From O'Halloran, D., Green, C., Harley, M., Stanley, M. & Knill, J. (eds), 1994,
Geological and Landscape Conservation. Geological Society, London, pp. 391–395.

in the UK only by those smaller, self-owned units of the service sector.

There are, as in any widely practised activity, many faces to the tourist industry: the large corporate machines, groups and chains, project developers, local enterprises and casual contributors, each comprising individual decision-makers who will, in the first instance, wish to secure the profitability of their operation. Throughout the industry, however, the profitability factor has been augmented by the genuine desire, by most providers, to ensure quality of product. Personal recommendation and a good reputation being the key to continued success. Seen in this light, one could argue that these smaller units, often operating as family businesses, providers of accommodation, catering, the visitor attractions, have long practised what is now fashionably termed 'sustainable tourism'.

Although international tourism enjoys a high profile and is a major economic and cultural component of modern life, the domestic market is by far the most significant in terms of movement volume and total tourism receipts. For the British, the foreign destination continues to exert a fascination, but the traditional UK holiday has held its own and is, indeed, itself in synthesis. (BTA 1992).

Forecasting for the future of the industry demonstrates that there will continue to be growth. There will be increases in leisure time, both through improved living standards and enforced through redundancy or early retirement. Forecasting also confirms that the industry will continue the trend towards the taking of more frequent holidays, extended weekend breaks for example, with the emphasis on more specific pursuits and away from the packaged 'mono-culture' for mass-tourism. (BTA Research Services 1990).

Crucial in this evolving definition of the tourist and central to the theme of this paper, is the fact that there is a tide of increasing expectation. The travelling public has matured. Many now combine the tourist experience with the development of specific skills; most recognize the educational value of travel, with a growing population using travel as part of the formal learning process.

As an expression of this increasing expectation, the volume of visitors actually sharing a leisure experience at any one time, will become a critical factor, both in the perceived quality of the experience and in the ability of a destination to sustain its tourist resort status.

There has, in fact, been a convergence of ideology: the consumer of the contemporary tourist product, qualitative and educational, with those concerned with heritage and nature conservation. The tourism industry is, therefore, well placed to act as mediator between those concerned primarily with conservation and those influencing or effecting change or development. These agencies have tended, traditionally, to be in opposition. The emergence of a new ethos in the tourism industry, the recognition that the resource base, and not just the market, is finite and sensitive, has placed the industry firmly mid-way between these old adversaries.

Qualitative teaching and public perceptions

The threat to our natural heritage is only as real as our perception of the problem. For many thousands of years, human beings have accidentally and deliberately brought changes upon the natural world. An awareness of this impact is a relatively recent phenomenon. Great Britain can hardly claim a physical landscape untouched by human hands. Indeed, the British perception of wilderness provides the Canadian humourist with much material.

The real threat is to the very basis for global survival brought about by unchecked abuse of our finite resource base. Although we are all affected, the issue falls outside the terms of this paper and this conference.

The pace of physical change, particularly within the past decade, has resulted in greater public awareness of the conservation question and a willingness to enter into the debate. Most will equate 'nature' with 'biological' conservation. The need to preserve the planet's biological heritage touches a wide audience and continues to enjoy a high profile. This is an emotive issue with direct relevance to our own quality of life and to our own survival. The heroes and villains are easily identifiable.

Geological or Earth science conservation, on the other hand, does not enjoy this profile. The need for conserving examples of our geological and geomorphological heritage, remains an academic debate. Philosophically and conceptually, this differentiation has no foundation. Section 52 (3) of the Wildlife and Countryside Act 1981 (Hansard 1981) states that: 'references to the conservation of the natural beauty of any land shall be construed as including references to the conservation of its flora, fauna and geological and geophysical features.'

There is no overnight solution to this dichotomy. Children are the future custodians of the world that we live in. Our prime duty must, therefore, be to provide the next generation with the quantitative and qualitative resources they will need in their task. Part of this task will be to reconcile the emotive imbalance in the perceived significance of biological and landscape, or geological conservation.

There is a world of a difference between the truth, defined in terms of empirical fact; by nature, second-hand and impersonal; and the truth perceived through the emotion of personal experience. The practical limitations of our current education system result in too little emphasis placed upon the 'learning experience' and a persistence of the need to isolate and compartmentalize subject material.

This concentration upon the teaching of subjects in isolated units has been at the obvious expense of an understanding of the inter dependence of subject material. Encouragement must be given, wherever possible, towards a more holistic approach to learning.

Truth cannot simply be an expression of fact. The answer to such persistent questions as 'Why are we here?' cannot be punctuated with a fullstop. It can, however, be felt; a sensation, an emotion, a moment of revelation, cementing an experience with undeniable affirmation, the very stuff of which teachers' dreams are made.

The essence of learning is embodied in the direct, or emotional experience as much as it is within formal classroom teaching. There must, therefore, be greater emphasis placed on an aesthetic, or emotional component within the teaching framework.

Two decades of discussion have been invested in the development of landscape assessment as a discipline. Most authorities agree that any methodology will require the subjective, as well as the objective component (Countryside Commission 1988).

A symposium of collected experiences, published in 1992 reports the results of recent attempts to introduce undergraduate students to various qualitative teaching models (Lee 1992). It is a sad reflection of educational priorities that the following aphorism, cited in the opening comments of the symposium, should assume such significance so late in the educational process: 'Tell me, I forget. Show me, I remember, Involve me, I understand.'

Current opportunities within education

Children enjoy only limited access to their local environment during formal education. Nature walks and field trips constitute the traditional response and are isolated and rare occasions confined to well-tried locations. And yet, the local environment is of paramount importance to any school and represents significant curriculum opportunities in the humanities, the sciences and the arts.

The daunting task of monitoring the many hundreds of sites with different designations, currently carried out by the various statutory and voluntary agencies presents a logistic problem of some magnitude.

With reference to multidisciplinary sites and, in this context, to Earth science sites or features of geological significance, proposals for a national campaign encouraging schools to 'adopt' their local Regionally Important Geological/geomorphological Sites (RIGS) have been developed. The scheme will be publicized through the magazine '*Watch*' and is to be co-ordinated at the local level between the RIGS groups, school, educational tour operators and, where appropriate, the local authority.

The majority of heritage and conservation bodies have programmes for education to which they each commit human and fiscal resources. In view of the tightening constraints under which each seeks to fulfill its objectives, there is scope for a more consolidated approach. With careful consideration of the requirements of those teaching all key stages of the National Curriculum, a combined strategy, comprising the individual expertise of each body, could result in greater penetration into the classroom.

Closer co-operation could also simplify the information on, and implementation of, the various funding and support initiatives provided by these agencies and a more 'user-friendly' image across the wide, but somewhat inaccessible spectrum of environmental and cultural opportunities currently on offer. Traditionally, the county and district resources have been freely available and, to varying degrees, integrated into the local educational system. The move, by many schools to seek grant-maintained status, with the associated withdrawal by the education authority as the funding agent, suggests that these one-time partners in an educational infrastructure may come to face each other in the commercial market place.

There are opportunities for private sector operations to bridge this gap. Marketing expertise, often lacking within those sectors of local government now facing a commercial future, and private sector infrastructure can re-open the dialogue that may have ended as a result of political change. Many ex-council facilities could thus adapt to free-market conditions, many will have the potential to attract interest from beyond local authority boundaries.

Tourism opportunities

There is no need to throw overboard many thousands of years of social evolution in preference of a subsistence culture, in order to re-establish an economy that is in balance with its natural resource base. Key factors underlying the new ethos in industry are scale and place. Management requires an acute awareness of the optimal scale and the locational significance of its operation.

The Historic Touring concept, piloted in the Heart of England region this year, is designed to focus this new ethos firmly within UK tourism. An educational package, developed in conjunction with the greater and lesser cultural and academic institutions within the study area, its tour content can be structured to complement most fields of historical, natural, cultural, archaeological or geophysical study. It also aims to address specific environmental issues, including effecting a re-distribution of the tourism load throughout the study area. It also provides expertise and infrastructure to the local school community under the cross-curricular programme entitled 'Landscape as the springboard to inspirational learning'.

At the heart of this initiative lies the emphasis upon the inspirational, the emotive quality of information dissemination. It is thus possible to create a point of interest, out in 'the field' well away from what the tourist industry refers to as the 'honey-pot'. Any community has its inherent tourist potential. The landscape, its physiology and history, along with the local flora, fauna, micro-culture and tradition shape the character of an area and, hence, the local tourist product.

Each has its own unique population with family, social and commercial links with the population of other communities. Taken together, these physical and human factors constitute those basic but largely unsung ingredients that form the natural base from which any conceived tourism strategy should and could spring.

Any strategy or development must be in sympathy with the host environment. Superficially, this requires physical (environmental) study in order to establish capacity levels, access, design criteria and so on. In the fuller context of sustainable tourism, a tourist development would, in itself, be environmentally or ecologically sound.

Any initiative should grow from the natural assets of the defined locality and, wherever possible, out of a consensus of local opinion. Research into prevailing social attitudes, currently being carried out amongst the rural communities of Herefordshire and South Shropshire, seeks to establish the community viewpoint and the willingness for, or against, potential tourism projects (Rural Surveys, Heart of England Tourist Board, in progress).

Central to this definition of sustainable tourism is the fact that the host community and environment should benefit from tourism revenue and/or qualitative improvement.

Following the publication of the government White Paper 'This Common Inheritance', the then Secretary of State for Employment, Michael Howard, set up the 'Tourism and the Environment Task Force'. The recommended 'visitor management strategies', to be funded by private sector partnerships, are well intentioned and politically expedient exercises in 'trouble-shooting'. Since the publication of the report, in 1991, tourism has been transferred from 'employment' to 'heritage' and government concern for environmental issues has been an obvious casualty of the economic recession (ETB 1992a,b).

The increasing awareness of environmental issues presents a double-edged sword. Whilst welcoming the trend towards a greater understanding of, and concern for, the environment, there has also been a downside. There has been a commercial response seeking to capitalize on the marketability of things 'green' and little evidence to support the fact that those tourist organizations advertising 'the green tour' operate under an environmentally sympathetic code of conduct.

The Association of Environmentally Sympathetic Tourism Organizations (AESTO) was founded to promote sustainable tourism and to address the following key issues.

- To establish a national forum for those tourism organizations whose operation may be considered truly environmentally sound. This forum will establish a unified code of good practice, seek to encourage industry at large to adopt these principles and form an effective lobby.
- Many of the smaller units of the tourist industry, referred to earlier as protagonists of a sustainable industry, are least well placed in the front line of current marketing opportunity. AESTO creates an effective means by which groupings of 'low-key' tourism initiatives can market their product.
- Recognizing that longer-term safeguards for the health and viability of the tourist industry and its resource base lie in education, AESTO seeks to foster an even closer relationship with the education establishment (DCC 1987; NCTT 1993).

Tourists have always been attracted to places that are seen or perceived as having that special quality. The industry has inherited a legacy of tourist destination cliches and vast numbers still tread the well-trodden path. In our own national context we have the Tower of London, Shakespeare, the Loch Ness Monster and Stonehenge, to name just four. It has already been suggested that consumer demand is beginning to exert an influence upon this narrower perception of the tourist attraction.

This legacy, and the present structure and dynamics of the tourist industry, represents the highlights on what must be seen as a more extensive landscape of challenge and opportunity.

Commercial opportunities tend to include Earth science sites of interest in the multidisciplinary context, together with other parameters, including scenic appeal, accessibility, infrastructure and market. There are, however some obvious exceptions. The Giant's Causeway is one. As with diversity of landscape and heritage, Great Britain boasts a rich geological and landform legacy, almost unprecedented in terms of variety within the given area. This is, in itself, a unique resource with its own inherent potential.

Summary

The complex and often irrational nature of the bureaucracy in the UK all too often poses the greatest threat to the country's environmental fabric. The statutory conservators of landscape may have to join the ranks of those seeking a commercial dimension in order to perpetuate their activities. Commercialism and conservation are not totally incompatible. A rationalization of the current allocation of government resources should be made a priority.

There is much statistical information giving testimony to the qualities which continue to attract foreign visitors and to stimulate domestic tourism. These include Britain's heritage, culture and the arts. Geology does not rank amongst the current top ten. Heritage and culture, essentially human activities, derive much of what both reisdent and visitor perceive as special, from their setting, the physical context. Accidents of history have made some locations more successful as tourist destinations than others. The physical environment, the infinite variety of landscape, is the underlying denominator which has provided the stage upon which the people of these islands have, for many thousands of years, enacted their pageant.

This paper has sought to propose a new definition of tourism. It is intended to go some way towards commending the industry as a focus for resolving some of the problems facing those involved in the conservation, classification and evaluation of geological and geomorphological sites of interest.

A commercial agency that has yet to subscribe to the principles of sustainability will continue to pose an environmental threat. Many private sector operators now accept that adoption of these principles also go some way towards securing economic longevity.

Recent political decisions on government spending demonstrate this threat from within the ranks of those very organizations invested with the custodianship of the environment and society. As in any real-world situation, there will be individuals motivated by the desire to serve the greater good and those more concerned with their own advantage.

This conference will not change human nature. It may, however, go some way towards effecting a positive change in the attitude of those with a disregard for the need to protect and to promote the physical environment.

References

BTA. 1992. *Annual Report.* British Tourist Authority.
BTA RESEARCH SERVICES. 1990. *English Heritage Monitor.* British Tourist Authority.
COUNTRYSIDE COMMISSION. 1988. *A Review of Recent Practice and Research in Landscape Conservation.* Countryside Commission, Cheltenham.
DCC. 1987. *The Tarka Experience – Devon Action for Tourism and the Environment.* Devon County Council.
ETB. 1992a. *Tourism and the Environment: Maintaining the Balance.* English Tourist Board.

—— 1992b. *Local Area Initiatives.* English Tourist Board.
HANSARD. 1981. Wildlife and Countryside Act, Section 52(3).
LEE, R. (ed.) 1992. Symposium on the teaching of geography in Higher Education. *Journal of Geography in Higher Education*, **16**, 123–151.
NCTT. 1993. *Green Tourism in Northumbria – Conference Proceedings.* Northumbria Centre for Travel and Tourism, Newcastle-upon-Tyne.
NEDC. 1991. *Competing for Growth.* National Economic Development Council, London.

The La Crosse River Marsh: development vs. preservation

ROBERT G. WINGATE

Department of Geography/Earth Science, University of Wisconsin–La Crosse, La Crosse, Wisconsin 54601, USA

Abstract: The La Crosse River Marsh separates the city of La Crosse, Wisconsin (population 51 153) into halves – the North Side and the South Side. Historically, several plans have been proposed to develop the marsh to connect the city. Plans have included industrial, transportation and residential development. Conservationists and environmentalists have fought for the preservation of the marsh as a unique geological and hydrological environment within an urban setting. The major development plan over the last two decades has focused on constructing a four-lane highway through the 1535 acre marsh to connect the central business district with an outlying shopping mall. In 1988, the Marsh Coalition was formed as a citizens' action group to oppose the throughway. Four groups including the City Council, Department of Natural Resources, Department of Transportation and La Crosse County Board were commissioned to work out a marsh land-use plan. In 1992, the Marsh Coalition succeeded in defeating major marsh development plans including the proposed marsh highway. The marsh will be sold to the State of Wisconsin Department of Natural Resources for perpetuity. It will be developed into a recreational area to include hiking, biking, cross-country skiing trails on the abandoned railroad tracks that traverse the marsh, and into an outdoor environmental classroom with learning stations and observation platforms. The marsh battle is a victory for site preservation because of raising the public conscience for the value of conservation versus development. It also ensures that the marsh will continue to provide a flood buffer zone for the city.

For the last half century the citizens of La Crosse, Wisconsin have waged a war over the proper utilization of the La Crosse River Marsh. The two armies consisted of the developers and the environmentalists.

The battleground was the part of La Crosse River Marsh that was located within the city limits of La Crosse. The area consisted of approximately 1535 acres (2 square miles) of which the city owned 386 acres. The remaining acreage was under private ownership. Because of its east–west orientation, the La Crosse River Marsh has historically divided the city into two halves, North La Crosse and South La Crosse.

This paper will investigate the vicissitudes of the war between the two armies. It will discuss the various plans the developers have offered to better utilize the marsh for industrial and commercial enterprises, residential schemes, and transportation corridors. The paper will follow the environmental resistance to the developmental plans. Finally, the paper will discuss the compromise Land-Use Plans that will bring the war to an end by the middle 1990s.

Early battle plans

The La Crosse River Marsh has seen some development over the years. The major impact on the marsh has been in the development of infrastructure. Two roads cross the marsh to connect the city's northside and southside. A railroad presently crosses the marsh on the eastern side. One set of electric high tension power lines run through the marsh. They originate from a transformer substation located next to the La Crosse River. The power company owns the marsh land over which the lines run. Sewer and water lines run through the marsh although they cannot be seen because they are buried. Finally, several hiking-biking and cross-country ski trails criss-cross the marsh. The trail system was developed on top of dykes of former abandoned railroad track rights-of-way that traversed the marsh in the early 1900s. The trails built on fill and afford excellent pedestrian access to the marsh from both sides of the city. The majority of the marsh exists as wildlife and water foul habitat and as a flood overflow basin at times of periodic inundation. Because there is so much open water in the marsh, one of the arguments used by developers is that the environmentalists are 'pandering to ducks and mosquitoes'. The counter argument is that many congested cities would give their 'eye teeth' to have an urban marsh in the centre of their city.

The first significant marsh improvement plan was proposed by Dr Frank Hoeschler in 1938 and became known as the Hoeschler Plan

(Juneau 1969). The plan called for filling in 680 acres of marsh with Mississippi River dredge spoils to raise the marsh level by approximately ten feet. The new land would create 1657 residential lots. The plan would relocate the La Crosse River and create a new lake surrounded by residential lots. Part of the marsh reclamation would be developed into a new industrial part in the vicinity of Copeland Avenue (see Fig. 1). Hoeschler's plan was submitted for approval to the La Crosse Common Council on July 9, 1938. The Common Council approved the plan, but the plan also had to be approved by the citizens of La Crosse in a referendum in late summer of 1938. The plan was defeated. A second referendum was held, and the plan was soundly defeated by the La Crosse citizens a second time.

Three reasons were offered for the voter refusal of the Hoeschler Plan. Firstly, the voters would have had to take on a partial tax burden to help finance the project. During depression time, this would have been a hard sell to tax payers. Secondly, there may have been a mistrust of the profit motive of a land speculator even though Dr Hoeschler was a respected citizen of the community. Thirdly, the citizens of La Crosse may have begun to realize even then what a valuable asset the marsh had become to the history and heritage of La Crosse. Never again after the voters' rejection was such an ambitious plan to fill in the marsh offered. The future attacks by developers would be more 'piecemeal'.

A second attempt at marsh improvement came in 1945, when the city hired an engineering firm from Chicago, Illinois (Alvord, Burdick, and Howson Engineers) to study the feasibility of channelizing and floor-proofing the La Crosse River (Alvord et al. 1946). The plan was presented to the Common Council for approval in 1946. The plan was rejected, because it did not guarantee 'flood-proofing', and the council did not want to take on the plan's debt burden. The marsh was saved again.

A third attack on the marsh came in 1947 with the implementation of the Bemel Project. The plan was on a much smaller scale. Max Bemel proposed industrial and commercial development on private land adjacent to Copeland Avenue on land he owned. Bemel gained permission to fill in land on either side of the thoroughfare which connected the central business district with the near North Side. Bemel used dredge spill to elevate his marsh property above the flood plain. The plan took three years to complete. Bemel sold off much of the property to other businesses. He developed his remaining property into Max's Auto Salvage, the fancy term for a junk yard. It was poetic justice that the city purchased Max's land in the 1980s and cleaned it up for urban renewal.

Bemel's project succeeded where other plans failed, because marsh land removal was on private rather than public land. Bemel completely financed and planned the undertaking himself. New business on the reclaimed land included fast food stores, filling stations, car dealerships, lumber yards and similar retail businesses.

The transportation assaults

In the 1950s the La Crosse Engineering Department proposed local street expansion into the marsh. After re-evaluation in 1958, most of the plan was abandoned except for the Lang Drive portion. In 1965 the City Engineering Department began working with the Army Corps of Engineers and the State Department of Transportation (DOT) to design a plan to eliminate the flood hazard of the marsh (Army Corps of Engineers 1967). The firm of Candeub, Fleissing and Associates was hired to study the marsh. The firm recommended altering the La Crosse River course to enter the Black River north of the city rather than follow its present course. The reclaimed land would be used for industrial and residential development (Candeub et al. 1965). In 1967 the Army Corps of Engineers decided the flood-proofing plan was not feasible, and the plan was abandoned except for the transportation recommendations.

In July of 1978, a severe thunderstorm caused a flash flood in the marsh. The force of the flood broke Lang Drive in two places cutting off the main transportation route between the North and South Side. The flood episode forced the issue on the redevelopment of Lang Drive. It was redesigned as a four-lane expressway. A new bridge was constructed over the La Crosse River to handle potential future floods. The new road also met the need for projected increase in traffic for future years.

One of the recommendations of the 1965 Candeub plan was to construct an expressway through the marsh to connect the La Crosse central business district with Interstate 90. The section of Interstate 90 that passed north of the city was completed in 1965. The 1968 the city charged the Egineering Department to work with the Corps of Engineers and the DOT to come up with an expressway feasibility plan. This was the beginning shot in a battle that has since raged for a quarter of a century.

In the early 1970s the city tried to sell the citizens of La Crosse on the marsh expressway plan by calling it the Bicentennial Freeway to commemorate 200 years of freedom. The plan called for the freeway to continue through the southside of La Crosse after leaving the marsh. Approximately 450 southside homes would be razed to accommodate the proposed freeway. This aroused the citizens' attention. Not only would the city be split by the marsh, but now it would be 'quartered' into east–west sections on the southside.

Dr Richard Fletcher led the citizens' action group that opposed the Bicentennial Freeway. He joined the University of Wisconsin–La Crosse Biology Department in 1968 after receiving his PhD from University of California–Berkeley. Dr Fletcher had experience with the Sacramento River marsh. To fight the Bicentennial Freeway more effectively, Fletcher was elected to the La Crosse Common Council and served on the council in the early 1970s. Because of citizen resistance led by an active voice on the Common Council, the Bicentennial Freeway plan was defeated.

Battle of the 1980s

In the summer of 1980 a new shopping mall called Valley View opened. The mall had overlooked the La Crosse River valley. The mall was located at an intersection of Interstate 90 at the north edge of the city. To give better access to the new mall, Highway 16, which ran along the eastern edge of the marsh at the base of the bluffs, was expanded from two to four lanes. Approximate driving time for the six-mile distance to the mall from the centre of the city was twelve minutes.

The mall had devastating effects on the central business district. Several stores moved to the mall vacating the city centre. It was postulated by developers that if there were an expressway through the marsh connecting the downtown with the mall, shoppers would return downtown (convoluted reasoning is wonderful). Besides, in a society which eats with its fingers to save time, the travel distance measured in time between the two business centres would be decreased by five minutes to a seven-minute trip.

By the middle 1980s it was recognized that there was a need to resolve the conflicts between the developers and the environmentalists over the environmentally sensitive La Crosse River Marsh. In 1988, state agencies and local government agreed to undertake a comprehensive study of the marsh to develop a mutually acceptable land-use plan. The participants included the Wisconsin Department of Transportation (DOT), the Wisconsin Department of Natural Resources (DNR), the city of La Crosse and the La Crosse County Board. The key to the plan was that it had to be approved by all parties. This was also the first time that the group was to specifically address the environmental protection needs of the marsh as well as development and transportation aspects. The group was to provide an information base on which intelligent decisions could be made. The charge called for community input outside of the four participating agencies. The study was entitled the La Crosse River Valley Study (*Newsletter* 1989).

To help formulate outside input a citizens' action committee entitled the La Crosse River Marsh Coalition (LRMC) was formed. Dr Richard Fletcher and Dr Charles Lee, University of Wisconsin–La Crosse History Department, were instrumental in formulating the coalition. A newsletter entitled *The Egret* was published periodically to communicate with the membership (*The Egret* 1993). The Marsh Coalition offered input to the land-use study, testified at open hearings on the various plans offered by the study group, and vigorously defended the conservation, recreation and wildlife habitat aspects of the marsh.

The final battle

Since 1989, the La Crosse River Valley Study Group has offered six different land-use plans for adoption (Land Use Plans 1–6). The major objection to the initial plans was the proposed placement of the expressway through the marsh. Only after the expressway was removed from the marsh was Plan 6 agreed upon by all co-operating agencies. Plan 6 was a compromise and refinement of the first five plans developed over a five-year period. Public involvement in the comprehensive planning process was an important element of the decision-making process. All meetings were open to the public. Public hearings were held for each plan. Differences and recommendations were returned to the various members of the committee so that all parties would agree on the final plan.

The major components of the final plan (Plan 6) contained victories for each agency. Approximately 25 acres of the study area were designated for commercial and industrial development. The parcels consisted of marginal wildlife habitat adjacent to present commercial and industrial properties (Fig. 1).

The transportation component consisted of recommendations to expand traffic routes essentially outside of the marsh area. This was considered a major victory by the environmentalists. To meet future traffic needs, transportation plans for the city and county will continue to be

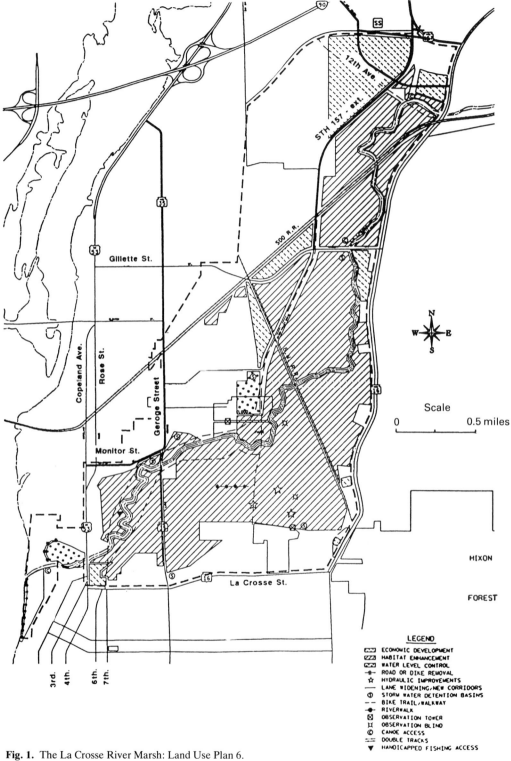

Fig. 1. The La Crosse River Marsh: Land Use Plan 6.

studied and developed over the coming years. However, future plans will not include new routes through the marsh for either rail or auto.

To safeguard the marsh, the plan called for selling the 386 acres of the marsh owned by the city to the DNR. A bonus of the plan is that the sale would raise an initial $167 000 for recreation development and would produce an additional $15 000 per annum in taxes. In the coming years the DNR would purchase marsh lands from private owners to ensure future marsh preservation. This would ensure a protected habitat for migrating birds and water fowl, plus the white-tailed deer, beaver, and muskrats that inhabit the marsh. To observe the wildlife, two observation towers and two blinds would be constructed at strategic viewing sites (Blum 1992). By selling the marsh to the DNR, it would be the DNR's responsibility to make the estimated $4.8 million improvements called for by the natural resource and recreation sections of Plan 6. Other natural resource development would include construction of sediment traps and detention basins, removal of certain dykes, noxious weed control, and reforestation in some areas of the marsh.

The recreation component consists of improvement of present bike/hiking/cross-country ski trails, connecting the existing state bike trails to the central business district, developing canoe access to the river, and providing handicapped fishing access (Blum 1993).

It is hoped by the end of 1993, the La Crosse River Valley Plan 6 will be a reality and by 1994 the improvements to the marsh mandated by the plan can be started.

After the war

Citizen improvement and action culminating in the formation of the Marsh Coalition helped preserve a valuable asset of La Crosse. By insisting on agreement on all components of the final plan by the participating agencies, all parties contributed to the greater good of the La Crosse community. If the developers would have won, the result would have been a permanent loss of a valuable recreational and environmental asset to the citizens of La Crosse. As in any war there are no clear winners and no clear losers. However, because of community compromise, despite minimum marsh development, the environmentalists have won the war to preserve the majority of the La Crosse River Marsh for the enjoyment of future generations of all La Crosse citizens.

References

ALVORD, BURDICK & HOWSON ENGINEERS. 1946. *Report of Marsh Fill*. Chicago, Ill.
ARMY CORPS OF ENGINEERS. 1967. *Report of the Army Corps of Engineers on the La Crosse River Marsh*.
BLUM, G. 1992. 'DNR May Need $167 000'. *La Crosse Tribune*, 14 December 1992.
—— 1993. 'City May Sell Part of Marsh'. *La Crosse Tribune*, 3 February 1993.
CANDEUB, FLEISSING & ASSOCIATES. 1965. *Report on Alternate Land Use Plans*. Chicago, Ill.
The Egret. 1993. Newsletter of the La Crosse River Marsh Coalition. La Crosse Wisconsin, **5** (1).
JUNEAU, D. 1969. *A Survey of La Crosse River Marsh Improvement Proposal 1938–1968*. Unpublished research project. University of Wisconsin–La Crosse.
La Crosse River Valley Land Use Plans 1–4. 1992. Supplemental Handout.
La Crosse River Valley Comprehensive Land Use Plan #5. 1992 LARS.
La Crosse River Valley Study Land Use Plan 6. 1992. LARS.
Newsletter. 2 February 1989. La Crosse River Valley Study Co-op Evaluating Agencies.

International support for conservation of geological, landscape and historical sites: the homeland initiative

GARRY McKENZIE

Geological Sciences, The Ohio State University, Columbus, Ohio 43210-1110, USA

Abstract: Site conservation requires financial support for site acquisition and maintenance. Many organizations, such as The Nature Conservancy in the United States, The National Trust in Scotland and World Heritage Sites, have developed creative ways (e.g. Nature Conservancy's debt-for-nature swops in the developing world) to conserve areas with natural (geological, ecological, geomorphological), historical, cultural, and industrial heritage value. The 'homeland initiative' is proposed as a model for educating the international public about local and global environmental change and the need for conservation and for building an international network of individuals and groups for collaboration on development of heritage sites.

For generations, overseas opportunities for a better life have resulted in emigrations from countries where resources and opportunities were limited. Often the emigrants have been successful and, although fully integrated into their new societies, they and their offspring have maintained an interest in the lands of their ancestors. It is proposed that this connection with their homelands be exploited to support the conservation of sites that are of special significance. A collaborative effort is needed between the scientific communities, preservation associations and cultural societies in overseas countries and the homeland, to select and evaluate candidate sites in the homeland that would be appropriate for acquisition.

A particularly appropriate country for testing this 'homeland initiative' might be Scotland, the source of many emigrants to North America, Australia and New Zealand. The clan system, clan maps, identification of many people of Scottish descent with their homeland areas (often as a result of tracing their ancestors) and the strong connections with Scotland provided by Highland Games and cultural societies (e.g. Scottish Heritage USA, Inc.) throughout the world, suggest that an opportunity exists to educate the public and to conserve the resources of Scotland that have international appeal.

Worldwide, protected and unprotected landscapes of special significance are under increased pressure from the human colony (e.g. Big Darby Creek, Ohio; Allan 1991). Conservation of these areas requires careful, long-term commitment of resources. Such resources will not be forthcoming unless there is widespread public agreement on the importance of conserving the landscape or there is adequate private funding. This paper presents the homeland initiative – a concept of international support for conservation of natural areas. This support is based partly on the cultural and historical ties of individuals and organizations to an established or a candidate conservation site. In the following discussion, it is suggested that there is a potential source of funds; a potential international network of scientists, conservationists, and governmental and non-governmental agencies/organizations; and an opportunity for Earth system science/global change education that goes beyond the current emphasis on tropical deforestation and world parks. The paper begins with a review of some reasons and options for site conservation that have a bearing on the concept proposed here.

Why protect geological sites?

There are many reasons why geological areas should be preserved. In North America, protection of geologically significant areas has its roots in the establishment of Yellowstone National Park. As with many areas that are now national, state or provincial parks, unusual landscapes and processes are the basis for the park. Often a geological feature (moraine, kettle, mountain) gives the park its name. The fauna and flora of such sites are related to these unusual geological substrates, geomorphic conditions, and micro-macrohabitats. As a result, biodiversity and aesthetics are additional reasons for preservation of such landscapes.

In addition to the above qualities, we must also consider a site's scientific, educational, political and historical significance. Natural landforms and geological sites are under attack from urbanization, agricultural development, or geological resource extraction. As long as world population continues to grow and planning agencies plan primarily for development rather than non-development, there will be a need to protect significant sites. These sites might be important because of their scientific value (type sections, characteristic landforms, important fossils, surface water quality or type, and groundwater or aquifer type), their educational value (best or last example of some geological or geomorphological feature in the region), or their historical value (the first place where a particular concept in the geosciences was proposed or understood). Additional reasons for protecting geological/geomorphological sites have been given by Toghill (1972) and others.

Some sites may have geopolitical significance. In the latter case, Ohio now has a state fossil and is in the process of designating a state soil type. Could landforms be far behind, and if one state does it will not others? Conceivably, the identified landform (e.g. kame) of a political area could become 'extinct' through urbanization and aggregate extraction, suggesting that the feature should be protected at some sites.

On-site changes in conservation areas can be expected through global environmental change. How these areas react to global change is another aspect of their scientific value. As we plan for these landscapes, we should be thinking about potential changes over the next 500 years as well as evaluating changes over the past several thousand years. We should also consider the educational and research value of these landscapes in a global change context.

Natural areas are not the only landforms that deserve and obtain protection. Burial and ceremonial mounds (anthropogenic landforms) made by natives in the Mississippi Valley of North America also have been protected, but not before many had been destroyed. Some have even been reconstructed as at the Mound City National Park in Ohio. Also, geologically significant landscapes with a prehistory component have been protected. One example is the Flint Ridge site in Ohio which contains numerous small quarries, the source of flint used throughout eastern North America. Today, abandoned surface mines are considered an eyesore and are being reclaimed. Should some of the modern human-made landscapes that are representative of the early mining and reclamation techniques of this century in North America be preserved (McKenzie 1979)? Maybe some modified landscapes that illustrate fluvial or coastal processes particularly well should be part of a conservation area?

Uniqueness, aesthetics, scientific, educational, geohistorical, and geopolitical factors all have a place in determining the importance of candidate geological and geomorphological sites. Because of this, our interests must reach beyond geological materials and processes. We should look at interactions with the biosphere and consider sites as components of the biogeosystem or Earth system. This expanded approach is particularly appropriate for the proposed homeland initiative concept and for sites where there is a strong potential for climatically driven on-site changes. As geoscientists, we seek to understand Earth history and can see many changes from the geological record. In setting up conservation sites, we should also take the long or geological view and be prepared to expect, and where there is scientific interest monitor on-site changes in the biogeosphere as a result of global climate change. The feasibility of a monitoring project would depend on the interest of the scientific community at the time a project is proposed. In some instances, atmospheric, biospheric or geospheric monitoring needs of the scientific community could lead to a request for local collaboration in establishing a heritage site. In any case, most of the monitoring would be done by a scientific group according to standard baseline and long-term monitoring criteria.

In Earth system science, we increasingly see the need to incorporate the human component as an agent of global and, of course, local change. Humans are geomorphic agents. The long-term evidence of this has been explored in southern Sweden in the Ystad Project (Berglund 1991) and in other parts of the world (Ponting 1991). Incorporating the history of humans at a site has the potential to expand public awareness of, interest in, and understanding of global change and landscape sites that we wish to conserve, mainly for their geological and landscape importance. Selection of sites on the basis of human history is not advocated; however, it is suggested that we utilize the history of sites, where appropriate, in selecting and supporting conservation sites. Such information would increase the 'ethnic attractiveness' of a site. A related concept – that geomorphological assets are cultural assets – has been promoted by Panizza & Piacente (1993).

Guidelines exist for evaluating a variety of candidate sites and determining which sites should be preserved first with the limited resources that we have available. It is beyond the scope of this paper to discuss such guidelines; however, they can be obtained from various organizations and government agencies.

Mechanisms for site conservation

Many mechanisms are recognized to conserve sites; some are more effective than others. They include purchase of land, conservation easements, land leases, management agreements and voluntary action on the parts of landowners. The debt-for-nature swops of The Nature Conservancy (Sawhill 1991) is one mechanism that has received much attention recently. It depends on the donations and income of the Conservancy for money to purchase debt of developing nations and to swop this debt for conservation bonds that will provide protection and management of conservation sites. All these mechanisms depend on individuals who are primarily concerned about the environment. The concept proposed here should reach those interested in the history and culture of a region, as well as those whose main interest is geological or environmental. By recognizing historical and culturally significant components of landscapes that are important for their geology and geomorphology, it should be possible to gain wider support of the national and international public for conservation of a site. At the same time we should be able to provide an opportunity for public education and long-time 'ownership'. We may be able to tap a different source of support than that which is focused on other very important environmental tasks around the world. With our focus on the environmental and human components, we should not disregard the potential for site-conservation support from those engaged in the extraction of geological resources. They, too, have a responsibility to protect special areas.

The homeland initiative

The concept of using established ethnic organizations throughout the world to support, promote and possibly adopt sites of geological significance was developed using Scotland as a model. It should be applicable in any area that: (1) produced numerous and/or wealthy emigrants; (2) has an overseas ethnic community with an interest in their place of origin; and (3) has an organized overseas ethnic community with an effective communication network. Countries such as Poland, Latvia, Ireland, Nigeria and Israel might also fit the model. In some cases, the international connections might be at the clan or tribal level, rather than solely at the national level.

Although there are only 5 million Scots in Scotland, many individuals around the world are of Scottish ancestry. Some of these recognize their Scottish roots, and even if they have not had a chance to visit 'their area' in Scotland, they are aware of the many publications that promote highland backgrounds, traditions and clan areas of Scotland (e.g. Moncreiffe 1981). There are magazines (published outside of Scotland) that promote Scottish traditions. Regular issues of *The Highlander* list about 125 Clan societies and more than 30 Scottish interest societies, while the annual directory issue lists more than 200 and 180, respectively (Anon. 1993). New developments in computer-generated genealogy charts and in data transfer, cater to the interests of those establishing their roots in Scotland. In addition, the Highland Games (>60 per year in North America) are a further draw for the individual interested in art, music, dancing and sports. This well-organized and often highly visible community is not just for the Scottish historian and amateur genealogist.

Feasibility

Research on the feasibility of the concept has just begun. The author is surveying selected clan societies and Scottish associations in North America and overseas. Responses are limited, but suggest that there are pros and cons to the idea.

According to the Great Lakes Commissioner of the Clan Donald (pers. comm.) which has more than 3000 members in North America, they are often faced with fund-raising requests for fine arts, scholarships, castle restoration and churches. Fortunately, wealthy benefactors were able to assist in acquisition of much land for the society and they now have a clan centre that attracts tourists and a land trust. Other societies are involved in forestry projects (e.g. MacNicol Society) and many are involved in castle restoration. According to the commissioner, the idea of doing something for the homeland often interests societies; however, they may take on projects that are too ambitious. The cost of removing unsafe portions of castles can be very expensive. Restoring them is another matter. And making money from the restoration for maintenance is difficult. According to a retired head of the National Trust in Scotland (speaking in Columbus Ohio, 1993) high maintenance costs are the reason the Trust has many of its properties. Most properties of the trusts make little money. In one case, they would be better off paying 200 000 tourists $2 each not to visit the site!

Thus there are problems with any site that requires upkeep and supervision. Natural or human-modified sites of geological significance might not be so expensive, but they would presumably all have some costs, even if only administrative. The actions of one relatively small group illustrates what can be done in this area. In the last 8 years, the Clan MacNicol Society (mainly in the USA) raised enough money to purchase 130 acres north of Portree on Skye, build a commemorative cairn, contribute to a restoration, and begin a reforestation project. They work closely with local authorities; a local clan group has formed a management committee (Murray Nicolson, pers. comm.). For a site of international scientific or other importance, particularly one in which a scientific or other easement had been obtained in exchange for financial assistance in site development, additional non-local representatives would be part of the management structure. The concept of ethnic support for sites appears to be reasonable in the Scottish example.

Advantages

To sum up, the main advantages of the proposed homeland initiative over more traditional approaches to supporting landscape conservation include: (1) international financial and popular support; (2) international network of professionals and amateurs, from many disciplines and non-profit organizations, with an interest in landscape conservation; (3) expanded interest in any on-site research and the research results by those in the history, anthropology, global change and genealogy communities; (4) worldwide publicity on a project, through radio, television and the press of ethnic groups abroad; and (5) expanded tourism as those overseas learn of ancestral sites of geological/geomorphological significance.

Disadvantages

Possible disadvantages of the initiative are (1) friction and misunderstandings between 'foreigners' and those who inhabit the country receiving support for the landscape project; (2) difficulties in selecting sites because of multiple interests and criteria of players; (3) unknown potential for funding from overseas; (4) possible disagreement between environmental organizations and trusts now operating in a country; and (5) loss of overseas support for other ethnic projects in the homeland and other national and international environmental projects.

Testing of the concept

There are several possible steps that could be taken to test the concept in a country that has the appropriate conditions described above. After establishing a directorate to determine who should be involved, a site evaluation committee would be selected. The next step would be selection of potential sites based primarily on landscape criteria, but including other criteria described here. Overseas input would be appropriate here. After consulting with organizations that have had fund-raising projects among the ethnic community overseas (in the Scottish example, e.g. see the Iona Appeal advertised in *The Highlander*), a fund-raising programme could be organized that would utilize the communications network of the international ethnic community associated with the proposed conservation site. Depending on the results obtained with a test of one ethnic group, modifications would be made for subsequent conservation projects targeting the same and other ethnic groups.

References

ALLAN, K. 1991. One of the last of the best. *Nature Conservancy*, **41**, 16–23.
ANON. 1993. Clan and Scottish societies. *The Highlander: the Magazine of Scottish Heritage*, **31**(2A), 1–15.
BERGLUND, B. (ed.) 1991. The Cultural Landscape During 6000 Years in Southern Sweden – The Ystad Project. *Ecological Bulletins*, **41**, Munksgaard, Copenhagen.
McKENZIE, G. D. 1979. Preservation of selected unreclaimed strip-mined lands in Ohio. *Ohio Journal of Science*, **79**(4), 170–173.
MONCREIFFE, I. 1981. (rev. ed.) *The Highland Clans*. Clarkson N. Potter, New York.
PANIZZA, M. & PIACENTE, S. 1993. Geomorphological assets evaluation. *Zeitschrift fur Geomorphologie, Supplementband*, **87**, 13–18.
PONTING, C. 1991. *A Green History of the World*. Penguin Group, New York.
SAWHILL, J. C. 1991. From the president: Using debt to save the rainforest. *Nature Conservancy*, **41**(1), 3 & 31.
TOGHILL, P. 1972. Geological Conservation in Shropshire (Letter). *Journal of the Geological Society, London*, **128**, 513–515.

Conservation through on-site interpretation for a public audience

PETER KEENE

Oxford Brookes University, Gipsy Lane Campus, Headington, Oxford OX3 0BP

Abstract: There is a perceived need to 'educate' visitors on-site at locations of Earth science interest. Visitors may be classified into four potential target audiences, each with their own needs. The apparent failure of much on-site education to reach the 'public' target audiences is related to the gulf between the aspirations of those who have a mission to educate and the interests of a majority of the public. Alternative approaches to win 'hearts and minds' include the adoption of less formal educational strategies, a more interdisciplinary approach to site conservation, and a more sensitive awareness of aesthetic landscape values. Earth scientists actively supporting the needs of conservation may see an advantage in strengthening their brief by widening their knowledge of the political, social and aesthetical considerations which so often determine the outcome of conservation issues.

This paper examines the assumption that one way of conserving Earth science sites is through the on-site education of adult non-specialist visitors. Many Earth scientists feel a sense of 'educational' mission. This may be driven by a wish to introduce others to the depth of reward that they have experienced through an understanding of rocks and landscape or it may perhaps be the belief in a more political imperative which runs something like this:

> 'There is a need to educate people about the environment both at an emotional awareness level and at a deeper level of understanding which equips people to participate in decisions affecting environmental conservation. In the hands of this educated democratic force hangs the long-term security of our valued environments.'

The public audience which is most likely to respond favourably to such overtures only clearly identifies itself when visiting sites. Where better to educate people than within the environment itself? Here the significance of what they see can be demonstrated in the landscape, a giant open-air laboratory. There is, therefore, a perceived need to educate the visiting public **on-site**, to help encourage deeper understanding and appreciation of our environmental heritage. Yet, 'serious' on-site Earth science interpretation for the general public is notoriously difficult. Is it impossible?

Failure to penetrate the public market

The degree of 'down-market' penetration from the academic end of the spectrum is, it seems, very limited. General audiences are resistant to what is on offer, even when sensitively interpreted and simplified. This is paralleled on the part of many specialists by a reluctance to take part in a process where the essential truth of concepts central to the understanding of Earth sciences can be viewed as being compromised by over-simplification in the name of popular explanation. In this process they see superficiality and trivialization undermining the very aims of achieving a level of understanding which is compatible with participation in environmental decision-making and conservation.

Not all academics would agree that this is an inevitable consequence of explaining sophisticated ideas to the uninitiated. Stephen Jay Gould, one of the few scientists who has managed to write very successfully for the popular market, states in the preface to his best-selling paperback *Wonderful Life. The Burgess Shale and the nature of history* (1989):

> I have fiercely maintained one personal rule in my so-called 'popular' writing. (The word is admirable in its literal sense, but has been debased to mean simplification or adulteration for easy listening without effort in return). I believe that we can still have a genre suitable for and accessible alike to professionals and interested lay people. The concepts of science, in all their richness and ambiguity, can be presented without any compromise, without any simplification counting as distortion, in language accessible to all intelligent people. Words, of course, must be varied, if only to eliminate a jargon and phraseology that would mystify anyone outside the priesthood, but conceptual depth should not vary at all between professional publication and general exposition.

Here, then, is a model worth emulating, but with the proviso that, if the comment for potential target audience Group D (see Fig. 1) is correct,

Visitors may be regarded as being part of a continuum spanning a broad spectrum of age and interests, but for convenience they can be grouped into four potential target audiences.	
A EDUCATION GROUPS Schools, Colleges, Universities, Adult groups In one sense this is a captive audience and one which readily accepts the idea of being educated in the environment, including the use of interactive interpretative techniques. The methodology of teaching field groups is relatively well understood, even if often not well implemented. 'Students' are used to structured, linear learning. They match the approach adopted by most academics and will often bring to the site a foundation knowledge which provides a tool kit with which to interpret seemingly novel environments.	B INTERESTED INFORMATION-SEEKING ADULT NON-SPECIALISTS Almost the ideal audience for the interpreter. Talking with such people is a rewarding experience. They are responsive, appreciative and interactive. Unfortunately, although their density at information points may be relatively high, this is a small (although influential) proportion of the adult general public. It is a group which needs little active encouragement. They satisfy the need felt by many earth scientists to introduce others to the depth of reward that they have experienced through an understanding of rocks and landscape. This group is committed and readily seeks out information and education.
C THOUGHTFUL ADULT NON-INFORMATION SEEKERS On occasion I count myself amongst these. They deliberately walk away from an information board fretting at the imposition of official graffiti which intrudes, trespasses, violates, encroaches, stands between, makes second-hand, the experience of place. This is the down-side of interpretation. There is a feel to landscape which is beyond meaning and reason. We must be wary of encouraging the destruction of environmental experience through intrusive, inescapable interpretive control. The resistance to being processed should not be confused with a resistance to education. The needs of this group should be addressed, not because of their numbers, which are not great, but because of their moral justification, and the powerful support they provide in creating a strong democratic lobby of conservationists.	D MASS OF GENERAL PUBLIC. (Only 4% of the public ever enter a bookshop) Sites of interest to earth scientists often serve other valuable functions and indeed the social role or aesthetic value of the 'open space' may be the principle reason why the site exists or is preserved. Many people visiting such sites are understandably completely unaware of the scientific value or interest of the place. Most have little interest in acquiring the sort of levels of understanding of the subject implied in the opening discussion. There does seem a particular resistance in the U.K. to mixing leisure or recreation with education - almost a resistance to education itself. Perhaps a subconscious recognition that what was taught at school did not really serve them well? Those directly involved in dealing with environmental interpretation for the public are well aware of a fundamental principle not faced by those dealing with a captive 'educational' market; that is the freedom of choice. To the extent that this market can vote with its feet, the whole strategy of environmental interpretation for the general public must be consumer driven.

Fig. 1. Potential target audience.

success is being measured in terms of 4% of the general public. One must conclude that much of what is attempted in the name of Earth science which is educational in its conception, is inappropriate for the main body of the public audience.

Interpretation which specifically addresses Earth science topics for the public at large can claim some success at stimulating an emotional (rather than an intellectual?) commitment to conserving valued environments. Interest is stimulated by an almost casual contact with the Earth sciences – a pretty fossil, a bizarre landform, an item of folklore. This haphazard approach is often dismissed by the professional Earth scientist as superficial to the point of encouraging completely erroneous views, often seriously inaccurate, descriptive rather than dealing with concepts or principles, and crammed with anecdotal, unconnected trivia!

Whatever one's views about Earth science education for the masses, there is clearly a yawning gulf between the educator and the public – a polarity which seems painfully more apparent in the Earth sciences than in other environmental disciplines such as archaeology, ecology or history.

Polarity examined

Polarity is a major metaphor in western culture. It implies either/or, right/wrong, black/white. It is commonly held today that we should seek to replace conflicting or opposing stances with an approach that emphasizes harmony.

It is a nice thought, but as a concept it is easier to articulate than to implement. Has it any relevance to our problem? The idea of replacing the notion of polarity with one of harmony or holism entered popular culture from the east, derived in particular from Zen and Taoist teachings. It is interesting that they also address polarity in the principle of Yin and Yang – opposites but equals. Consideration of the Yin and Yang principle can generate harmony or the 'middle way' by bestowing equal respect and sympathy to the seeming opposites. I do not wish to labour any philosophical discussion beyond that which might be regarded as straight common sense, but it does seem advantageous to look at the way the general public perceive, utilize and enjoy sites which we recognize as of significance to the Earth sciences and requiring conservation. The 'middle way' is also known as 'the watercourse way' – moving with the current?

When considering the gulf between mission and audience (Fig. 2), as an active geomorphologist, it is easy to identify with the 'mission driven' characteristics listed in the left-hand column. Yet how different is my approach to everyday living where much decision-making is unconscious, and where emotion is taken for granted as having a part to play in deciding what to do. Looking at the right-hand column ('consumer driven'), I can recognize that the characteristics listed there accurately reflect the way I behave, feel, make decisions, and enjoy myself when I am not busy being a geomorphologist. Visiting a ruin, a site of historical significance, I do not put historical reductionist blinkers on. I am open to all sorts of sensations, ideas and influences which impinge on my conscious and subconscious. Academics, it seems, work mainly with linear logic. 'People', think more like a computer with a random access memory. If they are inquisitive and seek information or interpretation, it is on a 'need to know' basis. This is how we normally learn about the world. When walking around a ruin, I may appreciate that understanding will be partial unless this monument is approached with a sense of historical perspective; yet, more important at the time may be the sense of place – the atmosphere overrides the need for interpretation. At one level I do not want to 'think'. I would like the option of not reading about it until afterwards.

Strategies for good practice

The failure of the 'academic strategy' to penetrate much beyond the educational audience suggests it is worth considering alternative

---

The more straightforward 'educational' markets (groups A and B) are generally receptive to the familiar structured academic approach. However, the identification of potential target audiences (figure 1) highlights the gulf which separates those with the mission to expand people's understanding of the earth sciences and the consuming mass of the general public (group D). It is useful to set in juxtaposition characteristics of this dichotomy, although such a list invites caricature and parody!

MISSION DRIVEN	▶ ◀	**CONSUMER DRIVEN**
EDUCATIONAL OBLIGATION	◀ ▶	MARKET DEMAND
INTELLECTUAL ACADEMIC ELITIST	◀ ▶	POPULAR
PROCESS	◀ ▶	DESCRIPTION
STRUCTURED PROGRESS	◀ ▶	RANDOM ACCESS MEMORY
ANALYTICAL SCIENTIFIC	◀ ▶	FEELING INTUITIVE
PYRAMIDICAL FOUNDATION	◀ ▶	'NEED TO KNOW' AVAILABILITY
LINEAR LOGICAL ORGANISATION	◀ ▶	RANDOM ACCESS
CONSTRUCTIONISM	◀ ▶	POST MODERNISM
LAWS	◀ ▶	PHENOMENA
AUTHORITY. ORDER	◀ ▶	ANARCHIC
PRECISION REDUCTIONIST	◀ ▶	GENERAL HOLISTIC
CONCEPTUAL UNDERSTANDING	◀ ▶	SENSE OF PLACE
ACCUMULATED KNOWLEDGE	◀ ▶	DIRECT EXPERIENCE
RATIONAL CONTROLLED	◀ ▶	ACCEPTANCE ENCOUNTER
BALANCED	◀ ▶	SENTIMENTAL
DISPASSIONATE	◀ ▶	EMOTIONAL
SERIOUS WORK	◀ ▶	TRIVIAL LEISURE
SOLEMN EARTH SCIENTIST	◀ ▶	HUMOROUS INTERPRETER
"I had an interesting time. I wish I knew more about rocks"		"It told me more about the Ipswichian Intergalactic than I wanted to know"

Fig. 2. The gulf between mission and audience.

strategies for the provision of Earth science interpretation in each of the target audience groups considered.

(A) Education groups

This group is comparatively well-served with interpretative material to reinforce the conservation message. A constraint on the effective use of this data, however, has been an inadequate mechanism for disseminating and sharing information concerning the location and availability of Earth science sites, or relevant publications and of the level of assistance available. This is likely to be one of the services developed by local RIGS (Regionally Important Geological/ geomorphological Sites) groups.

(B) Interested information-seeking adult non-specialists

This articulate group is predisposed to be sympathetic to conservation and has a voracious appetite for suitably presented environmental literature. However, much of the literature, even that supposedly seeking a wider audience, is still academic and heavy reading for the interested non-specialists. In that respect one cannot better the advice of Stephen Jay Gould (above). It is one of the encouraging signals for conservation that this group is growing in numbers and influence.

(C) Thoughtful adult non-information seekers

Let this group represent that part of us which seeks the experiential rather than the meaning. What does this mean in concrete terms? Wherever wilderness qualities survive, or where there is a sense of resonance within that place, then it should be axiomatic that intrusive interpretation which dictates the way that that site will be experienced should be avoided. It is too easy to package nature and sell it as a product – this can be done intellectually as well as physically. If we accept that the transmission of wonder is more important than on-site interpretation then the best conservation strategy here is to keep interpretation away, off-site, but available to those who seek it.

The pressure for on-site interpretation is increasing, not simply from those with an educational mission but from an expanding interpretation industry and those who have a commercial interest in leisure and tourism, for whom 'interpretation' may be seen as a planning lever to exploit otherwise protected sites.

(D) Mass of general public

It is now generally accepted that a wider public awareness is an important component in strategies to conserve valued Earth science sites. It is also clear that if the public are to be mobilized in this 'hearts and minds' campaign, then much more radical approaches are called for than have hitherto been the case. Such approaches might include community-based landscape art, poetry and drama which can enhance people's perception of a site. Similarly, multidisciplinary experiences could be encouraged which link rock, soil, ecosystems and history to heighten landscape awareness.

It has already been stressed that sites of Earth science interest may serve a variety of functions. These may relate to other disciplines or they may be connected to their recreational, open space or aesthetic landscape value. Just as the development of an integrated strategy for the conservation of such sites is sound environmentally, so an integrated approach to interpretation seems most likely to be effective in stimulating a body of opinion supportive of conservation strategies. Couched in these terms, conservation is seen by the visiting public, not as some obscure academic ploy with little relevance to their lives, but as a method of protecting a valued local environment.

Conclusions

Conservation involves the participation of an increasingly wide spectrum of the public. This welcome democratization of decision-making lies at the centre of the 'hearts and minds' debate, around which this paper revolves. However, the strategies suggested above offer no short cuts and at best may be regarded as models of good practice. Underlying these deliberations, I am driven by a deeper sense of unease. This relates to two possibilities:

(a) that public support won for conservation may yet prove to be an ephemeral enthusiasm, prone to fickle fashion and not yet reflecting the level of involvement which ensures the enduring, steadfast support of a sympathetic, knowledgeable public. Hence the need to reinforce the 'educational mission';

(b) that culturally we have experienced an undermining of some of the 'moral' certainties upon which an intellectual consensus depended and within which the Earth scientist and conservationist could find protection in arguing a case. The statement 'If we cannot preserve the landscape of the Burren then there is nowhere on Earth that is safe' is one which might find few dissenters amongst conservationists. Yet this debate today has political, social and aesthetic dimensions which many Earth scientists feel is beyond their chosen interest – and yet it is upon such considerations that conservation decisions are ultimately made.

'Good science is the ability to look at things in a new way and achieve an understanding that you didn't have before' (Hans Kornberg – biochemist). Earth scientists actively espousing the cause of conservation may see an advantage in strengthening their brief by widening their knowledge of the political, social and aesthetical considerations which so often determine the outcome of conservation issues.

National parks and geological heritage interpretation – examples from North America and applications to Australia

GABOR MARKOVICS

School of Natural Resources Management, Deakin University–Rusden Campus, 622 Blackburn Road, Clayton, Vic. 3168 Australia

Abstract: The benefit of interpreation in national parks in providing recreational, educational and management/conservational information has long been recognized. In Australia, during the 1970s and 1980s many aspects of parks were interpreted. The geological heritage of parks, largely escaped interpretation despite geology featuring as the centrepiece of the natural heritage in most parks.

In North America, the philosophy and practice of interpretation in national parks recognizes that geology is an integral part of a park's heritage and is included in the interpretative programmes.

This paper illustrates how geological heritage interpretation can benefit the community and public awareness of our natural heritage. Examples from the North Amercian continent will be examined and contrasted to that seen in places in Australia.

This paper addresses the lack of awareness and understanding of the geology of our national parks and the almost complete absence of interpretation of the geological heritage in Australian parks. In contrast to this, a recent visit to the North American continent showed that not only is the geological heritage interpreted to some extent in almost all parks, in many it is the focal point of the park (e.g. Grand Canyon National Park, Mt. St. Helens National Monument, Sunset Crater National Monument, Yellowstone National Park).

Current situation in Victoria

Victoria has 32 national parks (and many other state parks) of which 20 have geological heritage of significance. When one reads the general brochure on Victoria's national parks (Conservation & Environment, 1992) this aspect is obvious as most of the pictures are of breathtaking landscapes and the small paragraph on each park contains phrases like:

Wyperfeld NP	'The lake beds in peaceful Wyperfeld now fill only rarely.'
Grampians NP	'Victoria's largest national park is famous for its rugged sandstone ranges,'
Mt. Eccles NP	'Mount Eccles is a volcanic formation which last erupted about 7000 years ago,'
Port Campbell NP	'Spectacular coastal formation,'
Organ Pipes NP	'As well as 20 m high basalt columns,'
Baw Baw NP	'The granite Baw Baw plateau'

The above are only a few examples of the more obvious geological heritage in Victorian national parks. There are many state parks like Werribee Gorge and Tower Hill that exhibit the same heritage. In addition to the general brochure on our national parks, the Department of Conservation and Natural Resources (then the Department of Conservation and Environment) produces single information sheets on most of Victoria's parks. Some of these are quite old (greater than 5 years without revision), and only a few contain any information on the geology. For example, the Werribee Gorge Stage Park sheet (Conservation & Environment 1991) contains the following opening paragraph:

'Geological features, spectacular views, native flora and fauna and opportunities for bushwalking and rock climbing are the attractions of the 375 hectare Werribee Gorge State Park, about 65 km west of Melbourne. Five hundred million years of geological history – from ancient folded sea-bed sediments through glacial material to recent lava flows – have been revealed in the Gorge by the downcutting action of the Werribee river.'

The sheet continues to supply information (in sections) on location and access, facilities, plants and wildlife but no more information on the geology. This example is typical of most of the parks although some information on the geology can be obtained from the park rangers, particularly, where there are visitors' centres in a few parks (e.g. Wilson's Promontory NP, Grampians NP).

At the Department of Conservation & Natural Resources Information Centre in Melbourne, amongst the published literature, it was possible to locate 3 booklets available for purchase that had some geological interpretation relating to the parks:

(1) *Mt Buffalo NP* (1992) Conservation & Environment publication (1 page on geology and landscape of the park).
(2) *Wilsons Promontory NP* (1982) (6th edn, 1990) Conservation & Environment/ Victorian National Park Association (small sections on the geology along the walks).
(3) *Tower Hill* (1990) Victorian Geological Survey (booklet devoted to the geology of the park).

The remainder of the publications and brochures in the centre consist of a plethora on management plans (which do have small sections on geology in the more recent plans), plants, animals, salinity, water, whole farm planning, birds, shellfish, plant identification, agroforestry, tree diseases, whales, etc. This is scattered in amongst posters that highlight the spectacular scenery of the parks.

Current situation at Mount St Helens

In this paper it is not possible to show to the full extent how the absence of interpretation of the geological heritage in Victoria is strongly contrasted by what has been done in national parks in the United States and Canada. However, by using one park as an example (which is quite typical of all parks in the North American continent) the contrast may be shown. For this purpose, Mount St Helens National Monument in Washington State has been chosen. It is appropriate since it strongly reflects the US National Parks Service's (USNPS) attitude towards the interpretation of geology as an integral part of natural heritage. Unlike many of the other US parks Mt St Helens is a recent addition to the Parks Service, added to the list shortly after the 1980 eruption of the volcano.

Mt St Helens is an active volcano on the west coast of the USA. It is part of a belt of volcanic mountains (the Cascade Range) that formed because of movement of the Earth's crust during the last 150–200 million years. The Pacific plate has been slowly sliding underneath the western part of the North American continent and below this crust the rocks have been melting and then making their way to the surface and erupting as volcanoes. The eruptions in these volcanoes tend to be violent compared to the Hawaiian style of eruptions, and are therefore very dangerous. In 1980, Mt St Helens had been actively erupting throughout the early months and was quite a tourist attraction as it erupted ash into the atmosphere.

On May 18, 8.32 a.m., the northern side collapsed and left the landscape (and people) completely changed.

Mt St Helens is now a national monument that is jointly run by the USNPS and the US Forestry Service (USFS) . At the time of the May 18 eruption it was a recreational area centred around Spirit Lake, which is at the base of the volcano. The history behind the present visitors' centre near the I5 Highway turnoff to Spirit Lake is that 2 months after the eruption a caravan was set up as an information van. Shortly after, this was converted into a mobile unit while the centre was being built. The visitors' centre (which was completed in 1986) became the focal point of the park as access to Spirit Lake was destroyed during the eruption. The centre is a very impressive building and the interpretation of the actual event, the geological aspects being the event also became a focal issue in the centre and for the park.

The park (and centre) have the geological heritage interpreted in:

(1) many elaborate visitor centre exhibits;
(2) wayside exhibits (as you drive around the northern side of the volcano to get the present-day closest view of the blast area, some 10 km from the dome);
(3) a variety of audio-visual productions;
(4) an extensive publications area where one can buy many commercial and USNP/USFS publications;
(5) the *Volcano Review* – a visitors' guide newsletter (produced annually) that has information on trails, views, campsites, campsite programmes and interpreter-run programmes (at selected times throughout the seasons) on regeneration, the eruption, older features etc.

The interpretative exhibits in the visitors' centre are divided into four areas: a prehistory area, the May 1980 area, a recent history area and a present activity area. The exhibits range from:

- panels that supply information on volcanic events, why crustal plates move, to information about the actual eruptive event;

- a walk-through model of the volcano where inside you can view the magma chamber, see how the layers of rock build up, observe how volcanic events can be dated;
- two fully functional seismographs, one records the events of the volcano with the earthquake tremor data being readily observable, the other has the sensor just below your feet as you look at the machines and you can create your own earthquake pattern as you jump up and down or move about;
- displays that show aspects of volcanic dome growth, monitoring of the volcano, and the survivors of the blast;
- a series of computer quizzes that tests your knowledge about the features of the park and the eruptive event. The questions are drawn from the exhibits in the centre. This then can serve as a method of evaluating the effectiveness of the interpretative material.

Since 1990 the road to Spirit Lake has been reconstructed and they are building two more visitors' centres (the Coldwater Ridge Visitors' Centre and the Johnston Ridge Observatory Centre) closer to the volcano, a number of wayside exhibits and interpretative trail exhibits along the new road.

This example of one park, although very geologically orientated, demonstrates how interpretation is approached in the US parks. Without the interpretation of the park, visitors would find it difficult to appreciate the area of the park and the heritage that has developed as a result of the May 1980 eruption.

Involvement of other groups in geological interpretation

Another aspect of geological heritage interpretation that became obvious in the USA and Canada was the co-operative work by other agencies (government and private) in the documentation and interpretation of the geology. One such organization is the USGS (United States Geological Survey) which now sees its role extend beyond just geological research and information retrieval for academia and industry. The USGS produce resource material for various groups such as the Parks Service and co-operate with the USNPS to document and interpret the geological heritage.

This is starting to develop in Australia (although I don't think by any formal arrangement) by AGSO (Australian Geological Survey Organization, formerly the Bureau of Mineral Resources). This is an area that needs to be further developed along the lines of the US situation as AGSO has an enormous pool of documented geological heritage (albeit for other purposes) that covers many of our parks. This information can be used to help interpretation.

Interpretation can play a role in recreational, educational and management areas related to parks in USA and Canada. They both have a strong commitment to the area as seen by:

(1) the appointment of an Assistant Director for Interpretation (USNP);
(2) the development of interpretative plans (5 years with yearly reviews) by both USNPS and the CNPS;
(3) the development of surveys that supply information about the visitors to the parks. This information is then used in the planning of the interpretative material. In the US parks this was the Visitor Services Project (VSP) and in Canadian parks the Visitor Activity Management Process (VAMP);
(4) the establishment of a consultative body whose role is to supply expertise in the planning and development of interpretative material and facilities as well as the training of personnel in interpretation, e.g. the Harpers Ferry Center, USNPS.

Conclusion

In summary, we can learn from the North American experience when it comes to geological heritage interpretation. Particularly, when many of our parks have a strong geological 'theme'. Factors that would help are:

(1) a stronger Federal Government commitment to parks interpretation. This should encourage the development of natural historical societies that operate in parks. They have been seen to play a key role in the USA and Canada in interpretation;
(2) the development of co-operation between other government and private sector groups such as seen in the US by the USGS and the USNPS;
(3) a stronger commitment to interpretation by the various Australian Parks Services and a recognition that interpretation can play an important role not only in education, but also in recreation and management issues;
(4) the ability to produce interpretative material which requires:
- the development and implementation of a long-term interpretative plan that is regularly reviewed;
- an understanding and documentation of the heritage to be interpreted (i.e. the

establishment of a database of the heritage aspects);
- a knowledge of the people that are likely to use the interpreted material (the development of a survey that addresses visitor needs, expectations, background, demographics, etc.);
- a method of evaluating the 'effectiveness' of the produced interpretation.

References

Conservation & Environment. 1992. *Victoria's national parks – a brochure guide to Victoria's 32 national parks*. Department of Conservation and Environment, Melbourne.

—— 1991. *Werribee Gorge State Park – information sheet*. Department of Conservation and Environment, Melbourne.

Conservation system of geological sites in the old salt mine of Wieliczka (south Poland)

ZOFIA ALEXANDROWICZ & MALGORAZATA GONERA

Institute of Nature Conservation, Polish Academy of Sciences, 46 Lubicz St., 31 512 Krakow, Poland

Abstract: Wieliczka (near Cracow), one of the oldest European salt mines has been added to the World Heritage List. This is a type of museum illustrating geology and mining techniques and has been worked continuously since the eleventh century. Salt exploitation in the Wieliczka Mine is limited now and it will be stopped shortly.

The geology of the Miocene salt formation is complicated here. The deposit is situated close to the Carpathians and is partly covered by flysch nappes. Interesting tectonic and sedimentary structures, various type of evaporites and fossils are accessible in the mine. This site yields evidence useful for recognizing the salt deposition process.

Last year, numerous exposures in the mine were designated as documentary sites of inanimate nature. The conservation project comprises: (1) underground outcrops and sections recommended as educational areas; (2) particular localities protected because of their scientific value; (3) the nature reserve Crystal Caves assigned for professionals. These sites will be arranged by using appropriate conservation techniques. Crystal Caves protection required a special kind of conservation method for the large halite crystals (they are in danger of corrosion by humid air). We believe that the elaborate conservation model of subterranean outcrops is very useful for enclosing mines, which have now been given over to touristic and museum purposes.

Salt manufacturing in the vicinity of Wieliczka, one of the oldest salt working areas in Europe, dates back to the Neolithic period (around 3500 BC). At that time salt was obtained by evaporating water from salt springs. Salt mining began in the Middle Ages (about AD 1250) and continues today. During the seven centuries of mining, a system of nine levels was created under the city of Wieliczka, 64–327 m deep, with numerous galleries and chambers. This system is unique and incomparable to any other salt mine in the world. Some 2040 chambers from the period when salt was taken away by hand in large salt blocks have been preserved, 300 of which are classed as monuments. They are typical excavations and are unique examples of salt mining of the period between the eighteenth and early twentieth centuries. Some parts of the mine served for sight-seeing as early as the eighteenth century. Sculptures in salt and old mining equipment that have been preserved here are priceless cultural monuments spaced along the 2 km long tourist trail. Each year 600 000–800 000 visitors come to see them. The historical part of the mine embraces three upper levels, up to 104 m deep, and large parts of lower ones. In 1978 this part of the mine was placed by UNESCO on the First International List of the World Cultural and Natural Heritage.

The depletion of resources, periodic heavy flows of water into the mine (which have been recurring often during the last two years), subsidence of the bedrock and extensive damage to the city due to mining, all forced a decision to terminate exploitation and to concentrate on works aimed at preserving the mine as a tourist and educational resource. Accordingly, an increase of the mine's tourist and educational functions is planned. Until recently, calls for the preservation of geological exposures, making them accessible for educational purposes has not always been taken into consideration during mining activity. Thus, numerous important geological features were lost and the underground museum (located in one of the historical chambers where rock samples and illustrations were collected) is the only place where people can learn about the geology of the deposit.

Currently, a site management plan which will safeguard the important geological features of this site is being worked out. A basic selection was made and documentation of the geological sites of the historical mine was prepared. The preservation of the Crystal Caves requires a special approach. Research carried out here aims to determine the state of the halite crystals and the causes and scale of danger to the caves. As a result, the necessary safety activities will be determined and the possibility of making

From O'Halloran, D., Green, C., Harley, M., Stanley, M. & Knill, J. (eds), 1994, *Geological and Landscape Conservation*. Geological Society, London, pp. 417–422.

the caves accessible for sight-seeing will be considered.

Outline of the geological structure of the deposit

The saliferous formation of the Miocene (Mid-Badenian) extends as a narrow strip along the northern edge of the Carpathians. Near Krakow, it is exploited in the Wieliczka and Bochnia mines and by leaching the salt deposit in boreholes in other places. Towards the east, the emergence of salt formations widens to 8 km, and submerges beneath the Carpathians (Garlicki 1979). The development of saliferous sediments was linked with the creation of the Carpathian foredeep and the development of a shallow sea, temporally isolated from the open ocean, and with the last phase of the Carpathian flysch nappes (their folding, denuding and moving towards the north). Sedimentary and tectonic events that took place in the Mid-Miocene produced an unusually complex structure of salt deposits in Wieliczka. It was the subject of research by many generations of geologists, and studies by Niedzwiedzki (1892) and Zejszner (1844) can be considered as classics in the geology of salt deposits. The origin of the deposit still generates interest and discussion among geologists (Gawel 1962; Poborski & Skoczylas-Ciszewska 1963; Wiewiórka, 1974; Kolasa & Slaczka 1985).

A complex of the Miocene formations exposed in the Wieliczka salt mine is bipartite (Fig. 1). It consists of stratified deposits (about 50 m thick) and megabreccias (about 150 m thick). The stratified member consists of sandstones and silts with layers of gypsum and anhydrite, overlain by a complex of green laminated salt, followed by 'shaft' salt and 'spiza' salt with coarse-grained varities containing sandstone, mudstone, siltstone and anhydrite fragments. These deposits form three folds, strongly deformed and overturned towards the north. The megabreccia consists mainly of silt, marl and sand, containing debris of flysch rock and different sizes of salt (crystals, blocks, grains). Rock salt occurs in various types (Szybist 1975).

The characteristic sediments are known as the Wieliczka Beds representing the Wielician substage, and the formation exposed in the Wieliczka Salt Mine represents its stratotype (Luczkowska 1978). Many species of fossil plants (Aablocki 1930) and animals, e.g. Foraminifera (Reuss 1867), were found in them.

Strategy for the conservation of underground exposures

The securing of continuous access to underground deposits is the basis of conservation projects to protect scientifically and educationally important geological features located in the mine. The fragmentary nature of the exposures, resulting from the necessity to build protecting constructions, is a characteristic feature of underground conservation. The fragmentary nature of protection may also result from the variable geological structure of the deposits. Moreover, some exposures are not accessible for other reasons, including difficult, or dangerous tunnels or inadequate ventilation.

For these reasons, only those small fragments which meet the necessary criteria of a docu-

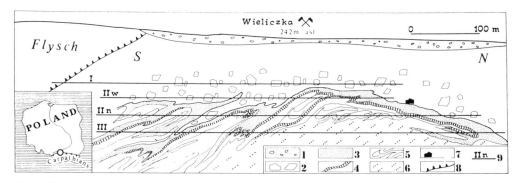

Fig. 1. Geological section of the Miocene salt deposit in Wieliczka (the vertical scale is the same as the horizontal). 1: Quaternary; 2: Salt Breccia Member. Stratified Salt Member – 3: "spiza" salts; 4: 'shaft' salts; 5: green salts. 6: substrate of evaporites (Lower Badenian? clays); 7: localization of Crystal Caves; 8: Carpathian overthurst; 9: levels of the mine and their depth below the surface – (I) 64 m, (IIw) 84 m, (IIn) 104 m, (III) 129 m.

mented site can be proposed for conservation and exhibition. Unique geological features occur in untypical deposits, and their origin is often controversial. In order to guarantee their effective preservation and the possibility of carrying out research, the protection of such places should be comprehensive and include particular zones of the mine; ranked as nature reserves.

The management plan for the preservation of the Wieliczka Salt Mine provides a model for underground environment protection methods. It may be implemented in active mines, but even more readily in worked out mines. The conservation system proposed rests on the following assumptions:

- The protection net is based on criteria of selection and scientific values of features which illustrate both typical and unique features of the deposits.
- Typical elements are scientifically documented and described in terms of the requirements of their legal protection as documentary sites of inanimate nature.
- Particular sites are connected by sightseeing trails which differ in the degree of their protection, presentation and accessibility (depending on their function) educational or scientific level.
- Unique parts of the mine or groups of features of special significance are designated as underground nature reserves to ensure their effective safeguard.
- Well-preserved nature reserves, appropriately protected and adapted, may be included in sight-seeing trails, especially for specialists with a limited number of visitors.
- There is no admittance to dangerous nature reserves. They are the subject of research aimed at determining the causes and scale of the danger and the conservation methods which should be applied.

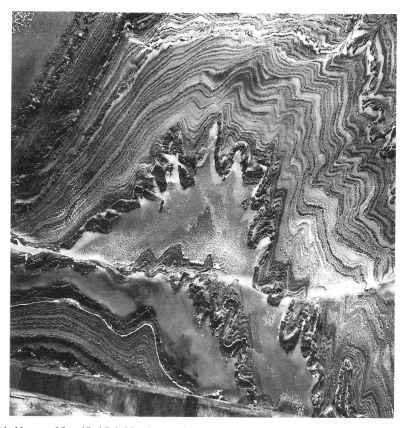

Fig. 2. Folded layers of Stratified Salt Member – spiza salt; level III of the mine (photo by Grzybowski).

- A museum is an important element of the proposed conservation system. It enables visitors to see the general geology of the deposit and methods of exploitation.

The conservation project realized in the ancient Salt Mine of Wieliczka

The plan for the conservation of geological values in the Wieliczka Salt Mine is concerned both with the creation of a network of sites documenting basic elements of deposits which can be used for education, and with the conservation of the Crystal Caves as a nature reserve which is especially endangered and currently not available for sight-seeing.

Documentation sites

Twenty-five sites, comprising of numerous exposures in the Wieliczka Salt Mine were proposed for protection as 'Documentation Sites of Inanimate Nature'. They were selected in terms of their value as the documentary basis for solving crucial geological problems of the Wieliczka deposits. The stratotype profile of the Wielician formation, an important stratigraphic unit of the Badenian in the Paratethys, also obtained a protected status. These sites represent deposits of both the stratified and megabreccia part of the orogen. They are located within two folds: the middle and the northern, and within neighbouring parts of the megabreccia. These exposures enable observation of a number of geological processes which are typical of the Wieliczka deposit. They include mineralogy and petrology, but also palaeontology and sedimentology, secondary salt crystallization and tectonic processes (Fig. 2). The exposures proposed for protection are in the walls of galleries and chambers located along the following trails:

(1) tourist trail in the historical part of the mine (six educational sites);

Fig. 3. Wall covered with halite crystals (the edges attain 20 cm length); Upper Crystal Cave (photo by Grzybowski).

(2) underground exhibition of the Krakow Salt Works Museum (four educational sites)
(3) the passage from the Kinga Shaft to the Crystal Caves (15 sites useful for scientific goals and high level education).

Other interesting outcrops in the mine are now being selected for conservation purposes.

The Crystal Caves

These are situated about 90 m below the surface, in the northeastern part of the mine (Pawlikowski & Wiewórka 1988). They developed in the zone of crevices at the border between the stratified and the magabreccial deposits (Figs 1 & 3). The Lower Crystal Cave ($706 m^3$) is a natural lenticular hollow, elongated upwards. Its upper walls and top are completely covered with halite crystals, usually about 10 cm cubes, sporadically 20–30 cm long. The Upper Crystal Cave ($1000 m^3$) is in the zone of primary crevices and lenses with partially preserved crystalline linings rich in very large crystals, up to 30 cm long, sporadically 40–50 cm. The Beautiful crystals exhibited in museums in Poland and abroad come mainly from here.

The Crystal Caves were discovered at the end of the last century and since then efforts to protect them have been made. The first regulation for their protection was issued in 1928 (Kreutz 1928; Müller 1928). Unfortunately, neither this nor subsequent regulations (from 1949) were legalized in the Nature Conservation Act. A new, currently prepared project for a nature reserve will soon gain legal status.

The condition of the cave is becoming worse, so visitors are not admitted. Processes of natural degradation are linked with the general condition of this part of the mine and result from the geological and hydrogeological conditions, the subsidence of the bedrock and an inadequate ventilation system. High relative humidity (74–80%) causes corrosion of the salt crystals and constitutes a direct threat to the crystalline nature of the caves. A potential threat is posed by numerous widened crevices which may cause sudden detachment of the crystal lining and undesirable transformation inside the caves. The results of continuous observations and measurements of these damaging factors will be the basis for drawing up a project to save the caves.

A further valuable part of the deposit – crucial for the explanation its origin – will soon be arranged as a nature reserve.

Geological museum

In the Maria Teresa Chamber III (level III of the mine) there is a geological museum. The exhibit contains rocks and minerals from the Miocene and Zechstein salt formations. There are salt crystals created by water saturated with salt and the fossilized remains of plants and animals found in the deposits. There is selected literature exploring the geology of the deposit, and maps, photographs and illustrations show the development of knowledge about the salt depsoits since the end of the eighteenth century. In addition, a geological exhibit is arranged in the renovated Wieliczka Castle of the Kradcow Salt Works Museum.

References

Garlicki, A. 1979. Sedimentation of Miocene Salt in Poland. *Geological Transaction, Polish Academy of Sciences*, **119**, 5–67.

Gawel, A. 1962. The Geological Structure of the Wieliczka Salt Deposits. *Prace Instytutu Geologicznego*, **30**(3), 305–327.

Kolasa, K. & Slączka, A. 1985. Sedimentary salt megabreccia exposed in the Wieliczka mine, Fore-Carpathian Depression. *Acta Geologica Polonica*, **35**(3–4), 221–230.

Kreutz, S. 1928. La Grotte à Cristaux de Wieliczka an tant que première reservation souterraine en Pologne. *Monuments de la Nature Inanimée de la République Polonaise*, **1**, 16–23.

Luczkowska, E. 1978. Holostratotypen der Unterstufen des Bardenian. *In*. Senes, J. (ed.) *Chronostratigraphie una Neostratigypen. Badenian*. Veda, Bratislava, 148–151.

Muller, A. 1928. Sur la découverte d'une Grotte à Cristaux dans les mines de Wieliczka. Situation de la Grotte. *Monuments de la Nature Inanimée de la République Polonaise*, **1**, 24–29.

Niedzwiedzki, J. 1892. *Zur Geologie von Wielizka*. Lwów.

Pawlikowski, M. & Wiewórka, J. 1988. The Crystal Caves in Wieliczka. *Protection of Nature*, **46**, 381–394.

Poborski, J. & Skoczylas-ciszewskae, K. 1963. Miocene in the Zone of the Carpathanian Overthrust in the Area of Wieliczka and Bochnia. *Annales Société Geol. de Pologne*, **33**(3), 339–348.

Reuss, A. E. 1867. Die fossile Fauna der Steinsalzablagerungen von Wielicza in Galizien. *Sitzber, k.k. Akad. Wiss.*, **55**, 1–116. Wien.

Szybist, A. 1975. Geological studies on brecciated part of Wielicka rock-salt deposis. *Przeglad geol.*, **7**, 428–431.

WIEWIÓRKA, J. 1974. Poziom najstarszej soli kamiennej w pokladowym zloźu solnym Wieliczki. *Studia i Mat. Dziejów Zup Solnych w Polsce,* **3**, 46–58.

ZABLOCKI, J. 1930. Tertiäre Glora des Salzlagers von Wieliczka. *Acta Soc. Bot. Polon.,* **7** (2), 139–156.

ZEJSZNER, L. 1844. *Geognostische beschreiburges Salz-Lagers von Wieliczka. Neves Janbuch fur Mineralogie,* Stuttgard.

A consensus approach: industry and geoscience at the National Stone Centre, UK

I. A. THOMAS & J. E. PRENTICE

National Stone Centre, Wirksworth, Derbyshire DE4 4FR, UK

Abstract: The National Stone Centre, based upon an important geological Site of Special Scientific Interest (SSSI), interprets the best teaching example of a limestone 'reef complex' in the UK and is probably now the most visited geological site in the country. Last year, 250 groups (mainly schools) were guided over the site and, in addition, large numbers of individuals used the self-guided trails and visited the indoor 'Story of Stone' exhibition. The aim of the centre is to educate the public in the important role that rocks and minerals play in everyday life. The intention is to promote greater understanding of the facts behind the debate on mineral extraction and the environment.

In geological circles, the argument for geological conservation has few dissenting voices. Outside our profession, whilst there is certainly a greater knowledge of geology now than in the past, very few of the general public have sufficient knowledge of geological dynamics or techniques to be able to appreciate the need for the preservation of geological sites. It is clearly imperative that there should be a wider understanding of geology and its methods, if the efforts of geological conservationists are to be properly understood outside the small circle of Earth scientists. It is against this background that we wish to describe the work of the National Stone Centre.

Essential dilemmas

In the first instance, the National Stone Centre was founded for reasons entirely different from those of geological conservation. The basic premise is a simple one. Very few people are aware of the important role that minerals play in their everyday lives – when told that each person in the British Isles 'consumes' on average 5 tonnes of bulk minerals each year, they are usually extremely surprised. Go on to explain the variety of minerals and their usage – in food, cosmetics, roads, plastics, etc. – and a vista, to which they have never given any thought, emerges. The next step reveals how far the mineral extractive industries contribute to the economy and to the quality of life. However, to get minerals you have to have quarries and mines and while everyone 'knows' that quarries can be dirty, smelly, noisy, no-one wants them in their backyard.

There is clearly a conflict here, between the vital need for minerals on the one hand, and the desire to preserve local amenities on the other.

If we are to resolve this conflict in a rational way, the facts about quarrying and mining need to be known – we need to understand how pressing is the demand for minerals, and how far their extraction can sensibly be tolerated. Part of this equation is the planning and implementation of appropriate after-uses, including of course geological conservation. At the Stone Centre, we take it as our function to provide the information on which there can be a better understanding of the extractive industry, leading to sound decisions by public authorities.

It must be made very clear that we do not see our role as providing a biased viewpoint in favour of the industry. Although we have been extensively funded by the extractive industry, and have industry representatives on our Council of Management, we have never sensed any pressure to present their side of the argument in a more favourable light, or indeed to modify our presentation in any way whatsoever. Equally we have wherever possible, sought a balance of funding from environmental and educational bodies, in order to underpin our ability to present a balanced and independent view.

Beyond this basic philosophy, however, there has been a more general aspect which leads away from, but is always connected to, the extractive industry. We have tried to present, in our exhibition and site trails established so far, and in our planned developments for the future, the interest of stone, in all its aspects. These start with geological origins, and go on through ecological themes of quarry regeneration, to building products and techniques, to architecture and to sculpture. We believe that if we convince the public that stone is an interesting subject in its own right, that there is interest and indeed fascination, in the material itself, its

From O'Halloran, D., Green, C., Harley, M., Stanley, M. & Knill, J. (eds), 1994, *Geological and Landscape Conservation.* Geological Society, London, pp. 423–427.

extraction, its use, and in its transport, then we will have fulfilled our aims.

The Stone Centre does, however, have other constraints which make it different from a normal conservation exercise. We realize that we cannot go on indefinitely relying on subsidies from the extractive and related construction industries and charitable trusts; we need to make ourselves financially independent. We are in no sense competing with our neighbours at Alton Towers; but we need to make the centre pay its way. At the moment, we can achieve this only be increasing our visitor numbers; this can only happen if the centre is attractive – that is to say, it is recreational as well as educational. Moreover, we need to maximize the number of people passing through the centre, if our national educational remit alone is to be fulfilled.

Where does educational conservation fit into this framework? In the first place, by showing that stone is interesting stuff, it will do much to help an understanding of the geologist's place in society. Secondly, by taking our 'geological trail' people see the way the evidence for tropical seas, for ancient reefs and for lagoons, etc. is studied – they see for themselves the way that geological reasoning builds up a picture of the past. By gaining this understanding of geological thinking, they will see the reasons for preserving the rocks in an accessible form. We believe that the Stone Centre has a unique role to play here, which is not filled by any other organization. The theme has by no means been fully developed on the site – we would like to see exhibits which take the visitor on from the geological trail to a display relating this to the need for geological conservation.

Turning to the site as now seen by visitors, the basic geological message is a complex one, largely outside the general experience or knowledge of most visitors. There are also many strands interconnecting with other fields, e.g. ecology, industrial history, the politics of mineral exploitation, all of which have a valid place. Standard checklists for good practice in interpretation (Page 1992) usually indicate that the target audience should first be identified and addressed accordingly. How can this be achieved, when an audience on a typical day is represented by a mix of under-five-year-olds on a birthday treat, oil company trainees, bored Hertfordshire teenagers and a reunion of octogenarian botany graduates? By way of context, it is also perhaps interesting to note that over half the population of England and Wales live within 80 miles (130 km) of the site) one of the factors influencing our choice of location in the first instance.

Towards solutions – the Rupert Bear approach

Is there then a way forward, without compromising scientific integrity or the well-being of the site? The national average reading age is currently the equivalent to that of an 11-year-old. This may be disconcerting for educationalists, but is comforting for those involved in interpretation in that it normalizes the language level required. Furthermore, specialist, even 'academic' visitors in 'leisure mode' appear to find short sentences, short text blocks and a straightforward approach much more digestible. Our tools are the indoor Story of Stone exhibition, in conjunction with site trails. The latter may be self-guided – using interpretative panels or the printed trail leaflets; for groups, it is possible to hire a guide. Where possible, we adopt the 'Rupert Bear' method of interpretation. Rupert Bear is a popular cartoon character – presented in picture form, accompanied by two-line captions, followed by a longer text at the foot of each page. Thus, the principle involves the object, i.e. an illustration or the view itself, a brief explanation and, if applicable, a longer text. This can be supplemented by paper guides or more expensive books as time, finance and subject matter permit.

After consideration in detail of the options available to us, we have adopted a basic theme which, using different language levels, can be adjusted to meet the needs of probably 90% of our visitors. For the remaining 10%, where feasible, we use guides with the specialist subject knowledge or awareness required, be they gardeners, archaeologists, quarry company trainees, sedimentologists or visitors facing particular learning difficulties. Our fundamental geological message rightly capitalizes on the features uniquely displayed at the site. The general geological theme, therefore, may be summarized as:

Fossil corals, sharks' teeth, large brachiopods and crinoids may be found on site. By analogy with modern environments, they constitute evidence of the past presence of tropical seas. By further extension, geologists can build up a detailed picture of the palaeoecology of the area – mainly filter-feeding animals in a complex of shallow lagoons (carbonate ramps), reef mounds (carbonate and build-ups) and off-shore submarine fringing screes. Lurking sharks and ash from a local volcano added to the excitement of this idyllic scene. Nearby, subsequent Namurian mudstones and sandstones were generated by a large fluviodeltaic system, which overwhelmed this hitherto attractive seascape (Thomas & Hughes 1993).

Building upon this central theme it is possible to tease out various elements, for example, how did this all happen when the site is now 78 miles inland, 200 m above the sea and with a climate which is rarely tropical? Cross reference can then be made to the issue of plate tectonics which is addressed in the Story of Stone exhibition at the Discovery Centre in the section dealing with geological processes – a stamp on the football being a metaphor for the crust on the Earth. Indeed, analogies are found to be the best means of portraying such messages – 'plate tectonics is rather like the skin on school custard'; 'Europe and America are parting at the rate at which our fingernails grow'. Another cross-subject link begins by looking at a dry stone wall – the effect of centuries of acid rain has etched out the fossils in the limestone. There is of course an opportunity here for a subsidiary conservation message – 'walls are one of the best places to find fossils but not the place to collect them'. Other uses of stone include lime burning (there is a small adjacent lime kiln and an allied message regarding pollution); lime in turn can be used in making mortar but the largest use of stone is in roadmaking so let us look at the road. The latter is made up of crushed igneous rocks, for which there are also related messages – Why not use local limestone? – How are igneous rocks formed? Where are they mainly found? – Hard rocks make good mountains and safer non-slip roads. From this a further theme can be developed relating to the conflict between mineral working and the environment – How can this conflict be resolved.

On the ground there are, of course, a number of potential conflicts between promoting a site for visitors and pure geological conservation – safety considerations may require fencing or declaration of 'no-go' areas, e.g. for research on a permit/indemnity basis. Interpretative panels require careful siting, so that they do not unduly detract from key viewpoints and can be viewed by the number of people anticipated at each stop. The ramifications of collecting policies and hammering are considerable – by way of compromise we permit limited collecting of material not *in situ* (subject to the centre having a prior claim on specimens) and we strongly discourage hammering for safety and conservation reasons. Pre-broken material is placed in visitor areas on the site for visitors to investigate further. However, knowing the centre's policies, on one occasion, an ardent volunteer, seeing a random large and fossiliferous block, decided to break it up for general distribution to visitors! He was completely unaware that a number of his geological colleagues who also guided parties, relied upon the same rock to demonstrate palaeoecological features. The search is on for a replacement specimen.

Developing consumer choice

The importance of canvassing visitors, e.g. about concerns, interests or misunderstandings, cannot be underestimated. This might include testing new trails, seeking visitor feedback, monitoring a visitors' book and questionnaire or interview surveys. Informal communication may be equally as valid in gauging whether or not people are enjoying their visit, are better informed on leaving, have changed their percep-

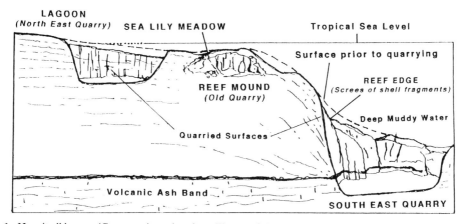

Fig. 1. How it all began. (Cross-section taken from Thomas & Hughes 1993.)

tions as a result etc. In an effort to inform the future planning of the centre we are co-operating with a researcher who is surveying many of these aspects. His initial findings are separately reported in this volume (Hose 1994). A related feature is is the Service Quality Audit. Annually we aim to assess through the eyes of a non-specialist volunteer new to the centre, aspects (including conservation issues) of the site which could be improved. The resulting confidential report is then used as an input to the maintenance and site works programme and the longterm planning of the centre. Interestingly, many of the inherent problems of portraying sound science to a lay audience are highlighted in these reports. As an example of our response, we are about to launch three new trail leaflets covering geology, history and ecology, the latter in particular following popular demand and comment. An experimental trail leaflet on 'Chemistry and the Environment' and publications for schools on setting up their own 'Geotrails' and on the use of geology in everyday life will follow. At a national level we are in the middle of a three-year programme, part-sponsored by the Department of the Environment, to produce information packs on the highly topical debate – Quarrying and the Environment. The first will be geared to the general reader, the second two to schools and, in particular, to national curriculum requirements.

In another field we are currently pioneering a scheme whereby quarrying companies, which have already built up a relationship with a school, are encouraged to part-sponsor class visits to the centre to provide a national context to reinforce their local experience of the industry. In researching the scheme, we specifically sought the views of schools on the ethics of industrial sponsorship – unanimously they considered that there would be no problems in this respect.

On a 'multi-disciplinary' site there are, of course, inherent potential conflicts between particular elements; even during the 1980s when the centre was being planned, the considerable development of vegetation began to obscure key geological sections. With growing interest in nature conservation, there has been some resistance to removal of this growth and the introduction of visitors, which we consider necessary in the name of geological enlightenment.

The future

What of the future? An exciting and ambitious 10-year plan for the centre to be launched shortly has, as an integral component, a strong geological conservation philosophy. The main commercial elements (the Geopark – a national trade and training centre for stone) are outside the designated Site of Special Scientific Interest (SSSI), the latter being devoted primarily to interpretation. Where development is envisaged in the six former quarries which constitute the SSSI, it is low key and compatible, e.g. stone masons' workshops, historic quarry plant and railways or gardens, and will not be allowed to compromise geological integrity. Of particular relevance outside the SSSI, is a proposal now in its initial stages, to develop the first major hands-on Earth science centre. The many science centres now operating in Britain are largely physics-based. Chemistry is probably only covered in depth at two centres (experiments usually being difficult to replicate and control without consuming material–; the biological sciences are of course admirably and widely served by nature reserves, field centres and visitor centres. In contrast, Earth science is clearly the poor relation with virtually no coverage. We intend to build further on our virtually unique capacity to blend 'indoor theory' with abundant field opportunities within metres of the building. Indeed, within 5 miles (8 km) of the site are excellent examples of vulcanicity, structural geology, mineralization, and carbonate, arid, paralic, deltaic and glacial sedimentation. The importance of this proposal to education is heightened by the inclusion of aspects of Earth science in the recent National Science Curriculum.

In the planning process therefore, we aim to respond to changing needs. In the geological field this is perhaps best exemplified by the comment of a 7– or 8-year-old visitor; when asked by a volunteer helper 'Do you like fossils lad?' The boy replied that he did, 'but was more interested in plate tectonics!' – a subject which is now being routinely referred to in our Junior Schools (to 9–11-year-olds) but which had not even emerged as a widely accepted sub-science in higher academic circles until the early 1970s.

Conclusion

The National Stone Centre, although not dedicated primarily to geological conservation, nevertheless furthers its aim in a variety of ways. By introducing a wide spectrum of the public, from primary school age to 'senior citizens', to the methodology of geological interpretation, it demonstrates the need for conservation of important geological sites. Because the site is traversed almost every day of the year by expert staff, the SSSI is given a degree of custodianship inconceivable in most geological sites. The extensive plans for future developments, although based on a variety of interests and studies, will serve to bring more people, with increasing diversity of interest, on to the site, and thus continue to spread conservation issues through an ever-widening circle.

References

HOSE, T. 1994. Telling the story of stone: assessing the client base. This volume.

PAGE, K. N. 1992. *Site Information Boards for Geological and Geomorpholigcal SSSI's*. Report by English Nature, Peterborough.

THOMAS, I. A. & HUGHES, K. 1993. Reconstructing ancient environments. *Teaching Earth Science*, **18**, 17–19.

Interpreting Earth science sites for the public

TIM BADMAN

Centre for Environmental Interpretation, The Manchester Metropolitan University, St Augustines, Lower Chatham Street, Manchester M15 6BY, UK
(Present address: Hampshire County Council, Countryside and Community Department, Mottisfont Court, High Street, Winchester SO23 8ZF, UK)

Abstract: Environmental interpretation is 'the art of explaining the meaning and significance of places of natural or historic importance to the people who visit them'. Over the last fifteen years, it has become a widely recognized and valuable part of countryside management and conservation in the UK, and it is continually being developed.

Environmental interpretation potentially offers one of the most powerful and effective means of raising public awareness of, and support for, Earth science conservation – by explaining the stores of geological sites in ways which are enjoyable, stimulating and informative. Work on interpretation of Earth science in the UK is in the early stages of development and its potential deserves to be more widely recognized.

This paper outlines a set of principles for the successful interpretation of Earth science, covering the selection of appropriate sites, identification of audiences, selection of themes, use of interpretative techniques and development of partnerships. We draw on examples of good practice from the UK and abroad, and on recent CEI work for English Nature.

This paper gives an introduction to the concept of environmental interpretation and sets out some key points of an approach to its application to Earth science sites.

What is environmental interpretation?

Environmental interpretation has become a recognized term to describe a range of activities carried out by managers of countryside and heritage sites. It can be defined as:

'the art of explaining the meaning and significance of sites visited by the public.'

Within this context three key elements are required:

(1) a specific site of natural, historic or cultural value or interest, which is being, or will be experienced at first hand by visitors;
(2) a visiting public, whether tourists, day visitors or local residents, who are making a recreational or educational visit;
(3) an interpreting organization or individual which aims to generate a concern for the site's conservation and/or an understanding of the processes and activities taking place (Barrow 1993).

In essence, what interpretation boils down to is the people who manage a site sharing their enthusiasm for its importance with the people who come to visit it.

The key aspects involved in developing effective site interpretation can be stated as follows:

- identification of sites where there are opportunities for interpretation;
- developing a clear message for each site which communicates what is particularly special or significant about it;
- identifying the audience(s) who visit the site, and producing interpretation which is aimed at communicating the message to them;
- imaginative use of the range of interpretative techniques available, related to accepted standards of good practice.

The value of interpretation to conservation and education about the environment is already widely recognized amongst conservationists and site managers in the UK. Nature conservationists have been interpreting since the early 1960s, when the first nature trails appeared. Such interpretation has helped raise the general level of awareness of nature in the countryside and strengthened the ecological conservation lobby.

Civic societies throughout the country have produced trails and tours through historic towns. Industrial archaeology groups have also taken up interpretation as a means of sharing their enthusiasm for derelict industrial landscapes, and of promoting their conservation. Most recently archaeologists, who were initially slow to get involved in interpreting their sites have realized its value, and have begun to enable the public to see beyond what often looked like muddy building sites, and to understand their significance and the reasons for the work that was put into excavating them.

Relatively little attention has been paid to interpretation by the Earth science conservationists, although it now seems a number of organizations and individuals are beginning to take an interest. Good examples of Earth science interpretation can be found in some other countries, for example on sites in parts of France and the USA.

Interpretation offers an effective method of increasing visitors' understanding and enjoyment of Earth science sites, and potentially their general attitudes towards Earth science. It also offers opportunities to influence their actions on site, and thereby contribute to site conservation. Essentially the idea is that if visitors enjoy and understand their visit, they are likely to value the site more, and support its conservation.

Points for action

Consideration of interpretation should be part of any scheme for Earth science site conservation. An effective approach to Earth science interpretation requires action on five key points.

1. Putting people on the agenda

A positive view is required towards providing for the visitors to Earth science sites.

We can be confident that a wide public will be interested in Earth science if it is explained in ways that are exciting and accessible to them. Certain aspects such as volcanoes and earthquakes, jewels, gemstones and precious metals never fail to capture their imagination. With Jurassic Park mania in full swing it is hardly necessary to mention dinosaurs and fossils. Whilst acknowledging that public interest in Earth science may not be on a par with other spheres of natural science at present, we can look to their success in engendering public awareness and support, and be confident we can make a similar impact.

2. Finding the right sites

This is not to say that we should be trying to interpret every geological or geomorphological site we encounter. Only a certain number of sites will lend themselves to effective interpretation and it is essential to identify and prioritize these. The following criteria are suggested as a basis for site selection (Centre for Environmental Interpretation 1992)

(a) Access
- Is the site readily and safely accessible to a suitable range of users?
- What are the visitor flow patterns on site – where do people go, and what do they do?

(b) Usage/audience
- Is there a clearly defined audience (or audiences) for interpretation? What are their needs or interests? What will they be able to understand?
- Is the scale of interpretation proposed appropriate to the level and type of use of the site.

(c) Site management
- Is the site well managed and maintained?
- Are there any potential management problems for interpretation (e.g. vandalism, erosion)?
- What are the objectives and wishes of the managing organization?
- Is the interpretation proposed right for the landscape setting of the site, and appropriate to other site management objectives?
- Does interpretation provision already exist on the site?
- Is it possible to link interpretation on the site with other facilities in the area? Is it possible to work in partnership with others?

(d) Geological interest
- Are the geological features and exposures readily visible and comprehensible to the public? Will the audience find them interesting and understandable?
- Is there a strong story to tell on the site? Are the features of the site particularly unique or special?

Visitor usage of a site should not be seen as in conflict with the conservation of its Earth Science interest. Surveys by English Nature have shown that damage by visitors to Earth Science sites is rare. A well planned programme of interpretation will bring benefits to the visitor management of our sites – it is an extra tool to help us in this work. It may also be a way to influence and educate new collectors.

3. Making the links

To ensure that the public will be interested in interpretation about geology it must demonstrate relevance to their own lives. This gives a basis for developing an explanation of the Earth science interest of the site.

The links between geology and people are there to be demonstrated, and the following checklist summarizes some of the most obvious.

Resources: The Earth is the source of our fundamental resources including minerals, coal, oil, gas and aggregates, and the foundation for soils and agriculture. Economic use of rocks has been made by humans for centuries.

Scenery and landscape: Geology provides the bedrock for our countryside, which has been shaped by geomorphological processes. The underlying geology is a key influence on past settlement, archaeology and traditional building materials.

Local themes: Geology has underpinned the development of local communities, providing building stones, determining the landscape, and giving the core to local economies reliant on mining and quarrying.

Historical/cultural themes: Geology has a long history, populated by interesting individuals and personalities. It is also the source of religious and poetic associations. W. H. Auden and Ted Hughes are amongst the British poets who have found inspiration in geological and geomorphological themes.

Magic and mystery: Earth science is full of mystery, and a rich source of stores of 'lands that time forgot' and magical association – particularly with caves and mountains.

International links: The continuity of rocks shows the past connections between now separate countries, and emphasizes the interrelatedness of the world. We can highlight the changing nature of the shape and location of modern lands through time.

Geological stories: Geological stories are fascinating in their own right – we have only to bring them alive! Telling these stories well means above all focusing on processes, emphasizing the dynamism of Earth science. We have to help people get inside geology. Recreating the appearance of past landscapes, and environments. Putting the movement into structural geology and metamorphism. Focusing on fossils as living animals. Looking at active process sites such as rivers, coasts, sand dunes. At all times emphasizing change and development, we can relate the past to the present and speculate about the future.

4. *Using our imagination*

There are a wide range of accepted methods of delivering effective interpretation, whose use is well established. A little imagination is all that is required to adapt them for interpreting Earth science.

A checklist of the principal techniques is as follows:

1. Guided walks and demonstration: Possibly the best way of finding out about Earth science is to go out with a skilled and enthusiastic guide, and look for yourself. There is no substitute for the personal contact of a talented interpreter. The quality of the interpretation rests on the abilities of the individual guide.

2. Interpretative panels: The opportunities for well designed and simple on-site panels are very significant. They do need to be sensitively located on sites which have been carefully selected. They also need to be well written using accessible language, and to avoid trying to say too much more – a limit of 200 words in total is a useful guide.

3. Publications: Landscape/nature trails are well tried and successful. Again design is crucial, as is ensuring that the finished product is easily available to the public or suitable local outlets. Themed leaflets can also be effective if clearly and well written. The challenge for Earth science interpreters here is to come up with exciting themes and effective and simple explanatory illustrations.

4. Permanent displays, audio visual, visitor centres: On a grander scale, it is possible to produce exciting and elaborate displays in visitor centres. These can be relatively expensive, but it is also possible to achieve very good results using simpler designs and graphics. Displays and exhibitions can be particularly effective if they are interactive and appeal to all the senses.

5. Travelling exhibitions: Travelling exhibitions are also effective, and can be useful in taking Earth science to the local community in schools and village halls.

6. Arts: Engaging audiences intellectually is only one way of going about our interpretation, and the use of performance and visual arts can sometimes be even more effective. Specially devised environmental theatre events are increasingly common, and there is no reason why these should not tackle Earth science themes. Sculpture in appropriate rock types can be an effective way of emphasizing the sense of place related to the local geology, whilst producing objects of beauty.

5. Developing partnerships

Finally, what is needed is the development of partnerships between Earth scientists and those involved in interpretation in order to develop interpretative projects. Countryside management organizations are probably the most important group for Earth scientists to be in touch with. Government conservation agencies, private landowners industry, as well as specialist advisors are also key contacts.

Earth science interpretation needs to be of high quality and Earth scientists should not be shy about making these contacts. Links should be particularly encouraged if they are developed through local geological groups or universities and colleges. Such partnerships may well be the way of combining the enthusiasm and knowledge of Earth scientists for their sites, with the communication and interpretative skills of countryside managers. They may also be the way to make the best use of resources and provide a basis for funding approaches.

There can be no doubt that Earth science as a whole would benefit from positive attempts to explain the subject to, and capture the imagination of, the public. Effective interpretative projects provide a way of addressing this issue. Above all what is needed is the will and the imagination to grasp the opportunities and to campaign for the resources to realize the ideas which are generated.

The author gratefully acknowledges the input of advice and ideas from colleagues at the Centre for Environmental Interpretation and staff at English Nature into the above paper.

References

BARROW, G. C. 1993. Environmental Interpretation and Conservation in Britain. *In*; GOLDSMITH, F. B. & WARREN, A. (eds) *Conservation in Progress*. John Wiley and Sons Ltd, Chichester.

CENTRE FOR ENVIRONMENTAL INTERPRETATION 1992. *Geological SSSI Interpretation Pilot Scheme Assessment*. Report to English Nature, Peterborough.

Further information

For further information about CEI's work on Earth science interpretation please contact:

- Gillian Binks and Gordon Rankmore, Centre for Environmental Interpretation, The Manchester Metropolitan University, St Augustines, Lower Chatham Street, Manchester M15 6BY, UK. Tel: 061 247 1067.

- James Carter, Centre for Environmental Interpretation (Scottish Office), University of Edinburgh, 20 Chambers Street, Edinburgh EH1 1JZ, UK. Tel: 031 650 8017.

CEI E-mail: (GeoNet) MCR1: CEI-EDINBURGH

Information signs for geological and geomorphological sites: basic principles

K. N. PAGE

English Nature, Northminster House, Peterborough, PE1 1UA, UK

Abstract: Earth processes and the rocks they create underpin and sustain all natural systems, yet despite this, the public profile and understanding of geology and geomorphology is appallingly low. Nevertheless, vast numbers of people are drawn to Earth science sites, such as coastlines, upland areas and urban open spaces in old mineral workings, without ever having picked up a geological textbook in their lives.

Once on a site, conservationists and Earth scientists have a golden opportunity to grasp the general public's attention and introduce them to Earth science and Earth science conservation. The medium of permanently mounted site interpretation panels is therefore a powerful tool in this respect, but should be supported by other techniques such as publications and by more general public relations strategies. Public interest in and support for Earth science issues is essential for effective conservation and indeed for maintaining funding of pure scientific research in Earth sciences.

The philosophy and principles behind the process of design and installation of site information signs at Earth science sites are discussed and three basic categories of sign presented, namely: (1) standardized site management signs; (2) site specific information plaques; and (3) interpretative site information boards.

Earth processes and the rocks they create are an integral part of all natural systems. Yet, despite this, the public profile and understanding of geology and geomorphology is appallingly low. 'Rocks are boring' is, as noted by Wray (1991, p. 27) an all too common cry. Few people realize their dependence on the geology of the planet that they live on. Geology nourishes them via the medium of the soil, shelters them in buildings constructed of its materials, and they use artifacts manufactured from processed minerals without ever directly acknowledging a geological source. They wonder at the beauty of mountains and seashores without appreciating the complexity of geological and geomorphological processes involved in landscape formation. Nevertheless, vast numbers of people are drawn to Earth science sites such as these – they may never have picked up a textbook in their lives, but they unknowingly and regularly make visits to geological localities. Earth scientists and conservation organizations then have a golden opportunity to grab the attention of the visitors and introduce them to Earth science and Earth science conservation. The medium of permanently mounted site-information boards is a powerful tool in this respect, but should always be supported by other techniques, such as publications, and by more general public relations strategies.

There are, of course, a small, but occasionally significant group of people who visit sites fully aware of any Earth science interest. The aims of these visitors may be educational (schools, colleges), scientific (researchers), economic (industrial geologists), recreational (amateur geologists) or commercial (specimen dealers). Members of these groups will have varying degrees of background knowledge and differing demands on a site. Informing these visitors of the nature and especially the conservation status of the site is often important, and some form of interpretation, suitably targeted, will often be very well received.

Nevertheless, certain activities, if unrestricted, can create problems for site conservation and some form of management instruction may be essential to aid visitor control. Signs erected primarily for management purposes need not necessarily include site interpretation, as geologically aware visitors are likely to know at least some aspects of the localities' interest. Certain types of visitor-induced management problems are widespread, for instance unauthorized entry onto private land or site damage by unscrupulous specimen collectors, and standardized signs or instructions may be an appropriate tool to aid conservation.

Site information signs have the additional value of beneficially raising the public profile of the implementing organisation or sponsor by demonstrating a commitment to environmental and heritage issues.

From O'Halloran, D., Green, C., Harley, M., Stanley, M. & Knill, J. (eds), 1994, Geological and Landscape Conservation. Geological Society, London, pp. 433–437.

The function of signs at Earth science sites

Signs and signboards have three basic functions at sites of importance for nature conservation (Page 1992b):

(1) to inform visitors of the conservation status of the site (i.e. legal status of protection, if appropriate);
(2) to control or manage visitors and therefore aid site conservation;
(3) to establish the role of the local or national organization responsible for the recognition, protection and/or management of the site.

Interpretative signboards, however, have a number of additional functions, as discussed by authors such as Aldridge (1975):

(4) enhancing visitor enjoyment in the belief that an understanding of the countryside increases the pleasure derived from visiting it;
(5) increasing the public understanding and appreciation of the countryside leading to a respect for it and an awareness of the need for its conservation;
(6) satisfying a visitor demand for information.

Prioritizing site signing schemes

A case could be made for identifying all important Earth science sites with some form of on-site sign, but the costs of such an operation could well be prohibitive for the initiating body. It is, therefore, necessary to prioritize which sites could or should be signed. Generally such sites will be:

(a) sites visible from, or within, tourist areas, viewing points or otherwise frequented by significant numbers of visitors;
(b) accessible sites in urban or densely populated areas;
(c) prime teaching sites; or
(d) sites which may be damaged or threatened by over-use, bad practice, ignorance or trespass.

Page 1992b

In certain areas, however, especially regions of high landscape value belonging to category (a), the very presence of signs and boards may interfere with the naturalness of the area and consequently actually decrease peoples enjoyment. In such cases it is generally considered that signing is not appropriate, except in very localized areas where visitors gather (car parks, information centres, etc).

Types of sign

Signs fall basically into two categories. Firstly, management signs, which do not include any form of interpretation and, secondly, interpretative signs which are designed to inform or educate visitors about the nature of the site.

Management signs are intended to simply inform visitors of the status of a site and will usually include some information or instruction which is intended to aid visitor management. Such signs are relatively simple and are therefore comparatively inexpensive to produce. These signs fall into two categories (Page 1992b):

(1) *Site-specific information plaques*, which include (i) the site name, (ii) a very brief statement of the interest of the site which may be tailored to the site or generalized (i.e. referring to a general geological or geomorphological interest), and (iii) visitor management information, if appropriate (e.g. 'please do not hammer the exposure'). In addition, a contact address for the implementing organization, or any other appropriate organization, is advisable as it provides a mechanism by which interested visitors can obtain more information.
(2) *Standardized site management signs* include a basic minimum of information which is not tailored to any specific site. Such signs are intended to be used as tools for site management and can include instructions appropriate for dealing with problems common to significant numbers of protected localities. Standardization considerably reduces the unit cost of sign production as larger print runs are possible. Ideally all management signs should include (i) a justification for the instructions (i.e. 'This site is protected for its geological [or geomorphological] features'), (ii) visitor management instructions (e.g. 'No tipping') and (iii) a contact address for additional information.

Interpretative signs are by their very nature more expensive to produce but their aim should be to involve a visitor with the site's importance, rather than simply instruct.

Interpreting geological sites

Interpretation is '... an educational activity which aims to reveal meanings and relationships through the use of original objects by first hand experience and by illustrative media, rather than simply to communicate factual information' (Tilden 1967). Environmental interpretation is therefore '... the act of explaining or revealing the character of an area through interrelationships between rocks, soil, plants and man to ... visitors in the field, with preparation and follow up usually in thematic or story form, to increase visitor awareness of the significance of the site visited and the desire to conserve it' (Aldridge 1975).

There a number of basic principles for interpretation which are widely applicable and quite relevant in the context of an earth science site. The following are taken from a useful Countryside Commission publication (1977/1987):

(a) 'Interpretation provides more than factual information, it should explain features and provoke a response.'
(b) 'Interpretation should present a complete picture, emphasising relationships between the parts.'
(c) 'The interpretive message is more likely to be understood if developed as a theme or story, rather than disjointed facts.'
(d) 'To achieve maximum understanding, interpretation should relate what is being displayed or described to something within the personality [or experience] or the visitor.'
(e) 'Visitors take part in interpretation from choice – it therefore must be enjoyable or they will switch off.'
(f) 'Interpretive programmes should be geared to specific age and interest groups.'
(g) 'Visitors are likely to gain a greater appreciation of the story if encouraged to interact' [models and other interactive objects may be inappropriate for site information boards, but encouragement to look at, or even pick up geological materials will promote visitor interaction].
(h) 'Subject specialists are not necessarily expert communicators – particularly to the general public and vice-versa. Therefore the preparation of an inter-active programme is best achieved by a combined exercise between subject and communications specialists.'
(i) 'Some sites do not need interpreting at all. They speak for themselves and to place a communicative medium between the feature and the visitor could diminish or destroy the experience.'

Site information boards are not, of course, the only technique available for interpreting Earth science sites. Many other techniques are available, and under certain circumstances (e.g. risk of vandalism, visual intrusion in areas of high landscape value, etc) certain methods may be more suitable. Examples include publications (leaflets and guide books), static displays and expeditions (in visitor centres and museums), guided tours and lectures, publicity in newspapers and magazines (advertisements, articles, etc.).

When planning an interpretative scheme at a site (having established that such a technique is appropriate and worthwhile there), a number of questions need answering and considerations built into the scheme (Aldridge 1975; Countryside Commission 1977; Page 1992b):

(1) Any scheme requires objectives, the basic question being: why provide interpretative facilities?
(2) What features could be interpreted? A number of themes and stories can be developed at most sites, but those actually employing an interpretative scheme should fit the objectives for the scheme more closely. It is not necessary to tell 'the complete story', which could actually confuse and deter visitors by its complexity, but simply to select one or two simple themes which can be told in a straightforward fashion with a minimum of text and simple diagrams.
(3) For whom are you interpreting? What sort of audience visits the site or what sort of audience would you like to visit the site? What does the audience itself want from the interpretation? These are important questions as the number, characteristics and needs of visitors will influence the depth of interpretation, its exact location and subsequent management.
(4) How should the site be interpreted? Is construction of an interpretation board really appropriate, or would some other interpretative technique (as listed above) be more appropriate?
(5) What, if any, other interpretative schemes exist in the immediate area? There is little more distracting and confusing to a visitor than a plethora of different signs and boards conveying different messages or instructions. If signs already exist on a site (for instance interpreting archaeology), it could be most appropriate to employ some other technique (such as leaflets supporting a self-guided geological trail). Alternatively, the new scheme could be incorporated with, or refer to, pre-existing facilities to provide an integrated approach to multidisciplinary interpretations of the site – in this case unification of design style and target

audience level is often advisable as this will emphasize interrelationships.
(6) What subsequent management is necessary? Implementing a site interpretation scheme is not the end of the process. The facility will need some form of monitoring to assess its effectiveness and, if necessary, the resources to repair or replace the board should damage occur.
(7) Are sufficient financial and staff/volunteer resources available to successfully implement the proposed scheme and initiate some level of management? Interpretation is an expensive process, if resources are not sufficient to design, produce and erect a suitable board, it may be necessary to consider an alternative, less expensive technique, for example a self-guided trail.

Implementing the signing scheme

Whatever level of signing is considered appropriate, from standard management sign to fully developed interpretation boards, there still remains the problems of production. Design should be a partnership between scientists (as experts on the site), environmental interpreters (as 'translators' taking the science and adapting it for the appropriate objectives and target audience) and professional designers (who will produce suitable artwork for the final sign itself). A fundamental part of the implementation phase is the selection of suitable materials to produce and mount the sign. Such materials should be chosen very carefully so as to:

(i) ease installation;
(ii) resist weathering and 'normal' wear and tear;
(iii) resist (as far as possible!) vandalism;
(iv) be expendable (i.e. to facilitate ease of maintenance or replacement); and
(v) conform to the budget available (without prejudicing installation and durability).

Allwood 1981

Materials suitable for signs and their mountings are many and varied but generally signs are printed on high impact plastic sheets or adhesive sheets which are then mounted on to a hardwood or aluminium backing. The mount, in turn, may form part of a self-supporting structure or be attached to or embedded into another structure such as a wall or cairn. Detailed descriptions of the range of available materials and production methods are provided by Allwood (1981) together with useful assessments of relative cost and durability. A brief summary of selected materials is also provided by Page (1992b), together with a detailed step-by-step procedural guide to the production and installation of interpretative boards at sites of Earth science importance.

Earth science interpretation in action

Despite the great potential that Earth science sites have for interpretation, very few signboarding schemes have so far been developed in Britain. The great stumbling block seems to be a lack of awareness by many local or national governmental, environmental and conservation organizations of any aspect of the sciences of geology and geomorphology. A few notable examples do exist in France, however, including

Fig. 1. An example of 'good practice' for site interpretation: Hunstanton Cliffs SSSI. The signboard interprets what the visitor can actually see, with the minimum of (non-technical) text and is designed to be 'attractive' to passers-by and not to interrupt the spectacular view. This is one of a series of signs designed by the Centre for Environmental Interpretation at the Metropolitan University of Manchester in conjunction with English Nature to promote Earth science and Earth science conservation at well-visited geological sites throughout England. Photo: C. D. Prosser.

the schemes in place at Digne and Thouars geological reserves in France (Stanley (1992) and Page (1992a) respectively). In these examples there is community involvement in the management of the sites, an interest and enthusiasm having been fostered by the local geological community.

In Britain, examples include Brownend Quarry SSSI in Staffordshire, where a project interpretation project was locally initiated, and with support from English Nature's predecessor (the Nature Conservancy Council (NCC)), the Staffordshire Wildlife Trust created a geological trail with fully developed on-site interpretation boards. Another British scheme at Upper Teesdale National Nature Reserve (NNR) however, formed part of a nationwide programme to provide interpretation for visitors to NNRs, that is reserves administered by English Nature (and previously by the NCC). Few other examples of interpretation of Earth science sites presently exist in Britain, but as part of an initiative by English Nature, several have been planned and two implemented – high-profile sites with high visitors numbers such as the coastal resorts of Scarborough and Hunstanton have been targeted (Fig. 1). The implemented schemes will have a high impact due to the large numbers of tourists frequenting the sites and some of these visitors will return home, having learnt a small but fundamental message – Earth science is part of our heritage, Earth science is interesting, Earth science is worth supporting.

References

ALDRIDGE, D. 1975. *Principles of countryside interpretation and interpretive planning. Guide to countryside interpretation* Part 1. HMSO.

ALLWOOD, J. 1981. *Information signs for the countryside. A guide to their production.* Countryside Commission, Cheltenham.

COUNTRYSIDE COMMISSION. 1977 (reprinted 1987). *Interpretive planning.* Countryside Commission advisory series, No. 2.

NATURE CONSERVANCY COUNCIL. 1990. *Earth science conservation in Great Britain – a strategy.* Nature Conservancy Council.

PAGE, K. N. 1992a. The French Jurassic connection. *Earth Science Conservation,* **30**, 11–12.

—— 1992b. *Site information boards for geological and geomorphological SSSIs.* English Nature Research Report 24.

STANLEY, M. 1992. The Digne Symposium – a personal view. *Earth Science Conservation,* **30**, 5–9.

TILDEN, F. 1967. *Interpreting our heritage.* University of North Carolina Press, Chapel Hill.

WRAY, J. 1991. Interpreting geological sites. *Earth Science Conservation,* **29**, 27–29.

Mineral collectors as conservationists

B. YOUNG

*British Geological Survey, Windsor Court, Windsor Terrace,
Newcastle upon Tyne NE2 4HB, UK
(Present address: BGS, Murchison House, West Mains Road,
Edinburgh EH9 3LA, UK)*

Abstract: Like most field-based sciences, mineralogy has derived great benefit from the activities of collectors, often amateur collectors. This remains true today. Indeed, field mineralogy is one of the very few scientific pursuits still to offer a true role to the amateur. An unprecedented growth in field mineralogy in recent years has been led almost entirely by amateurs. Mineral collecting in Britain is today probably more popular even than in the heyday of the Victorian collectors. However, the modern collector differs from his or her nineteenth century predecessor in two important respects. First today's collector is likely to have access to a wide range of sophisticated modern determinative techniques which enable many, often very significant, discoveries to be made. Second, the amount of raw material available in the field today is much less and is diminishing rapidly. These two connected factors are of vital importance in mineralogical conservation. More than ever before, sound curation, including accurate documentation, and conservation of collected material is vital. There is a need to cultivate a responsible approach to the wise use of scarce resources in the field. Donation of good representative and well-documented material to public collections, both local and national, will ensure preservation of material. Providing records and field observations to a local or national site documentation scheme and, when appropriate, publishing of findings in a suitable scientific journal, ensures that Earth science benefits fully from the results of much good field work. A number of collectors already well appreciate their responsibilities and follow some, or perhaps all, of these avenues. However, for whatever reasons, many remain to be persuaded of what must surely be seen as an obligation incumbent upon all who collect from Britain's rich but diminishing stock of mineralogical material.

Many of the raw materials which form the basis of civilization are minerals. Some understanding of minerals has therefore been essential since the beginning of human history. Initially an appreciation of which minerals or rocks could be used as tools or weapons sufficed. With the discovery of metals came the need to recognize essential ore and other minerals. The demands of increasingly sophisticated technology have developed the initially simple ability to recognize useful minerals into the complex science of modern mineralogy. It would be easy, therefore, to see mineralogy as an essentially utilitarian science serving the needs of industry and technology. It is clearly more than this.

From the earliest days of civilization humans have been fascinated by minerals not just as raw materials but as objects of interest and beauty. The use of colourful and durable rocks and minerals for personal adornment and for making objects of religious significance is perhaps as old as humankind itself.

As part of the natural world minerals, like plants and animals, excite curiosity and invite study in their own right. In common with the development of other branches of systematic natural history the emergence of mineralogy as a science has been possible only through the efforts of enthusiastic and dedicated collectors, many of whom, at least initially, had no particular motivation other than an almost instinctive urge to collect. The mineral collector has thus had a fundamental role in providing the material basis from which scientific mineralogy has evolved. Whereas the past importance of the mineral collector is thus undeniable, the value of his or her activities today are commonly underappreciated.

The mineral collector

The origins of mineral collecting and systematic mineralogy are clearly rooted in antiquity. Several Greek writers made reference to minerals and contemporary theories on their origin. Most notable of these was Theophrastus (372–287 BC) to whom we owe perhaps the earliest written works devoted to what today would be regarded as mineralogy. Four centuries later the Roman writer Pliny (Caius Pliny Secundus, AD 23–79) incorporated the mineralogical thought of his time into his

Natural History. The handful of texts on minerals published over the succeeding fifteen centuries contain little more than folklore with few useful facts. It was the publication in 1556 of *De Re Metallica* by the German physician Georgius Agricola which may be taken to mark the emergence of mineralogy as a science. The seventeenth and eighteenth centuries witnessed a continued growth in the science with the publication of important texts and the assembling of systematic collections, some of which survive in whole or in part. It was however, the nineteenth century which saw the remarkable growth of all aspects of natural history collecting, including minerals. The Victorian naturalist collectors are commonly credited with this upsurge of interest and there is no doubt that it is to them that many of the western world's great natural history, including mineral, collections owe their origin or their greatest period of growth.

Most collectors were amateurs in the sense of not deriving their principal income either from collecting or from the study of the collected objects. Such was the popularity of mineral collecting that a significant number of collectors became dealers supplying institutions and private collectors with specimens, often at high prices. Almost every major museum collection in the world today owes much to the activity of these early collectors and dealers.

The past few decades have seen a remarkable resurgence of interest in field mineralogy both in Europe and the USA. Mineral collecting is today probably more popular and followed by more people even than in the heyday of the Victorian collectors. The trade in mineral specimens continues with ever larger sums changing hands for a diminishing supply of high quality specimens.

The attraction of financial gain from collecting, or overcollecting, often illegally, has understandably earned mineral collectors, and dealers in particular, a bad name, especially with conservationists. Whereas the malpractice of the comparatively small number of such collectors deserves contempt they should not be overemphasized or allowed to overshadow the worthwhile contributions to mineralogy made by responsible collectors and dealers, most of whom genuinely understand and appreciate the need for conservation of sites, specimens and information. Over-collecting and financial greed are not by any means abuses only of modern collecting – they were well known to previous generations.

The modern collector differs from his or her predecessors in two important respects.

First, most of today's serious collectors own a good stereomicroscope and many enjoy access to a range of analytical or determinative techniques. When combined with a working knowledge of mineralogical and chemical principles, today's recreational mineralogist is well-placed to make significant contributions to the science. Indeed field mineralogy is one of the very few scientific pursuits still to offer a true role to the good amateur. It is noteworthy that in recent years many, or perhaps most, of the important finds of unusual or rare minerals in Britain are the results of dedicated amateur work, often of high scientific quality.

Second, the amount of raw material available to the collector at many sites is today much less than even a few years ago and is diminishing rapidly.

Both of these factors are of vital importance in mineralogical conservation.

Field collecting

The activities of collectors and dealers offer on one hand the potential for greatly extending our knowledge of a country's mineralogical heritage and on the other they pose potentially serious conservation problems.

Working mines and quarries or other temporary exposures frequently reveal interesting or important mineralogical material, perhaps for a limited period. In such cases a collector may perform an important function merely by rescuing material which would otherwise be destroyed. This potential is of course fully realized only if the collected specimens are adequately curated with full locality documentation. Ideally, representative examples should be donated, or at least offered, to a national or local museum with a clearly established mineral collection policy. Depositing records of field observations with a local or national site documentation scheme ensures the survival of vital locality information. Some form of formal or informal liaison between mining or quarrying companies, museums and interested individuals or societies can be very useful in this aspect of conservation.

Mineralized outcrops or sites under no immediate threat of destruction pose more of a conservation problem. Collecting which may reveal new finds or other information is desirable, though unrestrained removal of material cannot be sustained and a site may rapidly become exhausted. All collecting requires the consent of the land or mineral owner. Such owners may therefore be in a position, should they wish it, to exercise a considerable degree of control over collecting. However, many sites are remote and difficult to oversee. Ownership may be difficult to establish and many collectors simply do not attempt to obtain permission. Control solely by owners may, therefore, be unrealistic. Statutory protection by Site of Special Scientific Interest (SSSI) or Regionally Interesting Geological/geomorphological Sites (RIGS) status offers no real protection against overzealous collectors who, in any case, may be unaware of a site's status and even if they are, may simply ignore it.

A strong measure of self-discipline is essential in the modern collector. Standards of good field conduct, and collecting restraint, encouaged by many of the increasing number of reputable clubs or societies, have an important influence in this respect.

Localities threatened by damaging development or perceived as being seriously threatened by uncontrolled and destructive collecting may benefit from some form of rescue collecting. Material so collected can be stored by a responsible organization and made available to bona fide researchers on request. The recovery of a unique suite of tungsten and bismuth minerals from a vein outcrop in the Lake District jointly by the Nature Conservancy Council and British Geological Survey (BGS) (Nature Conservancy Council 1987, p. 36) and of minerals from mine dumps at Leadhills by the Royal Museum of Scotland (Jackson 1992) are rare examples of such measures applied to British sites. Collaboration with amateur collectors or societies, as in the Leadhills example, has been extremely valuable. Potential exists for more such collaborative rescue collecting at appropriate sites.

Curation, documentation and recording

Having been privileged to collect important, or perhaps even unique, material the collector must accept a duty to care for this either in his or her private collection or, as already suggested, by making specimens and accompanying information available to museums. However, even if such altruistic practices are widely followed the fact remains that, perhaps more than ever before, private collections represent an immense resource of very important specimens and information. It is thus desirable for the science that the collecting fraternity, perhaps through geological or mineralogical societies and clubs, makes known the whereabouts of such material. For the same reasons collectors should be encouraged to give serious thought to, and make some plans for, the fate of their collections upon their eventual demise. The responsible collector's appreciation of this conservation role for specimens may be epitomized by one collector known to this author who regards his specimens not as his property but as objects over which he has 'temporary custody'.

The conservation duty of the collector extends beyond the specimens themselves. With the determinative techniques available to many collectors today important mineralogical finds are being made increasingly often, paradoxically at a time when the amount of material available for collecting is diminishing rapidly. More than ever there is a need to place such discoveries on permanent record in the scientific literature. Club and society newsletters are generally unsatisfactory vehicles for such records: they are essentially ephemeral documents and the information is quickly lost or difficult to trace. For a number of years there has been a trend away from such topographic recording, especially in a number of long-established journals. This tendency is now reversing as the science again recognizes the need for this essential basic information.

A number of collectors already well appreciate the need to publish and having found that publication is not as daunting or as difficult as they feared, derive justifiable satisfaction from adding their findings to the mineralogical literature for future generations. The help and encouragement to record and publish findings given to amateur collectors by societies such as the Russell Society is particularly commendable.

Conclusions and recommendations

In today's increasingly conservation-conscious world collecting merely as an end in itself can scarcely be justified. Mineral collectors much surely realize that their freedom to collect is a privilege not a right. All privileges carry a weight of responsibility. Despite abuses of this privilege by some individuals, collectors have, on balance, served mineralogy well and can continue to do so. The price to be paid for continuation of this privilege must be an acceptance by all collectors of their responsibility to ensure, so far as possible, that sites, specimens and information are all conserved for the future. Failure to accept this must hasten the day when mineral collecting may be as unacceptable as collecting rare wild flowers or birds eggs.

It is suggested that the following points

should be the basis of a 'code of practice' for all collectors of minerals:

(1) Obtain consent to collect from the appropriate land or mineral owner.
(2) Exercise restraint in the amount of material collected.
(3) Record and retain full field observations and acquisition details with collected specimens.
(4) Exercise the highest possible curatorial standards for all collected specimens and accompanying information.
(5) Offer representative specimens of new finds to local or national museums with an established mineral collection policy.
(6) Provide information on new finds to a local or national site documentation scheme.
(7) Make known the whereabouts of important specimens held in private collections.
(8) Publish new or significant finds in appropriate established mineralogical literature.
(9) Made advance provision for the disposal of a private collection, or important parts of it, upon the death of the collector.

Mrs Sandra Clothier is thanked for preparing the typescript.

This contribution is published with the approval of the Director, British Geological Survey (NERC).

References

JACKSON, B. 1992. Working with others – the experience of the National Museums of Scotland. *In*: *Conserving Britain's Mineralogical Heritage*, (Conference abstract, University of Manchester 31 March – 1 April 1992).

NATURE CONSERVANCY COUNCIL. 1987. Rescue collection of rare Lake District minerals. *Earth Science Conservation*, **23**, 36–37.

The protection of Lower Palaeolithic sites in southern Britain

C. S. GAMBLE[1] & J. J. WYMER[2]

[1] Department of Archaeology, University of Southampton,
Highfield, Southampton S09 5NH, UK
[2] Wessex Archaeology, Portway House, Old Sarum Park,
Salisbury, Wiltshire SP4 6EB, UK

Abstract: Despite their importance nationally and internationally, the Lower Palaeolithic sites of southern England have remained, in management terms, a neglected resource. Statutory protection for these sites is almost non-existent. Conservation, as a routine aspect of the planning process, is rarely considered, while presentation leaves much to be desired.

These issues were highlighted in 1988 with an application to extract gravel from Kimbridge, in the Test Valley, Hampshire. These pits had previously produced almost one thousand hand-axes. Resulting investigations and discussions led to one positive outcome for, in 1990, English Heritage commissioned Wessex Archaeology to produce a research design for a survey of the Lower Palaeolithic sites south of the River Thames and Severn. The report emphasized the threats to the archaeology from current and future proposals for mineral extraction and many building developments or roadworks.

Out of this was born the Southern Rivers Palaeolithic Project. English Heritage are funding a full-scale survey lasting three years. The primary aim of the project is to assist management decisions in the planning process. A further aim is to establish the pattern and history of mineral extraction in the area. The survey involves detailed plotting of all the known find-spots of Lower Palaeolithic material, relating it to the Quaternary deposits in or on which it occurs and assessing its archaeological significance. The project is particularly important as providing a focus for co-operation between the mineral industry and archaeologists.

The archaeology of the earliest inhabitants of Britain is widely acknowledged as a world heritage resource. The sites of Swanscombe, Clacton and Hoxne are well known in both textbooks and general works devoted to the colonization of northwest Europe during the Middle Pleistocene period, from about half a million years ago. Key sites have recently been investigated at Boxgrove, Pontnewydd, La Cotte, High Lodge and Barnham. However, for all their importance, statutory protection for such sites is almost non-existent, although this situation has been redressed to some extent by the application of proposals set out in the Department of the Environment's publication of planning policy guidelines (DoE 1990). These give local planning authorities the responsibility for ensuring that the archaeological implications of planning decisions are properly assessed. These policies apply to all types of archaeological sites, from Lower Palaeolithic to post-medieval. Lower Palaeolithic sites such as those listed above are well documented and published so that, in the event of any threat by development to them, sufficient information is available for planners, with specialist advice, to make assessments and decisions. However, the great majority of Lower Palaeolithic sites are represented by little more than find-spots of flint artifacts. Fortunately, a very comprehensive gazetteer of these find-spots was compiled by Dr D. A. Roe and published by the Council for British Archaeology in 1968 (Roe 1968). Much of this information has been transferred into the relevant county Sites and Monuments Records but, in itself, is rarely suitable as a basis for making planning decisions if such sites appear to be involved. English Heritage was becoming acutely aware of this problem, especially in view of its Monument Protection Programme with its emphasis on preservation rather than excavation. Probably some 90% of Lower Palaeolithic find-spots are those made in Pleistocene river gravels and, with the great increase in mineral extraction since the last war, it was obvious that much palaeolithic archaeology was being lost. This realization that a valuable and limited archaeological resource was endangered by lack of information, sufficiently detailed to enable County Archaeological Officers, County Planners or Developers themselves to identify and evaluate the potential of sites, prompted English Heritage to take action. It announced in 1991 that it would fund a full-scale survey of the Lower Palaeolithic sites of Britain, south of the River Thames and Severn. It would last three

From O'Halloran, D., Green, C., Harley, M., Stanley, M. & Knill, J. (eds), 1994, Geological and Landscape Conservation. Geological Society, London, pp. 443–445.

years. This survey is referred to as the Southern Rivers Project and is now in its final year.

There are many important Lower Palaeolithic sites north of the Thames and the Severn, especially in East Anglia and along the middle and lower Thames Valley. The decision to restrict the present survey to the southern rivers was dictated by much less research having been conducted on the Palaeolithic period in that area in spite of it containing two counties with more find-spots than any others in Britain: Kent and Hampshire. It was also the area with the most active commercial quarrying of aggregate. The project is administered by the Trust for Wessex Archaeology at Salisbury. John Wymer is the project leader, with Andrew Lawson and Clive Gamble as management with academic advisors. Phil Harding assists with records and fieldwork, Karen Walker researches gravel extractions, past and present, and Linda Coleman is cartographer. The area is divided into six regions on the basis of major watersheds and a report is prepared annually on two of the regions. These reports will have a limited circulation, mainly for County Archaeological Officers, County Planners and certain commercial bodies. They are not confidential reports but expense prohibits more general circulation. A review team of Quaternary specialists meets regularly to monitor progress.

The plan is to prepare these reports so that those concerned can readily see where Lower Palaeolithic material has been found and in what context, and also identify likely areas where sites may occur. All known find-spots are listed in tabulated form giving details of provenance and national grid reference, circumstances of discovery, county sites and monuments number, geological sediment in or on which found, categories of finds made, quantities of and where preserved, major published references or sources of information, and the present state of the site. Obviously with the multitude of sites involved this information is much abbreviated and several items may be unknown. Perhaps the most important aspect of the work is the preparation of distribution maps with the location of sites superimposed on the Quaternary geology. These maps are on a scale of 1:50000 in most cases, but 1:25000 for especially prolific areas and town plans where necessary. Permission to use the resources of the British Geological Survey and publish the maps has been gratefully received from the Director of the BGS. Each site has a number relating it to the tabulated lists. It is thus possible for non-specialists to see at a glance which Quaternary deposits yield palaeolithic material and which do not and where sites are prolific or sparse. The maps also show find-spots of surface sites, generally at high levels above the Middle Pleistocene river terraces. In certain areas they are surprisingly prolific. Other information on the maps gives the areas of past and present major gravel extraction, and areas threatened by the current policies of planning authorities. Each of the six areas of the survey is subdivided into geographical regions, e.g. Medway Valley, the Solent, Sussex raised beaches etc., and the tabulated lists for each have brief texts summarizing the geology and history of discoveries and a separate section on current interpretations and assessments of critical sites.

Fieldwork involves visiting all the areas concerned and identifying as accurately as possible provenances hitherto vague or unlocated. Local knowledge, old editions of OS maps and even estate maps have allowed a good proportion of such sites to be given 'Accurate' or 'Estimated' six-figure grid references and reduced 'General' locations to a minimum. There has been excellent co-operation from archaeological units and local archaeologists.

With so much information to be conveyed on one map (find-spots with different symbols for accurate or estimated provenances, general provenances and critical sites, the drift geology, areas of past, present and threatened gravel extraction as outlined above) it was found impossible to do so without resorting to colour. All the details are superimposed on the relevant OS maps. The greater the clarity achieved by the cartographer, the greater became the cost of reproduction. The problem was resolved by the use of a Computer Aided Design System, purchased by Wessex Archaeology for this project and an investment for other and future use.

The first year's report is complete and published, covering the Upper Thames Valley, the Kennet Valley and the Solent drainage system. The second year's report is also now complete. This covers South of the Thames from Surrey to East Kent and the West Country. Work is now in progress on the Sussex rivers and raised beaches, and the extreme west of England.

In practice, as a working tool for the preservation of threatened Palaeolithic sites, the reports should enable county or consulting archaeologists to advise planners on recommended courses of action. This could vary from mere watching briefs of varying degrees of intensity, a site evaluation, possible excavation or refusal of planning permission. Conditions within a planning consent may entail a right to investigate certain features which may be exposed during the course of any work undertaken. Equally important, the information contained in the report could be the basis of discussion between planners and developers so that plans can be drawn up which avoid or minimize interference with sites.

Palaeolithic archaeology is often regarded by some professional archaeologists as something very different and mysterious to the archaeology of later periods. This is unjustified, but it has to be stressed that it is a branch of archaeology that does require the co-operation of several specialists in other fields of study. Quaternary geologists must align themselves with the archaeologists. They should take the opportunity to inspect all sections that have yielded palaeoliths and record and sample them accordingly. A good example of this co-operation is on the site at Dunbridge, Hampshire, where the Trust for Wessex Archaeology has a watching brief at the site as part of the planning conditions. The results of their archaeologist who specializes in the Palaeolithic and a consulting geologist has already clarified the composition of the fluvial sediment and the sequence of terrace deposits in the immediate area, apart from retrieving some artifacts. Ironically, this was the site which caused considerable debate and difference of opinion when planning permission was granted for gravel extraction as an adjacent pit had produced more hand-axes than any other site in Hampshire. In some respects the Southern Rivers Project was the outcome of the publicity it engendered.

It is hoped that in future there will be better co-operation between the mineral extraction industry, archaeologists and planners. The main aim of the Southern Rivers Project is to make the knowledge available, comprehensive and comprehensible, so that sites of the British Lower Palaeolithic will either be preserved or reveal the maximum information on the distribution, activities, age and environment of archaie *Homo sapiens* in this part of Europe.

References

ROE, D. A. 1968. A Gazetteer of British Lower and Middle Palaeolithic sites. *Research Report Council for British Archaeology*, **8**, 1–355.

DoE. 1990. *Planning Policy Guidance Note 16: Archaeology and Planning: (PPG 16). London, HMSO.*

Effective management of our Earth Heritage sites: the Earth Science Site Documentation Series in Scotland

C. C. J. MacFADYEN, J. A. McCURRY & S. KEAST
Scottish Natural Heritage, 2 Anderson Place, Edinburgh EH6 5NP, UK

Abstract: The identification of a network of key Earth Heritage sites is fundamental to most approaches to Earth science conservation. A comprehensive list of Earth science Sites of Special Scientific Interest (SSSI) have been identified in the UK through the Geological Conservation Review (GCR). The GCR site selection procedure was a purely scientific process and there was no scope to address the management of selected sites. Fundamental to the effective management of sites is a recognition that those involved in their management – the conservation officers, landowners, planners and developers – are non-specialists to whom the technical explanation of sites may be incomprehensible. For management to be effective it is vital to understand what it is one is managing. Consequently Earth science site conservation and enhancement is hindered by a lack of understandable information on each site. The Earth Science Site Documentation Series in Scotland was developed in direct response to this problem. The series provides a comprehensive and user-friendly report detailing information on each Earth science SSSI. This includes a plain English description of the Earth science interest and its significance, in conjunction with detailed management guidelines for each site. Extensive use is made of figures and photographs to help explain the interest and the latest available scientific information on the site is included to complete the package.

The identification of a network of key Earth Heritage sites is fundamental to most approaches to Earth science conservation (see Gordon 1993 for discussion). Sites are typically selected for their scientific and educational importance and are usually afforded a measure of legislative protection. In the UK the basic unit of protection is the Site of Special Scientific Interest (SSSI). A rigorous site selection procedure is important in establishing any network; however, as the Cairngorms Working Party recently noted 'SSSI designation in itself may not provide protection, the protection can only be assured when proper management is in place' (1992). In this paper we identify a major problem common to the 'proper management' of most Earth Heritage sites and present a strategy for their effective management. This strategy has been adopted by the recently formed government conservation agency Scottish Natural Heritage (SNH) and is presented here within this Scottish context.

The Geological Conservation Review

SNH inherited from its predecessor, the Nature Conservancy Council for Scotland, a Scotland-wide network of Earth science SSSIs, which have statutory protection under the Wildlife and Countryside Act 1981. The geological sites were selected in a UK-wide Earth science site selection process known as the Geological Conservation Review (GCR) initiated in 1977 and completed in 1989. This was an attempt to rationalize Earth science site coverage using strict scientific selection criteria based on the latest research (Nature Conservancy Council 1991, pp. 28–32). A total of 2200 Earth science SSSIs were identified, reflecting the remarkable geological diversity and rich scientific inheritance of the Earth sciences within the UK. Of these, 53 sites were identified in Scotland. Site selection is being followed up by publication of a series of subject-related volumes providing a detailed scientific account of each site. Each scientific account is written to the highest scientific standards and indicates why that site was of sufficient scientific importance to include within the GCR (e.g. Treagus 1992). Such a rigorous and rationalized basis to site selection, coupled with an explicit scientific justification for each site, is an essential basis for any network of sites selected for special protection on scientific grounds. Now that the network has been established it will require only minor modification through time to keep abreast of scientific advances in the Earth sciences.

The management of Earth science sites

Every Earth science SSSI when designated is notified to the landowner and occupier. The notification is in the form of a short scientific Statement of Interest derived from the GCR which is accompanied by a list of Potentially Damaging Operations (PDOs) and a map detailing the boundary of the site. The landowner/occupier must apply to the SNH Area Office for consent to carry out any activity on the site which occurs on the PDO list. This process ultimately results in discussions between the area staff and the landowner/occupier concerning the management of the site. However, it is often the case that area staff who are dealing with the notification and hence the management of these Earth science sites, are not Earth science specialists, having little or no geological training. Consequently, the technical Earth science terminology used in the scientific description which forms the core of the notification process, renders them incomprehensible to most area officers. Terms such as Silurian erypterid, protalus rampart, solifluction, harmotome mineralization and rhyodacite flows, mean little to non-geologically trained staff and landowners alike. A problem therefore arises where area staff may find themselves in the position of having to explain and justify particular sites, which they themselves do not understand, to landowners/occupiers who are perhaps less than sympathetic toward the designation. Unless this information is effectively communicated, the landowner/occupier may feel antipathy or indifference towards the interest and as a consequence the site may not be effectively conserved and managed. An understanding of the geology and geomorphology is not only a prerequisite to effective management of a site, but is an essential ingredient in nurturing the wider enjoyment of our Earth science heritage.

In order for staff to facilitate the effective management of sites and fulfil the SNH remit of having a devolved structure in which they can integrate effectively with the local community including giving advice on natural heritage matters, the Earth science interest has to be clearly communicated. The Earth Science Site Documentation Series described below achieves this aim.

The Earth Science Site Documentation Series: bridging the gap between the GCR and effective site management

The Earth Science Site Documentation Series, was initiated in September 1991 with the aim of describing and explaining the interest of each of Scotland's GCR sites in plain English for the benefit of SNH area staff who are directly concerned with the conservation and enhancement of sites at the local level. To date, 70 documentation reports have been produced within the Earth Science Branch of SNH's Research and Advisory Services Directorate based in Edinburgh. A large variety of geological and geomorphological sites have been documented, covering all regions of Scotland.

Sites selected for documentation typically have long histories of management problems or face major potentially damaging developments at present or in the near future. Field visits and literature surveys have been fundamental elements of each site documentation report since the project began, however a major additional element has been the consideration of site management.

The content structure of the site documentation reports

The Earth Science Site Documentation reports are comprised of five sections which are briefly outlined as follows.

(1) INTRODUCTION – an introduction to the report which outlines the scope of the project as a whole and provides details of site location and access to it. Major emphasis is placed on providing a clear description of the site boundary justifying its position.

(2) DESCRIPTION OF THE GEOLOGY – the most important component of the report, describing the Earth science interest in non-technical language. This is aided by a series of simplified geological/geomorphological maps, diagrammatic illustrations, palaeogeographic reconstructions and annotated photographs (the photographic coverage of sites, forms a base-line for future site monitoring). A precis of the information in this section is included in a sub-section entitled 'Significance of the Site'. This summary provides a plain English alternative to the Statement of Interest included in the notification details (see Fig. 1).

(3) SITE MANAGEMENT – is the site management section and it provides the key details for safeguarding the future of the site. Following an assessment of the current site condition an 'Interest Zonation Map' is produced in which the site is graded into areas of 'Crucial Interest' and areas of 'Context Interest'. Guidelines for the management of the site are provided, addressing current problems and anticipating future threats. For geomorpho-

> **A. STATEMENT OF INTEREST** (from notification document for Site of Special Scientific Interest)
>
> **GEOLOGICAL CONSERVATION REVIEW INTEREST: CALEDONIAN IGNEOUS**
>
> The Hill of Johnston undoubtedly provides the best examples of the later fractionation stages of the Inch Intrusion. In a small roadside quarry at its base the rocks are ferrogabbros and ferrodiorites, and these grade upwards to syenodiorite as alkali feldspar becomes relatively abundant. Higher up the hill the small natural exposures consist of syenodiorite and locally even more 'syenitic' material, although this may have been secondarily silicified. Syenitic rocks are also present as veins within the more basic members of the succession.
>
> **B. SIGNIFICANCE OF THE SITE** (from Earth Science Site Documentation Report in plain English)
>
> **GEOLOGICAL CONSERVATION REVIEW INTEREST: CALEDONIAN IGNEOUS**
>
> The Hill of Johnston rocks form part of the Inch Intrusion which is the largest of the six 'Younger Gabbroic Series' bodies exposed in NE Scotland. These igneous rocks crystallised from a molten magma which pushed or 'intruded' its way into surrounding Dalradian rocks about 489 million years ago. As the magma cooled a series of crystal fractions separated from it to produce a mineralogical layering in the rocks that subsequently formed. Compositionally these rocks varied from peridotites (an igneous rock enriched in the mineral olivine and containing no feldspar) to syenites (another igneous rock but one enriched in alkali feldspar).
>
> A network of Sites of Special Scientific Interest in north east Scotland have been selected which preserve the most important features of the mineralogical layering within the Inch Intrusion. The rocks at Hill of Johnston consist of gabbro (a coarse-grained equivalent of basalt) at the base, but grade upwards into syenite. The site is of special scientific interest because the syenites are a more evolved type of rock than the gabbros and shed light on the processes which helped to form the higher, more evolved, parts of the Inch Intrusion.

Fig. 1. A comparison between (A) the Statement of Interest (used in site notification document); and (B) its plain English analogue, the Significance of the Site (used in the Earth Science Site Documentation Report). This example is from the Hill of Johnston SSSI in NE Scotland.

logical sites, an assessment of geomorphological sensitivity to damage is routinely included.
(4) RESEARCH STATUS – this section provides a summary of the most relevant and up-to-date scientific research concerning the site. This provides the reader with an insight into the research status of the site and its potential for future research.
(5) REFERENCES – the reference section lists key scientific papers included at the back of the report.

Within the basic report structure there is abundant scope for flexibility to address the peculiarities and particular management problems of sites, consequently, no two reports are the same. Guidance given in the reports range from advice on minimizing damage where development is proposed, to defining best management practises. In the case of geomorphological sites, management often addresses issues of erosion and recreational access to upland SSSIs, such as the Cairngorms. The Earth Science Site Documentation Series provides the essential management advice to ensure effective protection.

It is seen as vital to promote accessibility to these reports. In addition to area staff, reports are available to the major co-partners in the management process, i.e. the site landowners/occupiers. They are also available on a consultative basis to planners and developers involved in any proposed site developments. Through demystifying the science of the sites, a greater understanding of our Earth Heritage is imparted to all in their management. The reports are also used as a means of identifying suitable sites for public interpretative facilities, in particular those sites which contain an easily understandable geological feature or landform such as the Arthur's Seat Volcano SSSI in Edinburgh (Site Documentation report number 41). The use of appropriate sites for interpretative and educational purposes is regarded as an important element in encouraging a conservation ethos to permeate public consciousness. The need for this awareness as part of an ongoing strategy in Earth science conservation is increasingly recognized (Ellis 1993; Magnusson 1993).

Conclusions

The GCR in Britain has identified an excellent core of Earth Heritage sites across the nation worthy of safeguarding on scientific grounds. To safeguard these sites there is a clear need to bridge the gap between their purely academic appreciation and the non-geological perspective held by of the majority of landowners/occupiers, managers, planners and the public. Decisions which have led to the destruction or damage of vulnerable sites are often taken without an appreciation of the effects of these actions. Consequently by sharing information about our key sites in an understandable manner through the Earth Science Site Documentation Series a greater awareness and enjoyment of our Earth Heritage is promoted. This is seen as fundamental to the effective management of our Earth Heritage sites.

References

CAIRNGORMS WORKING PARTY. 1992. *Common Sense and Sustainability: A Partnership for the Cairngorms*. Scottish Office, Edinburgh.

ELLIS, N. 1993. GCR Publications and Publicity. *Earth Science Conservation*, **33**.

GORDON, J. E. 1993. Geomorphology and conservation in the uplands: framework for a developing role in sustaining Scotland's natural heritage. *In:* STEVENS, C. *et al. Conserving our landscape and Evolving Landforms and Ice-Age Heritage*. English Nature, Peterborough.

MAGNUSSON, M. 1993. Making Rocks Talk. *Earth Science Conservation*, **32**.

NATURE CONSERVANCY COUNCIL. 1991. *Earth Science Conservation in Great Britain: A Strategy*. NCC, Peterborough.

TREAGUS, J. E. 1992. *Caledonian Structures in Britain: South of the Midland Valley*. Geological Conservation Review Series, **3**. Chapman & Hall, London.

Telling the story of stone – assessing the client base

T. A. HOSE

Faculty of Leisure and Environmental Management, The Buckinghamshire College, Queen Alexandra Road, High Wycombe, Buckinghamshire HP11 2JZ, UK

Abstract: Despite some twenty years of professional interpretative and conservation endeavour, little work has been undertaken on the nature and needs of visitors to geological attractions. An on-site visitor survey at the National Stone Centre in August 1993 sought to acquire such fundamental management data. It was found that the centre's visitors are mainly locally-based, casual arrivals as couples and small family groups with only a cursory knowledge of geology. It is suggested that the findings presented have a wider application to future developments in geological interpretation within the UK.

Site-specific geological interpretation, unlike many other forms of outdoor leisure and tourism provision, is not necessarily limited by seasonality; indeed for many locations it could be seen as a way of extending their tourism potential. Whilst opportunities abound to enhance the visitor experience, with the inclusion of geological interpretation, at numerous archaeological and industrial history sites little is done to capitalize on their Earth science asset. Undoubtedly, the development of an effective national geological interpretative strategy will ultimately provide the public pressure necessary to ensure the maintenance of the UK's rich geological heritage, for:

> 'It [interpretation] has the dual purposes of serving the best interests of the visitors who come to see and experience a site and also those of the place itself. Good interpretation will raise the value of a site in the eyes of those who come to visit; greater value will lead to greater conviction of the need to preserve and protect.
>
> Herbert 1989

Sadly, the national geological community has been slow in effective constituency-building at a time when its prime educational resource has been ravaged. The industrial base most under threat in the changing economic climate of the last decade had been responsible, in large part, for exposing the UK's richly available geological heritage through, for example, mining, quarrying and the excavation of railway cuttings. The service industries, such as retailing, leisure and tourism, which arose to replace the historic economic base were not so generous to the nation's geological heritage. Of course, much pressure to remove the industrial landscape came from the leisure and amenity lobby, always keen to find a recreational and especially sports use for reclaimed sites (Sports Council 1992) and often without regard to their inherent geological significance.

The rapid growth of interest since the 1960s in that period of the UK's historic past categorized today as our 'industrial heritage' forms a major component of the nation's second most important industry – tourism. Its growth (Cossons 1974) has contributed to the loss of many prime geological sites, such as mine tips, spoil heaps and quarries, through the widespread implementation of land reclamation schemes and associated amenity landscaping as the historic landscape is prepared for the tourist camera. Whilst the architecture, art and artefacts of the period are now the subject of much conservation concern, the basis on which that prosperity flourished is all but ignored as new, almost de-industrialized settings, are created as a tourist-acceptable backdrop. Interestingly, there is on record (Hewison 1987, pp. 93–94), to redress the balance, at least one example of amenity landscaping, at the North of England Open Air Museum, Beamish, County Durham which includes a reconstructed pit heap. This, in a county which has all but obliterated its mining heritage with recontouring and afforestation.

Whilst the displays and associated publications, at both national and provincial museums (such as Scunthorpe Museum, Humberside (Knell 1986; Grange 1993) often interpret the geology of their surrounding areas, there are still very few sites where the story of the rocks is explained *in situ* to the public. A variation on this theme is perhaps those guides to the building stones of either a structure (Purcell, undated) or a town (St Albans Museum 1990). At the National Stone Centre, an attempt is made to combine interpretation of both local geology *in situ* (Thomas & Hughes 1993) with a self-guided geological trail and national

From O'Halloran, D., Green, C., Harley, M., Stanley, M. & Knill, J. (eds), 1994, *Geological and Landscape Conservation*. Geological Society, London, pp. 451–457.

economic geology within the context of an on-site exhibition.

The National Stone Centre

The site and its facilities

The National Stone Centre is centred on an SSSI (Thomas 1984) and consists of six redundant limestone quarries. Its 20 hectare site is crossed by The High Peak Trail (a major public permissive path, developed and maintained by Derbyshire County Council, along the route of a disused railway) which separates the currently interpreted part of the site from the public access road and car parks; the link between the two is through the narrow arch of an old railway bridge. The National Stone Centre lies north of the town of Wirksworth, and to the south of the Peak District National Park – which is estimated to be within an hour's drive of some 20 million people (PPJPB 1988). Originally geared to cater for some 75 000 visitors annually, its location should indicate little difficulty in attaining that target.

Visitor context

The location of the Stone Centre is likely to result in comparable visitor behaviour and trends to that of the adjacent national park. There are strong seasonal and daily variations in the number of visitors to the park, with the holiday season concentrated in the summer months. It attracts some 500 000 visits in a typical summer week, with Sunday the busiest day seeing some 165 000 visits. There are about 18.5 million annual, of whom over 5 million are regular, visitors to the park, some 65% of day visitors to the park come from the surrounding conurbations. Over 95% of visitors have been to the area before; visitors tend to return to places they have previously appreciated. Some 40% of visitors set out to visit a particular area or site. 10% of visits to the Park are made by staying visitors, almost half of whom stay in the surrounding settlements. Further – which is significant for fee-charging attractions – over half of the summer overnight visitors either stay at campsites or caravan parks within the park. Visitors to sites and attractions tend to be in social groups of about three persons. Significantly, day visitors to the park are from Derbyshire, Nottinghamshire and the rest of the East Midlands and account for some 40% of the total visits (PPJPB 1988).

The survey

Challenges and approaches

The work presented here is the first British attempt to undertake a survey of visitors to a site of interpreted geological significance. Accurate visitor sampling is made extremely difficult by the site's multiple formal and informal access points. Consequently, with the limited survey resources available, it has not proven possible to accurately ascertain the numbers of persons entering the site. People enter the site for a variety of educational, social and recreational activities; these range from learning from the exhibition, to jogging and walking the dog!

The Discovery Building (housing the Story of Stone exhibition, shop and small cafe) forms a visual focus for the site and is its only permanent shelter; it was selected as the sampling point. Two surveys were conducted simultaneously during the period 7–16 August 1993 which was chosen to be representative of the normal summer visitor cycle to the centre. A respondent-completed questionnaire was issued over the entire period of the survey to all parties (defined as one or more visiting adults) purchasing tickets for the Story of Stone exhibition; there was some difficulty in issuing the questionnaire on the second weekend and consequently there is a gap in these data for a Saturday. However, just over 100 of the respondent-completed questionnaires were issued and some 50, representing a 48% response, were returned. The majority were fully completed and only two showed evidence of group duplication and were hence discounted.

Methodology

On eight days (chosen to avoid special events and to give an equal representative weekend and weekday sample), subjects exiting the building, and selected on a next-through-the-door basis from the shop, were given an in-depth interview; these were timed to take place from 11.00 to 1300 hrs and 1400 to 1600 hrs. To avoid unusually high refusal rates, and in an attempt to ensure that visitors had made their maximum intended use of the site, those carrying purchases from the cafe or intent on panning for gems (a 'hands-on' activity adjacent to the Discovery Building) with children were not approached. The interviews examined the material in the respondent-completed questionnaire, together with supplementary material to ascertain the respondent's geological knowledge and the use of the centre's facilities. Qualitative data on visitor perceptions were also sought and recorded.

During the course of the interview various text and illustrated prompt cards were used. The illustrated ones centred on common British fossils (Fig. 1) and the geological trail (Fig. 2). The fossils illustrated in Fig. 1 were selected on the basis that they were either found on-site or were likely to have been encountered by visitors in the shop, other museums or at another holiday destination. As such they served to discriminate between prior knowledge and that acquired on-site by the respondent.

Illustration 2 of Fig. 1 is identical, illustrations 4 and 8 are similar in many respects, and illustration 5 similar in some respects to those used in the geological trail guide leaflet available on-site. The bivalve illustrated in that leaflet is a Pecten and quite unlike that on the prompt card. Many respondents used common names for some of these fossils; for example 'sea lily' and 'mussel'. Of course, 2, 4, 5 and 8 could be readily found within the grounds of the centre.

The diagram in Fig. 2 was slightly modified, to clarify the points in question, from that found across the site on the interpretative panels along the geological trail.

Whilst these numbered points on Fig. 2 tested on-site acquired knowledge, they also served to indicate respondents' understanding; interestingly, a number thought the quarry itself was the actual lagoon, rather than the excavated site of its deposition products. A small number of respondents used geological terms such as 'strata' and 'crinoid beds'.

Responses

Some 76 in-depth interviews were conducted; there were 7 non-respondents, representing some 24 visits, giving a 92% response rate.

Unfortunately, due to the location of the sampling point and frequently inclement weather, a high number of respondents had not walked the geological trail at the time of the interview; this tended to be the last on-site activity of many visitors. In total, some 124 valid questionnaires, representing some 414 visits, were available for analysis from the two surveys.

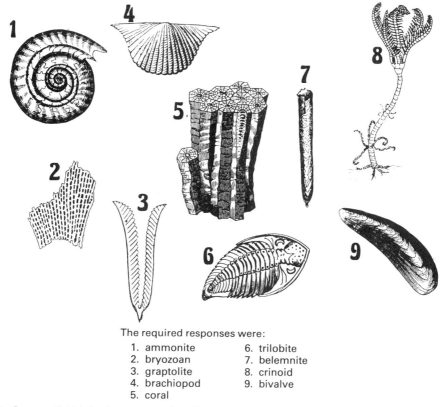

The required responses were:
1. ammonite
2. bryozoan
3. graptolite
4. brachiopod
5. coral
6. trilobite
7. belemnite
8. crinoid
9. bivalve

Fig. 1. Common British fossils prompt card used in the survey.

How it all began

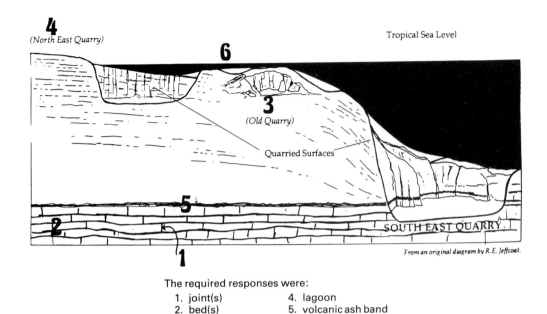

The required responses were:
1. joint(s)
2. bed(s)
3. reef mound
4. lagoon
5. volcanic ash band
6. sea lily meadow

Fig. 2. The geological trail prompt card used in the survey.

Throughout, the purpose was to assess the nature of the Stone Centre's client base in terms of origin, social groupings, general and specific geological attainment, micro-economics and preferences. However, it is interesting to speculate how many interviewers undertaking similar surveys elsewhere at heritage attractions are asked by their interviewees to identify everything from minerals in jewellery, to a 0.5 cm fragment of presumed Lower Carboniferous shark's tooth, and 10 kg lumps of limestone riddled with large Gigantoproductids lying on the floor of a camper van! Clearly, as in the Australian case:

'The results of this research [should] have significant implications for the marketing of fossicking, and for visitor satisfaction and management of the recreational experience.'
Jenkins 1992, p. 131

Major findings: social

The major results, relevant to a general examination of geolocial interpretation's client base, can be summarized in a series of tables.

The questionnaires were targeted at adults (defined as those aged 18 years and over) acting as party leaders/co-leaders. It seems clear that the most populous group are those in mid life-cycle with young families (Tables 1 and 2). Relatively large numbers were educated to 'A' Level standard and higher (Table 1), suggesting a high professional class component in the visitor population. Clearly geological interpretation has little appeal to those 19–29 years of age (Table 1).

Most visitors were accompanied; in the main by one or three other persons, often their children. Thus, couples and small family groups were dominant within the visitor population (Table 2). Clearly, there is a need to offer suitable activities for children since they are a major component of both group and overall visits. Again, the number of "adults only" parties indicates an interesting market segment with different needs and expenditure patterns to those of families.

There is clearly a large day-visitor component; much of this is of relatively local origin and depends upon the availability of private transport (Table 3). For those staying in holiday

Table 1. *Profile of all respondents*

Category	Percentage of respondents
Gender	
male	54
female	46
Age group	
15–19 years	2
20–29 years	7
30–44 years	44
45–64 years	38
over 64 years	9
Education	
left school at the minimum age	25
studied beyond 'O' level/GCSE	38
undertaken higher education	37
Work status	
full-time employment	57
part-time employment	13
not working	3
retired	18
househusband/wife	6
full-time student	3

accommodation the equal importance of staying with friends and relatives and camping/caravanning would suggest (accounting for 72%) a high proportion of middle-income visitors. However, some 26% of visitors use bed-and-breakfast premises and hotels.

Table 2. *Social grouping of all respondents: (a) size of group; (b) composition of group*

(a)

Number of individuals in visitor group	Percentage of respondents
1	2
2	35
3	16
4	30
5	13
6	3
7	1

Note: average group size is 3.3.

(b)

Composition	Percentage of respondents
Adults only	49
Adults with children	51

Note: 30% of visitors were children.

Table 3. *Residences and transport of all respondents at time of interview*

	Percentage of respondents
Accommodation	
permanent home	53
holiday home	45
camping/caravan site	35
hotel	13
bed & breakfast	13
youth hostel	2
other (e.g. self-catering cottage, staying with friends or relatives)	37
Location of accommodation	
Peak District National Park*	29.5
Ashbourne	10.7
Matlock	10.7
Derby connurbation	9.8
Nottingham connurbation	11.5
elsewhere in East Midlands	3.3
West Midlands	4.0
Manchester connurbation	4.0
elsewhere in UK	16.5
Transport	
car/van	94
public transport	2
bicycle	2
on foot	2

* Includes non-named areas within 5 km of the park boundary.

Major findings: site use

The overall size of the sample population makes accurate determination of the actual visitor pattern difficult. However, couples are a major component, and groups of three persons consistently account for about a fifth of group visits, throughout the week (Table 4).

The high number of first visits (Table 5) suggests that the Stone Centre has yet to establish itself as a regular-use facility by those from the nearby conurbations of Derby and Nottingham. Both the shop and cafe are clearly an asset welcomed by visitors. The exhibition, whilst having some popularity, is overshadowed by the geological trail; of course, this might be related to the entry charge to the exhibition and the free and open access, with or without the accompanying leaflet, to the geological trail!

Major findings: marketing

Interestingly, many respondents commented upon poor road signage for the site and yet it seems to be the most effective form of advertising (Table 6). More formal advertisements seem less effective than favourable comments from

Table 4. *Daily visitor pattern (based on all respondents)*

Day	Percentage of respondents within listed group size						
	1	2	3	4	5	6	7
Saturday	—	21	21	42	16	—	—
Sunday	—	35	12	24	18	—	12
Monday	—	23	31	31	8	—	—
Tuesday*	—	50	—	50	—	—	—
Wednesday	—	50	17	17	5	11	—
Thursday	6	44	12	31	6	—	—
Friday*	—	—	17	33	50	—	—
Saturday	—	55	22	22	—	—	—
Sunday	18	45	—	—	36	—	—
Monday	—	17	17	67	—	—	—

* Visitor sample too small for valid result.

Table 5. *Use of on-site facilities by respondents*

Use of facilities	Percentage of respondents
First-time visitors	84
Had made 2 or more visits during summer	15
Used the geological trail	57
Visited the exhibition	40*
Made a purchase at the shop	54
Used the cafe	42
Had a packed lunch on-site	22

* Estimated.

Table 6. *First knowledge of the site by respondents*

Means of discovery of site	Percentage of respondents
Tourist road signs	35
Leaflet	20
Word-of-mouth	15
Advertisement	8
Magaziene/newspaper	5

Table 7. *Daily newspapers regularly read by respondents**

Newspaper	Percentage of respondents
Daily Mail	13.9
Daily Telegraph	13.9
Daily Express	9.8
The Independent	8.2
The Guardian	7.4
Daily Mirror	6.6
The Times	3.3
The Sun	1.6
Other (mainly local papers)	25.4

* *Note*: 77.1% of respondents regularly read a daily newspaper; 22.9% do not.

Table 8. *Geological knowledge of interview respondents*

Category	Percentage interview respondents
Had never studied geology	61
Had studied geology as a hobby	17
Had studied geology at school	4
Had graduated in geology	5
Had studied geology as part of another subject (e.g. geography, mining)	9
Could name geological systems*	
Precambrian	9
Cambrian	16
Ordovician	8
Silurian	10
Devonian	5
Carboniferous	16
Permian	5
Triassic	12
Jurassic	96
Cretaceous	16
Miocene	3
Eocene	3
Oligocene	0
Pliocene	5
Pleistocene	8
Could identify from a drawing	
an ammonite	42
a bryozoan	3
a graptolite	5
a brachiopod	5
a colonial coral	9
a trilobite	20
a belemnite	14
a crinoid	13
a bivalve	8

* 20% could either name the system represented on-site or give an approximate age in Ma.

respondents' associates. The attractive full-colour leaflets are an obviously effective advertising medium.

The high readership of newspapers (Table 7), if not of the 'better quality' press, is indicative (given also the poor **admitted** showing of the tabloids) of the educational attainment of the visitor population. Significant in marketing terms was the high frequency of respondents taking regional evening, rather than metropolitan national, papers available in the East Midlands and Yorkshire and Humberside.

Major findings: geological knowledge

Of those people interviewed, 51.3% had either been or expressed an interest (weather permitting!) to go around the on-site geological trail. An attempt was made to assess both the general and trail-specific geological knowledge of those interviewed (Table 8).

The lack of formal study in geology is not surprising. Whilst the National Curriculum's inclusion of Earth sciences should boost children's interest and might encourage parents to seek out appropriate sites, it is too early to suggest that it will radically alter over the next decade the adult population's geological knowledge and understanding. However, at present, a significant minority of visitors claim some knowledge and interest. The view of Cossons (1974) writing on visitors to industrial heritage sites is apposite:

> '... the enthusiast visitor does represent a significant proportion of the total he is in a small minority when compared with educational parties or visitors with a general interest but little pre-existing knowledge.'

The results of the detailed questioning on geological knowledge suggest that the enthusiasts are some way off from achieving a degree of basic competence. The ability to name one particular geological system appears to be related more to media hype than basic knowledge; many respondents thought the Jurassic (in *'Jurassic Park'*) was a place!

Summary

The existing client base has been identified and categorized. The measured pre-existing and site-acquired knowledge of visitors suggests that there is a need to raise public awareness, knowledge and understanding of geological sites. Some means by which these can be achieved have been indicated: visitors prefer doing and seeing ('hands on'), to reading labels ('hands off') in exhibitions. In conclusion, as in the Australian example cited earlier:

> '... this study has highlighted the need for recognition of the needs and desired experiences of fossickers as a special interest travel segment... The results of analysis of visitor profiles and preferences will hopefully contribute to better marketing and management of fossicking opportunities.'
>
> Jenkins 1992, p. 140.

The findings reported in summary in this paper represent the early work in an on-going special interest tourism research project which aims to explore and develop a significant understanding of the micro- and macro-economic impact of site-specific geological interpretation within the UK. As more data are gathered and collated a clearer profile of visitors, existing and potential, to interpreted geological sites will emerge, enabling the development of a visitor and resource management model. Given the policies of the UK government, in relation to the funding of conservation generally and to access to the UK's rich geological heritage specifically, there is a clear need for such a model as a performance evaluative tool.

References

Cossons, N. 1974. The Conservation of Industrial Monuments. *Museums Journal*, **74**(2), 62–66.

Grange, W. 1993. The Lie of the Land. *Museums Journal*, **93**(3), 21.

Herbert, D. T. 1989. Does Interpretation Help? *In*: Herbert, D. T., Prentice, R. C. & Thomas, C. J. (eds) *Heritage Sites: Strategies for Marketing and Development*. Avebury, Aldershot, 191–230.

Hewison, R. 1987. *The Heritage Industry*. Methuen, London.

Jenkins, J. M. 1992. Fossickers and Rockhounds in Northern New South Wales. *In*: Weiler, B. & Hall, C. M. (eds) *Special Interest Tourism*. Belhaven Press, London, 129–140.

Knell, S. J. 1986. *The Natural History of the Frodingham Ironstone*. Scunthorpe Museum.

Peak Park Joint Planning Board 1988. *Visitor Survey 1986/7*. PPJPB, Bakewell.

Prentice, R. C. 1989. Visitors to Heritage Sites: a Market Segmentation by Visitor Characteristics. *In*: Herbert, D. T., Prentice, R. C. & Thomas, C. J. (eds) *Heritage Sites: Strategies for Marketing and Development*. Avebury, Aldershot, 15–61.

Purcell, D. (undated) *The Stones of Ely Cathedral* (2nd edn). Friends of Ely Cathedral.

Sports Council 1992. *The Amenity Reclamation of Mineral Workings*. Sports Council, London.

St Albans Museum 1990. *Discover St Albans Building Stones*. St Albans Museum.

Thomas, I. 1984. National Stone Centre. Geology and Physiography Section, Nature Conservancy Council, Information Circular, **21**, 11–16.

—— & Hughes, K. 1993. Reconstructing Ancient Environments. *Teaching Earth Sciences*, **18**(1), 17–19.

Keynote address

Reading the landscape

ALAN McKIRDY & ROBERT THREADGOULD
Scottish Natural Heritage, 2 Anderson Place, Edinburgh EH6 5NP, UK

Abstract: Scotland, for its size, has proved to be one of the world's most important natural laboratories for developing our understanding of Earth sciences. Many of the fundamental principles of Earth sciences were first developed in Scotland, and yet today the subject is one of the least understood amongst the wider community.
One of the key objectives of Scottish Natural Heritage (SNH) is to 'foster understanding and facilitate enjoyment of the natural heritage of Scotland', and, to that end, SNH is currently developing a series of initiatives, using the landscape as a basis for interpretation, to raise public awareness of the Earth sciences. Scotland's varied landscape is a reflection both of the underlying geology and the effects of weathering and erosion. This natural resource provides an invaluable platform from which to educate and inform the wider community in all aspects of the Earth sciences. The geological and geomorphological evolution of Scotland is an exciting and complex story which may be interpreted and brought to life using a variety of methods such as the publication of guides and booklets, development of visitor centres, on-site interpretation, public lectures and the media.

Not since Victorian times has the study of geology and geomorphology fired the public's imagination. Before the turn of the century, there were gentlemen clerics and field naturalists aplenty in Britain who were knowledgable about their local geology, and the study of rocks and landforms was integrated into their general understanding of countryside matters. It now seems, however, that the generalist has had his day and that the field of scientific endeavour is now largely the preserve of the specialist with a narrower view of the world. In this shift of emphasis from an holistic understanding of our surroundings to a more specialist approach, some subjects such as the Earth sciences have almost disappeared from public view.

In selling the need to conserve our Earth heritage to a wider audience than those who are professionally involved, we are faced with a number of challenges. Perhaps the most fundamental of these is the general lack of awareness of the subject in the population at large, so, in Britain, the sense of loss which would be felt if an important geological feature was damaged or destroyed, for example, would be minimal. Contrast this with the collective response of a local community to the loss of a much-loved hay meadow or bluebell wood. The response has a great deal to do with the value which the community ascribe to the various elements of their local environment. Most can intuitively appreciate the beauty and botanical diversity of the hay meadow and, therefore, when its existence is threatened it becomes a matter of concern to the community. In the case of a rock exposure, unless it is of cultural or historical significance, then it is unlikely that its loss would occasion much comment. It is against this background that the conservation of Earth science features has taken place in Britain in recent years.

The most obvious catalyst for change is for Earth sciences to have a more prominent place in the school curriculum. This would ensure that the next generation of decision-makers had a better understanding of Earth history and therefore the need to conserve its most important elements. To some extent, this change has already occurred in England and Wales, with the study of Earth history given a prominent position within the National Curriculum. In Scotland, however, that important change has yet to take place.

In any case, it will take time before the benefits of such changes are realized and to subsequently create a more receptive audience for the need to conserve geological and active process sites. The conservation agencies are a very important source of information and can help to create a climate in which effective conservation initiatives are more likely to flourish. Scottish Natural Heritage (SNH) are very keen to play their part in this process and we are encouraged in this endeavour by our founding legislation (Natural Heritage Act (Scotland),

1991). A number of initiatives are already at the planning or implementation stage and the following sections of this paper describe these projects in some detail.

Raising public awareness

The geological story we have to interpret in Britain is a fascinating one. Drifting continents, disappearing oceans, hypersaline lakes, volcanoes 30 km across and an ice age that lasted for two million years are all ingredients of the plot. Our role is to make these events which took place in the dim and distant geological past come alive, by presenting the information in an interesting way.

A recent survey demonstrated that most visitors come to Scotland because of the fine scenery and unspoilt countryside (Scottish Tourism Co-Ordinating Group 1992). This provides the ideal opportunity for SNH to build on this appreciation of the countryside by providing an informed commentary on the way in which the landscape has developed over geological time. The intention is that this technique of 'reading the landscape' will add a new dimension to the way in which people view our most precious natural asset and eventually lead to a fuller appreciation of the dynamic nature of the processes which fashioned the hills and glens as we see them today. It is also important that there is a much wider appreciation that processes such as river flow, slope movements and coastal change are sensitive indicators of landscape change. Understanding the dynamics of these geomorphological systems and their underlying controls is fundamental to the prediction of the effects of natural change or particular land management strategies.

Information for the general public can be delivered in a variety of different ways, such as:

- publication of leaflets, guide books and inclusion of geological information in more general tourist literature
- visitor centres ideally with audio-visual facilities
- on-site notice boards
- through the media

Although there are a number of partner organizations who are also involved in this area, such as local authorities and the museum service, it may become a primary function of SNH to ensure that this activity is adequately co-ordinated and delivered throughout the country.

1. Landscape fashioned by geology

In partnership with the British Geological Survey (BGS), SNH are currently producing a series of booklets which relate landscape to the underlying geology. This series is designed to appeal to a general readership and will be widely available. Geology and geomorphological processes are explained with reference to the familiar views and landscapes, so this series is the embodiment of the 'reading the landscape' approach to site interpretation.

The first two titles in the series, covering Edinburgh (SNH-BGS 1993a) and the Isle of Skye (SNH-BGS 1993b), have already been produced. Other titles which are planned include the Cairngorms and the Isle of Arran; both areas are popular tourist destinations.

2. Visitor centres

One of the most effective methods of raising public awareness in any natural subject is by way of a visitor centre. SNH is currently involved in developing a visitor centre at Knockan Cliff in NW Scotland, an area famous for its excellent exposure of the Moine Thrust and dramatic landscape. This part of Scotland has proved to be one of the world's most important natural laboratories for the study of geology. Many of the early breakthroughs in our understanding of thrust tectonics were made here by the pioneering geologists of the Geological Survey. In more recent times, geologists have continued to unravel the complex geological history of the area which includes 3-billion-year-old Lewisian gneiss, 1100-million-year-old shallow marine sediments (Moine Supergroup), 800-million-year-old alluvial and fluvial deposits (Torridonian sandstones) and 590–510-million-year-old shallow marine carbonate sediments (Cambro-Ordovician sequence); the foundations from which the unique and varied landscape of this part of NW Scotland was moulded some 10 000 years ago during the last ice age. These foundations are displayed in a recently developed interpretative wall (Fig. 1) which provides, in essence, a miniature geological cross-section through the area. To date, however, this understanding has largely been the domain of the academic researcher, or Earth science student, with the result that the majority of the wider community are largely unaware of this fascinating history.

It is envisaged at this stage that the centre will incorporate a main exhibition area, audio-visual theatre and 'ice-age' walkway. The 'ice-age' walkway, aided by interpretation boards, will

Fig. 1. Interpretative wall at Knockan Cliff. The wall represents a geological cross-section, in miniature, of the surrounding landscape.

use views of the Inverpolly National Nature Reserve to explain the role played by the last ice age in sculpting the landscape we see today. The Earth history will be further explained using audio-visual presentations, static and interactive displays. Our focus, however, will be wider than the immediate location of the visitor centre. It is also proposed that a series of 'satellite sites' will be used to illustrate the themes developed at the visitor centre. On-site noticeboards at each of these sites will provide an opportunity to describe the particular feature of interest, be it a rock face, landform or a panoramic view and also to refer the visitor to the centre where more information is available.

The centre, therefore, becomes the focal point from which the visitors can explore the surrounding countryside, providing an additional tourist facility for the area and the means by which information on the Earth sciences and landscape can be effectively delivered to an estimated audience of over 20 000 people each year. The primary benefit is that very few of these visitors will have had, or felt the need for, information on the Earth sciences.

In contrast to the remote nature of the Knockan Cliff site, a disused limestone quarry at East Kirkton, on the outskirts of the town of Bathgate, within easy reach by car from the major cities of Edinburgh and Glasgow, provides the ideal location for an additional visitor centre. The East Kirkton Quarry represents Carboniferous lagoonal or lake deposits which were associated with hot springs driven by nearby volcanic vents. Forested areas around the lake supported a considerable diversity of species. Fossil plants, invertebrates and vertebrates, including the remains of the famous Lizzie (*Westlothiana lizziae*), the oldest recorded fossil reptile, have been discovered at the quarry. The site, therefore, provides an ideal opportunity to raise public awareness within the urban environment; its position affording a ready and considerably sized audience within easy reach of the centre.

Both projects have a common requirement to involve the local community at every stage of the process, from planning through to implementation.

3. On-site interpretation boards

The use of on-site interpretation boards can greatly enhance the interest of a site or area to the passing tourist. One such example is situated on the northern side of the Firth of Forth. Looking southward towards the Edinburgh district, the board illustrates an artistic interpretation of the Carboniferous palaeoenvironment as it may have looked some 330 million years ago (Fig. 2).

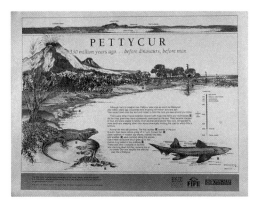

Fig. 2. Interpretative board at Pettycur, north side of the Firth of Forth. The board relates an artisitc impression of a Carboniferous palaeoenvironment with the landscape of the Edinburgh District.

4. Media

Television and radio are media that we have yet to exploit to any great degree, but considerable potential exists for the 'reading the landscape' approach to flourish. The public have an insatiable appetite for programmes on all aspects of natural history but, until recently, geology was considered to be too specialist for general viewers. This is now changing, as it is accepted practice for programmes on a particular habitat type of geographic area to be prefaced with an introductory explanation of the geological history of the region. Clearly, this is a useful beginning, but the next stage is for the primary focus of the programme to consider Earth history as a subject in its own right.

Conclusions

In promoting the Earth sciences, we have a great deal to learn from the way in which dinosaurs have been 'marketed'. This subject has been elevated from a rather esoteric scientific study to that of cult status, where almost every child in the country has a collection of plastic dinosaur models and an encyclopedic knowledge of the subject. From the less than promising starting point of a collection of some dusty old bones, approaching a 100% market penetration has been achieved! This success has largely been achieved by demystifying the science, allowing the subject to come to life. In the same way, SNH aims to demystify the wider aspects of the Earth sciences and use the concept of 'reading the landscape' as a window through which appreciation and understanding of the Earth sciences and Earth science conservation can be developed and brought to life.

References

NATURAL HERITAGE ACT (SCOTLAND). 1991. HMSO.

SCOTTISH TOURISM CO-ORDINATING GROUP. 1992. *Tourism and the Scottish Environment – A Sustainable Partnership.* Scottish Tourist Board, Edinburgh.

SNH-BGS. 1993a. *Edinburgh – A landscape fashioned by geology.* Scottish Natural Heritage, Edinburgh.

——. 1993b. *Skye – A landscape fashioned by geology.* Scottish Natural Heritage, Edinburgh.

Involving landowners, local societies and statutory bodies in Shropshire's geological conservation

PETER TOGHILL

School of Continuing Studies, University of Birmingham, Edgbaston, Birmingham B15 2TT, UK

Abstract: The production of two recent geological teaching trails in Shropshire has highlighted the importance of involving landowners, local councils, statutory bodies, and local and national societies. The trails through the Ordovician rocks of the Onny Valley, and Precambrian/Cambrian rocks of the Wrekin area, Shropshire show how a number of sites can be joined together to form an important educational project, as well as conserving the sites for the future. Without co-operation from a large consortium of interests neither of these two guides would have been published.

Landowners are often keen to become involved when geological parties can be more sensibly organized along a laid-out route, and a central permission granting system can be set up.

County Councils along with local district councils have a statutory role to play in providing countryside amenities, and may well wish to be involved in geological conservation, and management of sites. Other statutory bodies such as English Nature and the Countryside Commission can provide support and financial aid for fencing and signs, and the publication of a guide, and English Nature will always be keen to help with the practical conservation of its Site of Special Scientific Interest (SSSIs) and Regionally Important Geological/geomorphological Sites (RIGS).

Trails of this type pass through countryside with other features as well as geology, and the inclusion of other interests in the guide, e.g. ornithology, vegetation, soils, etc., is one way of involving a large group of organizations and people in geological conservation, and making a piece of landscape attractive to a wider audience.

Without co-operation between local and national organizations, many geological sites will continue to be uncared for and of little use, and may become unavailable.

The classic geological region of Shropshire (UK) contains a more varied sequence of rocks than any other region of comparable size in the UK (Toghill 1990). The county ranks high in the league table of visits from geologists and geological parties of all educational levels.

Although there was a great burst of geological conservation activity in the county in the 1970s, including the purchase of the famous Comley Quarry by the Shropshire Wildlife Trust in 1976, little has been done to provide information to the visitor about the county's geological heritage. Apart from British Geological Survey maps and memoirs, only four modern geological guides are available, written by professional geologists. These are:

(1) *Mortimer Forest Geology Trail* (Jenkinson 1991);
(2) *Onny Valley, Shropshire: Geology Teaching Trail* (Toghill 1992);
(3) *Ercall Quarries (Wrekin Area) Shropshire, Geology Teaching Trail* (Toghill & Beale 1994);
(4) *Wenlock Edge: Geology Teaching Trail* (Harley 1988).

The problem is that many Shropshire sites are isolated localities, and many of these, although of great historical importance, offer little to the casual visitor at present, e.g. Comley Quarry. The excellent, but now dated, Geologists' Association (GA) guide to South Shropshire (Whittard 1958; revised by Dean 1968) provides itineraries but little information about individual sites, and entries are often only a few lines. The GA guide was really an excellent way of guiding people to localities, but what the visitors then got out of the exposures was up to them.

However, a great deal of information is now available on Shropshire through academic journals and BGS files and maps, including the classic Church Stretton Memoir (1968). The Regionally Important Geological/geomorphological Sites (RIGS) scheme, supported by English Nature, is in its early days in Shropshire, but a large number of sites (several hundred) are already on file in records of the National

Scheme for Geological Site Documentation (NGSD), run by the Geological Curators Group and the Conservation Committee of the Geological Society of London, and on files in Ludlow Museum. All this information needs to be made available to the geological visitor.

It would seem appropriate to try and link a number of Shropshire sites into geology teaching trails which explain the geology of a small area, and link a number of sites which form a geological entity. Thus the author's (Toghill 1992) *Onny Valley, Shropshire, Geology Teaching Trail,* published as a GA Guide, with support from English Nature, set out to describe the Ordovician (Caradoc Series) rocks of the Onny Valley, and place them in the context of the Ordovician geology of Shropshire.

Landowners are often concerned about geological parties wandering over their land in a disorganized way and many would welcome the production and appearance of a guide which:

(1) puts the geologists on the right path;
(2) tells the visitor where to obtain permission;
(3) is published by a bona fide organization to which the landowner can appeal if things go wrong;
(4) allows tidying up of old quarries and exposures, and probably provides funds for new fencing, stiles, etc.;
(5) and finally provides a better relationship between geologists and landowners.

Consortia of interests

It is probable that a consortium of interests can become involved in the guides' production. County and local councils have a statutory interest in amenity provision, and can provide funding. English Nature has an obligation to look after its Sites of Special Scientific Interest (SSSIs) and engage in geological conservation. The Countryside Commission can provide funding, as it did in the Onny Valley, and in fact laid out its own Onny Valley Trail, along which the geological localities occur. The local Wildlife Trust will want to be involved, as the Royal Society for Nature Conservation (RSNC) has a geological remit, and the local trust may have geologists as members.

The local geologists as members of local geological societies can produce the text for the guide, but academics from more distant institutions may have to do this. If the area includes quarries, then quarry companies will probably be interested in being involved. Final publication of the trail guide by the GA or English Nature is a preferable way of getting the guide well known.

Derelict sites

Many classic Shropshire sites are overgrown old quarries which offer little to the visiting geologists and interested visitors. As each season passes, the nettles grow higher and the rubbish piles up. A good example of this is Meadowtown Quarry in the Shelve area, and it is probable that Comley Quarry would be in the same state if it hadn't been purchased by the Shropshire Wildlife Trust in 1976. Even roadside exposures like the famous unconformity at Hope Bowdler need regular maintenance to keep the grass and trees at bay.

Ercall Quarries

Ercall Quarries (Wrekin area) Shropshire (Toghill & Beale 1994) is an excellent example of what can be done, and is being done, in an area of large disused quarries (Fig. 1) when a consortium of interests is involved. Quarries of this type, when working activity ceases, soon become the focus of rubbish dumping, illegal motor cycle scrambling, and invasion of 'New Age travellers', as happened twice at Ercall Quarries in 1992.

These quarries contain both a geological and biological SSSI and are well visited by geologists who study the late Precambrian volcanics and early Cambrian sediments. The exposures are excellent for teaching at all levels, and research still continues (Wright *et al.* 1993) into the classic exposures centring on the Cambrian/ Precambrian unconformity.

The Ercall Quarries ceased working in 1985. At that time Shropshire County Council and the Nature Conservancy Council, along with advisers from the Shropshire Geological Society, asked the quarry owners to make the quarry faces safe, but to leave exposures for geological teaching, and to provide ramps and buttresses which, while making faces safe, also provided benches for access for geologists. This was all done admirably by Johnston Roadstone Ltd, and then the quarries and surrounding woodlands reverted to the complete ownership of the original landowner.

A serious problem arose in 1992 when the owner found the land becoming more and more open to public abuse, and he contemplated denying access to all, including geologists. However, since then a consortium of interests (Shropshire County Council, English Nature, Shropshire Wildlife Trust, Shropshire Geological Society, with possible support from the GA) has been meeting (with the owner's knowledge) to discuss ways of managing the area. Members of the Shropshire Geological

Fig. 1. Cambrian–Precambrian unconformity, Ercall Quarries, Shropshire.

Society, including the author, have been writing a GA Guide to the area, spurred on by the fact that the British Association will be visiting the quarries in August 1993. The guide (Toghill & Beale 1994) includes a geological trail covering all the well-known features of the quarries, and is aimed at all geological levels of knowledge. It is interesting to note that the County Council's involvement has a lot to do with a commitment to provide open space amenity reserves and a general education of the public towards an appreciation of countryside heritage, including geology.

It has been said that the sooner a management agreement is reached the better, since derelict land can soon be abused and take on an unattractive appearance. The geological exposures if not managed well soon loose their importance for teaching.

Conclusions

There is a need for a lot more co-operation between landowners, geologists, statutory bodies, and local and national geological societies, towards providing geology teaching trails. These should cover concise geological topics in areas capable of being visited in a day or less.

Sites in Shropshire which could be covered by geological trails and guides include: The Longmynd; Stretton Hills; Stiperstones/Shelve area; Hoar Edge/Chatwall area; Wenlock Edge area to cover more than the present Wenlock trail, and link in with the Ludlow Mortimer Forest Guide; Ironbridge area; Clee Hills; Carboniferous of NW Shropshire; and New Red Sandstone of North Shropshire plain.

References

GREIG, D. C., WRIGHT, J. E., HAINS, B. A. & MITCHELL, E. H. 1968. *Geology of the country around Church Stretton, Craven Arms, Wenlock Edge and Brown Clee.* Memoir of the Geological Survey of Great Britain, HMSO, London.

HARLEY, M. J. 1988. *Wenlock Edge: Geology Teaching Trail.* Nature Conservancy Council, Peterborough.

JENKINSON, A. 1991. *Mortimer Forest Geology Trail.* Scenosetters/Forestry Commission (based on an earlier edition – LAWSON, J. D. 1973, published by NCC, Peterborough.

TOGHILL, P. 1990. *Geology in Shropshire.* Swan Hill Press, Airlife Publishing, Shrewsbury.

—— 1992. *Onny Valley, Shropshire: Geology Teaching Trail.* Geologists' Association Guide No. 45. Geologists' Association, London.

—— & BEALE, S. 1994. *Ercall Quarries, Wrekin Area, Shropshire: Geology Teaching Trail.* Geologists' Association Guide No. 48. Geologists' Association, London.

WHITTARD, W. F. 1958. *Geology of some Classic British areas: Geological Itineraries for South Shropshire.* Geologists' Association Guide No. 27 (revised by DEAN, W. T. (1968). Geologists' Association, London.

WRIGHT, A. E., FAIRCHILD, I. J., MOSELEY, F. & DOWNIE, C. 1993. The Lower Cambrian Wrekin Quartzite and the age of its unconformity on the Ercall Granophyre. *Geological Magazine,* **130**, 257–264.

The geo-ecological education and geological site conservation in Romania

DAN GRIGORESCU

Faculty of Geology and Geophysics, University of Bucharest, Blvd. N. Balcescu, 70111 Bucharest, Romania

Abstract: The economic and social reform in the Central/Eastern European ex-communist countries placed environmental protection among the most acute priorities. Environmental protection should not refer only to the air, water, soil, flora and fauna, but also to places of 'spiritual value' such as geological sites with a particular scientific significance.

The understanding of the history of the Earth and the evolution of its equilibria should start from these places in order to achieve a solid ecological education of the young and of the public in general. Environmental studies in Romania have an increasing practical significance because of the growing impact of humankind's activities upon the environment. Such studies are inevitably interdisciplinary and cut across the traditional boundaries between sciences such as biology, geology, geography, chemistry, oceanography, law and engineering. The Romanian universities are extensively introducing environmental sciences into their curricula. Within the geological departments of the universities of Bucharest and Cluj a new specialization in geo-ecology is developing in close co-operation with universities from the UK, France and Greece. The objectives of this new speciality also include conservation of the several hundreds of geological sites spread throughout the country, from the Carpathian chain to the Danube Delta.

The impact of accelerated industrialization during the last half century is felt not only on the air, water, soil, flora and fauna of the Earth, but also on 'geological sites' – a term which designates those places that depict phenomena and events from the geological past with a special scientific and/or educational value.

The destructive effects on nature of these industries are more dramatic in the ex-communist Eastern European countries, including Romania, than in the western countries where these effects were diminished by a more or less efficient legislation. 'The conquering of nature by man', the communist slogan, has represented not only an extremist form of anthropocentric conception but also a way of action against nature, deliberately ignoring ecological equilibria.

Geological exploitation, for ores, coal, oil and gas, and for rocks in quarries, closely linked to industrial development, affects the 'vital elements' of nature by the pollution of air, water and soil, by massive afforestation, waste disposal, hazards to local populations and the 'spiritual elements' in which we include the geological sites and those of cultural and historical importance.

The aggressive attitude towards nature, which is specific for the communist society, has determined the quasi-absence of a pragmatic ecological education in schools and universities, and the lack of interest in training of professionals for environmental protection.

The geological heritage of Romania and its conservation

From a geographical and geological point of view Romania is a 'nature-gifted' country, with an equilibrated distribution of the three main geomorphological units: mountains, hills and plains, each of them roughly representing one-third of the total area of the country.

The Carpathian mountains arch across from the northeast to the southwest of the central part of the country, encircling the Transylvanian Plateau; the Black Sea and the Danube Delta (the largest delta in Europe) bound the southeastern side of the country.

From a geotectonic point of view Romania is included within the Alpine region of Europe, most of the country being consolidated during the Mesozoic and Cenozoic eras, but Proterozoic and Palaeozoic tectogenesis are also implied in the alpine basement; these facts are reflected in a wide chronostratigraphical range of the outcrops that cover most stages of the geochronological scale from the Early Proterozoic to the Holocene.

The variety of the petrogenetic conditions is reflected by the diversity of the rocks: igneous, volcanic, metamorphic and sedimentary. This geological variety is depicted by the 'geological sites' that can be grouped as mineralogical – petrographical, palaeontological (including palaeobotany, invertebrate and vertebrate sites), stratigraphical (stratotypes), structural–tectonic, geomorphological (impressive landscapes generated by water and wind erosion) and speleological (caves and other karstick phenomena). Currently, more than 100 sites have the recognized status of 'geological reserve'. These represent 0.2% of the total area of Romania (23 750 200 h). Generally, the protected areas currently encompass 4.8% of Romania's surface including three 'Biosphere reserves' of which the Danube Delta is by far the largest, 12 national and natural parks, 571 strictly protected areas, including the geological sites.

While Romania has been active in designating protected areas, the management of these areas is practically non-existent. This lack of management is partly due to the fragmentation of responsibility for identifying, designating and managing the protected areas between the central (governmental) and local authorities.

Also, the inefficiency of the general legislation for the protection of the environment (Law 9/1973 still in action), in spite of some recent added normative acts is proved by the lack of severe penalties for damage to the environment.

In practice, this law enabled the communist authorities to subordinate totally the environment to economic interests. Under the protection of Law 9/1973 huge quantities of Mesozoic limestones, some of them with a particular palaeontological and/or biostratigraphical importance were converted into cement and sold abroad at a dumping price; entire regions, especially in the hilly areas were entirely demolished, including villages for the exploitation of a low quality lignite (the case of the region south of Târgu Jiu in Oltenia) or of some poor bituminous schists (near the small town of Anina in Banat); and impressive peak of a volcanic mountain in the Calimani range was totally demolished for the exploitation of a small quantity of poor quality sulphur.

Of course, the economic and social development of a country cannot avoid environmental damage, but these can be diminished greatly by rational activity. A good example of two differing attitudes is given by two stages of great reconstruction within Bucharest. During the first decades of this century a large number of big buildings were raised in the central area of the capital. The foundations that were excavated uncovered hundreds of specimens of large mammals (mostly elephants, rhinos, bovids and horses). The builders, educated at the beginning of the century in the spirit of respect for science, considered it their civic duty to stop the digging everytime that fossil bones were unearthed, calling in palaeontologists. Today, these specimens are to be found in the collections of the University and of the Museum of Natural History in Bucharest.

Contrary to this, during the last twenty years when the Bucharest subway was built, a large part of the capital subsoil was excavated, but no fossils were reported! This reflects the new ideology, very pragmatic, implying a rush for 'efficiency' and a deliberate contempt for relics of the geological past. As a result, any bones were destroyed by the heavy machines. To stop the work for such trifles was regarded as a waste of time!

During the communist era interest for the conservation of geological sites was maintained only by small tourist and scientific associations or youth clubs which, lacking any legislative authority, were practically inefficient. They now have to resume their activities in the study and designation of places with a particular scientific importance. A large number of proposals for new reserves were submitted to the 'Commission for the Protection of Nature Monuments' by the Romanian Academy, but due to the lack of funds activity has been stopped at this level, preventing any concrete action for the management and monitoring of the reserves.

A new environmental strategy and the perspective for the geological site conservancy

After December 1989, Romania embarked on an ambitious macroeconomic adjustment programme to transform its economy from a centrally planned to a market economy. Among the major priorities of the reform during the transition period is the improvement of the quality of the environment.

The newly created Ministry of the Environment (since 1992 the Ministry of Water, Forests and Environmental Protection) started to build a new environment strategy for a short- and long-term perspective. The short-term strategy concentrates on localized environmental 'hot spots', where the level of the air, water and soil pollution is higher, mainly due to the toxic emissions and solid waste produced by chemical, mining and oil industries.

The long-term strategy will encourage economic growth in an environmentally acceptable manner. Different international bodies and organizations are co-operating in defining and financing the new environmental strategy, among which are the Operations Service of EC-PHARE, the Environment and Health Centre, the World Bank and US AID. Besides the Government, the non-governmental organizations (NGOs) are encouraged to contribute to environmental management.

The expected efficiency of the strategy depends essentially on the new general environmental legislation, which is still being debated by technicians from the different ministries involved, with a minor contribution from the NGOs. The drafted legislation for the environment contains a special section for the parks and protected areas, which includes the geological reserves. This chapter was criticized by the 'Society for the Conservancy of the Geological Heritage' (SCGH) in terms of definitions, responsibilities for the management of the geological sites and sanctions.

The new environment strategy launched by the Romanian government foresees an increase in the protected areas, including both animate and inaminate reserves from the current 0.64% of the total area of Romania to 1.2%.

The drafted environmental legislation divides the protected areas in accordance with their importance into **national** areas that come under government responsibility and **local** areas, that come under county responsibility; in both the NGOs might contribute with some specific activities.

As regards the geological sites, the only organization in the country that constantly deals with their study and protection is the SCGH, a non-governmental, non-profit organization, officially recognized since 1990. It continues the activity of a student research Circle for the Protection of Geological Monuments, established in 1978 in the Faculty of Geology at the University of Bucharest. The members of the circle were students and professors from the faculty. Their main tasks were:

- to inventory the physical conditions of the already declared geological reserves (80 at the time when the circle began its activities, out of which a dozen have been destroyed by natural erosion or by humans, due to the lack of an efficient control and protection management;
- to update scientific knowledge on the existing geological monuments and to make collections of representative samples from each reserve;
- to designate new geological sites to be protected.

The activity of the circle has to be resumed to sample in the field and for specific researchers (petrographical mineralogical, palaeontological) in the laboratory. The results of the studies were presented in the ordinary meetings of the circle and during the yearly sessions of the faculty; the best studies were printed in University publications or in the Academy Bulletin *Protection of Nature*.

In one way, one of the merits of the circle was the maintenance of interest in geological monuments as places representative of the 'Earth-memory', in a period when they were otherwise neglected by the authorities.

For geology students especially, such places also have an exceptional didactic importance; these students are the first who understand the irreversibility of the damage caused naturally or artificially, as well as having a duty for trying to avoid such damage as much as possible.

Immediately after December 1989 the circle was converted into a NGO, which has considerably increased the number of members (the SCGH with its headquarter in Bucharest now has branches in other university towns: Cluj-Napoca, Ploiesti, Baia-Mare); also the assumed activities were enlarged with curation, legislation and education.

The studies on the reserves are continuing in a more systematic way, the geosites (more than 100 now) are divided into districts, in order to enable a more efficient management and monitoring (Fig. 1). Using computer facilities all the information is recorded on standard geosite sheets, aiming to build a complete bank of data.

Some museums have undertaken responsibility for recording the geosites within their territory.

A draft legislation for the geosites (including the speleological ones) was submitted to the Ministry to be considered within the general law of the environment. The draft includes classification of the geosites in accordance with their type (mineralogical–petrographical, palaeontological, stratigraphical, tectonic, geomorphological) as well as with their importance (local, national, international; scientific, educational), aspects regarding the curation, organization of the collections, accessibility to the sites and collections etc.

The educational activity of the SCGH is developed through:

- informal conferences during the summer fieldwork presented by the members of the working groups in different areas for the local people, emphasizing the scientific/ educational importance of the geosites in

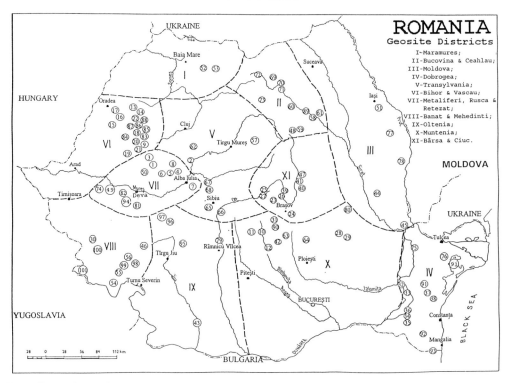

Fig. 1. Romanian geosites. 1, Dealul cu Melci; 2, Gârbova de sus; 3, Detunatele; 4, Piatra Corbului; 5, Pietrele Ampoiței; 6, Valea Mică; 7, Râpa Roșie; 8, Iezerul Ighiel; 9, Ghețarul Scărișoara; 10, Granitul de Albești; 11, Calcarul de Albești; 12, Suslănești; 13, Peștiș; 14, Valea Crișului; 15, Piatra Neamțului; 16, Tăsad; 17, Betfia; 18, Bohodei; 19, Izbucul Călugări; 20, Cetățile Ponorului; 21, Sighișel; 22, Cheile Crișului repede; 23, Măgura Codlei; 24, Purcăreni; 25, Carhaga; 26, Ormeniș; 27, Racoș; 28, Bădila; 29, Vulcanii noroioși de la Păclele Mari și Păclele mici; 30, Soceni; 31, Vama Strunga; 32, Hârșova; 33, Topalu; 34, Cernavodă; 35, Alimanu; 36, Seimeni; 37, Cheia; 38, Gura Dobrogei; 39, Aita Seacă; 40, Turia; 41, Valea Iadului; 42, Plaiul Domnesc; 43, Bucovăt; 44, Rateș; 45, Barboși; 46, Cheile Corcoaia; 47, Lacul Sf. Ana; 48, Lacu Roșu; 49, Lăpugiu; 50, Vâlcanu; 51, Dealul Repedea; 52, Chiuzbaia; 53, Creasta Cocoșului; 54, Svinița; 55, Bahna-Vârciorova; 56, Ponoare; 57, Lacu Ursu; 58, Cozla-Pietricica; 59, Munticelu-Cheile Șugălui; 60, Piatra Teiului; 61, Stânca Șerbești; 62, Cheile Turzii; 63, Plaiul Hoților; 64, Slănic-Prahova; 65, Cisnădioara; 66, Turnu Roșu; 67, Hășag; 68, Ocna Sibiului; 69, Pârâul Cilor; 70, Pojorâta; 71, Pietrele Doamnei; 72, Piatra Tibăului; 73, Doisprezece apostoli; 74, Rădmănești; 75, Dealul Bujoarele; 76, Agighiol; 77, Hulubăt; 78, Măluşteni; 79, Stăncioi; 80, Audreiașu de jos; 81, Sânpetru; 82, Cherghes; 83, Roșia; 84, Peștera Meziad; 85, Valea neagră; 86, Peștera Vântului; 87, Defileul Crișului repede; 88, Cornet; 89, Ceahlău; 90, Dealul Sasului; 91, Insula Popina; 92, Credința-Ciobănița; 93, Cotu Văii; 94, Rusca Monatană; 95, Peștera Muierii; 96, Peștera Tecuri; 97, Ohaba-Ponor; 98, Topolnița; 99, Domogled; 100, Beușnița; 101, Valea mare.

the area, the necessity for their conservation, the legislative aspects, etc.; the conferences are illustrated with slides;
- symposia on environmental protection and nature conservation organized by SCGH or other NGOs in different towns with the participation of scientists or other people with official responsibilities in these areas; usually such symposia are followed by videos of films in different, but related, subjects (the last symposium organized by SCGH took place on 22 April 1993, the 'world day of the Earth').

From next year the SCGH with the financial support of the 'Know-How fund' intends to organize summer courses for young people in geoscience conservation, that will also include field trips and participation in concrete management activities within the geosites. Hopefully, these courses will increase the number of volunteers for the protection and conservation of the geosites.

The recent re-introduction in Romania of geology in secondary schools as an independent discipline, which includes in its syllabus a special section dedicated to geosite protection, will

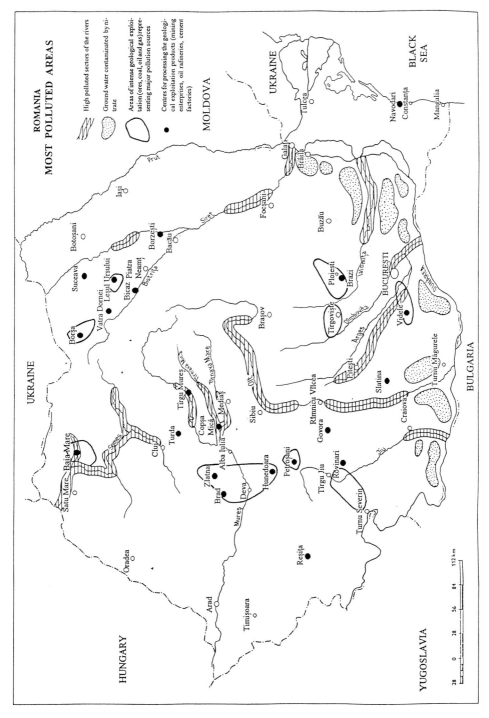

Fig. 2. Romania – state of the environment. Major pollution sources from geological activities. (Data for the pollution of surface and underground waters from the Environment strategy paper – Ministry of the Environment 1991).

probably attract more students for these activities.

Of great help in our endeavour to increase the educational sides of geosite protection is international co-operation. The participation of Romanian representatives in the meetings and symposia organized by the European Working Group for Earth Science Conservation offers a good opportunity for changing ideas and experiences as well as for initiating concrete projects in a regional framework in different aspects of geosite protection and conservation. Our society has already started a close collaboration with the Haute Provence Geological Reserve from Digne, France. The first group of Romanian students will participate during this summer in specific activities of management and monitoring in Digne; in exchange, a group of French students led by Dr Guy Martini, the Director of the Haute Provence Geological Reserve will visit geosites in Romania.

Our interest is to open geosites to the public, not to keep them under a sort of a 'glass bell'. For this reason the SCGH is preparing a booklet with touristic itineraries, crossing the geosites in the country.

Training of professional geo-ecologists

Managing the environment is a relatively new skill and, while the ability of Romanian professional and technical staff is high, there is a considerable need for training at all levels to develop a common approach for environmental management.

The geological domain with its practical activities (ore mining, coal, oil and gas exploitation and processing, quarries and related industries) represents a major pollution source, involving great waste disposals, pollution of surface waters, contamination of underground waters and radioactive effects on the environment (Fig. 2).

The future development of a European environmental legislation and policy will require well-trained specialists for management and monitoring of sites. Until December 1989, such activities were traditionally undertaken mostly by biologists and geographers, with a reduced efficiency due both to the lack of interest of the communist governmnet in such activities and to the low skill of the professionals.

With the aim of creating new specialists in geo-ecology, the Departments of Geology from two Romanian Universities (Bucharest and Cluj-Napoca) are collaborating within the framework of the TEMPUS programme (an EC programme for aiding the reform processes in higher education in Central/Eastern European countries, similar to ERASMUS and COMMET programmes in the EC) with universities from England (East London, Southampton), France (Paris VI) and Greece (Thessaloniki).

The main objective of the project is to build new curricula in geo-ecology, both at the undergraduate and postgraduate levels through intensive courses, study and practical placement periods for Romanian students in EC countries, as well as retraining of the professors.

I am thankful to Dr Marcian Bleahu, the former Minister of the Environment, for information regarding the state of the environment in Romania and the strategy for improving its quality. I thank very much Dr Des O'Halloran for the invitation to attend the Malvern conference and for all the facilities provided to make this possible.

Earth science conservation: the need for education and training

R.C.L. WILSON
Department of Earth Sciences, The Open University, Walton Hall, Milton Keynes MK7 6AA, UK

Abstract: The need to raise public awareness of the value of Earth science conservation (ESC) and to provide training materials to enable more people to become involved in practical conservation activities is considered in a UK context. Many local communities are already aware of their biological and cultural heritage, but ignorant of the geological and geomorphological features that underpin their sense of place. More needs to be done to educate many members of special interest groups (politicians, planners, landowners, developers, natural history societies and voluntary conservation organizations) about our local and national Earth science heritage. In both cases, raising awareness of the justification for ESC and its methods must be accompanied by a better knowledge and understanding of our science. Many professional Earth scientists are ignorant of the methods of ESC, as are biological conservationists. Professional users of Earth science sites take them for granted, so their awareness levels need enhancing in order to harness their expertise and energies to contribute to ESC activities.

To address this multi-faceted audience, the Open University is developing a training pack that aims to provide a broad scientific understanding that underpins an explanation of the needs for, and methods of, ESC. With appropriate face-to-face support, the pack will contribute to the education and training needs described.

The comments made in this paper arise from my involvement in the preparation of a training pack on Earth science conservation (ESC) for use in the UK. In the context of this paper, 'education' is defined as raising awareness and knowledge of ESC amongst a wide spectrum of the public and professions. 'Training' involves the development of skills, underpinned by knowledge, that enable the implementation of administrative and practical measures to conserve sites. Much still needs to be done to raise awareness about the need for ESC amongst the public, and most professional geologists and active biological conservationists are ignorant about such needs, let alone the methods of this relatively new area of endeavour.

In the UK, conservation of Earth science sites is justified for the following key reasons (Nature Conservancy Council 1990):

(1) Earth science sites have aesthetic, historical, cultural and wildlife value;
(2) our geological and geomorphological heritage must be preserved so that we may continue to seek new knowledge (i.e. sites are needed for pure and applied research);
(3) a network of sites is needed to contribute to the training of future Earth scientists, and to provide field facilities for schools.

Such justifications need to be promoted as widely as possible. The target audience in the UK is a disparate one, as shown in Table 1. There are many people who know nothing about Earth science or ESC. Others will know something about conservation, but nothing about ESC and vice versa. Only a minority will be familiar with both the science and the need for, and methods of, ESC. The current state of knowledge of the different categories by the potential audience shown in Table 1 is not intended to be critical or inflammatory about their level of knowledge, but is an attempt to be realistic about the wide range of needs education and training activities must address.

Education

Raising the awareness amongst three broad groups of people is essential to further the cause of ESC:

- local communities
- special interest groups
- site users

In the UK much has been done through the Regionally Important Geological/geomorphological Sites scheme (RIGS) to involve groups in large local government areas, but relatively little has been done to promote ESC amongst local communities, but they may change as English Nature refocuses its activities around the concept of **natural areas** (English Nature 1993). The

Table 1. *Potential UK audience categories conservation education training materials and the range of their prior knowledge (numbers are very approximate)*

	Approx. numbers in UK	Knowledge of:		Need for:	
		Geology	Earth science conservation	Education	Training
Specialists					
Non-geological staff of conservation agencies	1000	Poor/fair	Poor/fair	√	√
University Earth science teaching staff	1000	Good	Poor/fair	√	some
Academic and industry research staff	500	Good	Poor	√	some
Earth science new graduates	1500 p.a.	Good	Poor	√	some
Geologists in industry	5000	Good	Poor	√	some
Museum geologists	200	Good	Fair/good	√	some
Managers					
Landowners	10 000	Poor	Poor	√	some
Managers in business and commerce	?	Poor	Poor	√	
Local authority planning staff	>1000	Poor	Poor	√	some
Extractive site managers	?500	Poor–good	Poor	√	some
School teachers of science and geography	50 000	Poor	Poor	√	some
The public (part)					
County naturalist trust members	130 000	Poor	Poor	√	some
Amateur members of local geological societies	5000	Good	Poor/fair	√	some

character of such areas results from the interactions between their underlying rocks, landforms, soil, heritage, land use and other human activities. This mixture contributes to our sense of place. During the next year, English Nature will be piloting schemes for community education activities within two natural areas, and I hope that the lessons learnt should show how communities can be involved in local conservation activities.

Special interest groups include politicians, planners, landowners, developers, natural history societies and voluntary conservation organizations. Many of their activities already impinge on biological conservation, and some are sensitive to ESC issues, but more needs to be done to raise their general level of awareness. They need to be shown that in general ESC produces opportunities rather than threats.

Many users of Earth science sites take continued access to them for granted. More needs to be done to persuade them that such access is usually a privilege rather than a right, and that their behaviour, as teachers, students, amateurs or collectors needs to be such that features are not irrevocably damaged, or access to sites denied.

Users also need to be encouraged to become involved in conservation work, either by giving advice, or contributing to work at sites, or both. The RIGS scheme has great potential in this regard, and brings together users, special interest groups and communities.

Training

There is a need to train more people to apply the methods of ESC, but it is not on the same scale as that required to raise awareness. At present, the UK country conservation agencies need to train more of their staff (most of whom have a biological background) to undertake ESC, and this also applies to some staff employed by other commercial organizations (e.g. extractive industries, landfill operators, civil engineering companies) and voluntary conservation organizations.

Developing an education and training pack

Professional geologists should easily understand the justification for ESC rehearsed in the introduction to this paper, but the vast majority of the potential audience defined in Table 1 need to know some basic geological concepts in order to reach such an understanding. The materials currently being developed by the Open University (Wilson in press) are designed to give the target audience a very basic knowledge of geology which will be the foundation on which to build an understanding of the need for, and means of, ESC.

The training pack will follow the format of the OUs successful *Practical Conservation* series (funded by the former Nature Conservancy Council and other organizations). The style adopted is 'student active'. A variety of activities are used to break up the text and retain readers' interest, including looking at and interpreting field data, considering why actual sites need to be conserved, and how conservation of them can be achieved. Above all, the training material is designed to lead to **action** to support ESC activities. The material will be presented in three parts:

(1) *Earth Science Conservation: an overview*
Ten mini case studies show why sites need to be conserved, and introduce some of the basic geology needed to understand their importance. This introduction will be of interest to both geologists and non-geologists.

(2) *Introducing geology for Earth Science Conservation* This part will provide a general awareness of the subject; it will not produce expert geologists! The intention is to raise the user's geological knowledge to a level sufficient that they could converse with experts when needing to take advice on the value, utility, and conservation needs of an individual site under consideration. This part of the course would not be needed by staff already involved in ESC, or by many people working as professionals in the extractive industry, or by teachers of the subject.

A very simple introduction to minerals and rocks is followed by a basic introduction to igneous, sedimentary and metamorphic processes, with the discussion of processes including the origin of the major types of UK landforms. An introduction to geological time sufficient to understand terms on the '10 mile' geological maps of the UK is provided, and followed by an armchair guide to field work. Finally, a very 'potted' account is given of the major events in the geological evolution of what is today the British Isles.

(3) *Conservation in action* This part of the training pack discusses the need for ESC, the nature of sites, methods of selecting and enhancing them, and concludes with a chapter suggesting how public awareness could be raised concerning the need for and benefits of conserving geological and geomorphological site.

It is intended that with suitable face-to-face teaching support, the materials will support the wide range of education and training needs outlined in this paper.

References

ENGLISH NATURE 1993. *Natural areas: setting nature conservation objectives. A consultation paper.* English Nature, Peterborough.

NATURE CONSERVANCY COUNCIL 1990. *Earth Science Conservation in Britain: a strategy.* Nature Conservancy Council, Peterborough.

WILSON, R. C. L. (ed.) in press. *Earth Heritage Conservation.* Open University/Geological Society, London.

The changing nature of Earth science fieldwork within the UK school curriculum and the implications for conservation policy and site development

JOHN A. FISHER

University of Bath, School of Education, Claverton Down, Bath BA2 7AY, UK

Abstract: The introduction of the National Science Curriculum has changed the nature of fieldwork and thus the type of field site which is required by school teachers. Also there is a major shift in fieldwork practice from that where the teacher describes and interprets the geology (usually of large and prestigious sites) for pupils to learn and later reiterate to that where pupils are working within a wider science education framework which has different priorities.

This shift has two important implications. The first is that teachers need accessible and safe sites which are close to their school because opportunities for all educational visits are now severely restricted, on both cost and time grounds. Secondly, those sites must offer the opportunity for pupils to identify problems for themselves and to solve them using their expanding range of scientific concepts and skills. Thus the geology needs to be relatively uncomplicated and sites need to offer the potential for pupils to acquire knowledge and experience through progressive practical work which may be assessed as part of the National Curriculum. The learning objectives for the fieldwork need not be restricted to the science curriculum, however, but may impinge on wider environmental, social and aesthetic issues.

Traditional geology in the school curriculum

If we examine the place of geology within the education system of the UK up until about 1988, then we can understand how attitudes to, and perceptions of, geology as a school subject have become established and particularly how teaching methods have determined the role of field sites as teaching resources.

Geology for the majority of pupils, who left school at 16, has traditionally been a small part of their geography syllabus and thus taught by geographers in a way which was in accord with their needs and methods. Within that geographical framework, any identifiable geology or geomorphology had a largely service role, where information provided by Earth scientists was functional in understanding geographical issues. The scientific process by which geologists gained that understanding generally did not concern the geographer in UK schools.

When working in the field, descriptions of rock types and formations were almost always presented by teachers as a series of information sheets or field guides which pupils used to assist their comprehension of landscape features and these were often reinforced with field lectures.

Not only had the style and content of the teaching become stereotyped but also the field localities used by teachers. Large, well known, well recorded 'classic' sites were the ones which tended to be used because they were the ones in the textbook and, because of their academic significance, many had the status of 'Sites of Special Scientific Interest' (SSSIs). At certain times of the year school parties have had to take turns to see some of the smaller-scale features which they had travelled miles to see and the concentration of so many fieldworkers in such small areas exacerbated the problems of site misuse. The hammering of rock exposures and overcollecting of specimens along with the erosion made worse by so many student feet is familiar to all who work in the field and is a cause for genuine concern at many well known sites.

The policy of the Nature Conservancy Council to conserve sites by spreading the load gave rise to the publication *New Sites for Old* (Duff *et al.* 1985) in which alternative study areas in the Mendip Hills (County of Somerset, UK) are given a positive recommendation while not advertising those which were overused, small or vulnerable. For the experienced, the book was a valuable addition to their library, it gave enough information for the site to be found and put into context but assumed that the initiated would make their own evaluation and plan field activities. However, for those with little or no experience or expertise, the book failed to provide either enough information to explain the site or how it might be used as a teaching resource. Because it was not sufficiently 'teacher friendly' it probably contributed little to changing the attitudes or practice of fieldwork organizers. This is to be regretted because an evaluation of the effectiveness of that book, as an agent for change, could have provided an excellent pointer for future developments.

A time of curriculum change in UK schools

A process of major curriculum change began about eight years ago and the nature and direction of that change quickly made it clear that the established methods by which most subjects were taught were not going to serve the needs of school students in the future. Both the organization of the curriculum and the framework within which teachers work have changed and this has fundamentally affected the position of geology and its teaching in schools.

In January 1985 *The National Criteria* were published in order to shape developments of the General Certificate in Secondary Education, the new examination which was to subsume both the General Certificate in Education and the Certificate of Secondary Education; examination systems which had operated in parallel in most schools for about twenty years. The subject-specific, 'Science Criteria' defined, 'the essentials for all GCSE courses in the broad field of experimentally based science', and made clear that the criteria were to apply to examinations in geology (HMSO 1985a). Thus, for the first time, it was stated that geology must be an experimentally based course of study and conform to the same criteria for teaching and assessment as other sciences.

Later the same year, in March 1985, *Science 5–16: A Statement of Policy* was published (HMSO 1985b) which set out a philosophical framework for the development of the National Curriculum in Science which now determines the pattern of the education in science for all children in state schools in England and Wales (HMSO 1991).

Two of the ten 'Principles', which underpin that statement are those of 'Balance', which demands that all pupils be given opportunities to understand the nature of science through involvement in a 'Science Process', as well as to learn about science as a fund of knowledge. Also, 'Breadth', which ensures that pupils develop a wide conceptual understanding and knowledge through contact with all major branches of science and their applications.

The first of these Principles was to help determine the fundamental styles and approaches to teaching and the nature of the pupils' learning experiences. The second Principle requires, for the first time, that after years during which schools organized their science teaching around the separate disciplines of biology, chemistry and physics, the Earth sciences now have to be a substantial element of the science curriculum for all pupils.

Earth science within the school curriculum is thus in a state of flux. Small elements of geology, meteorology and astronomy remain in the geography curriculum where they retain much of their more traditional role and teaching methods. 'Geology' at GCSE and 'A' level has now to be taught as a science, but in many schools is still largely being taught from within the auspices of the geography department where there is not the experience of teaching through an experimental enquiry method which would place geology within a holistic model of science.

Conversely, most of the learning about Earth materials and phenomena is now being developed within science programmes which will be taught by those who have the experience and resources to teach science but who often do not have adequate specialist knowledge of this additional area of responsibility.

Experience with the National Curriculum is new but what is clear is that we are left with an outmoded legacy which shapes perceptions of geology as a school subject and this perception is firmly held by many science teachers who often regard it as a low level, inductive and descriptive subject. This attitude persists despite the fact that over the past 30 years the subject has undergone a conceptual and methodological revolution. Furthermore, at the time of the last major science curriculum developments during the 1960s, plate tectonics and the role of geology in solving economic problems had a high public profile but the status of geology in the education of young people did not change (Fisher 1992a). Because of neglect, geology is 30 years behind other areas of science in its evolution as a school science subject, thus the development and application of practical work in the field, which will have status equal with other areas of scientific investigation, is going to require considerable help and support from all who have a role to play.

The problem

The mismatch between traditional methods and organizational structures and the requirements of the National Curriculum creates a series of problems for course developers who have the task of preparing schemes of work and learning activities for the National Science Curriculum.

Furthermore, this problem extends to all those at a support level: organizations such as libraries, museums and environmental agencies.

The multidisciplinary nature of geology means that concepts can only advance satisfactorily in parallel with other areas of scientific knowledge, experience and skill. Conversely as the Earth sciences are so valuable in demonstrating applications, in providing interesting and useful extensions to other science topics and in contributing to a greater sense of coherence within course, it makes no sense to teach this area in isolation. Furthermore, the Earth sciences, in this multi-faceted role, are ideally suited to generating interesting starting points for valid and useful investigative work which should parallel and link with that experienced in the biology, chemistry or physics dimensions of the integrated curriculum (Fisher 1992*b*).

The statutory requirements will not only change the nature of any practical work carried out in the field but the type of field site which is needed by school teachers in order to meet the new requirements. Most significant is the major shift in fieldwork practice from that where the teacher describes and interprets the geology for pupils to learn and later reiterate, to that where pupils are working more independently while problem solving within a wider scientific framework.

This shift in approach has two important implications for those charged with identifying, developing and conserving geological sites. The first is that teachers need accessible and safe locations which are close to their school because opportunities for all educational visits are now severely restricted, on the grounds of both cost and time. Quick and easy access is also important if fieldwork is to become an integrated activity which takes place at strategic points throughout teaching schemes and this means that the field activity plus travelling time should not require more than about eighty minutes; that is, the fieldwork can be conducted within the normal lesson timetable.

Secondly, a group of sites should collectively offer for pupils opportunities to carry out investigative work which is in accord with the aims of the National Curriculum. Thus there is an urgent need for teachers, planners and conservation agencies to identify, develop and publicize sites which offer the potential for pupils to acquire knowledge and experience through progressively sophisticated practical work which they may then have monitored and assessed as part of the National Curriculum process.

For example, twelve-year-old children are now normally expected to demonstrate, during a scientific investigation that they can:

(a) ask questions, suggest ideas and make predictions, based on some relevant prior knowledge, in a form which can be investigated;
(b) carry out a fair test in which they select and use appropriate instruments to measure quantities such as volume and temperature;
(c) draw conclusions which link patterns in observations or results to the original question, prediction or idea.

(HMSO 1991)

If pupils are to demonstrate this level of scientific competence while working at a geological site, then two conditions must apply. The pupils must have sufficient background knowledge and understanding, as well as practice gained at other sites, for them to be able to ask, with the minimum of assistance, appropriate questions about the features being observed. Also the site must offer the scope to work at this level without presenting overwhelming or distracting complexities. The academic (specialist) significance of the geological features exposed are of less importance than how a site can stimulate curiosity, questions and thus hypotheses which may be tested by pupils using concepts and skills which are more universal within a broad and coherent science education.

The National Science Curriculum makes clear that pupils 'should develop the intellectual and practical skills which will allow them to explore and investigate the world of science and develop a fuller understanding of scientific phenomenon, the nature of the theories explaining these, and the procedures of scientific investigation'. It also clearly states what type of practical activity is appropriate to meet this aim across the whole range of the science curriculum.

Suitability of sites for such 'science investigations' is sometimes difficult to assess because, like all natural systems, they involve many variables. There is, therefore, a need for (at least) three questions to be asked when we evaluate sites as potential teaching resources:

(a) Is the range of the geological features to be observed at the site likely to encourage and enable pupils to focus on a few key issues and concepts which are appropriate for their age and ability?
(b) Can relevant information be clearly recognized, defined, collected and evaluated by pupils while they carry out scientific investigations at the site?
(c) Is the site accessible, safe, of appropriate size and scale; are the management and logistical problems reasonable?

A practical approach to matching field sites to teaching aims was put forward in 1988 in an NCC publication, *Earth Science Fieldwork in the Secondary School Curriculum* (Fisher & Harley 1988) which was produced in response to geology at GCSE level having to confirm to the science criteria. Although overtaken by events, this could still provide a model for future stages of site evaluation and development.

Not all children will have easy access to sites, much less those which are ideal for investigative work or the development of particular science skills; in this respect pupils from the counties of Avon and Cambridgeshire will not have equal opportunities, simply because of the nature and extent of the exposed geology. However, it must be remembered that those scientific investigations which become a formally assessed part of pupils' work can be carried out using the medium of any topic within the National Science Curriculum. Therefore, unless we are able to provide persuasive exemplars of high-quality, field-based, geological investigations, teachers will continue to be attracted to areas such as 'growing cress seeds' or 'making electromagnets' where they feel more confident and the notion of 'Breadth' in the pupil's learning experiences will be lost.

Of course, the learning objectives for the fieldwork need not be restricted to the science curriculum but may, as always, impinge on wider environmental, social and aesthetic issues. Feeling a sense of responsibility for a site and carrying out routine, small-scale conservation work while working there would be worthwhile in its own right, both from an educational perspective and for site maintenance and preservation.

The Site Adoption Scheme, which was an imaginative project initiated by the Nature Conservancy Council some years ago, may still provide a model by which this aim can be promoted and could encourage interested amateur groups, local authorities and teachers to co-operate in the task of site development and management (NCC 1984).

The establishment of the 'Regionally Important Geological/geomorphological Sites (RIGS) scheme (NCC 1991) also makes a valuable contribution to recording, preserving and making available more field sites at a regional level, but the development of a further layer or category in the network of recognized sites at a **district** level might enable local teachers to recognize and develop about 4 or 5 locations which are within a few miles of their schools.

Although conservation agencies recognize that school groups make up the largest percentage of organized field site workers and the concomitant need for appropriate site access and provision, the author believes that greater attention still needs to be given to the precise needs of schools and how their requirements might be used to influence site awareness, decision-making and conservation strategies. That is, we need to start from an understanding of the educational and logistic needs of teachers within a district and then do our best to provide sites which meet these requirements rather than starting with the geology and trust that this will satisfy the educational need; site development at this level should be educational-objective driven. For example, whether a site is located within easy walking distance of the school may be the factor which determines whether field investigations form part of the pupils' learning experience or not.

Many of those working on the RIGS scheme are teachers or those who have a good understanding of the school curriculum and they will be very aware that no experienced teacher would allow a particular resource to determine a scheme of work regardless of its suitability. Furthermore, all involved must give urgent attention to how the availability and potential of sites is going to be communicated to teachers bearing in mind that advice should now be in the language and style of the science educator rather than the academic geologist.

Sites which are to be conserved predominantly to meet local educational needs (as well as for their cultural and amenity value) could be identified as 'District Teaching Sites'. Many are likely to be known and documented by the RIGS scheme but sites which are classified using the priorities identified by practising teachers rather than geological criteria could mark the first stage in promoting them as educational resources.

The development of 'District Teaching Sites' needs to be done in conjunction with planners, local societies and landowners as well as teachers and educationalists and may not only require existing sites to be managed but fresh sites to be established. If there is a constructive dialogue between these parties then any opportunities which arise may be exploited. For example, if road works have exposed rock surfaces then, rather than the exposure be covered and landscaped they could become a feature of the local environment; a point of interest for the community as well as an available teaching resource.

If this process is not carried out at a local level and the teaching potential of the sites not well communicated to teachers, then practical work in the field will fail to become a routine part of Earth science teaching and ultimately a major reason for conservation will not exist.

References

Duff, K. L., McKirdy, A. P. & Harley, M. J. (eds) 1985. *New Sites for Old*. Nature Conservancy Council, Peterborough.

Fisher, J. A. 1992a. Geology: Earth Science or is it Geography. *School Science Review*, **73** (265), 141–145.

—— 1992b. National Curriculum Science – The Earth Science Dimension. *School Science Review*, **74** (266), 129–134.

—— & Harley, M. J. 1988. *Earth Science Fieldwork in the Secondary School Curriculum*. Nature Conservancy Council, Peterborough.

HMSO 1985a. *General Certificate in Education; The National Criteria – Science*.

—— 1985b. *Science 5–16: A Statement of Policy*.

—— 1991. *Science in the National Curriculum*.

Nature Conservancy Council 1984. *Practical Geological Conservation*. NCC, Peterborough.

—— 1991. *Earth Science Conservation in Great Britain – A Strategy*. NCC, Peterborough.

Communication of Earth science to the public – how successful has it been?

C. V. BUREK[1] & H. DAVIES[2]

[1] *Open University Tutor (NW & Welsh Regions), Newhaven, Church Street, Holt, Wrexham, Clwyd LL19 3JP, UK*

[2] *Open University Tutor (NW Region), Pharm House, Neston Road, Willaston, South Wirral L64 2TF, UK*

Abstract: Since 1971 the Open University (OU) has been using the medium of radio and television as a means of educating and attracting the adult population into education. Millions have had the opportunity to evesdrop on broadcasts involving Earth science and over 85 000 people have studied Earth science as part of their degree – a significant contribution to the raising of public awareness of Earth science. But how successful has it been?

In a small attempt to measure public awareness of geology, a survey was carried out in two villages either side of the River Dee where a Triassic sandstone outcrop forms a Site of Special Scientific Interest (SSSI) which is well interpreted with a public information board.

Analyses of the survey results are presented. They indicate that the general public still lacks awareness of geology. It was noted that those with school-age children in the family appeared to be more in touch with Earth science than the average. If this means that adults are educated through the younger generation then this gives a pointer as to where resources should be targeted.

Evidence drawn from enrolment data for the OU environment course points to another way in which Earth science education might be approached.

Finally, what can be achieved when the imagination has been fired will be illustrated by a look at a newly established geological trail round a disused quarry.

In its 1990 conservation strategy report, the Nature Conservancy Council (NCC) itemized six strategy themes of which theme 5 was 'raising public awareness'.

The mechanisms which were proposed by the NCC as likely to be most successful in promoting theme 5 were:

(i) utilization of the media especially television;
(ii) providing site information boards;
(iii) developing the Earth science component in field centres;
(iv) including Earth science conservation in curricula of school, university and adult education.

The Open University and communication of Earth science

The Open University (OU) is already well established as one of the leading educators of adults using the medium of television and radio. Since receiving its charter, the OU policy has been to provide educational opportunities for all adults who wish to study in their own homes in their own time. Through its Earth science course broadcasts, the OU has been able to make a significant contribution to raising public awareness.

Degree courses

The OU services over 100 000 students every year who are at various stages of their studies. Table 1 gives the numbers who have graduated or are in the process of graduating and have Earth science as some part (either major or minor) of their degree.

In the Science Faculty, Foundation Level (first year) teaching includes 25% Earth science, so all science graduates, in whatever branch of science, have a basic understanding of Earth science.

At Second Level, the Geology course (S236) is supported by two other courses (How the Earth works – the Earth's interior; Physical resources and the environment) which develop respectively geochemical/geophysical aspects and resource exploitation/landscape conservation aspects. Recent curriculum development means that students can link these courses to a more interdisciplinary approach given in courses such as 'Science matters' (an across the board

Table 1. *Cumulative totals of OU students with Earth science as a part of their degree*

Course	Total no. students
Foundation Level	
Integrated science 25% Earth science (S100, S101, 102)	85 819
Second Level	
Geology (S233, S236)	9 560
How the earth works (S237, S267)	5 611
Earth's physical resources (S238)	2 968
Science matters (S280)	1 237
Environment (U206)	3 776
Third Level	
Sedimentary processes and basin analysis (S336, S338)	2 049
Understanding the continents (S337, S339)	686

consideration of science that makes the news including Earth science topics such as global warming/rising sea-level and nuclear mineral mining and waste disposal) or 'Environment' (a cross-faculty course which includes natural hazards, sustainable energy resources, ocean resource exploitation).

Proceeding to Third Level a student can obtain an honours degree in Earth science through such courses as:

- Understanding the continents – tectonic and thermal processes of the lithosphere
- Sedimentary processes and basin analysis
- Oceanography
- Evolution

Associate and short courses

The OU also enrols students who wish to advance their knowledge in a subject without pursuing a degree. A student may enrol for one of the aforementioned courses or for the short-course programme which includes:

- Looking into the environment
- Practical conservation for land managers
- Changing countryside
- Remote sensing
- Renewable energy resources
- East Anglian studies: history of the environment

Impact on the general public

One of the most common routes for students into the OU system of study is by eavesdropping on a TV or radio broadcast. Degree courses are supported by regular television broadcasts which are compiled with the expertise of the BBC and are therefore attractively presented and equally appealing to student and casual viewer. It is almost impossible to assess just how many members of the public are receiving some Earth science education in this way but the BBC estimates 100 000 viewers per programme.

Measuring the success of communicating Earth science

It would require an enormous national survey to be able to come to any major conclusions on the impact that these OU contributions have made to the ambient level of Earth science knowledge in the community. However, by questioning visitors to a Site of Special Scientific Interest (SSSI), an attempt has been made to test a community's interest in Earth science.

The Farndon SSSI

For part of its length, the River Dee forms the boundary between England and Wales. Where it separates the Welsh village of Holt from the English village of Farndon, the river also cuts through an excellent section of the Triassic Chester Pebble Beds Formation of the Sherwood Sandstone Group. The beds dip 22–23° ESE exposing some 162 m of the succession including good sedimentary structures.

The section, shown in Fig. 1, is a designated SSSI and is supported by a interpretative display board which was installed in 1991 with finance from English Nature. The site is now managed by the Countryside Management Service of Cheshire County Council.

The public footpath by the river past the display board is a popular route for locals to walk with children and dogs, or to go just for an afternoon/evening stroll. It is also a popular fishing spot and a designated picnic site.

Holt and Farndon form a rural community of about 6000 people. The two villages house both

Fig. 1. View of the Farndon SSSI.

commuters to Chester and Wrexham and locally employed people but also have a significant number of retired people. With two large council housing estates and several large private housing estates, the villages contain a good mixture of all social groups.

Responses to the questionnaire

The survey of residents' reactions to the SSSI was conducted both during the daytime and in the evenings, both mid-week and at weekends. Everyone passing was stopped and no-one refused to answer the questions. Five percent of the total community participated. A disappointing 51% of those questioned had not even noticed the board. 2% of those who had noticed it had not gone to read it. This left 47% to respond to further questions. Of these, 84% said they found it informative. However, it was interesting to note that opinion was almost equally divided over the perception of who the board was really designed for:

- 26% thought it was primarily designed for the general public;
- 24% thought it was primarily designed for school children;
- 31% thought it was primarily designed for students;
- 19% thought it was primarily designed for professional geologists.

It was good to hear that no-one was hostile to the presence of the board and the fact that it might attract parties of students or schoolchildren to the site.

Of all those questioned, regardless of whether they had read the board or not, a disappointing 84% did not know what SSSI meant – disappointing because one-third of them were people who had read the board. A chi-square test of association between answers revealed a link (significant at $P = 0.1$ level) between those who had looked at the board and those who had children at school.

Conclusions

The survey results indicate that much more needs to be done in order to communicate Earth science to the wider population. BBC audience research data may indicate that many millions have eavesdropped on OU Earth science broadcasts but public awareness of the subject is still not great.

The chi-square test of association between the survey answers may hint at one way of effectively exploiting channels of communication. If children can be better informed about the significance of Earth science and Earth science conservation, they may indirectly bring their parents into the subject too.

Another pointer to effective communication of Earth science has been observed during the launch of the OU Environment course (U206). This course is an interdisciplinary study. In its launch year it attracted 1500 students. In the same year the Geology course only attracted 568

Fig. 2. Helsby Quarry Park.

students. The following year the Geology recruitment began to rise, reaching 830 students in 1993. It is regarded as significant that 20% of the additional intake on the Geology course were students who had studied Environment. U206 takes a holistic approach to the environment and sets Earth science in context showing the significance of its relationship to other subjects. Pleas for a holistic approach to Earth science conservation have already been made at the Malvern conference. On the basis of our experience we can only reiterate those pleas.

There is one other dimension to come out of a study of student recruitment to the Environment course and the Geology course. A significantly higher proportion of women students have been attracted to the Environment course as compared with the Geology course (50% vs. 39%) (Burek 1991). As women constitute over half of the general public with whom we are trying to communicate, any attempt at integrating Earth science conservation into a more holistic conservation approach is likely to draw more support from that section of society.

This paper has highlighted some areas where there is scope for improvement in communication of Earth science. However, there have already been some successes. A student who joined the OU in 1982 studied the Science Foundation Course in his first year and was so enthused by the Earth science component that he went on to include all the Earth science courses possible in his degree which he successfully completed in 1992. He immediately became aware of the general lack of knowledge of the subject and the importance of providing opportunities for school children and students to gain some knowledge locally without the necessity for expensive field trips. He identified an old abandoned local Triassic Sandstone quarry as being full of useful features (Fig. 2). He went on to persuade the local council, of whom he is an employee, to designate a geological trail which he has helped to write. He is currently overseeing the production of the interpretative display boards. This student has also gone on to become an active RIGS worker in Cheshire.

With such enthusiastic converts to the subject we can surely hope that the ambient level of knowledge of Earth science in the population will continue to rise and site conservation will be supported by an ever-increasing number.

References

BUREK, C. V. 1991. Great Britain's Open University: a new interdisciplinary course on the environment. *Symposium on Sustainable Development: Environmental Education.* International Federation of University Women, Geneva.

NATURE CONSERVANCY COUNCIL. 1990. *Earth Science Conservation in Great Britain – A Strategy.* NCC, Peterborough.

New Milestones: sculpture, community and the land

SUE CLIFFORD

Common Ground, 41 Shelton Street, London WC2H 9HJ, UK

Abstract: Common Ground has pioneered imaginative work on community involvement in local conservation, crossing boundaries between people, landscape, history, nature, culture, linking philosophy and practice, environment and the arts, and focusing on the conservation of cultural landscapes, through projects such as Parish Maps, New Milestones and Local Distinctiveness.

One recurring strand has been the importance to local people of the geology. The paper will demonstrate this through example.

Common Ground's New Milestones project is about what places mean to people who live in them and how to express that meaning in an imaginative and accessible way through sculpture. It hopes to stimulate the creation of small-scale works of the imagination, which express our sense of history, our love of place and of the natural world. It is involving local people in commissioning sculptures which will be valued and enduring features in the present and future life of the whole community.

Locality expresses itself in stone in so many ways. While nature does it best, culture, in England at least, often does it most.

Gravestones, stone walls, gate posts, sunken lanes, buildings, small quarries, railway cuttings are all part of the matrix of our cultural landscape, and taken so much for granted.

So often, particularity is being eroded. Stone walls, with their local ways of building guided by the nature of the stone, signal changes in the underlying rock. The toadstone (the local name for basalt) of Derbyshire proves its presence by showing dark in the pale limestone walls. Barbed wire does not have this quality – its effect is to homogenize.

Likewise the increased use of imported brick for building and marble for gravestones is pushing out local brick and stone, whose role of revealing the geology, playing host to local lichens and nurturing particular craftsmanship have reinforced local identity. People are being disinherited from their own local heritage and wisdom. If they have no direct link through work and everyday experience, no cultural relations with nature, how can they sustain an interest? If no one values what they know and value all around them, how can they be asked to care for special things for someone else?

Common Ground works through different projects to encourage local people to stand up for what they care about in their everyday landscape.

The New Milestones project

The New Milestones project is about our relations with nature and the land. It encourages local people to commission sculptors to help them explore and celebrate that relationship, and to leave modest momentos of positive feelings between people and places. In so doing, our hope has been that social sentiment about locality may focus discussion, and the works themselves continue to provoke people to consider the things so easily taken for granted.

Works have been commissioned by landowners, parish councils, farmers, local organizations, inner city tenants groups and are about the ordinary things and actions which make places what they are.

Our hope is that this process, of exploring and enhancing local distinctiveness and acting as a catalyst for heightening awareness of how the locality is valued, will further stimulate acknowledgement and active caring for the things which create a sense of identity.

Using Dorset as a pilot county, Common Ground started to try out the idea in 1985. The project is not about geology, but we have found that people's relationship with nature makes frequent reference to the land. Two examples must suffice.

Wayside carvings for the Weld Estate, Lulworth

The very first commission came through an enlightened landowner Wilfrid Weld who had just worked with the county landscape architect and archaeologist on an ecological and historical survey of the estate. One of his many concerns was to help visitors to enjoy the area and to lure the many visitors away from the most-used places. The chosen sculptor, Peter Randall-Page, worked with the help of the land agent, tenant farmers, estate workers and people from the pub, on a public footpath two miles from the nearest habitation. It is instructive to hear his evolving thoughts.

> 'The first time I made the spectacular walk from Lulworth Cove to Ringstead Bay, I felt quite overwhelmed both by the beauty of the place and by the sheer scale of the landscape. One feels very small in such places and I wanted to make work which would relate to the intimacy of human scale – something on which to refocus the senses before returning to the enormity of land, sea and sky.
>
> My objective which ran parallel with this was to make something which would strike up a resonance with the surrounding landscape by making a distillation of certain aspects of it. This area of the Dorset coast is famous for its abundance of fossils – in fact the chalk cliffs beneath this downland are literally made up of tiny fossils and the nearby Purbeck limestone comprises the fossilized remains of the gastropods, bivalves and ammonites etc. which once lived in an ancient sea. I liked the idea of making a kind of tribute to the ancient lives which now constitute our *terra firma*. I also wanted to incorporate something of the rhythms of the hills into the work – sweeping in broad rounded curves, tightening and plunging into deep gullies.
>
> I decided to make three carvings in Purbeck marble – not a true marble but a very hard local limestone, blue in colour and itself consisting entirely of tiny fossilized gastropod shells. Much prized by mediaeval carvers, Purbeck marble adorns many of our churches and cathedrals in the form of columns, foliage, fonts and other details of special importance. I have always thought Purbeck marble to be one of the most beautiful native stones and I felt that the look of preciousness that this material has would enhance the sense of intimacy I wanted to achieve. In the event I found that Purbeck marble is no longer quarried but was extremely lucky to find a small number of pieces which had been quarried over twenty years ago.'

The sculptures (Fig. 1), retire into their drystone niches in a hedgebank almost like wayside shrines, they start eddies of conversation and quiet contemplation. There are no plaques or signboards. In the village two miles away postcards and local knowledge will give glimpses into the background. But the real point is they are there for you to make of them what you will, to start reveries and musings, to provoke the imagination.

Chiswell Earthworks

The Isle of Portland, linked to mainland Dorset by the extraordinary Chesil Beach has given stone to England, and indeed the world, for centuries. One of its settlements, Chiswell, decided to celebrate its safety from flooding following the completion of a multimillion pound sea defence scheme: to signal renaissance.

The Portland Town Council and the Chesil Gallery approached Common Ground for help. Our suggestion was for a community-led project. It was agreed that a sculptor would be commissioned to help demonstrate local people's love of the place – for once stone would not leave the island to have value added to it elsewhere.

John Maine was chosen, and a small exhibition of his work and first ideas formed the focus of meetings with local people which helped expand horizons beyond 'a mermaid in the car park'. A subtle idea began to emerge, which John hoped 'would encourage people to look at Chiswell as a whole and perhaps act as a catalyst for other things to happen'.

Where the twenty-mile pebble beach reaches the island, a small triangle of sloping and slipping land – the West Weares – criss-crossed by paths, had been a dumping ground and had become a place to hurry through. John saw potential which the Borough Engineer also welcomed as echoing his ideas for stopping the landslips.

The sculpture would comprise five terraces, flowing waves undulating like the sea and making reference to the strip lynchets – the 4000-year-old stepped field systems of Dorset. But more than that, the land would be held back by dry stone walls representing the Portland Beds in stratigraphical turn, and each being worked in their traditonal ways.

The sculpture in effect would be a monument to the geology and to the masons: a powerful exposition of cultural intimacy with the land.

This ambitious work, begun in 1987 will be completed this autumn. The national failure of a government employment scheme robbed the project of 14 promised workers for a year. But much help has been forthcoming from local schools, the Masonry and Carving Course at Weymouth College, the Portland Environmental Team, the Devon and Dorset Regiment, the local Borstal, local masons, art students from all over the country, individuals and businesses, Portland Plant Hire, ARC who have given all the stone, as well as dogged commitment from

Fig. 1. Wayside Carvings by Peter Randall-Page, West Lulworth, Dorset (GR SY 784 811).

the artist, the Gallery owner, the Town Clerk and successive Mayors.

From the initial discussions, articles and letters in the local newspaper, Town Council Meetings, talking in the pubs and beside the evolving work, many people have been involved and offered their thoughts and criticism.

The sculpture is massive (Fig. 2) covering about half an acre and the longest of the five walls being 55 yards long. The lines of the paths have been respected, the walls dipping or stepping to let them through.

What has been created is a place with meaning for local people: something which they could not have accomplished without the vision of the artist, something which the artist could not have created without access to their intimate knowledge. Already people meet, play, sit and watch the sunset and the sea. Conversations begin; stories are exchanged; children bring visitors; tourists and questing artists who have heard this is worth a detour, seek out directions from the main street; the nearby inn does better some days; the Chesil Gallery hosts students and exhibitions which often connect.

Already people have visited from America, Australia and Japan. But more importantly, the everyday nature of this great building stone has been lifted out of the ordinary in its own place, the sculpture has helped make the everyday special.

The Chiswell Earthworks is a revelation of masonry techniques, and of the different Portland Beds, it demonstrates how each depends on the other. It will stand as a symbol of the culture of this place as well as adding to the cultural landscape in its own terms.

The quarries will always be a source of excitement to geologists. The whole island is a very particular place. One of its towns, Chiswell, of singular character, has initiated and demanded something unique to itself and has reinforced powerfully but subtley its own local distinctiveness. Its identity is now enhanced by a massive work of art, so integrated, it leaves visitors asking 'Where is the sculpture?' and local people being able to engage in more than ample reply.

Endpiece

Conservation and sustainability will remain elusive until we create a popular culture of wanting to care, and until we engage with the everyday landscape. Common Ground works to reclaim those difficult, intangible and elusive aspects of our relations with nature and places from the margins of professional activity, believing that we must embrace subjectivity, values, emotion, and we must include people. All of our projects from Parish Maps (which ask 'What do you value in your place?'), to Save Our Orchards and Local Distinctiveness focus on the cultural landscape and offer ways in to locally initiated action.

The New Milestones project is about what

Fig. 2. Chiswell Earthworks by John Maine 1986–93. Chiswell, Isle of Portland, Dorset (GR SY 684 732).

places mean to people who live and work in them, and how to express that meaning in an imaginative and accessible way through sculpture which will speak of the love of nature and place which people feel moved to express.

Rarely is attention paid to the commonplace and familiar aspects of our local surroundings. They are often overlooked or taken for granted but have great emotional value and meaning for the people who know them well. By recognizing and sharing their feelings about their place, it is hoped that people will find new ways to take an active part in caring.

With the recognition that local people have wisdom to offer, are good guardians and powerful allies, scientists would do well to help build imaginative bridges which will keep richness and particularity alive in the everday landscape.

References

KING, A. & CLIFFORD, S. 1987. *Holding Your Ground: an action guide to local conservation.* Wildwood House, London.

MORLAND, J. 1988. *The New Milestones project.* Common Ground, London.

Notes

Wayside Carvings by Peter Randall-Page can be found 2 miles along the inland path from Daggers Gate near Lulworth, Dorset. Grid ref. SY 784 811.

Chiswell Earthworks by John Maine are to be found beside the sea at Chiswell, Isle of Portland, Dorset. Grid ref. SY 684 732.

COMMON GROUND is a small non-governmental organization, a charity without membership. We emphasize the value of our everyday surroundings and the positive investment people can make in their own localities. We also forge links between the arts and the conservation of nature and our cultural landscapes. We offer ideas, inspiration and information through publications, exhibitions and projects (for example Parish Maps, Apple Day, Save Our Orchards, Tree Dressing Day, Local Distinctiveness).

Rock poems, rock music: using poetry and the arts to interpret geology

JAMES CARTER & TIM BADMAN

Centre for Environmental Interpretation, Manchester Metropolitan University, St. Augustines, Manchester M15 6BY, UK

Abstract: This paper suggests that conventional approaches to interpreting geology can only provide a partial solution to the work of raising the public's awareness and appreciation of the subject. Creative work offers another way of appreciating a place, and the paper discusses the role poetry and music can play in this.

Interpretation seeks to explain the significance of places to the people who visit them, and to encourage an awareness of conservation issues and values (Tilden 1957). Most geological interpretation follows the pattern of explaining the concerns and work of professionals in the field: it concentrates on scientific facts about the properties of various rocks, or on the geomorphology which has formed the landscape.

But if we are really trying to encourage people to take an interest in the subject, we need to take a wider approach than this. Very few visitors will want to pursue a scientific approach to the subject – they are more likely to find geology interesting if it is linked to their wider experience of a place, and most fundamentally to their emotional response to it (CEI 1990).

As an example, consider the process of erosion, which will be covered by many guidebooks and exhibitions. The better ones will communicate something of how vast that process is; of how much material the river has transported, and will try to put into context the length of time it has taken to shape the landscape. No guidebook yet written, however, can have the simple sense of drama and inexorability which Norman Nicholson achieves in his poem 'Beck'.

Not the beck only,
Not just the water –
The stones flow also,
Slow
As continental drift,
As the growth of coral
As the climb
Of a stalagmite.
                                (Nicholson 1982)

This expresses quite clearly the passage of time, and opens the reader's mind to a new vision of the landscape by introducing the idea of the stones flowing downhill as well as the water. The poem develops these ideas of inexorability and of a changing landscape, creating an image of tumultuous change happening in almost infinitely slow motion:

...top-heavy boulders
Tip over headlong,
An inch in a thousand years.

At the end Nicholson sees this process of erosion continuing as

An irresistible momentum,
Never to be reversed,
Never to be halted,
Till the tallest fell
Runs level with the lowland.
And scree lies flat as shingle
        (Nicholson 1982; extracts taken from 'Beck'
            in *Selected Poems* by Norman Nicholson,
                published by Faber & Faber Ltd)

Nicholson lives in Cumbria, and his writing is inspired directly by the landscapes of northwest England. 'Beck' mentions the hills of Ingleborough and Helvellyn specifically, and the poem could well be included in a guide booklet, or used on a panel on-site at either place. It would give the visitor an insight into the land which leaves a greater sense of awe, and a more deeply touching appreciation of geological time, than conventional interpretation can achieve.

Erosion has attracted the attention of other poets too. Ted Hughes, in 'Sugar Loaf', contemplates a stream running from a hill top to a pool, and establishes the water as something powerful and almost animate: 'The trickle *cutting* from the hill crown'; 'The water is *wild* as alcohol' (Hughes 1967, my italics).

This idea of water as alive is continued as he looks at the reflection in the pool:

I see the whole huge hill in the small pool's
    stomach.
This will be serious for the hill.
It suspects nothing.

Crammed with darkness, the dull, trusting giant
Leans, as over a crystal, over the water
Where his future is forming.
  (Hughes 1967; extracts from 'Sugar Loaf' in *Wodwo* by Ted Hughes. © 1960 Ted Hughes. Reprinted by permission of Faber & Faber Ltd and HarperCollins Publishers, Inc.)

This vision of geomorphology adds a sense of drama and hidden threat to the factual process: as in so much of his other work, Hughes links the natural world with deep-seated human responses and feelings. This linkage is surely one of the functions of interpretation. Both these poems have something to add to conventional interpretation: they take the standard approach of factual information one step further, to an emotional involvement which is essential if the public is to be captivated and inspired by the subject.

There are other possibilities too. Hughes' work recognizes that landscapes provoke emotional responses, and that these are directly linked to the geology of the place. This aspect of geology – its potential for shaping human thought and emotions – is explored by W. H. Auden in 'In Praise of Limestone'.

If it form the one landscape that we, the inconstant ones,
  Are consistently homesick for, this is chiefly
Because it dissolves in water.
  (Auden 1966)

Pedants may protest that only slightly acidic water will dissolve limestone. This is to miss the point: the poetic truth is to take the limestone landscape as emblematic of a particular state of mind and of a set of feelings, indeed as the source of those feelings. He links the underlying rocks directly to the character of those who live among them:

Watch, then, the band of rivals as they climb up and down
  Their steep stone gennels in twos and threes...
... accustomed to a stone that responds,
  They have never had to veil their faces in awe
Of a crater whose blazing fury could not be fixed...
  (Auden 1966)

and he provides wonderful characterizations of the landscapes which have attracted those who leave their native limestone:

..."Come!" purred the clays and gravels.
  "On our plains there is room for armies to drill; rivers
Wait to be tamed and slaves to construct you a tomb
In the grand manner: soft as earth is mankind and both
  Need to be altered." (Intendant Caesars rose and Left, slamming the door).
  (Auden 1966; extracts from 'In Praise of Limestone' in *Collected Poems* by W. H. Auden edited by Edward Mendelson. © 1951 W. H. Auden. Reprinted by permission of Faber & Faber Ltd and Random House, Inc.)

Clearly these interpretations of the character of landscape are personal, and not everyone will agree with them. But the link between rocks and the 'feel' of a place is indisputable, and deeply engrained in our culture even if it is not always recognized in interpretation: where else could the dark passions and tragedy of Emily Brontë's *Wuthering Heights* have been set but among the black gritstone walls and peat moors of the Pennines? This link is a fit subject for any interpretation which really seeks to get people thinking about and responding to the landscape they see.

There are many other opportunities for a creative or artistic approach to geology in interpretation. The poems discussed above are a small sample of the potential for involving literature; Sue Clifford's paper in this volume gives examples of sculpture projects which draw their inspiration from geology and recognize its place in the distinctive character of an area. With some imagination, any art form can be a vehicle for raising awareness of geology: at the Cévennes National Park in France, a temporary exhibition has used the musical qualities of stone. In addition to exhibits explaining how different rocks influence housing design in the various regions of the park, and how they were formed, the designers had included a 'lithophone': a sort of rudimentary xylophone, with thin bars of the various rocks suspended on cords. Mallets of wood and stone were provided, and visitors were invited to try out the different sounds made by the rocks. On one level, this is an amusing diversion. On another, it brings to our attention an important aspect of our experience of landscape, and hence of the geology which creates it: each rock type has a distinctive sound as we walk over paths and stones made from it.

Including ideas like these in geological interpretation does not mean abandoning the orthodoxies of scientific knowledge, nor the value of presenting them to the public. Rather it adds to them another dimension of experience, another way of understanding and appreciating the unique contribution of geology to our environment. If interpretation is about raising awareness, let us open visitors' hearts to new emotions and their ears to new sounds, as well as their minds to new facts.

References

AUDEN, W. H. 1966. 'In Praise of Limestone' in *Collected Shorter Poems 1927–1957*. Faber and Faber Ltd., London.

CEI (CENTRE FOR ENVIRONMENTAL INTERPRETATION) 1990. *Arts in Interpretation*. Environmental Interpretation, October 1990. CEI, Manchester.

HUGHES, Ted 1967. 'Sugar Loaf' in *Wodwo*. Faber and Faber Ltd., London.

NICHOLSON, Norman 1982. 'Beck' in *Selected Poems*. Faber and Faber Ltd., London.

TILDEN, Freeman 1957. *Interpreting our Heritage*. University of North Carolina Press.

Theme 5: International Convention

At about the same time that detailed planning for the Malvern Conference started in 1992, the proposal came forward quite separately within the Joint Nature Conservation Committee (JNCC) that there was scope to examine whether Earth science conservation on an international scale might be helped by an international convention. Other branches of conservation have gained from conventions such as Ramsar dealing with wetlands, Bonn covering migratory birds, and CITES regulating the international trade in endangered species. More recently, the Biodiversity and Climate Change conventions have placed those particular subjects on a political stage in a quite unprecedented manner. However, it was also recognized that the World Heritage convention, being site-focused, did cover some of the interests of an Earth sciences convention.

The initial discussions made it quite clear that there were significant differences of opinion about the objectives of such a convention and, indeed, whether a convention was needed. However, it was recognized that the subject could provide a focus for creative discussion within the Malvern Conference and out of this grew Theme 5, which provided a continuing topic for formal and informal discussion and debate throughout the meeting.

All delegates were sent, before the conference, a series of focused questions, which would form the basis of the two rounds of workshops, and summary sheets outlining existing international conventions related to environmental conservation.

The first session of Theme 5 consisted of three Keynote Addresses given by Chris Stevens, Professor Charles Holland and Dr Bernie Joyce, addressing, in their own way, the content, status and meaning of Earth science conservation.

On the following day, two hours were devoted to five parallel workshops (individuals were grouped alphabetically) which addressed the questions posed prior to the conference, exploring the possible content of, and so need for, a convention. One page reports from these meetings were prepared and circulated the next day.

There was a remarkable consensus. There was general support for a strategy or framework approach to international Earth science conservation, the objectives of which might be met by either an international convention or an internationally agreed strategy. Such a convention or strategy should self-evidently not jeopardize national sovereign rights. There was general acceptance of a core group of conservation interests covering geological, geomorphological and soil localities including ice, specimens, and geological information. It was also proposed that there should be two potential additions to such a menu, one of which recognized the interrelationship between biological and Earth science processes in the natural environment, and the other which covered the importance of the Earth's physical resources to humankind.

A summary of the workshop conclusions was prepared from the individual group reports and a further set of questions posed. The second series of workshops, with a new mix of membership but the same chairmen, then explored the scope of Earth science conservation in the context of the core plus the two potential additions, and sought views on the relevance to a convention or strategy.

These discussions left no doubt at the final plenary session of Theme 5, which immediately followed the second workshops, that there was a need for an international Earth sciences convention, but that the justification for, potential scope of, and objectives of, such a convention needed to be examined in depth. No-one failed to appreciate that the achievement of a convention would not be easy and could only follow from a long drawn-out and difficult exercise. If there was an easier route to achieve the same objectives then it should be adopted.

A draft Resolution was prepared following this concluding, plenary discussion and circulated for comment on the final morning of the conference. Comments were responded to, and an amended version was issued, prior to the closing session of the conference. The conference delegates endorsed both this statement and the establishment of an international task force (Dr P. Creaser [Australia], Prof. P. Jacobs [Belgium], Mr G. Martini [France], Ms A. Spiteri [Malta] and Mr C. Stevens [UK]) to pursue and report back on these ideas. The task force was charged with taking forward the Conference Resolution which followed on from the workshop discussions, facilitating the organization of a successor conference to Malvern, and expediting the creation of an international organization for Earth science conservation. The JNCC offered to provide this task force with a secretariat. The task force will write back to the delegates who took part at the conference, annually, until either a successor conference is held, or an international organization has been created.

There is no doubt that Theme 5 generated much interest and discussion, and many will remember this conference for the quality of the debate and discussion, both public and private on the convention issue. It was the talking point of the conference and helped ensure its success, adding value to its purpose.

The following text provides a record of the formal presentations and discussions held under Theme 5 during the conference.

Christ Stevens took the idea of an Earth science conservation convention to heart with his characteristic verve for a challenge. Nevertheless, he maintained a healthy scepticism that it might not be what is needed and that there could be simpler ways forward. He put a vast amount of time in before the conference helping the delegates to think about the issues beforehand, and he then contributed to the many discussions. He worked into the early hours of the morning ensuring that workshop reports were available to all delegates the next day. He was looking forward to the work of the task force with relish. That he is no longer with us to carry this challenge forward is difficult to appreciate. Those who had not met him before, will associate the Malvern Conference with Chris and it is fitting that the JNCC has formally decided that the conference volume will be dedicated to his memory.

Keynote Address

Defining geological conservation*

C. STEVENS
Joint Nature Conservation Committee, Monkstone House, Peterborough PE1 1JY, UK

Abstract: The paper explores three approaches to defining the scope of 'geological conservation' or 'Earth science conservation'. Its immediate purpose is to provide a background for discussion on definitions at the international convention workshops, and during other parts of the conference.

It is concluded that:

(1) Geological conservation is strongly linked to other cultural conservation areas: it is concerned with conserving the means of intellectual development, as opposed to economic conservation, which is concerned with out resources of oils, metals and aggregates. Geological conservation is particularly concerned with two aspects of the use of Earth science heritage: our geological understanding including the means to advance it in the future, and the aesthetic value of the heritage.

(2) Although the subject is generally regarded as distinct from conservation of the economic resources of the Earth, the latter is a closely related area with a complicated interface. Other closely related areas are wildlife and archaeological conservation. Most of the larger environmental issues of the day demand a holistic approach encompassing all these disciplines.

(3) A more informative title for the subject might be 'Earth heritage conservation' rather than geological conservation or Earth science conservation.

(4) A possible definition for the subject in the future is:

'Earth heritage conservation is concerned with sustaining the part of the physical resources of the Earth that represents our cultural heritage, including our geological understanding, and the inspirational response to the resource.'

I realized as soon as I tried to decide what to call this talk what a short straw I'd drawn. Should I call it 'geological', should I call it 'geological and landscape', should I call it 'geology and geomorphological', or 'Earth sciences', or should I go the whole way and call it 'Earth's resources'? All the definitions are fraught with implications and it is those implications that I would like to explore.

I would like to start roughly where all the national descriptions were at the very excellent International Symposium at Digne two years ago and I have drawn quite heavily on those and on our own experience in the past in this country to try and draw out the threads of the narrower of the definitions that we could come up with. A glacial esker, for example, may be threatened by quarries for sand and gravel; the aims of conservation must be seen to preserve its integrity as a landform for study, for enjoyment of landscape, as part of our heritage.

---

* Chris Stevens died a few weeks after the Malvern Conference. This text has been transcribed from a tape and is as close as possible to his spoken word.

Another range of issues concerns active process sites. Even on the most active process site, there can be a tracery of more or less delicate landforms which are too widely spaced to be damaged; such processes occur on the coast, on rivers and on slopes. And again a traditional focus has been to look at those processes and to say these should be allowed to continue, these are living laboratories, these are an integral part of our landscape, they are part of our natural heritage.

So that's a second theme within the traditional area and a third theme moves on both to the aesthetic and to the subject of outcrops. The enjoyment of scenery, the enjoyment of individual scenic elements and of scenery as a whole and the importance of the accessibility of rock exposures, whether they be small exposures in areas which are covered, as an awful lot of this country is, by agricultural land, urban land or whether they be exposures of the mountain-scale range which characterize some of the sites that were described very evocatively at Digne, in for example, the atlas.

And lastly, at what has been characterized

to be perhaps the site-based approach, the individual exposure for the purposes of study. And to me the unifying theme is very much not the concept of site nor of geography, but the concept of the use to which this was going to be put. A field visit may involve a society such as the Geologists' Association involved in amateur study, collecting and in scientific study, but it could equally well involve a group of international research scientists or school children.

Moving away from the site into the areas that are *ex situ* conservation, common themes very clearly spell out today on the subject of collecting, the importance of not only the ability to collect fossils to conserve the resource from which they came and all the information with them, but also to look after the actual samples themselves with their provenance data and to ensure that that part of the resource was safeguarded after it had left the ground, and an extension of that is to consider geological information. We are now moving into some areas which are only identified in some cases, but we had a case this morning when the geological information that could be taken from a site was recognized as a specific aim of Earth science geological conservation. And the last of those themes, that of soil conservation, spelt out in two ways, perhaps first of all from the stratotype approach, the soil as a record of the past, as a location at which study can be carried out and on the other hand touching for the first time on the economic aspects of soil as an agricultural and wildlife substrate and resource.

We, in Great Britain, had a definition of Earth science conservation along the lines that it was conservation of the Earth's resources and we were concerned that this definition was open to the accusation that so far we have not developed a strategy for conserving Great Britain's oil resources and a great many of its other resources. I am very indebted to colleagues in the country agencies for suggesting a very useful way of looking at this issue. It was suggested that the total physical resource of the Earth can be considered under two headings. The first of those, and the one that I have been trying to illustrate, could loosely be called cultural in the widest sense of the word because it was concerned with knowledge, data and understanding, with the way in which we got that with specimens, with science in short and with education, and also, because it was concerned with the inspirational side of the rather narrower view of what cultural means, the aesthetic values. Crucially it was associated with a value set – a heritage value set – and we believed that what unified all those themes that I have illustrated up to now, was a heritage value set, and that contrasted with an exploitive value set where the concern was, for example, metals, fuels, construction materials; some of the other less obvious of the Earth's resources like water and air and energy, and that we might as a provisional measure distinguish two sorts of conservation, Earth heritage conservation which is what I would, by preference, have called this talk when I started it off (except that I didn't want to give my punchline away beforehand) and extractive resources conservation.

Now this morning we have been challenged with what about the resources, never mind about the study and the trade in fossils, what about aggregates, what about groundwater, what about metal ores, what about energy and I believe that we have to address those questions in attempting a definition and I would like in this second part now to explore some of the issues and implications of tackling a much broader approach to the subject. I would like to start on the coast. Now the coastline contains two of Britain's Earth science heritage gems. First of all, the unrivalled exposure in the coastline in cliffs of a geological succession (our coastline is dotted with stratotypes); recognition of the importance of those sections and it is an unrivalled opportunity, especially for a country that is so urbanized and so dedicated to agricultural land and covered in soil, to seek the deeper geology. And secondly, it is a living laboratory of massive and powerful sedimentary systems. But driven by the energy of the sea, continued erosion and deposition are largely responsible for creating and underpinning all the habitats, all of the natural features and so natural areas around the coast which are instrumental to its attractiveness, also are instrumental to its economics. For example, in the case of saltmarshes, their health may be determined by the seawall height and of course requires an eroding cliff somewhere or other, or perhaps some china clay in to feed it. With the bottom situation where that salting has disappeared much more expensive and here economic arguments have been invoked and one of the things that has happened in Britain is a highly successful approach has been taken by bringing together all these arguments...

(tape runs out)

It has also stopped short of soils in general as productivity and it certainly stops short of groundwater. So in the question of groundwater, we have what is undoubtedly a natural resource and one that has not attracted our attention and so we are ever-widening the area that we can get involved with here. Aggregates is another one. The development of aggregates from coral reefs, which are, of course, ecologically very unsound places to obtain your aggregates, highlights very graphically the potential for conflict and if we continue, oil strategies are something that I have mentioned, coal and gas, and what about energy strategies? And Earth science conservation can continue in this way. Earth sciences have also made an enormous contribution to our understanding of the processes of global change and will continue to make an enormous contribution to the debate on global warming, but is it our area to get involved in? I think that if we continue this is what we will define Earth science conservation as. I would like to make a closing suggestion that we need to reorganize core areas, but we should move outside that core at every possible opportunity to participate with others in taking a holistic approach to the entire area. Earth heritage conservation is concerned with the part of the physical resources of the Earth that represents our cultural heritage, recognizing that that means both the scientific side of it and the inspirational side of it.

I believe that for convention we need to stick to the knitting for a definition and the issue at stake is 'will a convention focus on that area or will a convention move away and encompass a full approach?'. Both are possible with this definition. One of the issues is pragmatism and you saw this morning the questioning that Dr Scott received on the subject of resource allocations by government and that is a very good little taster of the immense and very political area that resources cover. And that is both a massive opportunity to go into that area and a signpost of the immense difficulty. I suggest that with whatever definition we carry forward, we should bear in mind the pragmatism of where it will lead us and lastly, to suggest that definition can serve for both.

Note: This text is a transcript of the recording of Chris Stevens' contribution. He did not leave a text and his words have received minimum editing.

Keynote address

Geological conservation: notion, necessity and nicety

C. H. HOLLAND

Department of Geology, Trinity College, Dublin 2, Ireland

Abstract: Of special significance among the very many and very varied geological sites of international importance are the boundary stratotypes defining (at their 'Golden Spikes') the Global Standard Stratigraphical Scale. Their particular role is briefly discussed. The probability of permanent conservation has to be assessed when these places are chosen. In these and other conserved geological sites there will always be problems of conflict between access and conservation. Some examples relating to the Silurian System are briefly reviewed. Other geological sites of such importance that the international geological community is entitled to expect some mechanism for conservation are some of the additional reference sections for the Global Standard Scale, the more famous lagerstätten of the world, localities of particular significance to the history of the science, and some others of unique geological character.

There are very many and very varied important geological sites. In 1989, the then Nature Conservancy Council listed about 2000 geological and geomorphological localities in Britain as of national or international importance for education and research, and thus meriting recognition as Sites of Special Scientific Interest (SSSIs). These had been selected from a resource of some 23 000. Many of the latter must depend for conservation upon local efforts, as for example the dedicated activities of the Malvern Hills Conservators or of the Shropshire Wildlife Trust.

In some cases it has to be said that there seems no reason to doubt that important sites will simply remain as such, and more or less accessible as such, without any early action being required. The splendidly fossiliferous Wenlock rocks of the Dingle Peninsula, County Kerry, Ireland may be regarded in this way. The extensive coastal sections here are reasonably remote and the sea continues to provide rich quantities of fresh rock.

At the far end of the scale of size there are whole landscapes: characteristic, often beautiful, sometimes unique, whose conservation is a matter for treatment as national parks, but whose geological component is nevertheless intrinsic. The karst country of the Burren in County Clare, Ireland is one of a spectrum of such landscapes which extends from northern England to mainland Europe. There is a sad history of commercial exploitation of such karstic surfaces for rockery stone. But the Burren has its own special character (botanical as well), where small cyclical changes of lithology are reflected in the terraced topography. Controversy has raged over the siting of an interpretative centre within such splendid country.

But the main thrust of the present contribution must be to explore the necessity and practicality of geological conservation in an international context. This applies above all to the so-called boundary stratotype sections in which, by international agreement, organized through the Commission on Stratigraphy of the International Union of Geological Sciences (IUGS), 'Golden Spikes' have been selected for definition of the bases of the various divisions of the Global Standard Stratigraphical Scale. I refer to the boundaries of systems, series, and stages. Matters have scarcely as yet progressed to the lower level of the chronozone, though this must be coming. Even the divisions higher in the hierarchy have proved difficult to settle. And it is sensible to note that some are inclined still to question the necessity for these Golden Spikes. They emphasize that what has so far proved most useful is the choice of the geological horizon at which a particular boundary is to be placed. I have expressed elsewhere (Holland 1978, 1986) what is now the orthodox view that the fixing of these marker points is not only necessary but urgent.

I have suggested that there are several reasons for this. First, the Golden Spike provides a standard, an anchoring device, analogous to the holotype in palaeontology. Secondly, once chosen, it can become the focus for additional work on various biostratigraphically useful fossil groups and physical attributes of the rock. An example of this was the micropalaeontological work of Mabillard & Aldridge (1985) on the boundary stratotype for the base of the Wenlock Series in Hughley Brook, Shropshire. Thirdly, there is a philosophical gain which I continue to believe is important. I have illustrated it by a number of versions of a diagram (e.g. Holland *in* Holland & Bassett 1989, fig. 1). It shows the unique property of the boundary stratotype that here, and here alone, we know by definition that a chosen point in rock represents a thus defined point in time.

As yet the most comprehensively defined of the systems is the Silurian and the accident of history, as well as the nature of the record, has resulted in the placing of most of the relevant Golden Spikes in Britain. The top of the Silurian System is, of course, defined by the base of the Devonian. This, the first such boundary to be agreed internationally through the approved mechanism of the IUGS, is taken in the Czech Republic.

The various criteria for selection of boundary stratotypes which emerged during the long search for a suitable base for the Devonian included the preservation of the section. Matters had of course been complicated here by previous grievous errors in correlation (Holland 1965, 1986, fig. 5; Bouček *et al.* 1966). Questions of accessibility and conservation were treated fully in the 'Guidelines and Statutes of the International Commission on Stratigraphy' (Cowie *et al.* 1986). The important point was made here that these are contrasting but complementary factors: 'Recent experience has shown that if access to an important outcrop is too easy and unrestricted then excessive collecting, even vandalism and plunder, may destroy the outcrop. Conservation and some restriction is therefore necessary in developed regions. Conservation in more remote regions may be easier but this depends on regional geological activity (with helicopters maybe) by outsiders.' The text also includes the following: 'There must be no insuperable physical and/or political obstacles for access by geologists of any nation; without great expense and ideally without much bureaucracy.'

Some Silurian examples will be examined with these matters in mind. Elaborate activity by the international Ordovician/Silurian Boundary Committee resulted in two prime candidates for selection of that boundary stratotype: the large island of Anticosti in the St Lawrence estuary, Canada and Dob's Linn, near Moffat, in the Southern Uplands of Scotland. The former is a marvellous area of simple structure and extensive exposure of top Ordovician and Lower Silurian rocks. The island is about the size of Belgium and is already very largely a fishing and hunting reserve owned by the Quebec Government. It was eventually rejected on the grounds that the conodont-defined Ordovician–Silurian boundary here was not easily recognizable about the world and that the graptolites, crucial in precise Silurian biostratigraphical correlation, are rarely present. Yet access here could be controlled and conservation of a suitable section (Barnes 1989) would have been simple.

Dob's Linn in the Southern Uplands of Scotland is situated in tectonically complex country, which has been affected also by glacial or periglacial processes. However, the dark shales abundantly exposed in continuous sections are exceedingly rich in graptolites. The area has undoubted historical importance in relation to Charles Lapworth's classical work. A rather tatty reminder of him remains on the wall of his cottage. The ground belongs to the National Trust for Scotland because of the scenic waterfall nearby. The site is, of course, an SSSI, but though a kind of on-site documentation was temporarily provided at the time of the original official examination by members of the international bodies in 1979, I understand that nothing has since been done.

The boundary stratotypes for the various Silurian series and stages are shown on Table 1. Some of them are on land belonging to the Forestry Commission; others are in private ownership. To the best of my knowledge they generally lack maintenance and documentation.

The base of the Devonian at Klonk in the Barrandian area (Prague basin), Czech Republic was chosen from four short-listed candidates after some 12 years of submissions, field visits, and discussions by the Silurian/Devonian Boundary Committee (McLaren 1977). The section at Ain Deliouine in the remote desert of southwestern Morocco was, according to one's taste, unsuitable or suitably inaccessible; but suffered from the effects of desert weathering on the critical graptolites.

In the basin and range country of Nevada there was much promise but comparatively little work outside discrete lines of section. The crucial graptolite *Monograptus uniformis* was not found here until after the visit by the Subcommission on Silurian Stratigraphy. I would emphasize that there is an urgency in these matters. Boundary stratotypes are required so that important and interesting aspects of geological research can continue within an agreed and rational framework.

Podolia in the Ukraine is a splendid area for

GLOBAL STANDARD STRATIGRAPHY			LOCATION OF BASAL BOUNDARY STRATOTYPE
SILURIAN SYSTEM	UPPER SILURIAN	PŘÍDOLÍ SERIES	(division into stages to await necessity)	BARRANDIAN (Pozary Section)
		LUDLOW SERIES	LUDFORDIAN STAGE	LUDLOW DISTRICT (Sunnyhill Quarry)
			GORSTIAN STAGE	LUDLOW DISTRICT (Pitch Coppice)
	LOWER SILURIAN	WENLOCK SERIES	HOMERIAN STAGE	WENLOCK DISTRICT (Whitwell Coppice)
			SHEINWOODIAN STAGE	WENLOCK DISTRICT (Hughley Brook)
		LLANDOVERY SERIES	TELYCHIAN STAGE	LLANDOVERY DISTRICT (Cefn Cerig section)
			AERONIAN STAGE	LLANDOVERY DISTRICT (Cefn Coed – Aeron Farm)
			RHUDDANIAN STAGE	SOUTHERN UPLANDS OF SCOTLAND (Dob's Linn)

Wenlock to Early Devonian stratigraphy, with impressive sections through little disturbed rocks exposed for many kilometres along the Dnestr River and its tributaries. In spite of the rich graptolitic and shelly faunas, there was an unfortunate absence of critical graptolites immediately below the chosen horizon. But I still believe that our good Russian colleagues were at that time swayed by the awesome problems of access.

The classical Barrandian area in Bohemia was finally chosen, backed by the rich collections of the National Museum in Prague. A monument in the foreground of the cliff section testifies to the importance the Czech authorities have given to the site. I understand it to be fully protected. There have been some complaints about permission to export specimens but these appear to have been resolved.

I have gone briefly through these examples to illustrate the kinds of places where conservation is required, some easily, some not so easily, achieved; and, above all, to emphasize what is insufficiently recognized: namely that an ideal site for any boundary is relatively unlikely. We must make do with the best available at the time, and, having gone through all the proper international procedures of selection, we are surely entitled to expect conservation.

The deliberations of the Subcommission on Silurian Stratigraphy resulted also in detailed descriptions of additional sections and regions, including those which had figured already in final processes of selection. Thus there is a whole web of additional reference sections (Holland & Bassett 1989). In some cases it is clear that conservation may again be necessary.

It is our business to see if some better way forward can be found for the conservation of all these sites. To give some idea of the scale of the problem, I once attempted to estimate the likely average number of stages employed in the post-Cambrian Palaeozoic systems and those of the Mesozoic. The figure was approximately ten, giving a total of eighty for these systems alone. However, it does seem to me that advice and decisions are needed for a list of internationally recognized geological sites longer than those already authoritatively chosen for the Global Standard Stratigraphical Scale and its additional reference sections. Surely there are other sites of sufficient international importance as to demand action. The kinds of places I have in mind include the famous lagerstätten such as Ediacara in South Australia, the localities in British Columbia for the Burgess Shale, and Solenhofen. There are also sites with a unique place in the history of the geological sciences, such as the Giant's Causeway in Northern Ireland; the nearby Portrush locality where the Plutonists claimed to see ammonites in igneous rock; and Siccar Point on the east coast of England, near the Scottish border, as one of the unconformities where James Hutton first glimpsed his 'succession of former worlds'. Finally, we should not neglect discrete sites of special scientific interest such as Shark Bay in Western Australia, where the tracks of an old cart through the field of stromatolites still testifies to their vulnerability; and the famous Lower Devonian plant and arthropod locality of Rhynie in Scotland, so vital to our understanding of the colonization of the land. Some of these are, of course, already well cared for. Rhynie, for instance, is presently one of the success stories of geological conservation.

In conclusion, there is a whole scale of localities requiring geological conservation, some more scientifically important than others, some more appreciated than others. Matters such as the conservation of data and the plundering of important fossils for commercial gain can all be related to particular sites. The problems are how to start the international procedure and where to close the list, how to organize, how to control, and how to pay.

References

BARNES, C. R. 1989. Lower Silurian chronostratigraphy of Anticosti Island, Quebec. *In*: HOLLAND, C. H. & BASSETT, M. G. (eds) *A Global Standard for the Silurian System*. National Museum of Wales, Geological Series, **9**, Cardiff, 101–108.

BOUČEK, B., HORNY, R. & CHLUPÁČ, I. 1966. Silurian versus Devonian. *Acta Musei Naturalis Pragae*, **22B**, 49–66.

COWIE, J. W., ZIEGLER, W., BOUCOT, A. J., BASSETT, M. G. & REMANE, J. 1986. Guidelines and Statutes of the International Commission on Stratigraphy (ICS). *Courier Forschungsinstitut Senckenberg*, **83**, 1–14.

HOLLAND, C. H. 1965. The Siluro-Devonian boundary. *Geological Magazine*, **102**, 213–221.

—— 1978. Stratigraphical classification and all that. *Lethaia*, **11**, 85–90.

—— 1986. Does the golden spike still glitter? *Journal of the Geological Society, London*, **143**, 3–21.

—— & BASSETT, M. G. (eds) 1989. *A Global Standard for the Silurian System*. National Museum of Wales, Geological Series **9**, Cardiff.

MABILLARD, J. E. & ALDRIDGE, R. J. 1985. Microfossil distribution across the base of the Wenlock Series in the type area. *Palaeontology*, **28**, 89–100.

MCLAREN, D. J. 1977. The Silurian–Devonian Boundary Committee: A final report. *In*: MARTINSSON, A. (ed.) *The Silurian–Devonian Boundary*. IUGS Series A, No. 5, Stuttgart, 1–34.

Keynote address

Identifying geological features of international significance: the Pacific Way

E. B. JOYCE

School of Earth Sciences, The University of Melbourne, Parkville VIC 3052, Australia

Abstract: If we agree that there is a need for an international convention to care for important geological sites and features, then we need to be able to generate lists of such sites and features, and we must be able to justify their significance at the international level. To do this, experience developed in Australia could be applied at the international level.

Geological and geomorphological sites, modern processes, and the landscape itself, should all be part of this effort at looking after our Earth heritage.

Is there a need for an International Convention on Geological Heritage?

The geology of the Earth contains within it not only an explanation of its own history, but also of the solar system and its place in the universe, and the development of the ecosystems of the modern Earth, and our place as humans in these systems.

The study of the past Earth is the province of Earth scientists or geologists, and those working in related fields such as astronomy, prehistory and ecology. The evidence of the Earth's history is in the geological and geomorphological sites and features, either at the Earth's surface and naturally exposed, or found by excavation and mining, drilling or geophysical techniques. To understand the Earth's history, this evidence must be looked at repeatedly, as new techniques become available, and as new concepts are developed.

Earth science mainly conducts its experimental work by setting up 'mental scenarios' of the Earth's history, and then attempting to confirm of falsify them by examining evidence in and from the field. The sites and features needed for such studies form a framework of reference areas in the field to which continuing access for study and sampling is needed.

All sites or features are unique in some way, but some are outstanding or of such rarity that they can be considered unique in the full sense, and so worthy of special care. Other sites and features may be grouped in ways which allow the selection of one or more representatives which can be given special care. To allow access and sampling by present and future Earth scientists, outstanding and representative sites and features need to be managed for protection against degradation, destruction or burial.

Sites and features forming our geological heritage may have significance at a range of levels, from international, through national and regional, to local significance only. International significance implies importance for geological study by scientists from other countries, and contributing to the overall geological world picture.

To maintain the value of sites and features of international significance, an international understanding of their importance should be reached, and an international agreement on how to look after them must be made.

Models for an international convention

The two previous speakers have set the scene (Holland, this volume; Stevens, this volume).

If we agree from my introductory discussion that there is a need to manage sites and features at an international level, how might this be done? I suggest two models for consideration:

(1) World Heritage
(2) National studies, such as those carried out by the UK and Australia.

World Heritage

World Heritage sites are generally large (and many major geological sites are small fossil areas, road cuttings, cliff sections – for example, see Fig. 1). World Heritage sites are generally selected with multiple values, such as biology, history, archaeology – this has been the case with Australian World Heritage sites (Table 1).

Fig. 1. Koonwarra Cretaceous fossil locality in Victoria, Australia. One of the great localities of the Mesozoic era, with very well-preserved flora and fauna, including plants, pollen, spores, fish, feathers and many insects, deposited in lake sediments 115 Ma ago. The site is an inconspicuous road cutting of limited extent. New road works led to the discovery of the 'Koonwarra Fish beds' in 1959, and a major collection of material is now housed in the Museum of Victoria (see inset of fossil fish). (Photo E. B. Joyce, 18 December 1969.)

World Heritage sites must be nominated by the government concerned, and supported by that government in future protection and management. Most countries are unlikely to support a purely geological site or a small site. For reasons of local support, e.g. by environmental lobby groups, these sites must have multiple values, and geology can only be one aspect of a nominated site.

About 100 World Heritage sites of natural significance (not all geological) have been recognized over twenty years. 'It seems possible that this number may rise to 200 by the year 2000' (Cowie 1992). Ultimately there will be probably only five or ten World Heritage sites for any one country, as each site requires a detailed and expensive submission, a detailed management plan and continuing support by the country concerned. A government is unlikely to be interested in say 10 sites of purely geological significance when the total list of World Heritage sites they might aim at for their country would only be some 5 or 6 at the most. So it seems likely that only three or four major geological sites will ever be listed as World Heritage for any one country. I argue that this will not be enough to provide a worldwide system.

Australia already has 10 World Heritage sites, a larger number than many other countries. These are all natural sites, and five in particular are geologically significant (Table 1).

A Task Force of the UNESCO Working Group on Geological (including Fossil) sites met in Paris in February 1991. Its report (Cowie 1992) provides a Global Indicative List of Geological Sites (GILGES), with 250 sites of possible World Heritage significance. This might provide a basis for a list of sites of international significance. GILGES includes 16 Australian sites (Cowie 1992) and the UK also has 16 sites in the listing. However, I would argue that most of the 250 sites in GILGES will never achieve World Heritage listing, for the reasons given above.

Table 1. *Number of sites of international significance for Australia*

Current World Heritage listings:	10 sites
Of these 10, those geological in major part are:	5 sites
Global Indicative List of Geological Sites (GILGES) Paris 1991: (See Cowie 1992)	16 sites
1986 Australian study: (Cochrane & Joyce 1986)	76
Number likely to be revised soon to:	c. 100
Number which might survive an international assessment:	c. 50
For comparison – RAMSAR for Australia:	c. 40 sites
An estimate of important geological sites for the whole world is:	c. 1000 sites

So the use of World Heritage listing for the selection and protection of a representative set of internationally significant geological sites is limited.

Experience of individual countries

Most countries are now involved in geological conservation, and each country has developed its own approach. Details of the results of this work are often difficult to obtain. A quick survey of conference reports and available publications suggests that many countries have prepared lists of sites ranging in number from 2000 or more (UK, Australia, New Zealand, Germany), 1000 or less (Ireland, Austria, Netherlands), down to 100 to 200 (Denmark, Romania).

Sometimes these numbers are for the total inventory for the country, of all levels of significance, and sometimes they are only the major sites, perhaps classified specifically as sites of international significance. From these data, it is probable that the number of international sites for many countries is likely to range from 20 to over 100.

UK and Australia studies have concluded that there may be 50 to 100 sites of international significance in each of these countries. These are countries with potentially many major sites, either due to the locally complex geology, as in the UK, or because of the extensive area involved in the case of Australia. Many countries of the world need only a smaller number of sites. Perhaps a total of 1000 sites might be required world-wide (Table 1). This is about the order of the number of sites governed by the RAMSAR Convention (The Convention on Wetlands of International Importance). How can such a large number of internationally significant sites be identified, documented, assessed and managed?

Firstly, a system is needed for each country to find and evaluate its major sites. This process will select and justify the importance of sites at levels of significance ranging from local, through regional and national, to international or world-level sites. Some countries have adopted a 'top-down' approach, in which the major sites are listed, generally because they are well known both inside and beyond that country. This may be because they are rare ('unique' is a term often applied). They may be of outstanding natural beauty or have aesthetic qualities (Iguazu Falls, Argentina and Brazil), be associated with other aspects of significance such as archaeology (Willandra Lakes, Australia), history of science (Hutton's unconformity at Siccar Point, Scotland), or biology (Galapagos Islands, Ecuador), or form part of areas reserved as national parks or similar generally large reserves (Hawaii Volcanoes National Park, USA).

The approach can be applied in the absence of any general assessment of features and sites in the country concerned. Perhaps time, expertise and funds for a general study were not available. In using this approach sites of possible international significance can be overlooked. However, this has been the approach used in the GILGES list developed by the Paris Task Force in its advice to UNESCO (Cowie 1992). A set of criteria (see Cowie 1992 p. 6), definitions of what constitutes sites of significance, and guidelines in the use of the criteria and definitions by the assessors are needed.

The national approach used by Australia A more rigorous approach is to carry out a full inventory of possible sites across the country concerned, and classify sites in a comparative way, with the sites of international significance emerging as the study proceeds. A definition of 'international significance' as well as criteria and guidelines are necessary with this approach also. This has been the process used in Australia by the Geological Society of Australia's state subcommittees (Cochrane & Joyce 1986; McBriar & Hasenohr 1991) and to some extent in the UK by the Geological Conservation Review. It can be summarized as the IDEM method (Joyce in press).

Sites are **I**dentified as the result of a general and, as far as possible, complete regional survey. Sites of any possible significance are then **D**ocumented to provide the information needed for them to be **E**valuated by a panel of experts, who will allot a level of significance, and may provide input for the final stage of **M**anagement. The criteria which might be used in assessing significance have been debated in several places, e.g. Davey (1984 and Joyce & King (1980).

For Australian geological conservation studies, a comparison has been made between the procedures used by the Geological Society of Australia, the Australian Heritage Commission, and UNESCO World Heritage in the approach to the identification and documentation, and in particular the evaluation of geological sites of international significance within Australia (see further discussion in Joyce, in press).

An Australian procedure used for site evaluation A standard procedure is carried out in Australia for all sites nominated to the Register of the National Estate maintained by the Australian Government. Nominated sites include not just geological and geomorphological sites, but other natural sites (botanical, zoological), as well as landscape sites, historical sites and archaeological sites.

To assess the nominations made to the Register, the documentation is examined by a group of experts in a wide range of fields (e.g. for the natural sites not just geology but also zoology, botany, geography, soil science, planning and administration of national parks). The Natural Sites panel in each state of Australia meets regularly to examine and discuss the documentation of each nominated site, argue about the significance of the site, discuss the suggested boundaries, and then make appropriate recommendations to the Australian Heritage Commission.

Such an assessment could not be realistically made by an individual or a small group, even if provided with clear definitions, criteria and checklists, and detailed assessment forms to follow. A panel with a breadth of expertise allows the information provided to be checked, and perhaps added to, and sites compared and finally evaluated. The assessment made will stand up to any later scrutiny by other scientific experts.

The Pacific Way The reference in this paper to the Pacific Way is to draw attention to the procedure often used to identify the geological significance of a site in Australia. In the Pacific Islands, and especially in the region surrounding Fiji, local meetings avoid confontation and the casting of votes along opposing lines, and instead try to reach an agreement acceptable to all those present (even if it takes a long time!)

In a similar way, a geological assessment panel, such as those of each state Division of the Geological Society of Australia' state Divisions, can argue about the attributes of the site, and reach a consensus (the Pacific Way) which each individual will feel able to justify in any future criticism of the significance level agreed on.

An individual geologist cannot be expected to be a competent judge in all aspects of geology, and so cannot completely evaluate all the information in the documentation of sites whose attributes extend beyond the individual's field of expertise. Thus a group of experts are needed, and ideally might include a stratigrapher, a palaeontologist, a geomorphologist, a field geologist who has worked in the area concerned (perhaps a geological survey or exploration company geologist), a geology teacher, and perhaps a geologist working in planning or the parks service, and a possible range of specialists such as a soil scientist, a volcanologist, a mineralogist, a historian of geology, and so on.

Whether the site has been shortlisted as a result of the 'top-down' approach (looking only for the best sites) or more rigorously by using the IDEM method, a process of geological evaluation or assessment will always be required, both within each country, as has been done by Australia and the UK, and internationally.

Towards an international convention

An international list

Generating a valid list for each country will require two conditions – a useable definition of international significance (as well as significance in general) and a group of local experts who can apply the definition and reach agreement on the significance of individual sites within each country. Once each country has prepared its own list, an internationally-based review process is needed to compare these listings and documentations and rationalize the list on a worldwide basis (agreed by the Pacific Way). Then an agreed list of sites, together with its back-up documentation, must be presented and publicized to an international audience.

An international agreement

What sort of international agreement is appropriate?

The range of possibilities have been discussed in Stevens (this volume), and will need further discussion in the workshops to follow.

The second speaker (Holland, this volume) has shown how geological conservation has traditionally been a site-orientated approach. This has been appropriate for the type of geological feature which we have been considering in the past. Many have been fossil sites, others have been sites representing stratigraphic and tectonic events e.g. stratotypes, major geological boundaries, unconformities. Mineral sites, and sites of historic interest, are also generally acceptable to those working in geological conservation; these and a number of other generally acceptable types of sites and features are included in a list of twelve types drawn up by Cochrane & Joyce (1986).

We can conclude that there needs to be an internationally-agreed convention, such as RAMSAR, to which each country will subscribe, which will both oversee the procedure of identifying, documenting and evaluating sites, and also require each country to maintain by local management the values of its sites of agreed international significance.

Two other aspects of geological conservation

However, I now want to extend this discussion of geological conservation into two further areas, by asking

- What about geomorphological sites?
- And what about landscape sites?

Some of the Task Force members of the Working Group at the UNESCO Paris World Heritage meeting initially had much difficulty in accepting geomorphological sites along with the more traditional fossil or stratigraphic sites, but as the result of discussion at the meeting it was agreed (Cowie 1992) that 'geomorphology is a valid and valuable part of geological science', although it was still suggested that such sites might fit better into another part of World Heritage.

Also as the result of the recommendations of a workshop held during the Paris meeting (Cowie 1992 p. 6) World Heritage criteria have recently been revised and now specifically refer to 'ongoing ... processes in the development of landforms' and 'outstanding examples representing significant geomorphic/physiographic features (i.e. landforms ...)'.

At a more local level, geomorphological sites were discussed at a recent conference where the conservation of active modern processes was recommended as a way of conserving (maintaining) actively developing landforms ('Conserving our landscape: Evolving landforms and ice-age heritage, 1992'). At this Malvern Conference, discussions in many papers assume without question that geomorphological sites and processes are aspects of geological conservation. So it seems that there is an increasing consensus on the inclusion of landforms and modern processes in the scope of geological conservation studies.

And finally landscape sites – these may be the wave of the future! This has been well-argued for recently (Robinson 1990). Landscape has been part of the title of this Malvern Conference, and again many authors have assumed that landscape is part of geological conservation. However, the overlap with the study of landscape by planners, by artists taking an aesthetic approach, and in Australia work by the National Trust and others (Schapper, J. pers. comm.) means we must tread carefully, and consider what has already been done in this area.

In Australia many of the geological sites now listed as World Heritage have a major landscape component. To give a further example, the Raak area of northwestern Victoria appeared on the GILGES list of 16 Australian sites by the Task

Fig. 2. The Raak and Boinkas of the Sunset County, Victoria, Australia. Unusual saline groundwater discharge depressions and gypsum deposits, of major late Quaternary palaeoclimate significance, forming an extensive and striking landscape amongst the Quaternary Mallee desert dunes of this semi-arid region in northwestern Victoria. (Photo A. J. Shugg, October 1979.)

Force of the UNESCO Working Group in Paris in 1991 (Cowie 1992), and is described as an extensive assemblage of landforms developed as a result of saline groundwater discharge during the late Quaternary history of a semi-arid region (Fig. 2).

The approach to landscape needs to be a geomorphological/physiographic approach, which will look to the assemblage of landforms, their genesis and development, current processes involved, and in particular their relationship to other aspects of the Earth's history, including the influence of underlying rock materials and structure. It will go beyond the approach used by the artist or planner, and will add a scientific explanation to the landscape and its human and pre-human history (Robinson 1990).

Summary

In this paper I have suggested an approach for selecting sites in each country, and bringing them to an international body for review. I have also argued for a broader view of geological sites, to include both geomorphological sites, and landscape sites.

I conclude that each country needs to develop, at the international level, a list of sites and features, both outstanding and representative, which to the best knowledge of local and other Earth scientists, contains the record of the history of the Earth within the borders of the country concerned.

An international convention on geological heritage should be developed which will call on all national governments, perhaps with help from UNESCO as the manager of World Heritage listing, to begin, continue or complete within their boundaries the processes of Identification, Documentation, Evaluation and Management (IDEM) which will guarantee the continuing availability for research, reference and education by Earth scientists of the major sites and features of our international geological heritage. In addition, all national governments should agree to meet and carry out the evaluation of such sites and features at an international level, so that the important sites and features of the Earth's geological heritage will be internationally recognized and managed for the future.

How do we now proceed? In the workshop group discussions, we might consider the following matters.

Conclusion – towards an international convention

We need to clarify our definition of what we are about, whether we call it 'geological conservation', as many countries have done for some years, or whether we begin using one of the newer terms suggested recently, such as:

- 'Earth science conservation' (used as the title of the major journal in the field, and also used by Stevens, this volume)
- 'Earth heritage conservation' (used by Stevens this volume)
- 'geological heritage' (as used in the title of the Digne meeting)
- 'heritage of the Earth' (as used by Martini at Digne)
- 'Earth science heritage'.

Whatever term we adopt, we will need a definition which will tell us and others what we are doing in our work to identify, document, evaluate and manage sites and features of geological and geomorphological significance. The definition may also include the relationship of this work to such aspects as Earth resources and sustainability, and to biological conservation including biodiversity.

We must consider how each country has pursued this work to date, and whether we can learn from each other, and identify and develop our future approaches within our own borders.

Then we must consider how this work can be extended beyond our borders, and become an international activity. How will we do this? If not through World Heritage (and I have suggested why I don't think this would work) then how do we begin working towards an international convention, and achieve a suitable result in the next few years?

Such international activity, and in particular the development of an international network of sites, will in turn provide a strong impetus for further work in countries which may be having difficulties with their geological conservation programmes, whether through lack of funding, political instability or other problems. It will also encourage the workers in these countries, and their governments, to appreciate the importance of looking after the geological heritage of the Earth.

References

COCHRANE, R. M. & JOYCE, E. B. 1986. *Geological Features of National and International Signifi-* *cance in Australia*. A report prepared for the Australian Heritage Commission.

'Conserving our landscape: Evolving landforms and ice-age heritage'. 14–17 May 1992, Crewe, Cheshire. Preprints of papers.

COWIE, J. W. 1992. *Report of Task Force Meeting, Paris, France, February 1991*. UNESCO World Heritage Convention, Working Group on Geological (inc. Fossil) Sites, IUGS Secretariat, Norway.

DAVEY, A. G. (ed.) 1984. Evaluation criteria for the cave and karst heritage of Australia. Report of the Australian Speleological Federation 'National Heritage Study'. *Helictite*, **15**(2), 1–41.

JOYCE, E. B. (in press) Assessing the significance of geological heritage sites: from the local level to world heritage. *First International Symposium on the Conservation of Our Geological Heritage, 11–16 June, 1991, Digne-les Bains, France*.

—— & KING, R. L. 1980. *Geological Features of the National Estate in Victoria. An inventory compiled for the Australian Heritage Commission*. Victorian Division, Geological Society of Australia Incorporated.

MCBRIAR, M. & HASENOHR, P. 1991. Some Australian examples of Geological Conservation. *TERRA Abstracts* Suppl. 2 to *TERRA Nova* **3**, 1.

ROBINSON, E. 1990. Seeing with new eyes. *Earth science conservation*, **28**, 24–26.

Workshop 1

Issues put to delegates prior to Workshop 1

What format could an Earth science convention take and what could it contain?

Possible issues relating to an Earth science convention you may wish to consider are:

(1) Could it be **strategic**, using the mechanism of national strategies to promote conservation internationally and foster research, training and education? (Biodiversity and Climate Change conventions are examples of this model.)
(2) Could it be **site-focused** and concentrate on conservation and management of internationally-important localities such as stratotypes? (Ramsar and World Heritage conventions are examples of this model, and the latter is the only existing convention explicitly covering Earth science features.)
(3) Could it be **issue-related** and deal with issues such as the international trade in fossil and mineral specimens? (The CITES and Bonn conventions fit this model.)
(4) Could it emphasize the dynamic interaction between humankind and the Earth, and focus on issues of **sustainable use** of Earth science features? (This is a feature, for example, of the Biodiversity convention.)
(5) Could Earth science issues be looked at **in isolation** or could a **holistic approach** – developing links with wildlife and archaeological conservation for instance – be adopted?
(6) Could it confine itself to **inter-country issues** or could it seek to promote **national strategies or action**?
(7) Are the following issues part of what you understand as Earth science conservation, or should some be excluded?
  - conserving localities for scientific research;
  - conserving localities for education and training;
  - conserving localities for amateur study, collecting or tourism;
  - conserving wilderness areas;
  - conserving geological sites or geomorphological features because they are attractive or aesthetically satisfying;
  - conserving museum collections of rocks, fossils and minerals;
  - conserving site maps, borehole records and other field data;
  - conserving our resources of oil, coal, gas, aggregates and metals;
  - conserving active geomorphological systems because of their economical value as, for example, coastal defences.

Reports of Workshop 1 groups

Group A–D

(1) The group split into three sub-groups to brainstorm the pros and cons of a convention based on (1) a strategic approach, (2) a site-focused approach, and (3) an issue-related approach. Each sub-group debated one of these and during the plenary discussion it emerged that those present strongly favoured the development of a 'hybrid' convention based on points 1–4 on the 'issues' list.
(2) Concern was expressed that a strong case for the need for a convention had yet to be made in the form of a statement summarizing the issues that should be addressed. There was agreement that the existence of a convention would help raise the standards of Earth science conservation and encourage compliance of existing and new legislation where this existed. But to obtain a convention, governments have to be convinced of the need: **we have to show that an Earth Heritage convention will be something new** which addresses issues not covered by existing conventions.
(3) The group felt that a convention should make links to wildlife and archaeological conservation where appropriate. The inclusion of Earth resource conservation was rejected for tactical reasons, as it was considered that broadening the scope of a convention in this way would drastically reduce the chances of international agreement on its scope.
(4) The group were not sure of the meaning of the question on the worksheet concerning inter-country issues versus national strategies (issue point 6). There was agreement that a convention must address international issues, and that this would require national strategies to be developed.
(5) Little time was spent discussing the scope of Earth science conservation summarized on the worksheet (issue point 7). The group agreed that all the issues listed should form

part of the activity, but for the reason stated in (3) above, felt that resources issues should not be included in a convention.

(6) In summary, the group concluded that more work was needed to define the issues a convention should address and that it should be a 'hybrid' i.e. strategic, site-focused and issue-related.

Group E–G

The meeting agreed unanimously that progress towards an international convention is desirable, and that this convention should embrace both geological and geomorphological interests.

Potential issues to be addressed in the convention were identified as follows: landscape, stratotypes, other geological sites, geomorphological features (including caves), soils, energy minerals, non-energy minerals, trade in fossils and minerals (including meteorites), geological data. The meeting considered this list should be addressed in a convention.

(1) *Soils*: while it was recognized that conservation of type profiles might be a relevant issue, the general conservation of soil as an economic resource was not.
(2) *Trade in geological specimens*: it was considered that no completely effective policy could be devised because of the problems of enforcement. Meeting agreed that a site protection policy would be most effective in controlling undesirable trade in specimens, and that legislation to preclude international movement of specimens would not be in the interests of science.
(3) *Energy and non-energy minerals*: it was agreed that it would be inappropriate to seek to include in the convention national economic resources.
(4) *Data*: the importance of data conservation as the key to scientific use of sites and specimens was recognized, but it was felt to fall within 'good practice' in relation to sites and specimens, rather than forming a separate issue.
(5) *Landscape, stratotypes, sites, geomorphological features*: an approach to the convention based on sites and with reference to their associated specimens was preferred – but within a framework of 'Geodiversity' having affinities with that underpinning the Biodiversity Convention. The need for national committees to nominate relevant sites was agreed. Key criteria for selection should be uniqueness and integrity. Nomination for World Heritage status was recognized as a possible complementary procedure.

Group H–N

Scope of Earth heritage conservation (EHC) The group accepted this as set out in point 7 of the 'issues' worksheet, but with some amendments. The scope should explicitly include water, soil, ice, climate change issues and the use/abuse of geological materials (including waste products). 'Sustainable use of' was preferred as a term to 'conservation' in some cases. The notion of 'wise use' (to include recycling) should be prominent, the link with impact on terrestrial processes especially those which affect biological systems, should be highlighted, and the reference to 'aesthetically satisfying' should link to biota too. The last bullet point of issue 7 should be expanded to cover the link between geomorphological processes and biological systems, and should also cover social value.

Form of international convention There was broad consensus that a convention would be desirable, but a number of cautions were registered about timing (was it wise to go for one at a time when the world economy was depressed?), about remembering the variation between the First and Third Worlds, and about the need to take a long view. It was seen as important that we build on the foundations prepared by the Digne Declaration. We need to consider who it is for; is it for Earth scientists or for the general public's benefit? Should we take the moral high ground as professional Earth scientists, and take this opportunity to create something that really gets to grips with the big issues, and not get bogged down in a long wish list?

The group agreed that a **strategic convention** (issue 1) was the best way forward, and that it should be structured in 3 parts:

- Principles
- Strategies to achieve the principles
- Main goals and operational programmes.

Principles Three main principles were agreed, as a series of formations which are similar to those developed at Digne; they should be enabling rather than site-specific:

- we should focus on a whole Earth approach which integrates the geosphere and biosphere;
- we should concern ourselves with the evolution of the Earth and its processes, noting that we have only temporary custody;

- we should focus on the understanding of Earth processes and resources, their sustainable use, and global understanding of these.

Next workshop We agreed that the following issues would be most useful to discuss at the next workshop:

(1) How do we drive forward these principles?
(2) What exactly do we want to achieve in tangible terms?
(3) How do we persuade governments to accept and implement these?
(4) We should produce a short paragraph on each component of EHC to flesh out what we mean by each of them. The Biodiversity Convention is a good model for this.

Group O–S

Definition of Earth heritage conservation (issue 7) Museum collections – these are inextricably linked to sites and their conservation, but need not be specifically mentioned in the definition.

- Soils could be included, as could water.
- Extractive industries/resources are not be to included (no clear agreement on this).
- Rewrite the definition to include 'Research, reference, education and aesthetic appreciation' and geological sites include stratotypes and fossil and mineral sites (etc) and landforms (geomorphology) and active geological/geomorphological processes.
- Do not involve 'cultural' issues.

Preamble to international convention An introductory statement for the general public and politicians, drawing on the Digne statement, but less specific is needed – use item 8 but rewrite it to be less emotive (e.g. 'slightest depredation' 'mutilate'). Call it the 'Malvern Convention' (Professor Prasad).

Type of convention There was confusion using the one-page hand out of the issues – needs to be re-written to show progression.

(1) Strategic (Biodiversity) = (framework?).
(2) Site-focused (Ramsar, World Heritage).
(3) Holistic.
   and side issues:
   - fossils and mineral trade;
   - sustainable use.
   See above for groups comments on these.

Positions regarding the type of convention ranged from strong support for (3) above, including Lovelock/Gaia supporter, and for (1) above as a start. There was general support for a convention – but feelings that it should ignore sustainability.

Conclusion We need an aim, need to work out first what we want to achieve. We should expose and try to use existing systems (UNEP for Mediterranean countries). Each country may have very different needs – how to we help each country to pursue Earth heritage conservation?

Group T–Z

(1) All participants supported the concept of an international convention on Earth science conservation.
   - The convention should take the form of a set of objectives, guidelines/general framework which can be used to influence national authorities in their implementation of sound Earth science conservation.
   - It may take the form of a report along the lines of the Bruntland initiative.
   - Concern was also expressed over the over-use of the phrase conservation/conserving as a potential area of marginalization, particularly with respect to attempts to influence the private sector/commercial enterprises etc.
(2) The convention should take paragraph 8 of the International declaration of the Digne Conference as a starting point and emphasize that we are custodians of the Earth's resources.
(3) Para. 8 of the Digne statement should also take into account the socio-economic factors of particular nation with relation to development (sustainable). The guidelines should seek to influence national policy rather than be prescriptive.
(4) An integrated/holistic approach should be adopted, utilizing biological, engineering and planning considerations.
(5) The educational benefits of Earth science conservation should be stated:
   - i.e. the relevance of Earth science conservation to the public;
   - technology transfer should be encouraged to enable less-developed nations in their implementation of Earth science conservation policies.
(6) There is a need for an international organization of Earth science conservation:
   - which must be operational;
   - and which must draw on geological surveys and other national authorities in order to professionalize the subject.

Conclusions from Workshop 1

(i) There is general support for a strategy or framework approach to international Earth science conservation, the objectives of which might be met through an international convention or an internationally agreed strategy.

(ii) Any such international strategy or framework must not jeopardize national sovereign rights.

(iii) Workshops were not unanimous over the total scope of any strategy or framework, but there was consensus for inclusion of at least the Core defined below. Possible additions to the Core – A and B below – were favoured by some workshops:

CORE: Scope would relate to the conservation of geological, geomorphological and soil localities, to active geomorphological processes including ice, to specimens, and to geological information.

ADDITION A: Scope would include the core above and in addition would recognize the interaction between biological and Earth science processes in the natural environment, and thus include genetic diversity and the natural habitats and species that underpin it.

ADDITION B: Scope would include the core above and in addition recognizes the importance of all the Earth's physical resources to humankind. Nations would commit themselves to a plan that would allow the economic and heritage potential of their resources to be realized.

A further option would consist of the Core plus Additions A and B.

Workshop 2

Issues put to delegates prior to Workshop 2

(1) Confirm that paragraphs (i) and (ii) of the Conclusion to Workshop 1 reflect general agreement within your workshop.
(2) Confirm that the Core of the Conclusion to Workshop 1, with Additions A and B, encompasses the total potential scope of Earth science conservation defined within your workshop.
(3) Considering the Core of the Conclusion to Workshop 1, Addition A and Addition B, identify the benefits to Earth science conservation of each of:

- Core,
- Core plus Addition A,
- Core plus Addition B,
- and Core plus Addition A plus Addition B.

(4) Place the four options in a priority order.
(5) Comment on the achievability of your highest priority option (only) in terms of (a) a strategic/framework convention and (b) a strategy agreed by the Earth science conservation community at an international level.
(6) Do you think that (a) the achievement of a convention or an international strategy should be carried forward by an action group with international membership, and (b) if so, should we initiate it at this conference?

Reports of Workshop 2 groups

Chairman's report (Dr K. L. Duff)

Six questions were posed to Workshop 2 groups, as set out on the worksheets issued to all participants. This report is set out to deal with each of these questions.

1. Confirm conclusions from Workshop 1. This group accepts and supports Conclusions (i) and (ii), but believed that much more work needed to be done to work up the justification and need for a convention. They saw the long-term aim being a strategic convention, but felt that a working group needed to be set up to develop strong and specific arguments to demonstrate the need for a convention. It should be original thinking, and should not just be an attempt to build a geological analogue to a biological convention. Geology has something special to offer, and we should do our best to promote this. It was felt vital that national sovereign rights were not jeopardized by whatever emerged from this.

2. Confirm the total potential scope of Earth heritage conservation. The Group confirmed that the Core, plus Additions A and B encompassed this. There was a suggestion that we needed to consider whether the atmospheric sciences should also be included in the scope.

3. Benefits of each option. Before considering the benefits of each option, the Group discussed the general issue of need for a convention or strategic framework, and identified a number of possible benefits, as listed below. However, the group still felt that these by themselves did not provide adequate justification for a convention, and that a working group was still needed to work this up in much more detail.

Benefits (General)

- A convention might provide an international 'label' for sites of very high importance, which may assist with efforts within countries to safeguard such sites. The group concluded that the World Heritage Convention was probably not the best solution here, since it was too elitist, the sites needed to be very large, and the scope was far too wide, encompassing cultural sites, archaeological sites, etc.
- Earth science sites would be seen to have equivalent status to biological sites.
- Mechanisms for controlling international trade in specimens might be provided, and could be selective (as is CITES). The question of need for such controls is still unanswered.
- An international information bank could be created, and would assist in control of trade, and in providing comprehensive data on Earth history. This might also cover Earth science knowledge which would help us to understand and manage other processes.
- An international forum would be created.

Benefits for the 'Core'.
These were adequately covered by the benefits listed under 'General'.

Benefits for Core plus 'A'.

- Opportunity for stronger links with biology, palaeobotany and biospeleology, which would make Earth science more relevant, and more likely to be heard.

From O'Halloran, D., Green, C., Harley, M., Stanley, M. & Knill, J. (eds), 1994, *Geological and Landscape Conservation*. Geological Society, London, pp. 519–521.

- Habitat formation is often controlled by Earth processes, for example marshes and wetlands, and closer links would be mutually beneficial.
- Better integration of land-use management planning, in ways which brought together biological and geological issues.

Benefits for Core plus 'A' and 'B'.
- This would provide a powerful vehicle to influence resource management, but might run the risk of diverting attention away from a focus on key sites.
- Might provide international good practice guides which would create a level playing field on environmental controls for multinationals.
- This option was seen as a longer-term objective.
- Both this option and the previous one might provide Earth scientists with major opportunities to demonstrate the relevance of what we do, especially if we present it as being directly related to quality of life issues.

Benefits for Core plus 'B'.
The group did not feel that this could usefully be separated from the other options.

4. Priority of the four options. The group considered the relative priority of the options, and concluded that they had an almost unanimous preference for a convention which covered only the Core plus Addition 'A', at the same time a group should be established to explore the feasibility and justification for extending this to cover Addition 'B', but no action should be taken to implement the latter until there had been much more debate and consideration.

The group's second preference was for action solely on the Core plus Addition 'A'.

5. Feasibility and achievability of preferred option. The group felt that there was likely to be governmental scepticism about the need for a convention, and that no approaches should be made until powerful justifications had been developed and agreed. These must be clear, understandable, and relevant. They must also be very well focused, and will need to be marketed effectively. In the meantime, we must not lose sight of the need to maintain the site protection focus. There was a suggestion that conservationists would be likely to support a convention, but it was also clear that there is very considerable variation between geologists at the international level. In some countries most geologists are aware of, and support, Earth heritage conservation, whilst in others there are very low levels of understanding and support.

Some delegates thought that the current high levels of environmental awareness created a situation where it was timely to go for a convention, and believed that we may get higher levels of public support than we expect; personally, I am not so convinced.

6. The way ahead. The group believes that an action group should be set up to take this forward, and develop strong and convincing justifications before we go public. The action group needs to be truly international, and not just be restricted to First World delegates. The action group should be charged with the production of a detailed document which sets out justifications and need, and proposes the way forward in the form of an action plan; costs should also be assessed. This document should then be used as the basis for detailed discussions with the international community of Earth heritage conservationists, and as the basis for seeking funds for this work.

Chairman's report (Professor R. C. L. Wilson)

1. The group confirmed that paragraphs (i) and (ii) of the Conclusions to Workshop 1 were acceptable, but felt that care would be needed to ensure that an international convention/framework/strategy would not be perceived as a threat to national sovereignty, even though it was not intended to be.

2. The group confirmed the scope of the proposed Core, and agreed that it and Additions A and B represented the total potential scope of Earth science conservation.

3. It was felt that much more work was needed to be done to define the needs for and benefits of an international convention/strategy/framework. The group warmed to the idea of viewing the need to preserve a diverse Earth heritage under three value sets:

- Knowledge-based
- Cultural
- Amenity-based.

Each of these could be applied to a consideration of the need for conservation as applied to **specimens**, **information**, **sites**, **processes** and **resources**, but as far as the last named area is concerned, the group were against including control. We felt that the Core plus A and B should be embraced by a convention or framework but focus strongly on the issues raised by the Core and Addition B (but little progress was made in defining the issues more clearly).

4. The group accepted the need for an action group, and urged that it should co-opt members in order to incorporate a world view. We also urge that existing networks be used to disseminate the information.

Conclusions from Workshop 2

There was agreement to the following conclusions from Workshop 1:

(i) that there was general support for a strategy or framework approach to international Earth science conservation, the objectives of which might be met through an international convention or an internationally agreed strategy.
(ii) any such international strategy or framework must not jeopardize national sovereign rights.

The workshops agreed that the scope for such a convention or strategy, including the preamble, should consist of:

- a core which would include the conservation of geological, geomorphological and soil localities, active geomorphological processes including ice, specimens, and geological information.
- the scope would further recognize (i) the need for an understanding of the interactions between biological and geological processes in the natural environment, and thus the natural habitats that underpin it and (ii) the importance of the Earth's physical resources to humankind.

The workshops further recognized that a strategy agreed by the Earth science conservation community at an international level will be a necessary first step towards the achievement of an international convention.

Malvern Conference on Geological and Landscape Conservation 1993: Resolution endorsed by closing session of conference

Resolution (*amended* version)

The Malvern International Conference 1993

- believes that there is need for an international Earth science conservation convention;
- recognizes that the **justification for**, **potential scope of**, **and objectives of**, such a convention should be examined in depth; and
- supports the establishment of an international task force which will pursue, and report back on, these propositions.

Action Plan

(1) The following delegates at the Malvern Conference have agreed to form a task force, for a period of two years in the first instance, with the purpose of:
  - progressing the Malvern International Conference 1993 Resolution taking account of the conclusions of the Workshops at the conference;
  - facilitating the organization of a successor conference to the Malvern Conference 1993;
  - expediting the creation of an international organization for Earth science conservation which will, on formation, take over the functions of the task force.

  Dr P. Creaser (Australia)
  Prof. P. Jacobs (Belgium)
  Mr G. Martini (France)
  Ms A. Spiteri (Malta)
  Mr C. Stevens (United Kingdom)

(2) The international task force will seek the support of delegates and others in carrying forward its report, and widen its membership as necessary.

(3) The Joint Nature Conservation Committee of the United Kingdom (Monkstone House, City Road, Peterborough PE1 1JY, UK) will provide a secretariat to support this task force for a period of two years after which time it is anticipated that a permanent international organization will be in place.

(4) The task force will report back annually in writing to the delegates who attended the Malvern International Conference 1993 until either a permanent international organization is in place or a successor conference is held, whichever is the earlier.

22 July 1993
Malvern

Index

Agassiz Rock 1
Agenda 21 33, 43
aggregate
  consumption 34, 39, 268
  quarrying and mining impact 228, 288–90
    planning controls 34–7
  Quaternary study resource 39–40, 41, 87–91
Aggtalek 250, 251
agricultural industry conflicts 229
Aguilón 331–2
Alpine Fault 263
Alport 125
ammonite conservation project 383–5
Anarjokka National Park 153–4
Anina Mts National Park 160
apatite mining 154–5
Apuseni 160, 222–5
archaeology and Earth science 88–90, 198, 202, 203, 443–5
Arctic environments *see* Svalbard
Area of Outstanding Natural Beauty (AONB) 191
art forms, role in landscape of 487–90, 493–4
Arthur's Seat 1
Askja 228, 229
Aukstaitija National Park 273
Aurajoki river 145
Australia
  national conservation 69–70, 76–7, 319, 508–10
  state conservation
    Queensland 303–7
    South Australia 319–21
    Victoria 105–9, 413–16
australites 69
Austria, cave conservation 213–14

ball clay 80, 85–6
Banff National Park 279
Bartonian Stage stratotype 125
Baumanns Cave 175
bauxite mining 224
Bavarian Pfahl 119
Baw Baw National Park 413
Bayrischer Wald National Park 177
beach management 84, 193
Berchtesgaden National Park 177
Bicaz-Hasmas National Park 159
biology and Earth science conservation 201–2, 203
Black Country Geological Society 353–7
Blackford Hill 1
blanket bogs 18
blockfields 165
Blue Grotto Arch 207
bogs 17, 20, 198
Bohlen Wall 119
Bollin, River 192, 194
boulder beds 40
Brecon Beacons National Park 279

Brewin's Canal SSSI 356
Broads National Park 198
Brownend Quarry 437
Bruntland Commission 7, 18–19, 205
Bucegi National Park 159
building stone 347
Bükki National Park 251
Bulgaria 247–8
bulk fill material 84
Burgess Shale 76, 506
Butenai Interglacial 261–2

calcium carbonate uses 80, 86
Caliman National Park 159
Canada 64, 76
  Canadian Heritage Rivers System (CHRS) 127–31
Carboniferous Limestone Natural Area 125
Castle Rock, Edinburgh 1
catchment planning 194
cave conservation
  Austria 213–14
  Bulgaria 247
  Estonia 240
  Hungary 252
  Northern Ireland 337–8
  India 256
  Romania 224–5
  Russia 299
  USA 209–11, 387–90
Ceahlau National Park 159
Central Pinnacle 113
Cerna Valley National Park 160
Cetatile Ponorului 160
Cévennes National Park 494
chalk landscape Natural Area 125
chalk uses 80, 86
chemical pollution problems 13
Chengjiang Fauna Reserve 244
Chesil Beach 28
Chihsing Shan 114
China, conservation programme 243–5
china clay *see* kaolin
Chingshui Cliff 114
Chiswell Earthworks 488–90
Christmas Cove 321
Cioclovina National Park 160
Clacton Channel 88
Clara Bog 20
clint 238
coal mining, landscape impact of 93–6
Coalbrookdale Formation 366
coastal conservation 125–6, 193
Coldwater Cave 210
Colliford dam 84
Commission on Geological Sciences for Environmental Planning 23–6

525

Committee on Land Utilisation in Rural Areas 1
community involvement 328, 383, 436–7, 461, 464, 473–4
compaction and soil conservation 13, 171
computer software for site recording 371–2
　design features 272–5
　development 375–6
　future 379–80
　standardization 376–9
conservation of Earth science sites 1
　approaches 500–1
　practicalities 328, 404–5, 499–500
　reasons 403–4, 473
　see also International Convention on Earth Science Conservation
construction industry conflicts 228, 268
Convention Concerning the Protection of the World Cultural and Natural Heritage see World Heritage Convention
copper mining 100–1, 171, 247
Cornwall and Devon china clay industry 80–5
country sports, role in landscape of 279–81
Countryside Council for Wales xii
Cozia National Park 159
Cuilcagh Mt 337–41
Culm Measures 122
Culm River 142
cultural heritage and landscape 487
Czech Republic 93–6, 505–6

Dalyan 271
Dark Peak 125
Dartmoor 79, 199, 280
De Zåndkoele 324–6
Deccan Volcanics 256
Dee Cliffs SSSI 484
Denmark, palaeontological conservation 64
Derwent River 142
Devon and Cornwall china clay industry 80–5
Dibdale/Burton Road project 361–2
Digne ammonite conservation project 383–5
Dino-Park, Germany 118
Dinorwic pumped storage scheme 170
Dinosaur Provincial Park 76
District Teaching Sites 480
Dobrogea Mt 160
Dob's Linn Silurian type site 504
documentation see computer software
Dolgarrog aluminium smelter 170
Downton Castle Sandstone Formation 369
Dudley urban conservation projects 350–2
　Black Country Geological Society work 353–4, 356, 366
　Dibdale/Burton Road project 361–2
　museum recording scheme 365, 366–9
　Wren's Nest NNR 350, 366
Dulyn Cwm 165
Dunbridge Gravel 88
dune conservation 247
Dwejra dolines 207
Dyffryn Mymbyr 165

dynamic landscape management 187, 188, 191–4
Dzukija National Park 275

Earth Science Site Documentation Series 448–50
Earth Summit (Rio 92) 26, 43
East Kirkton Quarry 335, 461
Ediacara Fossil Reserve 319–20
Edinburgh 1
education
　approaches to 473–4, 477–80, 483–4
　importance of 10, 244–5, 252, 392–3, 407, 410, 459
　resource use in 194, 424–6, 463, 464, 474–5
　see also interpretation also information signs
Eire see Ireland
Elan River 142
Elton Formation 366
emotion, role in landscape interpretation 493–4
engineering works, role of 359–62
English Nature xii, 313–15
environmental interpretation see interpretation
environmental pressure evaluation 153
Environmentally Sensitive Areas 280
Ercall Quarries 464–5
erosion problems, soils 13–14
erratic conservation 240, 262
Estonia 237–40
European Soil Charter 54
Evesham, Vale of 125
Exmoor 198, 280
exploitation management principles 7–10
Eyjafjallajökull 229

Farndon SSSI 484
fens 18, 125, 198
Fertô-Lake National Park 251
Feshie River (Scotland) 147–50, 201
　flood management 150–2
Ffestiniog 170
Ffynnon Llugwy, Cwm 165
Finland, river conservation 145
fire clay mining, Romania 224
Five Peaks 113
flood management and river conservation 150–1
Flow Country 197
fluvial conservation see river conservation
footpath erosion 171
footprint conservation 250
forestry and conservation problems 167–70
Forth, Firth of 461
fossil conservation see palaeontological conservation
Foundation for Aggregates and Landscape 41–2
France 383–5, 494
Fraser Island 304
French River (Canada) Provincial Waterway Park 127, 129
Frog Beds 257

Gardermoen 47–50
geoenvironmental indicators 25–6

Geological Conservation Review (GCR) 2, 447
Geological Reserves Sub-Committee 1
geological societies, role of 309–10, 353–7
geomorphology, conservation of 511–12
  classification 187, 191
  management 186–7, 188, 192–4, 202
    application to river systems 133–7, 147–52
geopark concept 179
geothermal power 228, 267
geotope concept 28–9, 117–18
Germany 64, 118–19, 175–9
Ghar Dalam cave 206, 207
Ghar Hasan cave 207
Ghetarul de la Scarisoara cave 60
Giant's Causeway 506
GILGES 71, 72, 119, 508, 509
GIS (Geographic Information Systems), role in recording of 366–9, 379
glacial striae 1
Glasgow 1
Golden Spikes 503–4
Graianog, Cwm 165
Grampians National Park 413
Grand River (Canada) 129
gravel *see* aggregate
greenhouse effect 14
Guilin National Park 244
Gullet Quarry 1

Harz-North-Rim-Fault 119
Hayes Cutting SINC 356
Hekla 229
Helgoland 119
Helsby Quarry Park 486
Hengchen Peninsula 115
Hepste River 142
heritage aspects of conservation 500
Heunesäulen 118
Highland Water 142
Hochharz National Park 177
Höger complex 27, 28
Hogland 240
Holm Fen 125
Holocene sediments 201–2
  see also peatlands
Holywell Coombe 89
Holzmaden 119
homeland initiative 403, 405–6
Hominid sites 76, 256
Hong Kong 291–6
Horse Cave (Kentucky) 387–90
Horse Gully 321
Hortobágyi National Park 251
Hsiukuluan Shan 113
Hsueh Shan 113
Huangshan, Mt 243
Humpleu Cave 160
Hunaglong 243
Hungary 249–52
Hunstanton Cliff SSSI 437
Huoyen, Mt 115

Hurst Castle Spit 194
Hutton, James 1
hydro-electric schemes 224
hydrogeology, peatlands 20

Iceland 227–9
Idwal, Cwm 164, 165
Il-Hofor dolines 207
Il-Maghlaq fault 207
Il-Maqluba sinkhole 207
Ilumetsa craters 240
India 255–7
industrial heritage 451
information signs 434, 436
interpretive approaches 434–7, 461
integrated resource management (IRM) 206–7
International Convention on Earth Science Conservation 30, 507, 510
  development of 497–8, 515, 519
  resolution 523
  site classification problems 507–12
  workshop reports 515–18, 519–21
International Council for Local Environmental Initiatives (ICLEI) 43
interpretation of the environment
  approaches to 430–2, 435–6, 493–4
  defined 429–30, 434–5
  on site 407–11
  use of boards 434, 436
Ipolytarnòc 250
Ireland 17–18, 215, 503
iron ore mining 154–5
IUGS GEOSITES 71, 72
Izvorul Tàusoarelor cave 223

Japan, ammonite cast display 384–5
Jasmund National Park 177–8
Jasper National Park 279
Joint Nature Conservation Committee (JNCC) xii
Juizhaigou 243

Kaali Meteorite Crater 240
Kaiserstuhl 119
kaolin extraction 80–5
Kärdla Impact Structures 240
karst *see* limestone
Katla 229
Kenting Coral Reef Coast 115
Kimbridge Farm Quarry 88
Kiskunsági National Park 251
knoch land system 169
Knockan Cliff 336, 460
Kodalen 154–5
Koonwarra fossil site 508
Köycegiz Lake 271
Krafla 228
Kruger National Park 279
Kullana wavecut platform 207
Kurnool Caves 256

La Crosse (Wisconsin) marsh conservation project 397–401
Laach, Lake 119
Lahemaa National Park 239
Lakagígar 229
lake conservation 247
Lake District 125
landfill 88
landowners, problems of site responsibility 464, 465
Land's End granite 79
landscape conservation 145, 460, 487, 511–12
lava tubes 306
Lawn Hill National Park 304
Laxá River 228
lead mining 171
Leadhills 441
Lechuguilla Cave 210–11
Lepidodendron stumps 1
Lieth salt dome 119
limestone/karst landscape 205–7, 214, 221, 222–5, 247
  pavements 215–20
  see also caves
lithium uses 85
Lithuania 259–62, 273–7
Llydaw, Llyn 170
local government, role in sustainable development policies 43–4, 350–1, 354, 355, 356
lochan land system 169
Lugg River 142
Lulworth wayside carvings project 487–8, 489
Lüneburger Heide 118, 119, 179

Maesnant River 142
Malta, limestone conservation project 205–7
Malvern Hills 1, 125, 287–90
Mam Tor 125
mammal faunal sites 257
management approaches to conservation 447–8
Marble Arch Caves 337–8, 340
Marble Gorge 113
Marchlyn Mawr, Llyn 170
media, geological coverage by 462
Meilte River 142
Melynllyn, Cwm 165
Mendip Hills 79
Merkers salt mine 119
Messel 119
metal ore mining 99–102, 153–5, 171
meteorites and craters 69, 240, 297, 299
mica uses 84
Micklefield Quarry 349
Mineralogy of Wales Network 99
minerals
  economic mineral extraction 34–7, 41–2
    history conservation 79–80, 99–102
    sustainability of 33–4
  specimen conservation 69, 247, 299, 439–41
minerotrophic mires *see* fens

mining and conservation problems
  coal 93–6
  metalliferous 99–102, 153–5, 170–1, 247
mires *see* peatlands
Moel Tryfan 171
Moine Thrust 336, 460
moraine conservation 27–8, 247
Morrison, Mt 113
Mt Eccles National Park 413
Mt St Helens National Monument 414–15
mud volcanoes 115
Murgon Fossil site 76
Müritz National Park 178
music, use in communication of 494
Mývatn, Lake 228

Nanhuta Shan 113
Nant Ffrancon 164, 165
Nant y Benglog 164
Naracoorte 76–7
National Nature Reserves (NNR) 121
National Parks
  Poland 181–2
  UK 134, 191, 198–9
National Stone Centre (UK) 423–6, 452–7
Natural Areas concept 122–6
natural monument concept 117, 247
naturalness evaluated 283, 285–6
Nature Conservancy Council 1, 3
Nature Reserves Investigation Committee 1
Netherlands 17, 323–8
New Milestones project 487
New Zealand 263–8
Newer Volcanic Province (Australia) 105–8
nickel ore exploitation 153–4
non-renewability problems 7, 8
Nördlinger Ries 119, 179
North Esk River 142
Northern Ireland 337–41
Northumberland National Park 198, 279
Norway 153–5
Numinbah Valley 304

Oberharz National Park 179
Odertal National Park 178
Oldhamstocks Burn 142
Olduvai Gorge 76
ombrotrophic mires *see* bogs
Open University Earth sciences 475, 483–4
Organ Pipes National Park 413
Ozark Underground Laboratory 209, 210

Padarn, Llyn 165
palaeoclimate analysis 197–8
palaeontological conservation 63–4, 67, 76–7
  Australia 69–70
  Bulgaria 247
  Canada 64
  China 243, 244
  Denmark 64
  France 383–5

Germany 64
India 256, 257
Romania 64–5
Russia 299
South Africa 65
Spain 57–61, 329–33
UK 65
USA 65
Parys Mt copper mine 99–103
Peak District National Park 199
peatlands 17–20, 197–9, 201–3, 340
Peking Man fossil site 243, 244
Pembrokeshire Coast National ark 198
Penghu's Columnar Basalt 113–14
Penmaenmawr 171
Peris, Llyn 165, 170
Pestera Vîntului cave 224
Pettycur 462
phosphorus resources 154–5
Piatra Altarului Cave 160
pliosaur conservation 70
pocket valley 221
poetry, use in communication of 493–4
Poland 181–3, 417–21
Polesie National Park 182
pollution problems, soil 13
polygons 165
Popigai crater 297
Port Campbell National Park 413
Portile de Fier National Park 160
Portland earthworks 488–90
Portrush 506
Potentially Damaging Operations (PDO) 2–3, 448
Pouk Hill Quarry 354
power generation and conservation problems 170, 228, 267
Pozary Silurian type site 505–6
precautionary principle 9–10
precipitated calcium carbonate 80
Pricopan granite 160
protalus ramparts 164
pumice mining 228

quarrying and conservation problems 170–1
Quaternary features 119, 185, 261–2, 443
   see also aggregate

Raak and Boinkas 511
raised bogs 18
Rammelsberg 119
rare earth elements 85
recording methods see computer software
Reeves Point 320
Regionally Important Geological/geomorphological Sites (RIGS) xii, 3, 134–5, 311, 480
   launch 313
   work 314–16, 335–6, 343–5
resource conservation 7–10, 500
Rhynie 506
Ricla 331–2
Rio Conference (Earth Summit) 26, 43, 205

river corridors 194
river conservation 133–7, 145, 147, 151
   see also Canadian Heritage Rivers System also Feshie River
Riversleigh 76–7
road construction and conservation 171, 268, 336
roches moutonées 169
Rodna National park 159
Romania 64–5, 157–60, 222–5, 467–72
Rotorua geyser conservation 267
Roztocze National Park 182
Rudabánya 250
Rügen coast 119
Russian Commonwealth 297–300

Sächsische Schweiz National Park 178
Ság-hill volcanic cone 251–2
St Andreasberg mines 119
St Austell granite 80
Salisbury Crags 1
salt mines conservation 119, 417–21
sand and gravel see aggregate
Saxon Sandstone Mts 119
Scàrisoara ice cave 222
Scientific Committee on Problems of the Environment (SCOPE) 24
Scotland 185, 186, 198, 215
   RIGS work 335–6, 343–5
   river conservation 147–52
   SSSI management 447–50
Scottish Natural Heritage xii, 460–2
sculpture and landscape 487–90
Severn Vale 125
Shamao Shan 114
Shanwang Fossil site 243, 244
Shark Bay 506
Shropshire geological herigage 463–4
Siccar Point 1, 506
Siebengebirge 119
signs see information signs
Silurian type sites 504–6
Site adoption schemes 328, 480
site maintenance, problems of 328, 464
Site of Special Scientific Interest (SSSI) 2, 121, 134, 191, 447–8
skills development in conservation 474
Skjaldbreiour 229
slate quarrying 170–1
Small Yehliu Coast 114–15
Snaefellsjökull 229
Snowdonia National Park 161–3
   management
      conflicts in land use 167–72
      conservation strategy 166–7
      constraints 166
      future development 172
   Snowdon range 163–4
   Y Carneddau 165–6
   Y Glyderau 164–5
soils 8, 11–15, 53–5
solifluction lobes 165

Solnhofen-Eichstätt limestone 119
Songshan National Park 244
South Africa 65
Southern Rivers Project 90, 443
Spain 57–61, 329–33
Species Recovery Programme 122
Spey Islands 152
springs 221
Standing Conference on Problems Associated with the Coastline (SCOPAC) 193
Steinheim Basin 119
stone stripes 165
Stores Cavern 367–8
stratigraphical aspects of conservation 299, 503–4
Sturt Gorge 320
subrosion 118
Suhua Coast 114
sustainability 7–10, 43, 47, 186–7, 205
  planning policy 34–7, 43–4
  resources considered
    minerals 33–4
    peatlands 18–20
    soils 54–5
Suursaari 240
Svalbard 283–6
swallet 221
Switzerland 27–9, 39–42
synthetic filler 83–4

Tafen Pinnacle 113
Taishan, Mt 243
Taiwan geomorphology 113–15
talc mining 154
Tanzania 76
Tapa Pinnacle 113
Taroko Gorge 113
Tata museum 251
Taupo Volcanic Zone
Tawu Shan 113
Tenterfields LNR 356
Ter River 142
Tertiary volcano sites 119
Thorne Moors 201–3
Three Pinnacles 113
tin 85
titanium 85
toadstone 487
Tongariro Volcanic Centre 264
topaz uses 85
Torphin Quarry 335, 343, 345
tourism and land-use pressure 171–2, 207, 225, 229, 305, 391–2, 394–5, 460–1
training needs in conservation 474
Trakai National Park 274
trampling and soil conservation 171
Tumbling Creek Cave 209–10
tungsten 85
Turkey 271
TV, geological coverage by 462

UK
  approaches to conservation xii–xiii, 313–15
  limestone pavements 215–20
  palaeontology 65
  peatland 197–9
Ukraine (Western) 232–5
Undara National Park 306
United Nations Conference on Environment and Development (UNCED) 26
Unteres Elbetal National Park 179
Upper Teesdale NNR 437
uranium 85
urban geology 231–5, 347–9, 350–2, 353–6
USA 414–15
  cave conservation 209–11, 387–90
  marsh conservation 397–401
  palaeontological conservation 65–6

Victoria lines fault 207
Victoria Park, Glasgow 1
Vik Talc Quarry 154
visitor pressure *see* tourism
volcano conservation 119, 227–9, 247
volunteer conservation workers, role of xii, 309–12, 314–16, 335–6, 353–4, 356, 359
Vorpommersche Boddenkindschaft 178

Wales xii, 198, 215
  see also Snowdonia National Park
Water End 142
water resources and conservation problems 170
waterfall conservation 247
Wattenmeer National Park 178
wayside carvings project 487–8, 389
Werribee Gorge State Park 413
Westlothiana lizziae 63–4, 335, 461
Wieliczka salt mines 417–21
Wildlife Enhancement Scheme 122
Wilsede, Mt 119
Wolterholten 326–8
World Commission on Environment and Development (Bruntland Commission) 7, 18–19, 205
World Heritage Convention 71, 72, 75
  Australian sites 76–7, 508
  scope of geology in 507–9
Wren's Nest NNR 350, 366
Wulingyuan National Park 243, 244
Wyperfield National Park 413

Yanmingshan National Park 114
Yehliu 144
Yorkshire Dales National Park 198, 279
Yu Shan 113

Zechstein history 119
Zigong dinosaurs 243–4, 245
zinc mining 171